INTERNATIONAL SERIES OF MONOGRAPHS IN
PURE AND APPLIED BIOLOGY

Division: **ZOOLOGY**

General Editor: G. A. Kerkut

Volume 22

BIOSPELEOLOGY

NOTE TO READERS

Please note that throughout the text of this book the spelling 'Biospeology' has been adopted, following the French edition (see discussion on this point in the Introduction). However, since the spelling 'Biospeleology' is the more correct version it has been used for the main title of this book.

BIOSPELEOLOGY

The Biology of Cavernicolous Animals

BY

A. VANDEL

Faculty of Science, Toulouse
Director, Subterranean Laboratory,
Centre National de la Recherche Scientifique

TRANSLATED INTO ENGLISH

BY

B. E. FREEMAN

Lecturer in Invertebrate Zoology,
Wye College, University of London

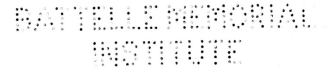

PERGAMON PRESS

OXFORD · LONDON · EDINBURGH · NEW YORK
PARIS · FRANKFURT

Pergamon Press Ltd., Headington Hill Hall, Oxford
4 & 5 Fitzroy Square, London W.1

Pergamon Press (Scotland) Ltd., 2 & 3 Teviot Place, Edinburgh 1

Pergamon Press Inc.,122 East 55th Street, New York 22, N.Y.

Pergamon Press GmbH, Kaiserstrasse 75, Frankfurt am Main

First edition 1965

Library of Congress Catalog Card No. 63–21137

This is an authorized translation of the original volume
Biospéologie—la biologie des animaux cavernicoles
published in 1964 by Gauthier-Villars, Paris

CONTENTS

PART 2

A LIST OF CAVERNICOLOUS SPECIES

CONTENTS   ix

CONTENTS   xi

PART 3

GEOGRAPHICAL DISTRIBUTION AND ECOLOGY OF CAVERNICOLES

PART 5

THE BEHAVIOUR OF CAVERNICOLES; SENSITIVITY AND SENSE ORGANS

Part 6

The Evolution of Cavernicoles

PREFACE

THIS book would be incomplete but for the kindness shown by numerous biologists in informing me of recent research and publications which would otherwise still be unknown to me. These workers are:

Aellen, V., Geneva	Hadži, J., Ljubljana
Aljančič, M., Ljubljana	Henrot, H., Paris
Angelier, E., Toulouse	Juberthie, Ch., Moulis
Barr, Th.C., Lexington	Juberthie-Jupeau, L., Moulis
Bellard Pietri, E. de, Caracas	Lanza, B., Florence
Bole, J., Ljubljana	Leleup, N., Pretoria
Bolivar Y Pieltain, C., Mexico	Machado, A. de Barros, Dundo
Bouillon, M., Moulis	Matjašič, J., Ljubljana
Cassagnau, P., Toulouse	Meštrov, M., Zagreb
Caumartin, V., Moulis	Motas, C., Bucharest
Chodorowski, A., Warsawa	Orghidan, Tr., Bucharest
Chopard, L., Paris	Petrowski, Tr., Skoplje
Christiansen, K., Grinnell	Pretner, E., Postojna
Coiffait, H., Moulis	†Remy, P., Paris
Condé, B., Nancy	Richards, A., Sydney
Cotti, G., Lugano	Riedel, A., Warsawa
Delamare-Deboutteville, Cl., Nancy	Rioja, E., Mexico
Dudich, E., Budapest	Rouch, R., Toulouse
Español, F., Barcelona	Ruffo, S., Verona
Fage, L., Paris	Strinati, P., Geneva
Franciscolo, M.E., Genoa	Strouhal, H., Vienna
Gueorguiev, V.B., Sofia	Uéno, S.I., Kyoto

Professor René Jeannel and the late P.A.Chappuis generously gave their extensive biospeological library to the subterranean laboratory of the Centre National de la Recherche Scientifique. I have used these precious collections extensively in compiling the present work.

I owe a great debt of gratitude to Alex Townsend, Librarian of the British Museum (Natural History), who made available photocopies of publications unobtainable in France. Finally, the figures illustrating this book have, for the most part, been reproduced in London with the aid of A.Townsend. I wish to thank him for all the help he has given me during this work.

† Deceased

INTRODUCTION

A. SPELEOLOGY

The aims of simple sciences such as geometry and crystallography may be perfectly and unambiguously defined.

But in more complex sciences the boundaries between them are always rather uncertain, because these sciences are so intimately connected that they can be separated only in an arbitrary manner. This is the case in the majority of the natural sciences, of which geology, oceanography and meteorology are good examples. Speleology belongs to this type of complex science. It is a "science synthétique", as justly described by Racovitza at the time when he founded l'Institut de Spéléologie de Cluj.

The term "Speleology" was derived by Emile Rivière from two Greek words: σπηλαῖον, a cave, and λόγος, a discourse. L. de Nussac in his *Essai de Spéologie*, published in Brive in 1892, proposed to replace the former term with the more euphonious one "Spéologie". This term was adopted by E.G. Racovitza and R. Jeannel in their writings. Nevertheless, E.A. Martel, remarked on the first page of his well-known book *Les Abîmes* that the term spéologie is less exact than spéléologie, because the Greeks used the word "σπέος" for artificial excavations, in particular the tombs in which the Egyptian Pharaohs were buried. Therefore for reasons of priority and correctness the term Spéléologie should be retained.

Speleology is the "science of caves", but the term taken in its true etymological sense is too restrictive. The cave is an anthropomorphic idea (B. Condé). We would indeed agree with Racovitza that the science of speleology is the study of the subterranean world. If this definition is accepted, it is clearly evident that speleology constitutes one of the sciences of the earth, which also include geology, physical geography and geophysics. The phenomena of weathering and erosion, etc. are part of the science of physical speleology.

In the present study we shall not be concerned with the physical aspects of speleology in themselves, rather we shall be considering the subterranean world as a habitat for living creatures.

B. BIOSPEOLOGY

Armand Viré (1904) proposed the term "biospéologie", to refer to the study of subterranean life. This science really came into being in the mid-nineteenth century. Since that time it has quickly developed owing to the efforts of many workers in exploring the subterranean world. Thanks to their work we are now in a position to draw up a complete list of subterranean species. Hundreds of memoirs have been devoted to the systematics of cavernicolous species. The structure of these animals and of some of their most remarkable anatomical peculiarities such as their modified eyes, are now fully understood.

Nevertheless, the great majority of publications devoted to cavernicolous animals are of a purely descriptive nature. Biological data on these creatures are qutie fragmentary. As for the experimental studies on the cavernicolous animals they are very few in number and most of them have been carried out only recently.

The reason for this can be found in the conditions which have presided till now over biospeological research. The study of cavernicolous animals has, to begin with, been carried out by amateurs with, for the most part, inadequate resources. When professional zoologists entered the field they have been, by their training, more interested in systematics and morphology than in biology and experimental research. It is only in the last few years that adequate laboratory facilities — in particular caves fitted out as laboratories — have been placed at the disposal of biospeologists.

The creation of underground laboratories has opened the era of experimental biospeology. Already the culture of cavernicolous animals and experimental work on them has taught us a great deal. Thus, newly acquired data permit us to see the origin and evolution of the cavernicolous animals in a new light. The reader will find this aspect of the subject discussed in the latter part of this book.

BIBLIOGRAPHY

MARTEL, E. A. (1894) *Les Abîmes*. Paris, Delagrave.
NUSSAC, L. DE (1892) *Essai de Spéologie*. Brive.
RACOVITZA, E. G. (1907) Essai sur les problèmes biospéologiques. *Biospeologica*, I. *Archiv. Zool. expér. géner.* 4, **VI**.
RACOVITZA, E. G. (1926) L'Institut de Spéologie de Cluj, et considérations générales sur l'importance, le rôle et l'organisation des Instituts de Recherches Scientifiques. *Trav. Inst. Speol. Cluj*, **I**.
VIRÉ, A. (1904) La Biospéologie. *Compt. rend. Acad. Sci. Paris;* **CXXXIX**.

PART 1

Biospeology

CHAPTER I

THE SUBTERRANEAN WORLD

THE STUDIES of zoologists and botanists have from the beginning been concerned mainly with the surface of the earth, including, of course, its atmosphere and its freshwater and marine environments. It was not until comparatively recent times that their attention became attracted to the subterranean environment, an immense new world of great diversity.

But it would be naïve to believe that there is clear-cut demarcation between the subterranean and surface environments. Today it appears with increasing clarity that the subterranean fauna is united with that of the surface by closely related intermediate forms. While this is true of terrestrial forms its validity is even more evident in aquatic life; but this will be considered in greater detail later. Cavernicolous animals, completely subservient to the subterranean environment, constitute a very small assemblage, which may be considered to be the end point of subterranean evolution.

The biospeologist must not ignore the intermediate groups of animals and plants which exist between the surface of the soil and the cave itself, because it is these which contain the majority of subterranean forms. However, the subterranean world presents so diverse a picture that a universal description might well be questioned. From the outset it is advisable to subdivide it.

To Racovitza (1907) must go the distinction of having worked out the first rational classification of the subterranean world. Although this classification has been improved and made more precise, its basis is still that suggested by Racovitza. The subterranean world may be divided up in the following manner:

> A. Solid media   (a) Rocks
>   (b) Earth
>
> B. Liquid media   (a) Subterranean streams
>   (b) Ground water

This classification is necessarily relative since in the subterranean world the limits between different media are much less evident than they are on the earth's surface. Not only may it be difficult to establish the demarcation between a subterranean stream and the ground water, but even the separation of solid and liquid media may become arbitrary underground.

For example, because the atmospheric humidity in caves is always near saturation, aquatic forms are able to spend long periods out of water; inversely, terrestrial forms can withstand prolonged submergence without harm. This will be further considered later on.

A. SOLID MEDIA

(a) Rocks

(1) *Natural Cavities.* Subterranean cavities are more usually hollowed out in calcareous rocks. However, caves have been reported in rocks of various composition (granite, basalt, lava, laterite, sandstone, etc.). Caves are found in rocks of various geological ages but occur more frequently and are of a larger size in Devonian and Carboniferous limestone (late Palaeozoic) and in the calcareous deposits of the Jurassic and Cretaceous (Mesozoic).

Depending on the processes of speleogenesis, the subterranean cavities may be present in the form of horizontal tunnels (grottoes and caves) or vertical perforations (shafts, gulfs, chasms, swallow-holes, etc.). At the same time as the subterranean cavities are being hollowed out they are continually receiving deposits of various sorts: ice, calcium carbonate and clay, for example. These cavities finally fill up through evolution.

The inhabitants of the subterranean cavities are the true cavernicoles.

(2) *Crevices.* Although the large cavities are accessible to man, there are very many others of so small a size that one can only observe them via their openings. These may be termed crevices. The latter may be recognised as "cracked seams", formed between stratigraphical layers, "diaclases" (Daubzée) resulting from rock breaks and arranged in a more or less perpendicular direction; and faults formed by the dislocation of strata. These crevices may be hollowed out to form finally large caves.

Many biospeleologists have wished to attribute to these crevices an essential role in the population of the subterranean world. This point will be considered in detail later, when evidence will be given to show that their importance in this respect has been somewhat exaggerated.

(3) *Artificial Cavities.* Man has frequently excavated rocks; firstly as a means of establishing communication (road and rail tunnels, canals for irrigation and navigation); secondly for the exploitation of mineral or fuel deposits; and thirdly to obtain materials for construction (quarries and catacombs). For a long time it was believed that artificial cavities were of little interest to the biologist because they were thought to contain only commonplace species of recent introduction attracted by the darkness and the humidity. Today it is known that although these conclusions are certainly correct in many cases, in others, particularly where rock masses already containing subterranean life have been penetrated, it is erroneous.

These species find in the artificial tunnels excavated by man conditions similar to those which exist in natural caves. The artificial tunnels constitute a trap for cavernicoles which are attracted by ligneous debris and remains of food and excrement always present in these situations.

Tombs may also provide shelter for true cavernicoles, an example being that of the collembolan *Sinella cavernarum* collected by Folsom (1902) from a tomb in Pennsylvania.

(b) Soils

The soil is a complex mixture of the products of decomposition of bed rocks and of external deposits. The latter may be of inorganic origin (fallen or wind-blown material and alluvium), or of organic origin (decomposing vegetation and the bodies and excrement of animals). The study of the soil is the true science of pedology.

The animals which live in the soil are generally quite different from cavernicoles. However, as is so frequently the case in biospeology, it is difficult rigidly to separate the two groups. Numerous examples can be cited of species which live both in the soil and in caves.

This is why the biospeologist must not ignore soil animals. However, a detailed treatment of these animals would be out of place in a book of this nature. Only the essentials will be dealt with here.

Soil animals may be divided into several ecological categories.

(1) Muscicoles and Humicoles. These animals are not even true soil animals, let alone cavernicolous ones. However, certain tropical humicoles show characteristics comparable with those of cavernicoles. These are "cavernicoles *en puissance*". This point will be considered in detail in the last part of this book.

(2) Endogeans. The endogeans are the true inhabitants of the soil. (The term "endogés" was proposed by Georges Pruvot and published by Racovitza.) Some entomologists bring the endogeans into the group of insects classified as "found under stones", because during periods of heavy rain certain true endogeans move towards the surface of the soil, and come to rest under large stones, a situation where they may be easily captured. However, this is not a general characteristic of endogeans and is, moreover, a temporary one. The interstices found within the soil itself constitute the true habitat of the endogeans.

In the small crevices within the soil the atmospheric humidity is near to saturation, and thus similar to that in caves. These biotopes seem to be intermediate between aquatic and terrestrial habitats. The external fecundation which is characteristic of aquatic animals, is known in terrestrial arthropods only in the Symphyla. which are true endogeans (Juberthie-Jupeau). Although the endogeans are different from the cavernicoles they often show characteristics analogous to them. For example, their integuments are frequently depigmented, and their eyes are frequently reduced, atrophied or even completely absent.

Many endogeans live in the entrances of caves, so that in this case the distinction between endogeans and cavernicoles becomes particularly delicate. In addition it will be suggested later (p. 8) that certain lines of cavernicoles have probably evolved from endogeans. For this reason biospeologists should pay attention to this group of animals.

Several examples of endogeans will now be given.

Protista: Flagellates; Amoebina Nuda (*Amoeba terricola* and *A. verrucosa,* etc.) and Amoebina Testacea are present in the soil.

Platyhelminthes: A number of terrestrial planarians are endogeans, occupying a similar niche to the earthworms, while others feed upon earthworms themselves. *Rhynchodemus (Microptana) humicola* Vejdovsky, is found in central Europe, it is milky-white and completely devoid of pigment. This species possesses anteriorly a pair of very small, pigmented eyes. Other endogeous planarians are not only quite lacking in pigmentation but are also eyeless, for example *Geoplana typhlops* Dendy, of Tasmania and *Geobia subterranea* Fr. Müller, of Brazil. The latter species feeds by sucking the blood of the earthworm *Lumbricus corethrurus,* a feature observed by Fritz Müller in 1857.

Nematoda: Many free-living nematodes are found in the soil.

Annelida: Oligochaetes belonging to the families Enchytraeidae and Lumbricidae are typical endogeans.

Mollusca: The gasteropod, *Caeciloides acicula,* on which an excellent monograph has been written by Wächtler(1929), is a true endogean. This species is unpigmented, is without eyes and has reproduction of the cavernicolous type (p. 358). *Testacella* is a genus of slugs which possesses a very small shell. They live up to a metre down in the soil where they feed upon earthworms. They still retain their eyes, however.

The Arthropoda constitute the majority of endogeous animals.

Crustacea: This group is represented by the terrestrial isopods. There are numerous members of the families of the Trichoniscidae, Squamiferidae and Armadillidiidae which are typically endogeous forms.

Myriapoda: Some Chilopoda (e.g. Geophilidae, *Crytops*), and Diplopoda (e.g. *Gervaisia*) and all the Symphyla, are endogeans.

Arachnida: The primitive opilionids of the genus *Siro* are typical endogeans.

Many Acarina belonging to the groups Parasitiformes, Trombidiformes and Oribatidae, spend an endogeous life.

Apterygota: The orders Collembola and Diplura (Japygidae and Campodeidae) contain many endogeans.

Isoptera: Most termites may be regarded as endogeans.

Orthoptera: The mole crickets with their reduced eyes and limbs modified for digging, are well-known endogeans.

The Stenopelmatidae belonging to the tribe Gryllacridoidea contain several endogeans which are, above all, remarkable for their fossorial adaptations. *Stenopelmatus talpa*, an apterous American species, has reduced antennae and its anterior legs are armed with spines. Another species of this family is the curious *Oryctopus prodigiosus* (Fig. 1), which was

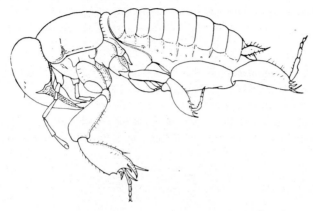

FIG. 1. *Oryctopus prodigiosus* (after Chopard).

discovered by Père Pantel at Madurai in India. The females are apterous, the anterior legs are modified for digging; the antennae are very much reduced, and the eyes are composed of only a few omatidia. The ovipositor is rudimentary.

FIG. 2. *Cylindrachaeta* (after Chopard).

The genus *Cylindrachaeta* of the family Cylindrachaetidae lives in the soil in Australia. Their habits are similar to those of the mole cricket (Tindale, 1928). They have a cylindrical body and the fore legs are fossorial. The eyes are reduced to a single omatidium (Fig. 2).

Homoptera: Typhlobrixia (Cixiidae) is a Madagascan genus. They are endogeans but sometimes are found in caves (Synave, 1953).

Hymenoptera: Worker ants of the genus *Lasius* (*flavus* group), and also of the genus *Solenopsis*, are depigmented and have reduced eyes. The strictly endogeous genera *Epitritus* and *Strumigenys* are blind.

Coleoptera: Numerous Coleoptera inhabit the soil. In this habitat the families best represented are the carabids, the pselaphids, the scyd-maenids, the catopids and the curculionids.

Coiffait (1959) has made a detailed study of these often microscopic beetles. They bury themselves in the soil to a depth of 50–60 cm which makes their study particularly arduous. He gave the name "Edaphobies" to these creatures. Coiffait gave a long list of characteristics belonging to them. They are different from the cavernicoles but show a parallel evolution to them; for example, depigmentation of the cuticle, and a tendency towards a reduction of the eyes and wings.

Vertebrata. There are many terrestrial vertebrates which are truly endogeous.

Amphibia: Gymnophiona.

Squamata: Amphisbaena (South America), *Blanus cinereus* (Spain), *Chirotes lumbricoides* (Mexico) etc.

Ophidia: Typhlopidae.

Marsupialia: Notoryctes typhlops.

Insectivora: Talpa (Europe and Asia), *Condylura* and *Scalops* (America) and *Chrysochloris* (Africa).

Rodentia: Spalax and *Geomys* (America).

(3) Myrmecophiles and Termitophiles. The distinction between myrmecophiles and termitophiles on the one hand, from endogeans on the other, is a delicate and often artificial one. For instance, the terrestrial isopods belonging to the family Squamiferidae and the genus *Platyarthrus* contain typical myrmecophiles. Such an example is *P. hoffmannseggi,* but the endogeous members of this genus are found only exceptionally with ants, for instance *P. costulatus.*

We cannot treat the extensive subject of the biological commensalism here. It is sufficient to say here that certain myrmecophiles show characteristics which are unquestionably those of the cavernicolous; beetles of the genus *Claviger* are testaceous and lack eyes, and further modified are the members of the genus *Platyarthrus,* which are perfectly white, deprived of pigment, and blind. Among the termitophiles the terrestrial isopods of the genera *Phylloniscus* and *Termitoniscus* which resemble *Platyarthrus* but which are more primitive members of the sub-order Oniscoidea may be cited.

(4) Pholeophiles. Burrows and underground nests give cover to a rich commensal fauna to which Falcoz has given the name "pholeophiles". This varied fauna has caught the attention of many naturalists (see bibliography), and is mainly composed of arthropods. These animals are often depigmented and blind.

B. LIQUID MEDIA

(a) Marine Caves

Marine caves contain quite a varied fauna but these animals have not usually undergone noticeable modification and have but a slight interest to the biospeologist. However two gobiids from the coast of California live in crevices *(Lethops connectens)* and in holes made by crabs *(Typhlogobius californiensis)* and have reduced eyes. But here it will suffice to refer the reader to an excellent review by Arndt (1936).

The only important marine cave which is unquestionably of interest to biospeologists is the Jameo de Agua situated on the island of Lanzarote (eastern Canary Island). This cave contains a subterranean lake which communicates with the sea. The influence of the tide is significant here, but the water of this lake is cooler than that of the sea and is also less salty, owing, to the infiltration of fresh water. The lake contains two depigmented crustaceans with reduced eyes; a galatheid, *Munidopsis polymorpha;* and a mysid, *Heterompsis cotti.*

(b) Subterranean Rivers

It has been appreciated for a long time that the hollowing out of subs terranean cavities is mainly due to the action of water. This is why cave- are traversed during their early existence by streams, while shafts and swallow-holes contain waterfalls.

Subterranean water courses are sometimes merely surface rivers which have buried themselves where they meet calcareous beds. In the majority of cases, however, subterranean rivers have their origin within caves. They result from a flowing together of oozings (vadose waters) which, after having passed through permeable strata, collect together to form streams or rivers.

Subterranean rivers have a tendency to sink lower and lower into the ground. The old stream beds do not, however, remain completely dry. During floods the old courses may be temporarily restored, or the seepage of water may give rise to underground lakes and pools, or to little ponds surrounded by walls of stalagmites to which one gives the name "gours". In these lakes, pools and gours, the aquatic subterranean fauna lives.

(c) Lake Bottoms

The profundal regions of lakes, and in particular those of the great alpine lakes, constitute an environment similar to the subterranean media. Here there is continual darkness and constant low temperature; an absence of seasonal rhythm; there is silt providing a similar environment to that of the mud deposited in the bottom of the subterranean pools. There is also a complete absence of green vegetation and in consequence, of herbivores. Nutriment is derived from deposits which sink from the surface of the lake. The fauna consists of limivores and carnivores.

The fauna of the depths of the great lakes below the Alps contain several species which are true cavernicoles, *Niyhargus foreli* and *Asellus cavaticus foreli.*

B 2

Lake Ochrid contains in its profundal zone many troglobia and troglophiles; in particular the two amphipods *Niphargus foreli ochridanus* and *Synurella ambulans longidactylus*, and the copepod *Ceuthonectes serbicus*, may be mentioned.

(d) Interstitial Medium

The name "interstitial medium" is given to the aquatic environment present in the minute interstices between grains of sand. The fauna which inhabits this environment has been termed the interstitial fauna.

Many types of interstitial medium may be distinguished:

(1) The sandy beaches which border the sea (salt water).
(2) The sandy beaches of lakes (fresh water).
(3) The sands of estuaries (brackish water).
(4) The sands deposited by underground streams.
(5) The medium constituted by ground water.
(6) The parafluvial subterranean nappes.

These nappes surround streams and extend under their banks to a distance of 1–3 m. The water which traverses the parafluvial nappes underflow is not always of the same origin. If the stream passes through an impermeable terrain, the parafluvial nappes are fed only by water from the stream. But if the banks extend into an alluvial plain then the parafluvial nappes are fed from the stream and otherwise by ground water.

Only the last three types of interstitial media can be described as hypogeous. Sandy deposits of the surface of the earth differ from the underground phreatic nappes notably because the former contain green vegetation, more especially algae, which are completely absent in the latter. However, it is quite impossible to set precise limits to the different groups of interstitial media. In the interstitial habitat the freshwater, brackish and marine faunae mix one with another in such a way that it is impossible to say where one begins and the other ends.

In the study of cavernicolous animals the marine interstitial fauna should not be neglected because it is closely related to the interstitial brackish and freshwater faunae. Most of the organisms of the interstitial biocoenose are euryhaline. We can be sure that many animals living in the fresh water interstitial medium have evolved from some marine environment. However, in this book, only the fauna of phreatic and parafluvial nappes will be considered.

The animals which populate phreatic nappes have been given the name "phreatobies" (Motas and Tanasachi). The term "phreaticole" is a barbarism. The study of this biological medium constitutes the particular science of phreatobiology (Motas, 1958). The fauna which occupy the parafluvial nappes have received the name "hyporheic fauna" (Orghidan, 1959).

It is convenient to include in the interstitial media a new biotope which has recently been proposed by Milan Meštrov (1962). This is the "hypotelminorheic medium". This medium has been formerly recognised by P. Remy (1926). One may observe in mountainous regions deposits of silt or clay, covered with dead leaves, humus, and more especially grass. These deposits are traversed at a depth of some centimetres by tiny streams of water. This biotope contains many representatives of the fauna of springs and of streams, such as the molluscs *Pisidium* and *Bythinella*, and also insect larvae. But species may also be found which have formerly been regarded as true cavernicoles (*Niphargus*, *Stenasellus*, *Pelodrilus leruthi*, and many of the light-avoiding planarians). These observations have led Meštrov to regard the aquatic "cavernicoles" as types originally inhabiting hypotelminorheic media. These species may well have been carried into underground caves where they have been able to persist, providing they found an environment broadly similar to their original one.

The interstitial phreatic, hyporheic and hypotelminorheic media contain so numerous and varied species, which are otherwise similar to those found in caves, that several biospeologists have denied the existence of true cavernicoles. "Es besteht keine Höhlenfauna" stated Karaman (1954). So extreme a position is certainly an exaggerated one. Many cavernicoles, including the Crustacea *Caecosphaeroma*, *Monolistra*, *Sphaeromides*, and *Troglocaris*, and most vertebrates, for example, *Amblyopsis*, *Caecobarbus*, *Anoptichthys* and *Proteus*, have certainly never populated the interstitial medium at any stage in their evolution. They passed directly from the waters of the earth's surface to the subterranean rivers and lakes.

These examples drawn from the aquatic subterranean fauna show, nevertheless, how vague and imprecise are the limits of the superficial, interstitial and cavernicolous faunae. But the biospeologist will discover in the imperceptible transitions between the hypogeous forms, the conditions on which to base an understanding of the origin and evolution of subterranean life.

The faunae which inhabit the interstitial media have been the subject, during the last 20 years, of a considerable number of studies carried out in many countries. The principal workers are given in the table below:

England	A. G. Nicholls
France	E. Angelier; P. A. Chappuis; Cl. Delamare-Deboutteville; E. Fauré-Fremiet; L. Hertzog; J. Y. Picard; P. Remy
Belgium	R. Leruth
Switzerland	C. Walter
Italy	S. Ruffo
Germany	E. Haine; S. Hussmann; W. Noll; A. Remane; H. W. Schafer; H. J. Stammer

Austria	A. Ruttner-Kolisko
Hungary	L. Szalay
Rumania	C. Motas and collaborators
Yugoslavia	St. Karaman; M. Meštrov
Poland	J. Wiszniewski
Russia	D. N. Sassuchin
United States	R. W. Pennak; C. B. Wilson
Madagascar	R. Paulian.

(e) Springs

The ground water and the underground streams come to the surface as springs. The outlets of large underground rivers are termed resurgents. Both contain particular faunae which have been the subjects of many interesting studies.

As in the case of the interstitial media, springs are an intermingled environment with characteristics intermediate to those of the subterranean and epigeous environments. They contain a mixed fauna of epigeous and cavernicolous species.

The environmental characteristics of springs recall those of the subterranean habitat, for there is only slight variation in the temperature of the water. Here, similarly, light is often reduced by overhanging rocks or vegetation and dead leaves.

CONCLUSIONS

This rapid review of subterranean life has of necessity been very summary. Many volumes would be necessary to describe and study in detail the different biotopes with which we are going to deal. This review fulfils a double object however. Firstly it allows us to make an appraisal of the extent and diversity of the subterranean world. On the other hand it gives us information concerning particular faunae inhabiting different subterranean biotopes, faunae which are, however, so intimately related to one another that it is difficult precisely to separate them. A critical examination of the hypogeous animals clearly shows that the cavernicolous fauna are neither exceptional nor the result of some accident, a view subscribed to by some ill-informed authors. They are completely surrounded by related forms, to which they are united by a succession of closely allied species. We discern already the multifarious sources from which the cavernicoles have taken their origin.

BIBLIOGRAPHY

The Subterranean World

RACOVITZA, E. G. (1907) Essai sur les problèmes biospéologiques. *Biospeologica. I. Archiv. Zool. Exper. Gén.* (4), **VI.**

Artificial Cavities

ALTHERR, E. (1932) La faune des mines de Bex. *Rev. Suisse Zool.* **XLV.**
BALAZUC, J., DRESCO, E., and HENROT, H. (1951) Biologie des carrières souterraines de la région parisienne. *Vie et Milieu*, **II.**
BUTTNER, K. (1926) Die Stollen, Bergwerke und Höhlen in der Umgebung von Zwickau und ihre Tierwelt. *Jahresber. Ver. Naturk. Zwickau.*
FOLSOM, J. W. (1902) Collembola of the grave. *Psyche*, **IX.**
HARTUNG, W. (1931) Über die Tierbevölkerung von Bergwerksschächten im Vergleich zur Höhlentierwelt. *Sitzb. Gesell. Naturf. Fr. Berlin.*
HNATENWYTSCH, B. (1939) Die Fauna der Erzgruben von Schneeberg im Erzgebirge. *Zool. Jahrb. Abt. System.* **LVI.**
HUSSON, R. (1936) Contribution à l'étude de la faune des cavités souterraines artificielles. *Annal. Sc. Nat. Zool.* (10), **XIX.**
KOLOSVARY, G. (1934) Recherches biologiques dans les grottes de pierre à chaux de la Hongrie. *Folia zool. hydrobiol. Riga*, **VI.**
LANDOIS, H. (1896) Pflanzen- und Tierleben in den Bergwerken. *24. Jahresbericht d. westf. Provinz. Ver. f. Wiss. Kunst. Münster.*
MARCHAL, C. (1899) Aperçu sur la faune vivante des Mines. *Bull. Soc. Hist. Nat. Autun*, **XII,**
MASCHKE, K. (1936) Die Höhlenfauna des Glatzer Schneeberges. 5 Die Metazoenfauna der Bergwerke bei Märisch-Altstadt. *Beitr. Biol. Glatz. Schneeb.* 2. Heft, Breslau.
MÜHLMANN, H. (1942) Die rezente Metazoenfauna der Harzer Höhlen und Bergwerke. *Zoogeographica*, **IV.**
VIRÉ, A. (1896) La faune des Catacombes de Paris. *Bull. Mus. Hist. Nat. Paris*, **II.**

Endogeans

COIFFAIT, H. (1959) Contribution à la connaissance des Coléoptères du sol. *Vie et Milieu, Suppl.* No. 7.
DELAMARE DEBOUTTEVILLE, C. (1951) Microfaune du sol des pays tempérés et tropicaux. *Vie et Milieu, Suppl.* No. 1.
FENTON, G. R. (1947) Essay review: the soil fauna, with special reference to the ecosystem of forest soil. *J. anim. Ecol.* **XVI.**
FRANZ, H. (1950) *Bodenzoologie als Grundlage der Bodenpflege.* Berlin.
JACOT, A. P. (1940) The fauna of the soil. *Quart. rev. Biol.*
KEVAN, D. K. MAC E., 1955 *Soil Zoology.* London.
KÜHNELT, W. (1950) *Bodenbioloie.* Wien.
SCHALLER, F. (1962) *Die Unterwelt des Tierreiches.* Berlin–Göttingen–Heidelberg.
SYNAVE, H. (1953) Un Cixiide troglobie découvert dans les galeries souterraines du système de Namoroka *(Hemiptera-Homoptera). Natural. Malgache.* **V.**
TINDALE, N. B. (1928) Australian mole crickets of the family Gryllotalpidae (Orthoptera). *Rec. South Austral. Mus.* **IV.**
WÄCHTLER, W. (1929) Anatomie und Biologie der augenlosen Landlungenschnecke *Caecilioides acicula* Müll. *Zeit. Morphol. Ökol. Tiere.* **XIII.**

Myrmecophiles and Termitophiles

DONISTHORPE, H. (1927) *The Guests of British Ants.* London.
WASMANN, E. (1894) *Kritisches Verzeichnis der myrmekophilen und termitophilen Arthropoden.* Berlin.
WASMANN, E. (1925) Die Ameisenmimikry, ein exakter Beitrag zum Mimikryproblem und zur Theorie der Anpassung. *Abhandl. z. theor. Biol.* **XIX.**

Pholeophiles

DELAMARE- DEBOUTTEVILLE CL. et PAULIAN, R. (1952) *Faune des nids et des terriers en Basse Côte d'Ivoire.* Paris.
FALCOZ, L. (1914) *Contribution à l'étude de la faune des microcavernes; faune des terriers et des nids.* Lyon.
HESELHAUS, F. (1914) Über Arthropoden in Nestern. *Tijdschr. v. Entomo!.* **LVII.**
JEANNEL, R. (1945) Mission scientifique de l'Omo. VI. Zoologie No. 57 Faune des terriers des Rats-Taupes. *Mém. Mus. Hist. Nat. Paris.* N.S. **XIX.**
MARIÉ, P. (1930) Contribution à l'étude et à la recherche des Arthropodes commensaux de la Marmotte des Alpes. *Annal. Sc. Nat. Zool.* (10), **XIII.**

Marine Caves

ARNDT, W. (1936) Fortschritte unserer Kenntnis der tierischen Bewohnerschaft der Meereshöhlen. *Mitteil. Höhl. Karstf.*
CALMAN, W. T. (1904) On *Munidopsis polymorpha* Koelbel, a cave dwelling marine crustacean from the Canary Islands. *Annals and Magaz. Nat. Hist.* (7), **XIV.**
FAGE, L. and MONOD, TH. (1936) La faune marine du Jameo de Agua, lac souterrain de l'île de Lanzarote. *Biospeologica*, **LXIII.** *Archiv. Zool. exper. Gen.* **LXXVIII.**
HARMS, W. (1921) Das rudimentäre Sehorgan eines Höhlendecapoden, *Munidopsis polymorpha* Koelbel aus der Cueva de los Verdes auf der Insel Lanzarote. *Zool. Anz.* **LII.**

Lake Bottoms

EKMAN SVEN (1917) Allgemeine Bemerkungen über die Tiefenfauna der Binnenseen. *Intern. Rev. gesamt. Hydrobiol.* **VIII.**
FEHLMANN, J. W. (1911) Die Tiefenfauna des Luganer Sees. *Intern. Rev. gesamt. Hydrobiol. Suppl.* **IV.**
FOREL, F. A. (1901) *Le Léman.* Tome III; Lausanne.
MONARD, A. (1919) La faune profonde du lac de Neuchâtel. *Bull. Soc. neuchatel. Sc. Nat.* **XLIV.**
STANKOVIČ, S. (1960) *The Balkan Lake Ochrid and its living World.* Den Haag.
ZSCHOKKE, F. (1911) *Die Tiefseefauna der Seen Mitteleuropas.* Leipzig.

Interstitial Media

(for a complete bibliography, see Delamare-Debouteville, 1960)
ANGELIER, E. (1962) Remarques sur la répartition de la faune dans le milieu interstitiel hyporhéique. *Zool. Anz.* **CLXVIII.**
DELAMARE-DEBOUTTEVILLE, CL. (1960) *Biologie des eaux souterraines littorales et continentales.* Paris.
KARAMAN, ST. (1954) Über unsere unterirdische Fauna. *Acta. Mus. Maced. Sc. Nat.* **I.**
MEŠTROV, M. (1960) Faunističko-ekološka i biocenoloska istraživanja podzemnih voda savske nizine. *Biološki Gla..nik.* **XIII.**
MEŠTROV, M. (1962) Un nouveau milieu aquatique souterrain: le biotope hypotelminorhéique. *Compt. rend. Acad. Sci. Paris.* **CCLIV.**

Motas, C. (1958) Freatobıologia, o nouă ramură a Limnologiei. *Natura* **X.**

Motas, C. (1962) Procédé des sondages phreatiques. Division du domaine souterrain. Classification écologique des animaux souterrains. Le Psammon. *Acta. Mus. Maced. Sc. Nat.* **VIII.**

Motas, C., Botosaneanu, L. et Negrea, St. (1962) *Cercetari asupra Biologiei Izvoarelor si Apelor Freatice din Partea Centrala a Cimpiei Romîne.* Bucuresti.

Orghidan, R. (1959) Ein neuer Lebensraum des unterirdischen Wassers. Der hyporeische Biotop. *Archiv. f. Hydrobiol.* **LV.**

Picard, J. Y. (1962) Contribution à la connaissance de la faune psammique de Lorraine. *Vie et Milieu* **XIII**

Remy, P. (1926) Les Niphargus des sources sont-ils des emigrés des nappes d'eaux souterraines? *Comte. rend. Conr. Soc. Sav.*

Ruffo, S. (1961) Problemi relativi allo studio della fauna interstiziale iporreica. *Boll. d. Zool.* **XXVIII.**

Springs

Beyer, H. (1932) Die Tierwelt der Quellen und Bäche des Baumbergegebietes. *Abh. Westf. Prov. Mus. f. Naturk. Münster* **III.**

Bornhauser, K. (1912) Die Tierwelt der Quellen. *Intern. Rev. gesammt. Hydrobiol.* Suppl. **V.**

Thienemann, A. (1924) Hydrobiologische Untersuchungen an Quellen. *Archiv. f. Hydrobiol.* **XIV.**

Vandel, A. (1920) Sur la faune des sources. *Bull. Soc. Zool. France* **XLV.**

CHAPTER II

THE CAVERNICOLES

THOSE animals which live underground have received the name "hypo-gean"†; this term is in contrast to "epigean" which categorises animals which live at the surface of the soil.

In the preceding chapter the heterogeneity of the hypogean population has been shown. In this book only two categories of hypogean will be dealt with: the cavernicoles and the phreatobia.

A. CLASSIFICATION OF THE CAVERNICOLES AND THEIR NOMENCLATURE

The cavernicoles are an essentially heterogeneous assembly and the first task of the biospeologist is to erect a classification and establish a nom-enclature.

From the first studies carried out on the underground environment, it was quite clear that some cavernicoles occupied different regions of the caves. An ecological classification was therefore established.

The first attempt at classification was by J.C.Schiödte in 1849, who dealt with this problem in his *Specimen Faunae Subterraneae*. This classi-fication, which today appears to be so crude, divided the cavernicoles into four categories: animals of shade, crepuscular animals, animals of dark regions and lastly, animals living on stalactites. Joseph (1882) revived Schiödte's classification but removed the last category.

Another, more rational, classification was proposed by J.B.Schiner (1854), in a paper devoted to the cavernicolous fauna dealt with in an extensive work by A.Schmidl, on the caves of the Adelsberg region. He distinguished those forms strictly adapted for life in caves, which he called "troglobies"; others which were frequent in caves but not totally confined to them called troglophiles; and finally the occasional cavernicoles. Schiner's classi-fication has been generally accepted by biospeologists. Racovitza (1907) has proposed the name "trogloxene" to replace the "occasional guest" category proposed by Schiner.

The same classification has been accepted by ecologists who have only

† Racovitza (1907) opposed "hypogean" not with respect to "epigean" but to "caver-nicole". This concept is in contradiction to the etymology of the term hypogean.

16

adopted different terms. Thus R.Hesse in his classic Treatise of Zoo-geography, substituted the terms eucaval, tychocaval and xenocaval. The conchyologist C.R.Boettger, used the same nomenclature. A.Thienemann in his work on the waters of the continent created the terms stygobie, stygophile and stygoxene, terms which were adopted by the hydracaro-logist C.Motas.

Dudich, in his monograph on the "Baradla" cave, proposed a slightly altered classification which does not have much advantage over the classical classification:

Classical Classification	Classification of Dudich
Troglobie	Eutroglobionte
Troglophile	Hemitroglobionte
Regular trogloxene	Pseudotroglobionte
Occasional trogloxene	Tychotroglobionte

Two Italian biologists, a botanist R.Tomaselli, and a zoologist M.Pavan, elaborated a more complex classification consisting of seven different categories.

Cavernicoles	By chance	With intolerance	No reproduction	Eutrogloxene (1)
			Reproduction	Subtrogloxene (2)
		With tolerance	No reproduction	Aphyletic Trogloxene (3)
			Reproduction	Phyletic Trogloxene (4)
	By choice	Facultative	No reproduction	Subtroglophile (5)
			Reproduction	Eutroglophile (6)
		Obligatory	Reproduction	Troglobie (7)

This classification is extremely theoretical and its application is beset with many difficulties, mainly due to our very fragmentary knowledge of the reproduction of cavernicoles.

It is necessary to understand that the terms troglobie (= troglobian), and especially troglophile and trogloxene correspond to heterogeneous groups. But they are extremely convenient to use, and for this reason they will be adopted in this book.

B. CHARACTERISTICS OF THE CAVERNICOLES

Omitting the troglophiles and trogloxenes for the time being, the subject of this section will be the true cavernicoles, that is to say the troglobia.

We shall regard the troglobia as those living creatures which inhabit underground caves and waters, while their presence in the surface habitat is quite exceptional. This ecological criterion is the only one which has

B 2a

universal validity and does not lead to any doubt. However, many bio-speologists have attempted to complete this definition by the addition of other considerations. It is certain that cavernicoles possess a special physi-ology and this subject will be developed in the fourth part of this book. Most biospeologists maintain that cavernicoles also show a morphology of their own. We will examine this statement in accordance with the facts.

(a) Size

Biospeologists generally agree that troglobia are larger than the epigeans belonging to the same group. According to Coiffait (1959) the sizes of endogeous Coleoptera are generally less than those of related epigeous species, whilst those of troglobia are usually greater. The latter conclusion is certainly justified in many cases. The terrestrial cavernicolous isopods belonging to the genus *Titanethes* are the largest of the Trichoniscidae. The cavernicolous crayfish, *Cambarus bartoni tenebrosus* and the amphipod *Crangonyx gracilis*, are larger than the related epigeous species (Banta, 1907). *Heteromerus longicornis*, a truly cavernicolous species from the caves of Herzegovina is the largest of the Collembola (Absolon, 1900). The cavern-icolous campodeids are mainly larger than the epigeous species (Con-dé, 1955). The cavernicolous members of the genus *Koenenia* are larger than the surface forms (Peyerimhoff, 1908). The cavernicolous beetles belonging to the Trechinae and Bathysciinae are the largest species in these two sub-families (Coiffait, 1955).

This rule is, however, far from being general. The cavernicolous isopods of the families Parasellidae and Anthuridae, the Syncarida belonging to the family Bathynellacae, and generally speaking all the members of the interstitial fauna, are of microscopic size.

In fact these two manifestations, giantism and dwarfism, are opposites only in appearance. They both demonstrate the unbalanced states of growth which so frequently show themselves at the end of the evolution of ancient phyletic lines.

(b) Form of the Body and its Appendages

All books concerned with biospeology state that cavernicoles possess a more slender body with more slender appendages than do related epigeous forms. This statement is certainly exact where it applies to interstitial forms which crawl between grains of sand, and also to certain groups such as the Orthoptera, Trechinae (Carabidae), and the opilionids (Fig. 3). But it can be easily seen that the cavernicoles only exaggerate the mor-phological characteristics of the groups to which they belong, and which are manifested, even if to a lesser degree, in epigeous forms.

On the other hand, cavernicoles related to stout epigeous types with short limbs, give no impression of slenderness, and they possess normal appendages.

This question will be examined again in Chapter XXXI of this book.

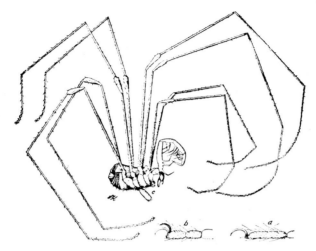

FIG. 3. The cavernicolous opilionid, *Phalangodes armata*, from caves in Kentucky (after Hubbard).

(c) Physogastry

It has often been stated that physogastry is characteristic of certain groups of cavernicoles. In fact, it appears only incidentally in the troglobia. It has been reported in certain female Diptera such as *Allopnyxia patrizii* (Fig. 45).

With respect to the swelling of the posterior part of the body in certain Bathysciinae such as *Leptodirus* (Fig. 53), this is due to a false physogastry, caused by a swelling of the elytra while the abdomen remains normal.

(d) Apterism

Insects that are troglobia are generally apterous, whilst the troglophiles are frequently brachypterous. The reduction or disappearance of wings is associated with a regression of the alary muscles. However, the nymphs of apterous and cavernicolous insects generally possess wing cases or pterothecae, for example *Geotrechus* (Coiffait), Bathysciinae (S. Deleurance). In contrast, elytra are never atrophied in cavernicoles (Jeannel, 1926).

Apterism is a manifestation not confined to cavernicoles, but largely widespread in epigeous insects. For example, the muscicolous species of the genus *Trechus* often possess small wings, sometimes reduced to vestiges similar to those of the cavernicolous species of *Aphaenops* (Fig. 4). Apterism represents the manifestation of a phyletic regression. Jeannel wrote with much truth: "L'aptérisme se présente comme une orthogenèse, qui survient chez certaines lignées, indépendamment des conditions de vie et particulièrement du climat".

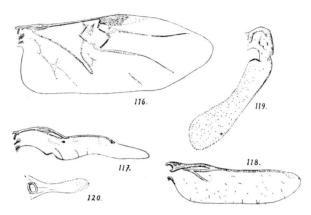

FIG. 4. The reduction of the wings in the *Trechinae*. *116*, *Trechus quadristriatus; 117*, *Trechus obtusus 118*, *Trechus rufulus; 119*, *Trechus distigma; 120*, *Aphaenops cerberus*. (After Jeannel).

(e) Depigmentation

The relations between pigments and cavernicolous life will be fully discussed in Chapter XXV. For the moment it will be sufficient to point out that the great majority of troglobia are partially or totally depigmented forms. However, it is known that a few species retain normal pigmentation, even though they live permanently in underground caves.

On the other hand, depigmentation is not totally confined to cavernicolous species. Thousands of humicoles, lapidicoles, endogeans, myrmecophiles and termitophiles have been recorded which are partially or totally depigmented.

(f) Reduction of the Eyes (Anophthalmy)

The reduction of the eyes is a phenomenon frequently encountered in cavernicoles. This regression may take many forms. In all cavernicolous groups a series of reductions extending from barely perceptible regression to the total disappearance of the eyes and associated structures, may be observed. This topic will be dealt with in Chapter XXVI.

However, many epigeous species have eyes which are reduced or absent, so that it would be false to state that the reduction of the eyes is correlated with underground life only.

In conclusion it may be said that no single morphological criterion can be taken as strictly characteristic of the cavernicoles. All that may be stated is that certain manifestations, in particular depigmentation and the loss of eyes, are statistically more frequent in cavernicoles than in epigeans.

BIBLIOGRAPHY

Classification of the Cavernicoles

BAGGINI, A. (1961) Sulla distribuzione della Fauna cavernicola Italiana nelle categorie biospeologiche. *Rassegna Speleologica Italiana*, **XIII.**
DUDICH, E. (1932) Biologie der Aggteleker Tropfsteinhöhle "Baradla", Ungarn. *Speleol. Monographien*, **XIII.**
HESSE, R. (1924) *Tiergeographie auf ökologischer Grundlage.* Jena, Fischer.
JOSEPH, G. (1882) Systematisches Verzeichnis der in den Tropfsteingrotten von Krain einheimischen Arthropoden nebst Diagnosen der vom Verfasser entdeckten und bisher noch nicht beschriebenen Arten. *Berliner Entomol. Zeit.* **XXV.**
PAVAN, M. (1944) Appunti di biospeologia. I. Considerazioni sui concetti di troglobio, troglofilo e troglosseno. *Le Grotte d'Italia*, (2), **V.**
PAVAN, M. (1950) Observations sur les concepts de troglobie, troglophile et trogloxène. *Bull. Trim. Assoc. Spéléol. Est.* Vesoul, **III.**
PAVAN, M. (1950) Consideraciones sobre los conceptos de Troglobio, Troglofilo y Trogloxeno. *Spelaeon*, **I.**
PAVAN, M. (1958) Relazione sulla classificazione biologica degli animali cavernicoli. *Rassegna Speleol. Ital. e. Soc. Spel. Ital.* (4), **II.**
RACOVITZA, E. G. (1907) Essai sur les problèmes biospéologiques. *Biospeologica, I; Archiv. Zool. exper. gener.* (4), **VI.**
SCHINER, J. R. (1854) Fauna der Adelsberger, Lueger und Magdalener-Grotte. *In* A. SCHMIDL, *Die Grotten und Höhlen von Adelsberg, Lueg, Planina und Loos.* Wien, Braunmüller.
SCHIÖDTE, J. C. (1849) *Specimen Faunae subterraneae. Bidrag til den underjordiske Fauna.* Kjöbenhavn.
THIENEMANN, A. (1926) Die Binnengewässer Mitteleuropas. *Die Binnengewässer*, Bd. **I,** Stuttgart.
TOMASELLI, R. (1955) Relazione sulla nomenclatura botanica speleologica. *Archiv. Bot.* **XXXI.**
TOMASELLI, R. (1956) Relazione sulla nomenclatura botanica speleologica. *Rassegna Speleol. Ital. e. Soc. Spel. Ital.* **III.**

Characteristics of the Cavernicoles

ABSOLON, K. (1900) Über zwei neue Collembolen aus den Höhlen des österreichischen Occupationsgebietes. *Zool. Anz.* **XXIII.**
BANTA, A. M. (1907) The fauna of the Mayfield's Cave. *Carnegie Inst. Washington* No. 69.
COIFFAIT, H. (1955) Les fossiles vivants du sol. *Notes biospéologiques;* X.
COIFFAIT, H. (1959) Contribution à la connaissance des Coléopteres du sol. *Vie et Milieu,* Suppl. No. 7.
CONDÉ, B. (1955) Matériaux pour une Monographie des Diploures Campodéidés. *Mem. Mus. Hist. Nat. Paris. N. S. Ser. A. Zool.* **XII.**
JEANNEL, R. (1926) Monographie des *Trechinae* (première livraison). *L'Abeille,* **XXXII.**
JEANNEL, R. (1943) *Les Fossiles vivants des cavernes.* Paris, Gallimard.
PEYERIMHOFF, P. DE (1908) Palpigradi (ière série). *Biospeologica,* **VIII**; *Archiv. Zool. exper. gener.* (4), **IX.**

CHAPTER III

THE ORIGIN AND DEVELOPMENT
OF BIOSPEOLOGY

INTRODUCTION

While the animals which inhabit the seas, the rivers and the land have captured the attention of man for many centuries, it was not until comparatively recent times that naturalists began to be interested in subterranean life.

The frequentation of the caves is, however, an old human custom. Since the beginnings of our ancestry, *Australopithecus* and *Sinanthropus* used caves for shelter. Much later Neanderthal Man and afterwards the Man of the Upper Palaeolithic made caves their normal habitation. The remarkable animal artists belonged to the Magdalenian culture and reproduced with admirable talent the horse and the bison, animals which they hunted each day. However, one of them, with whom one must not contest the title of the first biospeologist, faithfully reproduced a cavernicolous insect which the eminent orthopterist Lucien Chopard, has unhesitatingly attributed to the genus *Troglophilus*. It was carved on a piece of bone discovered by Count Bégouen in the Grotte des Trois-Frères, at Montesquieu-Avantès (Ariège). (Fig. 43).

The ancient Greeks believed that the caves, which are numerous in Greece, particularly in Attica, were the abode of the dead. It is probably this legend which for a long time led to the belief that caves could not contain life. Professor Jeannel wrote: "For two centuries it was thought impossible that caves could favour the development of life. Also, the discovery of living animals, without eyes, in the depths of caverns, was taken as a surprising anomaly. For some years the cavernicolous fauna was always considered as a very exceptional accident in the world, simply a biological curiosity". We know today that nothing could be further from the truth.

A. BEGINNINGS OF BIOSPEOLOGY

It was *Proteus* which firstly, because of its size, attracted the attention of man, and also the first to receive a vernacular name, the olm. It is probable that the inhabitants of Carniola and Dalmatia had known of it for many years. It was reported for the first time in a book printed in 1689 by Baron Johann Weichard Valvasor, in a study which he devoted to his native country, *Die Ehre des Herzogthum Crain* (Laibach), in which he speaks of a "Dragon", in a spring near Laibach (Ljubljana); the "Dragon" could only be *Proteus*. But it was only 80 years later in 1768, that this amphibian was described by Laurenti, under the name *Proteus anguinus*.

In 1799, Baron Alexander von Humboldt accompanied by the French botanist Bonpland visited the famous cavern in the Caripe valley in Venezuela, known as the Cueva del Guàcharo. He described a cavernicolous bird, which had been known for a long time by the Indians, and called guàcharo, but which was new to science. He named it *Steatornis caripensis*.

One cannot, however, speak of biospeology at this time, but only of natural history. *Proteus* represents a veritable giant amongst the European cave fauna, because it reaches a length of 30 cm. Most European cavernicoles measure only a few millimetres in length. It is for this reason that they were not discovered until informed explorers with a knowledge of zoology and entomology visited the caves.

It was in 1831 that Count Franz von Hohenwart, who had already found again *Proteus* in the caves of Carniola, collected the first cavernicolous insect in the immense Adelsberg cave; this was one of the more extraordinary cavernicolous insects by reason of its physogastry and the slenderness of its appendages. This curious beetle was described in the following year by Fernand Schmidt under the name *Leptodirus Hohenwarti*.

In 1845, the Danish zoologist, J.C. Schiödte, visited caves in Carniola. He collected a varied fauna consisting of insects (Coleoptera and Collembola), Arachnida (an araneid and a chernetid), and Crustacea (an amphipod and an isopod). Four years later he published the first work on biospeology: *Specimen Faunae Subterraneae -- Bidrag til den underjordiske Fauna*, Kjöbenhavn, 1849. In this book he reported the species which he had discovered, and he drew them and described them with great care. This important work marked the beginning of biospeological research, and at the same time the birth of biospeology.

The same year an Austrian scientist, Dr. Adolf Schmidl, undertook the systematic exploration of the underground caves in the region of Adelsberg. In 1854 he published an important speleological monograph which rapidly became a classic: *Die Grotten und Höhlen von Adelsberg, Lueg, Planina und Lass*, Wien. This work also contained an article on cavernicolous fauna by Dr. J.R. Schiner and a botanical contribution by Dr. Alois Porkorny.

From that time, biospeological research was pursued actively in many European countries and also in North America.

In the middle of the last century L. Miller and J. von Frivaldsky described the cavernicolous fauna of Hungary; and V. Motschoulsky (1840–1869) dealt with the first insect cavernicoles captured in the caves of Caucasia.

The first discovery of cavernicolous insects made in France was in the Pyrenees. In 1857 Delarouzée and then Linder, explored the famous Grotte de Bétharram and described its remarkable entomological fauna. In the same year Charles Lespès visited the caves of Ariège and discovered numerous cavernicolous Coleoptera. The Pyrenean caves were surveyed the following year by a pleiad of active, enthusiastic entomologists: Abeille de Perrin, de Saulcy, Piochard de la Brûlerie, de Bonvouloir, Valéry-Mayet, Mestre, Marquet, etc. Starting in 1870, Eugène Simon explored the caves in Southern France and found the first cavernicolous Arachnida and Isopoda.

In North America during the years 1842–1844, de Kay, Wyman and Tellkampf described the first cavernicolous fish, *Amblyopsis spelaeus*, caught in the immense Mammoth cave. Several years later F.Poey discovered two remarkable cavernicolous fish belonging to the family Brotulidae (*Lucifuga subterraneus* and *Stygicolus dentatus*), in the caves of Cuba. However, American biospeology did not advance until the beginning of the 1870's with the remarkable studies of E.D.Cope and A.S.Packard.

In 1866 the zoologist, Dominik Bilimek, Curator of the Museum of Mexico and friend of the Emperor of Mexico, Ferdinand Maximilien, visited the cave of Cacahuamilpa, situated not far from the capital, and described the first representatives of the Mexican cavernicolous fauna, one of the most interesting in the New World.

B. BIOSPEOLOGICAL RESEARCH THROUGHOUT THE WORLD

After 1870 biospeological research increased rapidly and publications became so numerous that it is impossible to treat them systematically in a review of this nature. But we shall try to isolate the different factors which brought about the development of biospeology.

(a) Lithological Factors

Biospeology can expand only in countries possessing numerous caves. Although subterranean cavities can be found in rocks of diverse natures the immense majority of them are hollowed out in calcareous beds. They are particularly numerous in regions having undergone karstic evolution.

(b) Geographical Factors

It is not sufficient that regions should contain numerous caves for them to possess an interesting cavernicolous fauna. Geological evolution and climatic factors have created strongly different conditions which will now be enumerated.

(1) The zones of the Earth which were covered by glaciers during the Quaternary, such as Europe and North America, possess only a sparse

cavernicolous fauna. Terrestrial cavernicoles are lacking, or at least they are extremely rare. The aquatic cavernicoles, whose means of dispersal are more effective than those owned by terrestrial forms, were able to colonise the zones previously occupied by glaciers. However, the cavernicolous faunae in these regions are less rich in species and less abundant in individuals than those of more southerly countries. All the north of Europe has passed through such conditions and one can understand why biospeology has not been developed in Scandinavian countries.

However, aquatic cavernicolous faunae are found in the caves of south and west Germany as well as in the Tatras (Chodorowski, 1959).

(2) The regions richest in cavernicoles and highly specialised subterranean forms are the karstic regions possessing a Mediterranean climate.

The region of the world possessing the most abundant and varied cavernicolous fauna is the karstic zone bordering the Adriatic, that is to say Istria, Carniola, Croatia, Dalmatia, Herzegovina, Montenegro and Albania. It is not surprising that Karst itself has been the cradle of biospeology.

Relatively recent lists of the subterranean fauna of the Balkans have been prepared by Karaman (1935, 1954) and by Hadži (1957).

A rich and varied fauna is found in the south-east of Europe, that is, Transylvania, Bulgaria, Macedonia and Greece. A list of the Bulgarian cavernicolous fauna by V. B. Gueorguiev and P. Beron has recently come out (1962).

Finally, the remaining sections of the circum-Mediterranean fauna populate the caves of the Crimea and Transcaucasia. They have been described by Birstein and Lopachov (1940) and Birstein (1950).

The western part of the Mediterranean is equally rich in cavernicoles, in particular Portugal, the Spanish Levant, Cantabria, the Pyrenees, the Southern Alps, Sardinia (Patrizi, 1956, 1958) and Puglia, in the southern part of Italy (Ruffo, 1955).

Cavernicoles have been collected in North Africa (Algeria and Morocco) but the hypogean fauna is much less rich and varied than in the northern Mediterranean.

In the Mediterranean regions of Asia there are many places to interest biospeologists, notably Syria, the Lebanon and Iran.

In Japan there are many caves which have been hollowed out from calcareous or volcanic rock. However, Japanese biospeology was not developed until quite recently, in fact in 1915. Its short history has been reviewed by S. Uéno (1957).

In North America, the karstic regions rich in caves occur in the central and eastern regions of the United States (Indiana, Kentucky, Virginia, Missouri, Tennessee, Alabama and Texas). Their fauna has been accurately listed by Brother G. Nicholas (1960, 1962), the prime-mover in a group of biospeologists from Notre-Dame (Indiana).

Mexico and also the island of Cuba are places of great interest for biospeologists. These two regions have been investigated for many years but there is still a great deal to discover. In Guatemala there is the immense Lanquin Cave, whose fauna is still known imperfectly.

The countries of the Southern Hemisphere occupy an analogous situation to those of the Mediterranean region. There are many caves in South Africa, Australia and New Zealand, which are now beginning to attract the attention of biospeologists, for several of them contain a quite original cavernicolous fauna.

(3) From a biospeological point of view, the tropical regions are sharply contrasted with the Mediterranean countries. The species which populate their underground cavities differ only slightly from their epigeous congeners. Thus the tropical caves contain no true cavernicoles, or at least, terrestrial cavernicoles, because amongst the aquatic hypogeous types (in particular in the Crustacea and in the fish), many species showing characters which are incontestably those of cavernicoles are found.

Four regions of Africa are particularly interesting to the biospeologist: Somalia, investigated by Italian zoologists; East Africa (Kenya and Zanzibar) where the caves have been explored by Sjöstedt, and then by Alluaud and Jeannel; the Congo, where the subterranean cavities have been studied by Heuts and Leleup; and lastly Madagascar, where there are immense caves which have been visited by Decary, by Millot and by Paulian. It may be added that Hiernaux and Villiers (1955) have collected the fauna of some caves in Guinea which have been hollowed out in sandstone and laterite. Mr. N. Leleup is preparing a general review of the cavernicolous fauna of tropical Africa.

In the tropical regions of Asia there are many caves. A faunal list of these caves has recently been drawn up by Lindberg (1960). The two most interesting regions for the biospeologist are Assam and the Burma/Malaya area.

In southern China to the south of the Yang-Tse-Kiang there lies a vast karstic region which is rich in caverns. This area has been recently explored by the Hungarian speleologist Balazs (1961, 1962), who has given an account of the cavernicolous fauna. Similarly, Formosa possesses a number of caves containing fauna.

Further south, Tong King, as well as the islands of the Indonesian Archipelago (Sumatra, Java, Borneo, and the Philippines), contain caves but very little is known of their fauna (Stadler, 1927). Finally, the Japanese zoologist H. Torii has investigated the caves in the archipelagos of Micronesia including the Mariannes and the Carolines.

South America remains the *terra incognita* of biospeology. Thanks to the efforts of Bellard Pietri (1954, 1956), a preliminary list of the Venezuelan caves and their fauna has been drawn up. The Cueva del Guàcharo, near Caripe, is one of the best-known caves in Venezuela (Bordon, 1959; Bellard Pietri, 1960). There is a cavern of the same name in Trinidad. Finally a number of cavernicolous fish have been discovered in Brazil.

(c) The Human Factor

A factor which is at the root of all human activities is the personal factor. Biospeology has progressed in those countries which have been fortunate enough to be the birthplace of active and enthusiastic workers who have been the leaders in their field and who have become associated with many disciples each of whom has wished to improve upon the work of his more famous contemporaries. A list of various prominent workers appears below, although this should not be regarded as exhaustive:

France: E.G.Racovitza; René Jeannel; Louis Fage
Belgium: Robert Leruth
Portugal: Antonio de Barros Machado
Germany: Hans Stammer; A.Remane
Hungary: Endre Dudich
Czechoslovakia: Karl Absolon
Yugoslavia: Stanko Karaman
Rumania: E.G.Racovitza; P.A.Chappuis; C.Motas
Bulgaria: Ivan Buresch
Russia: J.A.Birstein; E.V.Borutzky
Japan: Masuzo Uéno and Schun-Ichi Uéno
United States: A.S.Packard; Carl Eigenmann
Mexico: L.Herrera; Candido Bolivar y Pieltain
Venezuela: Eugenio de Bellard Pietri
New Zealand: Ch.Chilton; A.M.Richards

C. THE ORGANISATION OF BIOSPEOLOGICAL RESEARCH

Up until the beginning of the century biospeology was the work of amateurs, good entomologists and true scientists to be sure, but they worked alone. During half a century these researches have become organised, the stage of the individual plan has now passed to that of systematic organisation.

Biospeologica

In the first place mention must be made of a most important organisation whose foundation is due to the initiative of the Rumanian zoologist E.G.Racovitza, who in so doing made a great contribution to biospeology. Racovitza and the outstanding entomologist René Jeannel, surrounded themselves by a group of associates amongst whom may be mentioned Louis Fage. These biologists founded in 1907 an association which received the name Biospeologica. This organisation which was originated in France at the Arago Laboratory in Banyuls-sur-mer, was continued in Rumania,

when the Cluj Institute of Speology was founded in 1920, under the direc-
torship of E. G. Racovitza. Upon its foundation this association proposed
a threefold aim:

(1) The exploration of caves and research on cavernicolous animals.
Many thousands of caves have been investigated by the collaborators
of *Biospeologica*, most of them in Europe but many in Africa and
N. America as well. These caves have been described in the *Enumera-
tions*, of which nine series have appeared. The first volume was published in 1907 and the last in 1962.
(2) The identification of the material collected. This has been entrusted
only to the most qualified specialists. All the groups containing
cavernicoles have been reviewed in this way, not only animals, but
also fungi.
(3) The publication of the results obtained in memoirs, the whole collection of which has been given the name *Biospeologica*. This collection
appears in the French zoological periodical *Archives de Zoologie
expérimentale et générale*. At the present time the collection of *Biospeologica* contains 81 memoirs.

The Work of Robert Leruth

Robert Leruth followed in Belgium his organisation and research in
keeping with the spirit of *Biospeologica*. But his efforts, meritorious as
they were, did not produce a work comparable with that of *Biospeologica*.
This was largely due to the paucity of the terrestrial cavernicolous fauna
of Belgium, but also because Leruth was almost alone to collect material,
and lastly because he died at the early age of 28, on the 11th June 1940,
following wounds received during the German invasion of Belgium.

During the course of his short career Leruth published the results of his
research in 3 series of publications.
(1) "Exploration biologique des cavernes de la Belgique et du Limbourg
hollandais" (1931–1936), a series of memoirs appearing in various
scientific journals.
(2) "Notes d'Hydrobiologie souterraine" (1935–1939), published in many
different periodicals, but mainly in the *Bulletin de la Société Royale
de Liège*.
(3) "Etudes biospéologiques" (1937–1941), which appeared in the *Bulletin du Musée Royal d'Histoire Naturelle de Belgique*.

The Cave Research Group

In England the Cave Research Group led by Brig. E. A. Glennie and
Miss M. Hazelton have assembled important material collected principally

in British caves. Five volumes published between 1946 and 1960 give an account of the results obtained.

The Work of Karl Absolon

The Moravian zoologist, Karl Absolon, of Brno, spent almost half a century exploring the caves of Karst which border on the Adriatic. The results of these collections were grouped into a series of monographs which received the general title *Studien aus dem Gebiete der allgemeinen Karstforschung, der wissenschaftlichen Höhlenkunde, der Eiszeitforschung und den Nachbargebieten*. These monographs were published at Brno (Brünn), between 1932 and 1942.

D. THE PRESENT STATE OF BIOSPEOLOGY

Biospeology originated about a hundred years ago. If the numerous memoirs which have appeared during that time are reviewed, they can be divided into three categories:

(1) Numerous caves have been visited and described and their fauna collected. The results obtained have been recorded in publications which may be categorised as "Speleological and biospeological monographs". These monographs deal with investigations of indispensable value. This task must be continued although it is already far advanced in Europe, the Mediterranean region and in North America. Much less information is available for other parts of the world.

(2) The material collected from the caves has been examined by specialists, and has thus become the subject of numerous publications which may be classed as "Systematic Monographs". This task is as indispensable as the speleological research because precise and searching systematics are more necessary to biospeologie than to any other natural sciences, for the isolation of the different evolutionary lines of cavernicoles in separate cave systems, almost certainly without genetic interchange between them, leads to the production of hypogean forms divided into numerous small populations of species, subspecies and races.

These systematic studies have allowed the precise geographical distributions of cavernicolous species to be evaluated. These distributions are often remarkable in the sense that they frequently correspond to the configuration of ancient land masses, quite different from that of the continents today.

A knowledge of the morphological and biogeographical contributions to biospeology allows the evolutionary history of the principal cavernicolous groups to be traced with some accuracy. A book by Professor R. Jeannel (1943), *Les fossiles vivants des cavernes* demonstrates the state of biospeology some twenty years ago.

(3) Biospeological studies often include ideas concerning the origin of the cavernicoles, the factors which have determined their evolution, and the significance of their morphological characteristics. These considerations are usually only weak hypotheses, for they rarely rely on precise observations and even more exceptionally, on experiments.

Thus, biospeology, although it has been in existence for over a century, is still a young science, almost totally founded on simple observations of facts.

Until a very recent date the biology of the cavernicoles remained almost unknown. That which is known of the ecology and ethology of hypogeous animals proceeds from a few, generally imprecise observations. Their reproduction, development, longevity and physiology remains unknown.

Finally, until the last few years, a few scattered experiments have been undertaken on the cavernicoles, but the results of these have been tentative and subject to caution (as in the case of the experiments reported by Viré and Kammerer).

The reasons for the immense gaps in our knowledge arise from the extreme difficulty of breeding cavernicoles in ordinary laboratories. This culture work could not be successfully undertaken until specially designed subterranean laboratories were constructed. More will be said of these laboratories in the next chapter. It was not until such time as zoologists were able to use such laboratories that the progress of biospeology became very rapid. It is to record the results which have been obtained since that time that this book has been written.

It has become forcibly apparent that the culture of cavernicoles demands minute attention to detail, and great perseverance. Reproduction in *Proteus* and in *Aphaenops* was obtained only after many years of work marked by unfruitful experiment and failure.

Regarding experimental research upon cavernicoles, it can here be affirmed that it will be bristling with difficulties. It requires the efforts and ingenuity of many generations of workers. That is to say, the development of biospeology is past the stage where the contributions of individual isolated workers are of prime importance. For its continued development biospeology must become collective work, carried out by teams of research workers using the facilities provided by specialised underground laboratories.

BIBLIOGRAPHY
Historical

BÉGOUEN, H. (1929) La Grotte des Trois-Frères. *Mitteil. Höhl. Karstf.*
CHOPARD, L. (1928) Sur une gravure d'insecte de l'époque magdalénienne. *Compt. rend. Soc. Biogeogr.* Paris, **V.**
JEANNEL, R. (1943) *Les Fossiles Vivants des Cavernes.* Paris, Gallimard.

World Research in Biospeology.

BALAZS, D. (1961) Die Höhlen des südchinesischen Karstgebietes. *Die Höhle* **XII.**
BALAZS, D. (1962) Beiträge zur Speläologie des südchinesischen Karstgebietes. *Karszt ès Barlangkutatàs.* **II.**
BELLARD PIETRI, E. DE (1954) Hacia un Atlas espeleologico de Venezuela. *Bol. Soc. Venezol. Cienc. Nat.* **XV.**
BELLARD PIETRI, E. DE (1956) La Espeleologia en Venezuela; Flora y Fauna hipogea. *Bol. Soc. Venezol. Cienc. Nat.* **XVII.**
BELLARD PIETRI, E. DE (1960) La Cueva del Guácharo. *Bol. Soc. Venezol. Cienc. Nat.* **XXI.**
BIRSTEIN, J. A. and LOPACHOV, G. V. (1940) Erforschungen der Höhlenfauna der USSR in den Jahren 1935–1939. Biospeologica Sovietica. **I,** *Bull. Soc. Natural. Moscou.* **XLIX.**
BIRSTEIN, J. A. (1950) Peščernaja fauna zapadnogo Zakavkazja. *Zool. Zurn. Moskwa.* **XXIX.**
BORDON, C. (1959) Breves Notas sobre la fauna entomologica de la Cueva del Guacharo. *Bol. Soc. Venezol. Cienc. Nat.* **XXI.**
CHODOROWSKI, A. (1959) Les études biospéologiques en Pologne (Biospeologica Polonica, II). *Speleologia,* I.
GUEORGUIEV, V., and BERON, P. (1962) Essai sur la faune cavernicole de Bulgarie. *Annales. Spéléol.* **XVII.**
HADŽI, J. (1957) Fortschritte in der Erforschung der Höhlenfauna des dinarischen Karstes. *Verhandl. Deutsch. Zool. Gesell. Graz.*
HIERNAUX, C. R. et VILLIERS, A. (1955) Speologica africana — Étude préliminaire de six cavernes de Guinée. *Bull. Inst. fr. Afrique Noire. Ser. A.* **XVII.**
KARAMAN, ST. (1935) Die Fauna der unterirdischen Gewässer Jugoslaviens. *Verhandl. Intern. Ver. f. theor. angew. Limnologie.* **VII.**
KARAMAN, ST. (1954) Über unsere unterirdische Fauna. *Acta Mus. Maced. Sc. Nat.* **I.**
LINDBERG, K. (1960) Revue des recherches biospèologiques en Asie moyenne et dans le Sud du continent asiatique. *Rassegna speleol. ital.* **XII.**
MOTAS, C. (1957) Die speläologischen Forschungen in Rumänien. *Geologie.* **VI.**
MOTAS, C. (1960) Speologia in R. P. Romina. Problemi, realizari, perspective. *Natura,* Bucuresti. No. 4.
MOTAS, C. (1961) La Speologia in Romania. *Rassegna Speleologica Italiana.* **XIII.**
NICHOLAS, BROTHER G. (1960) Checklist of Macroscopic Troglobitic Organisms of the United States. *Amer. Midl. Nat.,* **LXIV.**
NICHOLAS, BROTHER, G. (1962) Checklist of troglobitic Organisms of Middle America. *Amer. Midl. Nat.* **LXVIII.**
PATRIZI, S. (1956) Nota preliminare su alcuni risultati di ricerche biologiche in Grotte della Sardegna. *Atti d. VII Congresso Naz. Speleol. Sardegna* (1955). Como.
PATRIZI, S. (1958) Nuovi Reperti sulla Fauna cavernicola della Sardegna. *Atti d. VIII Congresso Naz. Speleol.* (1956). Como.
RUFFO, S. (1955) Le attuali conoscenze sulla fauna cavernicola della regione pugliese. *Mem. Biogeogr. Adriatica.* **III.**
STADLER, H. (1927) Fortschritte in der Erforschung der tierischen Bewohnerschaft der Höhlen Südasiens und Indonesiens. *Mitteil. Höhl. Karstf.*
UÉNO, S. (1957) Blind aquatic beetles of Japan with some accounts of the fauna of Japanese subterranean waters. *Archiv. f. Hydrobiol.* **LIII.**

BIOGRAPHIES

ABSOLON: STROUHAL, H. (1961) In memoriam Univ. Prof. Dr. Phil. Karl Absolon. *Die Höhle;* **XII.**
PRETNER, E. (1961) Karl Absolon. *Nase. Jame.* **II,** (1960).
CHAPPUIS: JEANNEL, R. (1960) Pierre-Alfred Chappuis (1891–1960). *Annal. Speleol.* **XV.**
VANDEL, A. (1960) L'œuvre scientifique de P.-A. Chappuis. *Annal. Speleol.,* **XV.**
ROUCH, R. (1960) Liste des travaux de Monsieur P.-A. Chappuis. *Annal. Speleol.* **XV.**
HUMBOLDT: JIMENEZ, A. N. (1960) Humboldt, Espeleologo precursor. INRA, La Habana.
KARAMAN: KARAMAN, Z. (1960) Dr. Stanko Karaman. *Crustaceana,* **I.**
LERUTH: MARÉCHAL, P. (1942) A la mémoire de Robert Leruth, biospéologiste (1912–1940). *Bull. Mus. R. Hist. Nat. Belgique,* **XVIII.**
RACOVITZA: GUIART, J. et JEANNEL, R. (1948) Emile Georges Racovitza. *Archiv. Zool. exper. gener.* **LXXXVI.**
RADU, V. GH. (1948) Le Professeur Emile G. Racovitza. *Bull. Soc. Stii. Cluj.* **X.**
CODREANU, R. (1948) A la mémoire du Professeur Émile Racovitza. *Notationes biologicae. Bucarest.* **VI.**
MOTAS, C. (1960) Emil Racovitza (1868–1947). *In Figuri de Naturalisti.* Bucuresti.
SPANDL: PESTA, O. (1927) Hermann Spandl. *Speläol. Jahrb.* **VII–IX.**
MOTAS, C. (1948) Professorul Emil Racovitza. *Anal. Acad. Rep. Pop. Romane. Mem. Sec. Stiint- Ser. III;* **XXIII.**

CHAPTER IV

BIOSPEOLOGICAL MEANS AND METHODS

THREE successive tasks must be undertaken by the biospeologist:

(*a*) Collection of material.

(*b*) The culture of cavernicoles so that they may be observed at leisure, and in order that experiments may be carried out upon them.

(*c*) Publication of results obtained.

A. COLLECTING TECHNIQUES

There is no place in the present work to explain the techniques of speleology. In any case these have already been fully discussed in many papers and books, but it is appropriate here to give a few indications of the methods of collection used in biospeology.

These methods were developed at first by "hunters of cavernicoles". These people were sometimes only countrymen, but their lack of education did not hinder their excellent observation of Nature. Pierre Manaud and Jean-Marie Brunet who lived in the Pyrénées du Couserans, were such people. Some of their techniques have been improved and developed by biospeologists. The essentials of the subject will be summarised here, but the reader who wishes to obtain fuller information is referred to the following publications:

CAVERNICOLOUS FAUNA

Publications in French

Delherm de Larcenne (1875); Racovitza (1913); Bettinger (1922); Jeannel (1943); Colas (1948); Chappuis (1950).

Publications in German

Dohrn (1866); Dombrowski (1906); Winkler (1912); Wettstein (1920); Kuntzen (1924); Chappuis (1930).

Publications in Italian

Pollonera (1899); Alzona (1905); Cotti (1957).

Publications in Spanish

Balcells (1959).

Publications in Portuguese

Machado (1945); Sousa (1959).

Publications in English

Barber (1931); Hazelton and Glennie (1953).

Publications in Russian

Birstein Y. A. and Boroutzky, E. V. (1956).

INTERSTITIAL FAUNA

Leruth (1939); Chappuis (1942); Delamare-Deboutteville (1960, Chapt. II).

(a) Collecting at the Mouths of Caves

Collecting must begin well before the mouth of the cave is entered. In fact, the entrances of caves are a particular biotope where members of the "parietal fauna" may be discovered. There are no special methods for collecting this fauna. The entrances of caves including those of deep shafts or pits, provide shelter for numerous humicoles and endogeans.

Humicoles may be collected either by sifting the dead leaves, wood fragments and other diverse debris in which they live, or by treating such material in one of the various modifications of the Berlese apparatus. In the latter case the material to be sorted is brought back to the laboratory in a large canvas sack, and the total fauna it contains may be extracted.

The endogeans may be taken from under rocks which may be turned over with the aid of an entomologist's pick. They may be extracted from the earth by the method of washing. The debris which floats to the surface of water is removed and dried upon metal gauze which rests over a hollow container.

(b) Collecting in Caves

After having inspected the entrance of the cave, the biospeologist should wear strong overalls and rubber boots for collecting in the cave itself. Lighting is most important and requires particular attention. The speleologist frequently uses electric light but this is inadequate for the needs of the biospeologist. An acetylene lamp is required for successful collecting. The lamp must not be carried in the hands, since both hands are needed for other purposes in biospeological collecting, and it is unwise to rest the lamp on the ground since there is then the liability of its being turned over and extinguished. Finally, a lamp carried in the hand prohibits the scaling and climbing of ladders. It is most convenient to attach the burner to the helmet and to connect this to a generator by means of a rubber tube. The generator may be worn round the waist on a stout leather belt. The generator must be able to be rapidly detached when the biospeologist is obliged to worm his way through narrow openings (châtière).

The biospeologist should carry a small sling-bag to contain collecting instruments, specimen tubes, a torch, candles, matches in a water-tight box, prickers for the acetylene jets, spare burners, a reserve of carbide and also a notebook and pencil, and some paper for making labels. If it is intended to investigate previously unexplored caves, a nylon rope (hemp ropes are dangerous because they may rot), and a light alloy ladder about 10 metres in length should be taken.

The method of investigation pursued by the biospeologist is quite different from that of the speleologist. Unlike the latter, he must proceed very slowly. He must examine the walls, stalactites and any humid surfaces, with meticulous care. Clayey deposits must be inspected and large stones turned over. Scrupulous attention should be given to any strange objects, pieces of wood, and bits of paper, masses of dead leaves, fallen wax from candles, human faeces and the droppings of rodents and of bats, any of which may attract cavernicoles.

Fallen material should be examined to some considerable depth, because the larvae of cavernicolous insects, as well as endogeans may live at a depth of half a metre or more. The stalagmitic floor must be broken up with a hammer in order to give access to the layer of clay upon which it has been built up.

Lakes, "gours", even the smallest pools of water, and seepages, must be carefully examined.

Interstitial fauna is usually collected outside caves on the banks of rivers. A large hole about a metre in depth should be dug in sand or gravel. The water which seeps into and rapidly fills the hole, contains the phreatobious organisms which may be captured and removed with a fine net.

(c) Collecting Equipment

Because of the small size and agility of cavernicoles the biospeologist finds it convenient to use collecting equipment similar to that of the entomologist, for instance the aspirator, fine flexible forceps made with watch springs and a fine brush or a woodcock's feather.

A small net of strong cloth and a silk net are indispensable for obtaining aquatic fauna, and a pipe may be necessary if the water is deep.

(d) Traps and Bait

A cave cannot be explored thoroughly during a single visit. It is advisable to make several excursions to the same one, where possible at different times of the year, so that the total fauna of the cave may be captured. Under such conditions it is advantageous to set up traps in the cave. This method is indispensable if one wishes to collect a large number of specimens so that they may be cultured in the laboratory.

(1) Bait. The most simple and efficient bait consists of a piece of cheese (Roquefort cheese, made with sheep's milk, is particularly recommended) placed under a stone. More simply, the surface of a rock may be rubbed with cheese and this area then covered with a slab of stone. Tainted meat may also be employed, or even crushed snails, but cheese is preferable. After a few days the cavernicoles appear in large numbers around the bait and can be easily collected. Masses of dead leaves are also good for bait, but are not effective until a few months have elapsed.

(2) Traps. The use of permanent traps such as those invented by Barber (1931) and ones derived from them, is to be strongly discouraged. These traps consist of tubes or flasks with the bait placed above them and contain a preserving fluid (glycol is most frequently used). These traps function for many months or even years and herein lies their danger. If neglectful entomologists forget to remove such traps the fauna of the cave in which they have been set may be totally destroyed to no purpose.

For the capture of aquatic animals and particularly Crustacea a particular type of flat net may be employed. This device consists of a disk of netting mounted on a frame to keep it rigid, and is suspended by four strings. A piece of meat is tied in the centre of the disk to act as bait, and the trap is weighted with a pebble. The trap is then placed on the bottom of a subterranean pool or stream and left for a number of hours. It may then be retrieved and the Crustacea (isopods and amphipods) easily collected.

(e) Treatment of Cavernicoles after Capture

(1) If it is wished to keep the specimens alive for breeding experiments, it is important to maintain them at a temperature as close as possible to that of the cave in which they were captured. Terrestrial cavernicoles should be kept in a container where the humidity of the air approaches saturation point.

The terrestrial forms may be placed in specimen tubes which have been partly filled with moistened plaster of Paris and are stoppered with cotton wool. The tubes are finally placed in a large-mouthed Thermos flask. The aquatic forms may be placed directly in a Thermos flask.

(2) If it is wished to collect samples for systematic study or to place in a collection it is convenient to kill the animals immediately. The methods used vary with the group in question.

Insects. From the aspirator the insects are transferred to a killing bottle containing wood shavings soaked in ethyl acetate.

Crustacea, myriapods, arachnids and small insects. These animals are conveniently preserved directly in 75% alcohol.

Planarians and Oligochaetes. These groups require the use of special fixatives. For the planarians, which must be fixed when extended so that they can be serially sectioned, de Beauchamp has recommended the following mixture:

90% alcohol – 6 parts
40% formol (commercial formalin) – 3 parts
Glacial acetic acid – 1 part.

The biospeologist should carry in his bag a metal bottle containing the usual fixatives.

B. ATTEMPTS TO TRANSPLANT FAUNA

Direct observations made on cavernicoles in the actual caves in which they are living are very useful, and biospeologists must always be encouraged not to neglect them because they furnish valuable information on the ecology of these animals. However, their scope is limited. In particular it is difficult to affirm in the absence of experimental data that the extremely stable climatic conditions that are found in caves are indispensable for the physiological equilibria and the reproduction of the cavernicoles.

Before resorting to direct experimentation in the laboratory one may use field experiments. These consist of transplanting fauna from one cave to another situated in a different locality.

It is well known that such experiments should only be undertaken for scientific purposes, and all necessary precautions should be observed. If they are carried out by amateur entomologists out of malice or idle curiosity they may give rise to regrettable confusions, as in the case of *Troglodromus* in the Alpes-Maritimes (Bonadona, 1955, 1956).

The results of these transplantations are variable. Some of them fail. Ginet (1952) transplanted populations of two aquatic crustacea and a species of bathysciine beetle from various caves in the department of Ain to the Grotte de la Balme (Isère). None of those species became acclimatised to its new habitat (Ginet, 1960).

Other attempts have been more successful. Banta (1907) reported that the species of cavernicolous fish which inhabited the caves of the Mississippi valley, *Amblyopsis spelaeus*, had been successfully introduced to Mayfield's Cave (Indiana) by Dr. C. Eigenmann, where it became acclimatised and reproduced.

Patrizi (1956) introduced in 1952, specimens of *Bathysciola derosai* from a cave in Monte Argentario to another in Central Italy. In 1954 these beetles had survived in their new habitat and had multiplied.

The culture of the bathysciine cavernicole *Troglodomus bucheti* from the Alpes-Maritimes was carried out twelve years ago at the Moulis Cave

Laboratory. Some of them escaped. These individuals started the colonies which now flourish in many parts of the cave.

Dr. Henrot explored for several years the Grotte du Lapin which opens on the Causse du Larzac near Le Clapier (Aveyron). He was surprised to find this cave almost devoid of life in spite of the apparently favourable conditions. Only two campodeids and a non-cavernicolous staphylinid were captured. H. Coiffait and M. Bouillon, attached to the Moulis Subterranean Laboratory, hearing of Dr. Henrot's experience, planned a transplantation experiment. On the 27th March 1957 they introduced into the Grotte du Lapin various cavernicoles captured in the Pyrénées ariègeoises. They were:

> 75 specimens of *Aphaenops cerberus*
> 170 specimens of *Speonomus diecki*
> 30 specimens of *Speonomus stygius*
> 400 specimens of *Antrocharis querilhaci*
> 450 specimens of *Paraspeonomus vandeli*
> 110 specimens of *Typhloblaniulus lorifer*.

During two visits made on the 3rd June 1960 and the 4th March 1961 examples of *Paraspeonomus vandeli* and *Typhloblaniulus lorifer* were captured. No specimens of the other introduced species were observed. It seems that selective discrimination had become established in this heterogeneous population. Certain species had become adapted whilst others were rapidly eliminated.

C. BREEDING OF CAVERNICOLES

Informative as transplantations may be, they cannot replace breeding methods. The breeding of cavernicoles is essential to acquire data on their physiological requirements, including food, and upon their behaviour, breeding habits, development and longevity. Finally, it is most important to culture cavernicoles in order to provide specimens upon which to carry out controlled experiments.

(a) Different Stages of Breeding

The meaning of the word breeding should be clearly understood. Breeding is composed of several successive stages:

(1) Keeping the captured animal alive for some time is the first and easiest stage that has been realised with greater or lesser success for all cavernicoles.

(2) To obtain reproduction of cavernicoles in captivity, that is, copulation and production of eggs.

(3) To trace the development of the egg and the embryo up to the point of eclosion; this is a delicate stage since the requirements of the egg and the embryo are frequently different from those of the adults.

(4) The fourth stage is post-embryonic growth, or the larval stage and metamorphose in insects, which finally results in the production of the adult.

(5) Although the individual cycle of development is completed, the task of the breeder is not over. He still has to maintain his stock of specimens for several successive generations in order to confirm the complete success of his experiments. Such conclusive experiments, however, are still rare. A few examples will now be given.

De Beauchamp (1935) found that the light-avoiding planarians are simple to keep in captivity and can be maintained up to sexual maturity. However, cocoons are seldom obtained and even more rarely do the young hatch and develop. Such conditions are, however, easy to obtain in the epigean planarians. In the culture of *Dendrocoelum (Polycladodes) album* the adults of the second generation reach sexual maturity but do not reproduce. A culture of *Dendrocoelum (Dendrocoelides) collini* reached the third generation, but then all the specimens died.

The culture of *Niphargus virei* carried out by R. Ginet at the Moulis Subterranean Laboratories reached the third generation (assuming the first generation corresponds to field specimens).

Only one genus of cavernicolous fishes, *Anoptichthys*, will reproduce normally in aquaria. One species of this genus *A. jordani*, which comes from the caves of Mexico, has become a commercially sold aquarium fish. But this is not a very specialised form, being nearer to the epigean species *Astyanax mexicanus*.

(b) Breeding Methods

The culture of cavernicoles has been attempted on several occasions. The most successful attempts have been made by Leonida Boldori (1933–1959). This worker has bred a variety of cavernicolous arthropods (Coleoptera, Myriapoda, Isopoda, Chernetidae, Collembola). This biospeologist has kept several species in culture for five years and has observed reproduction in many of them.

S. Glaçon and G. Le Masne (1951) were successful in maintaining breeding cultures of several species of bathysciine Coleoptera, by keeping them in "Janet nests" in refrigerated containers, and by maintaining suitable micro-climatic conditions. However, the more delicate species, as the species of the genus *Aphaenops,* although they survived for several months in the "Janet nests" did not breed.

It seems, therefore, that the real solution is to carry out breeding work in natural caves, that is, to convert suitable caves into laboratories.

D. SUBTERRANEAN LABORATORIES

"Les cavernes et les abîmes sont des laboratoires naturels
tout prêts pour d'innombrables et curieuses études".

E. A. Martel, *Les Abîmes*, 1894, p. 565

The sentence quoted above demonstrates that the idea of transforming a cave into an underground laboratory is not a new one. But such a project requires considerable financial support, such as would be beyond the resources of the private individual. Such an enterprise can only be undertaken as a collective scheme. Although there have been about a dozen attempts of this kind very few have reached the stage of the attainment of a fully experimental laboratory.

These various attempts will now be reviewed country by country, an order which seems preferable to the chronological one.

FRANCE

The Laboratoire des Catacombes

At the time when E. A. Martel was carrying out his spectacular speleological explorations, Armand Viré concerned himself with the subterranean fauna. When he came to Paris to continue his studies at the Muséum National d'Histoire Naturelle, he thought of using the Catacombes which extend under the Jardin des Plantes as a laboratory. He submitted his plan to Alphonse Milne-Edwards, director of the Museum, who approved it and supplied the necessary means for it to be carried out. The work was started in 1896 and finished a year later.

The Laboratoire des Catacombes undoubtedly was the first biospeological laboratory to be set up in a subterranean cave (Viré, 1897). Viré cultured the cavernicoles collected in the Catacombes themselves, and also those brought from various caves in France, Corsica, Italy, Karst, Austria, Algeria and the United States (Viré, 1901). He also carried out experiments intended to show the influence of darkness upon surface living forms and, inversely, of the action of light upon cavernicoles (Viré, 1904). Most of the results obtained are open to some doubt, at least as far as their interpretation is concerned. However, the observation that *Proteus* becomes pigmented and assumes a violet-black colour on prolonged exposure to the light, is true.

Viré estimated that the experiments he had initiated should be continued until the twenty-first century. In fact the life of the Laboratoire des Catacombes was much shorter. It was destroyed by the great flood of the Seine in 1910, and has never been reconstructed.

The Laboratoire de Spéléobiologie Expérimentale de Saint-Paër

Henri Gadeau de Kerville in 1909 and 1910, prepared a disused quarry, within the town boundary of Saint-Paër (Seine-Inférieure) and installed a subterranean laboratory. He called it the "Laboratoire de Spéléobiologie Expérimentale". It was inaugurated on the 10th July 1910 (Gadeau de Kerville, 1911).

It seemed, however, that the laboratory was not used for experimental purposes. At least, the only indication which can be obtained from the scientific records, refers to the case of a *Proteus* which lived for $14\frac{1}{2}$ years in the Saint Paër laboratory, and which, as it did not receive any food for the last 8 years, finally died from starvation (Gadeau de Kerville, 1926).

The Laboratoire Souterrain du Centre National de la Recherche Scientifique

The Laboratoire Souterrain du Centre National de la Recherche Scientifique was instituted by a special decree made on 11th February 1948 (Vandel, 1954). It was constructed at Moulis in the Pyrénées ariègeoises, one of the most celebrated areas of biospeology. It includes a cave laboratory installed in the first section of the Grotte de Moulis, a surface building consisting of living quarters, laboratories, a library and a workshop, etc. (Pl. I, II, III). The laboratory was inaugurated on the 26th June 1954. This laboratory is specially equipped for the breeding of cavernicoles, and for experiments on them. A summary of early results obtained has been given by Vandel (1959). Reference will often be made in this book to research carried out in the Moulis laboratory.

The Grotte d'Antheuil

The Grotte d'Antheuil is situated some 40 km from Dijon, above the valley of the river Ouche (Côte d'Or). It was fitted out as a climatological and ecological station. The culture of cavernicoles is planned (De Loriol, Tintant and Rousset, 1959).

BELGIUM

The Laboratoire Souterrain de Han-sur-Lesse

The Fédération spéléologique de Belgique established a subterranean laboratory in 1959 at Han-sur-Lesse, very near to the famous Han grotto. It has been called the "Laboratoire Souterrain E. de Pierpont". An extensive research programme is planned which includes the study of many aspects of the subterranean world (hydrology, mineralogy, climatology, biology, radioactivity, etc.) (Liégois, 1958).

B 3

PLATE I. Moulis (Ariège). At the centre of the plate an anticline whose strata are vertical. This anticline is made up of Jurassic dolomite. This is the type of rock in which the cave laboratory has been excavated. The anticline is bordered on the west (to the right of the photograph) by schists of the Upper Lias (woodland coomb) and the calcareous rocks of the Lower Lias (Hettangian) (visible on a level on the quarry). To the extreme right and high up, can be seen the steep slopes of the axial zone of the Pyrenees. To the right and down below is the surface laboratory. To the left is the village of Moulis. Just to the right of the church steeple can be seen the entrance to the cave laboratory (Photo by Barbé).

PLATE II. A front view of the subterranean laboratory of the Centre National de la Recherche Scientifique (Photo by Carrère).

PLATE III. One of the rooms in the cave laboratory (Photo by Carrère).

ITALY

Stazione Biologica Sperimentale Sotteranea di Napoli

An ancient cave hollowed out in the sub-soil beneath Naples was fitted out, in 1954, by Dr. Pietro Parenzan, as the "Stazione Biologica Sperimentale Sotterranea" (Parenzan, 1954, 1956).

YUGOSLAVIA

The Postojna Biospeological Station

One of the best situated subterranean laboratories is undoubtedly that which has been installed at the famous cave which has been called the Adelsberg Cave under Austro-Hungarian rule, Postumia Cave under Italian rule, and Postojnska jama, under Yugoslavian rule.

The plan for the installation of the laboratory was conceived by the "Istituto Italiano di Speleologia," which had its headquarters at Postumia. The work of building the laboratories was carried out under the supervision of the director, the great speleologist G.A.Perco, between 1930 and 1931. The biospeological station was installed in a side gallery called the "Grotta dei nomi nuovi". It is made up of three sections, one devoted to the study of the hypogeous flora, another to aquatic fauna and the third to terrestrial subterranean animals. Professor E.Dudich (1933) has given an excellent description of the laboratory (cf. also Jeannel, 1936); and he compiled an extremely well-thought-out programme of research to be carried out at the Biological Station. However, it does not seem that the exceptionally favourable conditions offered by the Postumia Station have always been taken advantage of by biospeologists.

The Biospeological Station was used during the last war as a fuel depot, and was seriously damaged by fire, but the laboratory has now been repaired and cavernicoles can now be cultured there.

Other Yugoslavian Stations

The right gallery of the cave, Podpeška jama in Slovenia, was set up as a laboratory for the purposes of studying the ecology of the cavernicolous fauna. Apparently, only meteorological studies have been carried out there (Kenk and Seliškar, 1931).

Also, Dr. Marko Aljančič of the Ljubljana Faculty of Medicine in 1960 set up a subterranean laboratory for biological work in Tular Cave at Kranj, which is about 25 km north of Ljubljana.

HUNGARY

The Biospeological Station at "Baradla", a Cave Near Aggtelek

Hungary has the distinction of possessing one of the largest caves in Europe (9 km of galleries). This is "Baradla", which opens near Aggtelek in the north-east of Hungary. Professor Endre Dudich has dealt with this vast subterranean system in a magnificent monograph (Dudich, 1932).

It was to be expected that Hungarian speleologists would use this famous cave for the construction of a subterranean laboratory. This was started in 1959. The Baradla Subterranean Laboratory is attached to the Institutum Zoosystematicum of the University of Budapest, and is under the direction of Professor Endre Dudich. A section of the cave called the Ròkalyuk (Fox Hole) was fitted out as a biospeological laboratory. The building of a surface laboratory and also of a museum is also planned (Dudich, 1960, 1962).

RUMANIA

The Institutul de Speologie de Cluj

This institute possesses no cave-laboratories. However, it is a very old foundation and therefore worthy of mention here. It is a University Institute attached to the University of Cluj. It was founded in 1920 by Emile-Georges Racovitza who, after living for 30 years in France, returned to his native country at the end of the First World War. He was at first assisted by Dr. René Jeannel, and later by P.A.Chappuis.

According to its founder the Institute was intended to carry out that work called "Biospeologica", which had been defined 13 years previously by E.G.Racovitza. That is, the Institute was specially concerned with research into cavernicolous life, the identification of species, and their description. It was also proposed to set up a list of caves visited (Racovitza, 1926; Chappuis, 1948).

The Institutul de Speologie Emil Racovitza

The death of Racovitza in 1947 followed by the departure of P.A.Chappuis in 1949, temporarily slowed down the progress of speleology in Rumania. However, there are now two Institutes of Speleology in that country, one at Cluj and the other at Bucharest.

They were combined into a single Institute in 1956, which was placed under the direction of Professor Constantin Motas, and called the "Institutul de Speologie Emil Racovitza". The director has given an excellent report of the speleological activities carried out by this Institute (Motas, 1961). The reader is advised also to consult a magnificent book entitled *Grottes de Roumanie.*

Finally, a laboratory of experimental biospeology is to be set up at Closani, in the southern Carpathians.

RUSSIA

The Biological Station at Kossino

Built near to the caves of Kutais, at Rion in Georgia, this station considerably helps the work of biospeologists in that country (Boroutzky, 1930).

UNITED STATES

The Cave Farm, University of Indiana

In 1903 the University of Indiana purchased, on the advice of C. H. Eigenmann some 182 acres of land located near Mitchell (Indiana). On this site "Farm Cave", also known as "Donaldson Cave", opens. It is crossed by a river in which the blind fish *Amblyopsis spelaeus* lives. F. Payne (1907) and C. H. Eigenmann (1909) devoted a considerable amount of detailed research to this species. Some breeding aquaria were installed in the cave although it was not, apparently, fitted out as a real subterranean laboratory. The cave is no longer used by biospeologists today.

A proposal for the construction of a proper subterranean laboratory (Cave Research Centre) in the well-known Mammoth Cave is being considered at the time of writing.

E. BIOSPEOLOGICAL PUBLICATIONS

(a) Treatises and General Works

ENGLAND

HAZELTON, M. and GLENNIE, E. A. (1953) Cave fauna and flora, in C. H. D. Cullinford, *British Caving*. London.

FRANCE

VIRÉ, A. (1900) *La faune souterraine de France*. Paris.
RACOVITZA, E. G. (1907) Essai sur les problèmes biospéologiques. *Biospeologica, I. Archiv. Zool. expér. géner.* (4), **VI**.
JEANNEL, R. (1926) *Faune cavernicole de la France*. Paris.
JEANNEL, R. (1943) *Les Fossiles vivants des cavernes*. Paris.

BELGIUM

LERUTH, R. (1939) La biologie du domaine souterrain et la faune cavernicole de Belgique. *Mém. Mus. R. Hist. Nat. Belgique*, No. 87.

GERMANY

HAMANN, O. (1896) *Europäische Höhlenfauna.* Jena.
LENGENSDORF, F. (1952) *Von Höhlen und Höhlentieren.* 2. Auflage, Leipzig.

ITALY

RUFFO, S. (1959) La Fauna delle Caverne, in *Conosci l'Italia III. La Fauna*, Milano.

UNITED STATES

EIGENMANN, C. H. (1909) Cave Vertebrates of America – A Study in degenerative evolution. *Carnegie Inst. Washington. Public.* No. 104.
PACKARD, A. S. (1886) The Cave Fauna of North America, with remarks on the anatomy of the brain and origin of the blind species. *Mem. Nat. Acad. Sci. Washington,* **IV.**
PAYNE, F. (1907) The reactions of the blind fish *Amblyopsis spelaeus* to light. *Biol. Bull.* **XIII.**

(b) Works Devoted to Subterranean Aquatic Fauna

CHAPPUIS, P. A. (1927) Die Tierwelt der unterirdischen Gewässer. *Die Binnengewässer*, **III,** Stuttgart.
SPANDL, H. (1926) Die Tierwelt der unterirdischen Gewässer. *Speläologische Monographien*, **XI,** Wien.

(c) Reviews Devoted to Aquatic Cavernicoles

GRAETER, E. (1909) Die zoologische Erforschung der Höhlengewässer seit dem Jahre 1900 mit Ausschluß der Vertebraten. *Intern. Rev. Gesammt. Hydrobiol.* **II.**
REMY, P. (1953) Eaux souterraines. In: Revue bibliographique de l'hydrobiologie française (1940–1950). *Annal. Stat. Hydrobiol. appl.*

(d) Catalogues

(1) General Catalogue

WOLF, B. (1934–1938) *Animalium Cavernarum Catalogus.* Three volumes. S'Gravenhage.

(2) Regional Catalogues

FRANCE

BEDEL, L. and SIMON, E. (1875) Liste générale des Articulés cavernicoles de l'Europe, *Journal de Zoologie*, **IV.**
JEANNEL, R. (1926) *Faune cavernicole de la France.* Paris.

GERMANY

ARNDT, W. (1940) Die Anzahl der bisher in Deutschland (Altreich) in Höhlen und im Grundwasser lebend angetroffenen Tierarten. *Mitteil. Höhl. Karstf.*

AMERICA

NICHOLAS, BROTHER, G. (1960) Checklist of Macroscopic Troglobitic Organisms of the United States—in Symposium: Speciation and Raciation in Cavernicoles. *Amer. Midl. Nat.* **LXIV.**
NICHOLAS, BROTHER, G. (1962) Checklist of troglobitic Organisms of Middle America. *Amer. Midl. Nat.* **LXVIII.**

(e) Periodicals

There are very few periodicals devoted entirely to biospeology. On the other hand, speleological publications are very numerous. We shall mention only those which regularly publish articles of biospeological interest.

ENGLAND

The Transactions of the Cave Research Group of Great Britain. 5 volumes have appeared since 1947. This periodical also publishes *Biological Supplements.*

FRANCE

Biospeologica. 81 parts have appeared between the years 1907 and 1962. This publication is included in the *Archives de Zoologie expérimentale et générale.*

Notes biospéologiques. 13 volumes appeared between 1947 and 1958.

Annales de Spéléologie. From Volume **XIV** (1959), the *Annales de Spéléologie* have become a publication of the Centre National de la Recherche Scientifique, and represents the official journal of the subterranean laboratory of the C.N.R.S. It is in two parts: one devoted to physical speleology and corresponding to the old *Annales de Spéléologie;* the other to biospeology and which is a continuation of the *Notes biospéologiques.*

Sous le Plancher, journal of the Spéléo-Club of Dijon, publishes biospeological articles, many of which are devoted to Chiroptera.

SPAIN

Speleon. Review edited by the Instituto de Geologia Applicada, Oviedo. Each volume contains biospeological articles.

ITALY

Italy possessed a biospeological periodical: *Proteus, Rivista internazionale di Biologia sotterranea.* This periodical finished publication at Volume **III,** 1905. It should not be confused with the Yugoslavian periodical *Proteus* published in Ljubljana.

However there are several speleological journals published in Italy, which contain articles on biospeology: *Le Grotte d'Italia*, Rivista dell'Istituto di Speleologia. This revue is edited by the "Sezione dell'Istituto di Geologia dell'Università di Bologna". It is the oldest Italian speleological publication, dating from 1927. The third series is now in the course of publication.

Rassegna Speleologica italiana. This journal is edited by Salvatore del'Oca, Como; by 1960 twelve volumes had appeared.

Studia Spelaeologica. This is the "Organe ufficiale del Centro Speleologico meridionale e della Stazione biologica sperimentale sotterranea di Napoli".

Two other Italian periodicals contain articles on biospeology.

These are: *Bollettino della Società Entomologica Italiana*, Genova. In 1960 the collected bulletin contained 90 volumes. The Society also publishes the Journal called *Memoria.*

Doriana, Supplemento al *Annali del Museo Civico di Storia Naturale* "G.Doria", Genova.

BELGIUM

The *Etudes biospéologiques* were produced under the direction of Robert Leruth. There are 35 numbers produced between 1937 and 1952, which appear in the *Bulletin de l'Institut Royal des Sciences de Belgique* and in the *Bulletin du Musée Royal d'Histoire Naturelle de Belgique.* Publications ceased on Leruth's death.

SWITZERLAND

Stalactite, official journal of the "Société Suisse de Spéléologie", contains papers by Swiss biospeologists.

GERMANY

Mitteilungen über Höhlen- und Karstforschung, Zeitschrift des Hauptverbandes Deutscher Höhlenforscher, Berlin, (this journal later changes its title to *Zeitschrift für Karst- und Höhlenkunde*). This periodical which contains a large amount of biospeological information appeared between 1925 and 1943.

AUSTRIA

Die Höhle, Zeitschrift für Karst- und Höhlenkunde, Wien. By 1960, 11 volumes of this journal had been published.

CZECHOSLOVAKIA

Studien aus dem Gebiete der allgemeinen Karstforschung, wissenschaftlicher Höhlenkunde und den Nachbargebieten, Brünn. This journal edited by Dr. Karl Absolon, contains a "Biologische Serie" of which nine parts have appeared between 1932 and 1942. This series relates to studies on the collection "Biospeologica balcanica" which has been assembled by Dr. Absolon.

HUNGARY

Fragmenta Faunistica Hungarica. Six volumes of this journal had appeared by 1943. They are changed later in *Opuscula Zoologica Instituti Zoosystematici Universitatis Budapestensis*, Budapest. By 1960 four volumes of the latter journal had been published and it contained a series of articles on biospeology, entitled "Biospeologica hungarica".

YUGOSLAVIA

Numerous Yugoslavian publications concerning speleology contain numerous articles on biospeology.
Acta Musei Macedonici Scientarum Naturalium, Skopje.
Fragmenta balcanica Musei Macedonicii Scientiarum Naturalium, Skopje.
Folia balcanica, Skopje.

RUMANIA

Buletinul Societatii de Stiinte din Cluj (Rumania). This journal contains numerous papers by workers at the Institutul de Speologie of Cluj.

However, the publications of the workers at the Cluj Institute can be found in a collection entitled *Lucrarile Institululi de Speologie din Cluj*. Nine successive volumes have appeared containing 128 papers. In 1960 publication of this periodical was resumed; it now also contains work carried out at the Instituts of Bucharest and Cluj.

BULGARIA

Mitteilungen aus den Königlichen Naturwissenschaftlichen Instituten in Sofia. This journal contains papers concerned with the study of biological material collected by Dr. Iwan Buresch, and his collaborators, from Bulgarian caves.

TURKEY

Istanbul Universitesi fen Fakültesi Mecmuasi, Istanbul. This journal publishes papers which are concerned with the material collected by Professor C. Kosswig and Dr. de Lattin from Turkish caves.

POLAND

Speologia, Warszawa. This periodical appeared in 1959. It publishes biospeological papers entitled "Biospeologica Polonica", of which five have already been produced.

RUSSIA

"Biospeologica Sovietica" appears in the *Bulletin of the Society of Natural Sciences of Moscow, Biological Section*. Ten papers appeared between 1940 and 1948.

We must also mention the *Spelaeological Bulletin of the Institute of Natural Sciences M. Gorky University of Molotov*. Edited by D. E. Charitonov and G. A. Maximovich. No. 1 of this series (1947) contains two papers by the editors. No subsequent numbers of this bulletin seem to have appeared.

UNITED STATES

Three American speleological periodicals contain papers of interest to biospeologists. They are:

Bulletin of the National Speleological Society, Alexandria, Virginia.

N. S. S., National Speleological Society, Arlington, Virginia.

Cave Notes. Publication of Cave Research Associates, Berkeley, California.

Two other American periodicals devoted to the Natural Sciences regularly publish biospeological papers:

American Midland Naturalist, Notre Dame, Indiana.

Journal of the Elisha Mitchell Scientific Society, Chapel Hill, N. C.

MEXICO

Ciencia, Mexico. This periodical contains papers published by Mexican biospeologists.

Anales del Instituto de Biologia, Mexico. This journal contains many biospeological papers.

SOUTH AFRICA

The Bulletin of the South African Speleological Association, Cape Town.

AUSTRALIA

Journal of the Sydney University Speleological Society, Sydney. By 1960 six volumes of this journal had appeared.

NEW ZEALAND

The New Zealand Speleological Society, Auckland. The Society has published 35 numbers of its journal, dating between 1952 and 1960.

F. THE CONGRESS OF SPELEOLOGY

The first International Congress of Speleology was held at Paris in 1953. The third volume of *Comptes Rendus* which appeared in 1956 is devoted to biospeology. The second and third International Congresses of Speleology were held at Bari (Italy), and at Vienna (Austria), respectively.

The "Congressi Nazionali di Speleologia" which is held regularly in Italy, publishes reports in *Atti*, which contains numerous biospeological papers.

BIBLIOGRAPHY

Collecting Techniques

ALZONA, C. (1905) Brevi notizie sulle raccolte zoologiche nelle caverne. *Boll. d. Naturalista*. **XXIV**.

BALCELLS, E. (1959) El Estudio biologico de las Cavidades subterraneas. *Memor. Asemblea reg. Espeleol. Carranza (Vizcaya)*, Bilbao.

BARBER, H. S. (1931) Traps for Cave-inhabiting Insects. *J. Elisha Mitchell Sc. Soc.* **XLVI**.

BETTINGER, L. (1922) La faune cavernicole et la chasse dans les cavernes. *Bull. Soc. Hist. Nat. Savoie.* **CXCI.**

BIRSTEIN, Y. A. and BOROUTZKY, E. V. (1956) La méthodique de l'étude des eaux souterraines. *In La vie des eaux douces de l'URSS.* Tome IV; partie 1; Chapitre 43. Moscou — Léningrad.

CHAPPUIS, P. A. (1930) Methodik der Erforschung der subterranen Fauna. *In Abderhalden, Handbuch d. biol. Arbeitsmethoden.* Abt. IX; Teil 7.

CHAPPUIS, P. A. (1942) Eine neue Methode zur Untersuchung der Grundwasserfauna. *Acta Sc. Math. Nat. Kolozvar.* No. 6.

CHAPPUIS, P. A. (1950) La récolte de la Faune souterraine. *Notes biospéologiques.* **V.**

COLAS, G. (1948–1962). *Guide de l'Entomologiste.* Paris. 1 ère Edit. 1948; 2 ème Edit. 1962.

COTTI, G. (1957) Guida alla ricerca della flora e fauna delle caverne. *Rass. Spel. ital. e Soc. Spel. Ital.* Guide didattiche. **I.**

DELAMARE-DEBOUTTEVILLE, CL. (1960) *Biologie des eaux souterraines littorales et continentales.* Paris.

DELHERM DE LARCENNE (1875) Chasse dans les grottes. *Nouvelles Entomologiques.* **VII.**

DOHRN, C.A. (1866) Ueber den Fang der Höhlenkäfer. *Stettin. entomol. Zeit.* **XXVII.**

DOMBROWSKI, E. (1906) Einiges über das Sammeln von Höhlenkäfern. *Natur und Haus.* **XIV.**

HAZELTON, M. and GLENNIE, E. A. (1953) Cave Fauna and Flora. *In* Cullingford, C. H. D. *British Caving.* London.

JEANNEL, R. (1943) *Les Fossiles vivants des cavernes.* Paris.

KUNTZEN, H. (1924) Das Sammeln von Höhlenkäfern. *Mitteil. Höhl. Karstf.*

LERUTH, R. (1939) Notes d'Hydrobiologie souterraine. **VII.** Une méthode intéressante pour l'étude de la nappe phréatique. *Bull. Soc. R. Sc. Liège.* No. 2.

MACHADO, A. DE BARROS. (1945) Instrucôes para a exploraçâo biologica das Cavernas. *Broteria; Ser. Cienc. Nat.* **XIV.**

POLLONERA, C. (1899) Un metodo par raccogliere i molluschi cavernicoli. *In Alto.* **X.**

RACOVITZA, E. G. (1913) Biospeologica. Instructions pour la récolte et la conservation des biotes cavernicoles et pour la rédaction des données bionomiques nécessaires à leur étude. *Edition des Archives de Zoologie expérimentale et générale.* Paris.

SOUSA, M. (1959) Breves notas sobre a colheita e conservacâo na Biospeologia. *Bol. Soc. Portug. Espel.* No. 2–3.

WETTSTEIN, O. (1920) Anleitung zum Sammeln von Tieren und Pflanzen in Höhlen. *Speläol. Jahrb.* **I.**

WINKLER, A. (1912) Eine neue Sammeltechnik für Subterrankäfer. *Col. Rundschau.* **I.**

Attempts to Transplant Fauna

BANTA, A.M. (1907) The Fauna of the Mayfield's Cave. *Carnegie Inst. Washington Public.* No. 67.

BONADONA, P. (1955) Notes de biospéologie provençale. *Notes biospéologiques.* **X.**

BONADONA, P. (1956) A propos des *Troglodromus* Dev. *Notes biospéologiques.* **XI.**

GINET, R. (1952) Essai d'acclimatation de cavernicoles dans la grotte de la Balme (Isère). *Bull. Soc. linn. Lyon.* **XXI.**

GINET, R. (1960) Écologie, Ethologie et Biologie de *Niphargus* (Amphipodes Gammaridés hypogés). *Annal. Spéléol.* **XV.**

PATRIZI, S. (1956) Introduzione ed acclimazione del Coleottero Catopide *Bathysciola derosai* Dod. in una grotta laziale. *Grotte d'Italia* (3) **I.**

Breeding of Cavernicoles

BEAUCHAMP, P. DE (1935) Observations sur les *Dendrocoelum* obscuricoles en élevages. *Compt. Rend. Soc. Biol. Paris.* **CXX.**

Boldori, L. (1933) Animali cavernicoli in schiavitù. *Atti d. I. Congr. Speleol. naz. Trieste.*
Boldori, L. (1935) Animali cavernicoli in schiavitù. II. *Boll. Soc. entomol. Ital.* **LXVII.**
Boldori, L. (1938) Animali cavernicoli in schiavitù. III. *Natura.* **XXIX.**
Boldori, L., and Cerruti, M. (1959) Animali cavernicoli in schiavitù. IV. *Boll. Soc. Entomol. Ital.* **LXXXIX.**
Glaçon, S. and le Masne, G. (1951) Une méthode d'élevage et de transport des Coléoptères cavernicoles. *Compt. rend. Acad. Sci. Paris.* **CCXXXII.**

The Subterranean Laboratories

The Laboratoire des Catacombes

Viré, A. (1897) Le Laboratoire des Catacombes. *Bull. Mus. Hist. Nat. Paris.* **III.**
Viré, A. (1901) Liste des principales espèces étrangères entrées en 1900 et 1901 dans les collections du Laboratoire de Biologie souterraine du Muséum. *Bull. Mus. Hist. Nat. Paris.* **VII.**
Viré, A. (1904) Sur quelques expériences effectuées au laboratoire des Catacombes du Muséum d'Histoire Naturelle. *Compt. rend. Acad. Sci. Paris.* **CXXXVIII.**

The Laboratoire de Spéléobiologie Expérimentale de Saint-Paër

Gadeau de Kerville, H. (1911). Le Laboratoire de Spéléobiologie expérimentale d'Henri Gadeau de Kerville à Saint-Paër (Seine-Inférieure). *Bull. Soc. Amis. Sc. Nat. Rouen.*
Gadeau de Kerville, H. (1926) Note sur an Protée anguillard (*Proteus anguinus* Laur.) ayant vécu sans aucune nourriture. *Bull. Soc. Zool. France.* **LI.**

The Laboratoire du Centre National de la Recherche Scientifique

Vandel, A. (1954) Le Laboratoire souterrain du Centre National de la Recherche Scientifique. Editions du C.N.R.S., Paris.
Vandel, A. (1959) Les Activités du Laboratoire souterrain du Centre National de la Recherche Scientifique. *Annal. Soc. R. Zool. Belgique.* **XCIX.**
Vandel, A. (1962) Les Activités du Laboratoire souterrain du Centre National de la Recherche Scientifique. *Conférences du Palais de la Découverte.* No. A. 280.

The Grotte d'Antheuil

Loriol, B. de, Tintant, H. and Rousset, A. (1959) Antheuil; son site; sa grotte. *Sous le Plancher.*

The Laboratoire Souterrain de Han-sur-Lesse

Liégeois, P.G. (1958) Les activités de la Fédération spéléologique de Belgique et le Laboratoire souterrain de Han-sur-Lesse. Publications de la Fédération Spéléologique de Belgique.

Stazione Biologica Sperimentale Sotteranea di Napoli

Parenzan, P. (1954) Istituzione della Stazione biologica sperimentale sotterranea di Napoli. *Atti d. VI Congr. naz. d. Speleol.*, Trieste.
Parenzan, P. (1956) Istituzione della stazione biologica sperimentale sotterranea di Napoli. *Grotte d'Italia.* (3) **I.**

The Postojna Biospeological Station

Dudich, E. (1932–1933) Die speläobiologische Station zu Postumia und ihre Bedeutung für die Höhlenkunde. *Speläol. Jahrb.* **XIII–XIV.**
Jeannel, R. (1936) La Grotte de Postumia et sa station biologique souterraine. *La Terre et la Vie.* **VI.**

Other Yugoslav Stations

ALJANČIČ, M. (1961) Biospeološki Laboratorij v Jami Tular Pri Kranju. *Naše Jame*, Ljubljana. **II.**

KENK, R., and SELIŠKAR, A. (1931) Študije o ekologiji jamskih živali. I. Meteorološka in hidrološka opazovanja v Podpeški jami v letih 1928–1931. *Prirod. Razpr*, Ljubljana. **I.**

The Biospeological Station at "Baradla", a Cave near Aggtelek

DUDICH, E. (1932) Biologie der Aggteleker Tropfsteinhöhle "Baradla" in Ungarn. *Speläolog. Monogr.* **XIII.**

DUDICH, E. (1960) Das höhlenbiologische Laboratorium der Eötvös Loránd Universität. Biospeologica hungarica, **X.** *Annal. Univ. Sc. Budapest. Sect. Biol.* **III.**

DUDICH, E. (1962) Über das ungarische Laboratorium für Höhlenbiologie. *Karszt ès Barlangkutatàs.* **11**

The Institutul de Speologie de Cluj

RACOVITZA, E.G. (1926) L'Institut de Spéologie de Cluj, et considérations générales sur l'importance, le rôle et l'organisation des Instituts de Recherches Scientifiques. *Trav. Inst. Spéologie. Cluj.* **I.**

CHAPPUIS, P.A. (1948) L'Activité de l'Institut de Spéologie de 1920 à 1947. *Bul. Soc. Stiinte d. Cluj.* **X.**

The Institutul de Speologie Emil Racovitza

MOTAS, C. (1961) La Speologia in Romania. *Rassegna Speleol. Ital.* **XIII.**

SERBAN, M., VIEHMAN, I. and COMAN, D. (1961) *Grottes de Roumanie.* Méridiens-Editions. Bucharest.

The Biological Station at Kossino

BOROUTZKY, R. (1930) Expedition der Biologischen Station zu Kossino, zum Zwecke der Erforschung der Kutais-Höhlen am Rion, im Jahre 1929 (Transkaukasien, Georgien). *Mitteil. Höhl. Karstf.*

PART 2

A List of Cavernicolous Species

There is no place in a work of this nature to give a detailed list of all the animals and plants found in caves. This would merely reproduce Volume III of B. Wolf's *Catalogus Animalium Cavernarum* and require the space of some 2000 pages.

But it is advisable, so that the reader may orientate himself in the complex world of cavernicoles, to review the different groups of living organisms which inhabit caves, and to fix their respective places in the systematic classification.

Numerous cavernicoles have been discovered during the past ten or fifteen years. These have, of course, not been dealt with in the classic treatises. The reader will find most of these organisms mentioned in this book. This list stops at the end of 1962.

CHAPTER V

SUBTERRANEAN PLANTS

A. INTRODUCTION

The study of plants which live in the subterranean environment developed well before that of cavernicolous animals — with the exception of *Proteus* — because the first treatise on subterranean botany entitled *Plantae subterraneae, descriptae and delineatae,* was published in 1772. It was written by Johannes Antonius Scopoli, a naturalist from the Carniola, which is a classic area for speleology.

Kyrle (1923) created the terms speleobotany and speleozoology. The second term represents a true science with a valid basis, but the former term represents a non-existent science. This is clearly because there are no true cave-living plants. Maheu (1906) correctly stated "No underground species exists which has not an analogous form at the surface from which it differs only in alteration of colour or of general characteristics".

It is interesting to study the modifications which plants undergo when they are kept in darkness or when they develop in weak light. These facts are of interest to botanists and plant physiologists but fall outside the domain of biospeology.

An important bibliography relates to plants found in a subterranean environment. Amongst the more important books on this subject may be mentioned those of Pokorny (1853, 1854), Call (1897), Maheu (1906), Lämmermayer (1912–1916), Morton and Gams (1925), Tomaselli (1949, 1950), Tosco (1957–1958) and Mason-Williams and Benson-Evans (1958).

B. FUNGI

Lists of fungi found in caves have been compiled by Maheu (1906), Lagarde (1913, 1917, 1922), Negri (1920), Bailey (1933), Wolf (1938) and Mason-Williams and Benson-Evans (1958).

The fungi which grow in dark situations belong to several groups: Myxomycetes, Phycomycetes, Ascomycetes, Basidiomycetes, etc. Some are epizoic and others parasitic, and these will be referred to in Chapter XV. The other types develop on organic matter introduced into the caves: wood, paper, food, excrements, etc. Polluted clay deposits may be covered in moulds, but the natural clay of caves is always without them.

No fungi are exclusively cavernicolous, with the exception of certain insect parasites (see Chapter XV), but here it is rather a question of host specificity than of hypogean specialisation.

It is true to say that many new species of fungi have been collected in caves, and also new genera *(Corallinopsis, Mahevia)* but no one has proved that these new species are not to be found above the subterranean habitat. Most of the undescribed forms are parasites of insects, that is, those fungi which rarely attract the attention of mycologists,

Another reason to doubt the existence of true cavernicolous fungi lies in the radical changes which cryptogams undergo under the influence of the subterranean environments, and which are likely to lead mycologists astray (Klein, 1922). Although the smaller forms are relatively slightly modified when they develop underground, the higher fungi, particularly the Basidiomycetes, may show malformations or deformations under such conditions. "They may appear to be similar to another species, away from their original type, and thus specifically, and often generically, unrecognisable" (Lagarde, 1917).

The modifications of structure which the fungi undergo when grown in the subterranean environment are often far-reaching. They have frequently been described by mycologists. Special mention may be made of the "Rhizomorphs" studied by R. Schneider (1885, 1886), which seem to represent reduced and profoundly modified forms of the higher fungi. These phenomena are more relevant to plant physiology than to biospeology.

A study of the cave dwelling Aspergillales and Mucorales using a whole range of modern techniques has recently been resumed by Caumartin (1961 a and b). In subterranean media these moulds have a tendency to form cysts. These are frequently found in cave clay, often at great depths. Caumartin has reported that the formation of cysts is brought about by the presence of sulphur, resulting from the reduction of sulphates. On the other hand, in the presence of ferric salts, the growth of mycelia is considerably increased and sporulation occurs.

Biospeologists should not overlook the importance of sub-soil fungi because they form a source of food for cavernicolous animals; they can also supply them with the vitamins and the growth substances they require.

C. CYANOPHYCEAE

Cyanophyceae belonging to many different species and genera have been collected in caves. Their presence has been observed not only at the entry of the subterranean cavities but also in total darkness. It also appears to be well established that these very primitive plants are capable of synthesising their various pigments in the absence of light (Etard and Bouilhac, 1898; Magdeburg, 1933; Claus, 1955).

Magdeburg (1933) observed in Franconian caves some very friable tufa which contained many lower plants and in particular, Cyanophyceae. He considered that certain Cyanophyceae and especially Chroococcaceae *(Gleocapsa biformis* and *G. polydermatica, Aphanothece castagnei)* play a part in the precipitation of calcium carbonate and in the formation of calcareous tufa. Although the observations of this German botanist are interesting, his conclusions should not be accepted without confirmation.

D. ALGAE

For some considerable time it has been known that some algae *(Chlorella, Scenedesmus, Pleurococcus,* etc.*)* are able, unlike Phanerogams, to manufacture chlorophyll in complete darkness. This fact was first reported by Radais (1900) and confirmed by modern plant physiologists (Myers, Goodwin). The formation of chlorophyll and carotenoid pigments is, in fact, greater in cultures kept in the dark than in illuminated ones (Goodwin, 1954). Pigments of the symbiotic algae entering into association with fungi in lichens, may also be developed in the absence of light (Artari).

Therefore it is not surprising to find in caves, greenish or reddish beds formed by *Protococcus* and *Haematococcus.* But as the phenomenon of photosynthesis cannot occur in total darkness, the development of these algae must depend upon the presence of organic matter. Diatoms are often collected in subterranean waters, but they cannot be kept alive in total darkness. Living diatoms have been collected at the bottom of dimly lit shafts; but those collected from subterranean rivers are always dead and reduced to their siliceous frustules (Maheu).

E. PLANTS OTHER THAN CRYPTOGAMS

Mosses, liverworts, ferns and various Phanerogams are found at the entrances of caves. These are all shade-loving forms which grow in damp, poorly lit situations. Plants which germinate in partial darkness often develop in an atypical manner. These modified processes of development have been carefully described by botanists. At any rate it is clear that no higher plant penetrates deeply into the subterranean environment. Therefore the study of these forms is not relevant to biospeology.

BIBLIOGRAPHY

Introduction

CALL, E. (1897) Note on the Flora of Mammoth Cave, Kentucky. *J. Cincinnati Soc. Nat. Hist.* **XIX.**

KYRLE, G. (1923) Grundriß der theoretischen Spelaeologie. *Spelaologische Monographien* I, Wien.

LÄMMERMAYR, L. (1912, 1914, 1916) Die grüne Pflanzenwelt der Höhlen. *Denkschr. Akad. Wiss. Wien. Math. Naturw. Kl.* **LXXXVII, XC, XCII.**

MAHEU, J. (1906) Contribution a l'étude de la flore souterraine de France. *Annal. Sc. Nat. Botanique.* (9), **III.**

MASON-WILLIAMS, A. and BENSON-EVANS, K. (1958) A preliminary investigation into the bacterial and botanical Flora of Caves in South Wales. *Cave Research Group of Great Britain Publication* No. 8.

MORTON (FR.) and GAMS, H. (1925) Höhlenpflanzen. *Speläologische Monographien* V, Wien.

POKORNY, A. (1853) Über die unterirdische Flora der Karsthöhlen. *Verhandl. Zool. Bot. Ver. Wien.* **III.**

POKORNY, A. (1854) Zur Flora subterranea der Karsthöhlen. *In* SCHMIDL, A., *Die Grotten und Höhlen von Adelsberg, Lueg, Planina und Laas.* Wien.

TOMASELLI, R. (1949) Osservazioni di biospeologia vegetale. *Rassegna Speleol. Ital.* **I.**

TOMASELLI, R. (1950) Per un censimento della flora cavernicola italiana. *Rassegna Speleol. Ital.* **II.**

TOSCO, U. (1957–1958) Contributi alla conoscenza della vegetazione e della flora cavernicola italiana. *Le Grotte d'Italia* (3) **II.**

Fungi

BAILEY, V. (1933) Cave Life of Kentucky, mainly in the Mammoth Cave Region. *Amer. Midl. Nat.* **XIV.**

CAUMARTIN, V. (1961 a) La microbiologie souterraine: ses techniques, ses problèmes. *Bull. Soc. Bot. Nord. France* **XIV.**

CAUMARTIN, V. (1961 b) Le Comportement des moisissures dans le milieu souterrain. *Troisième Congr. Intern. Spéléologie.* Vienne.

KLEIN, G. (1922) Ein *Mucor* aus einer Dachsteinhöhle. *Verhandl. Zool. Bot. Gesell. Wien* **LXXII.**

LAGARDE, J. (1913) Champignons, 1ère Série. *Biospeologica,* **XXXII.** *Archiv. Zool. exper. géner* **LIII.**

LAGARDE, J. (1917) Champignons, 2ème Série. *Biospeologica,* **XXXVIII.** *Archiv. Zool. exper. géner* **LVI.**

LAGARDE, J. (1922) Champignons, 3ème Série. *Biospeologica,* **XLVI.** *Archiv. Zool. exper. géner* **LX.**

MAHEU, J. (1906) Contribution a l'étude de la flore souterraine de France. *Annal. Sc. Nat. Bot.* (9), **III.**

MASON-WILLIAMS, A. and BENSON-EVANS, K. (1958) A preliminary investigation into the bacterial and botanical Flora of Caves in South Wales. *Cave Research Group of Great Britain. Public.* No. 8.

NEGRI, G. (1920) Su un musco cavernicolo crescente nell'oscurita assoluta. *Atti d. R. Accad. Lincei. Rendiconti* (5), **XXIX.**

SCHNEIDER, R. (1885) Über subterrane Organismen. *Königl. Realschule z. Berlin. Real-gymnasium. Bericht über das Schuljahr 1884–1885.* No. 85; Berlin.

SCHNEIDER, R. (1886) Amphibisches Leben in den Rhizomorphen bei Burgk. *Math. Naturw. Mitteil. Sitz. k. Preuss. Akad. Wiss.* **VII.**

WOLF, F. A. (1938) Fungal Flora of Yucatan caves. *In* Fauna of the Caves of Yucatan, by A. S. Pearse. *Carnegie Instit. Washington Public.* No. 491.

Cyanophyceae

CLAUS, G. (1955) Algae and their mode of life in the Baradla Cave at Aggtelek. *Acta Botan. Hungar.* **II.**

ETARD, A. and BOUILHAC (1898) Présence des Chlorophylles dans un *Nostoc* cultivé à l'abri de la lumière. *Compt. rend. Acad. Sci. Paris* **CXXVII.**

MAGDEBURG, P. (1929–1932) Organogene Kalkkonkretionen in Höhlen. *S.B. Naturf. Gesell. Leipzig* **LVI–LIX.**

Algae

GOODWIN, T. W. (1954) Some observations on Carotenoid synthesis by the Alga *Chlorella vulgaris. Experientia* **X.**

RADAIS, M. (1900) Sur la culture pure d'une algue verte; formation de chlorophylle à l'obscurité. *Compt. rend. Acad. Sci. Paris* **CXXX.**

CHAPTER VI

THE FREE-LIVING PROTISTA

IN THIS chapter only the free-living Protista will be considered. Epizoic and parasitic forms will be mentioned in Chapter XV. It is advisable to distinguish within the free-living Protista those which are found in subterranean waters and those which are found in the clay deposits in caves.

A. PROTISTA OF SUBTERRANEAN WATERS

Protista are well represented in numbers and in species in underground waters. They have often been referred to in the past by many authors (Ehrenberg, 1860, 1862; Joseph, 1879 a and b; Vejdovsky, 1880; Schneider, 1886; Packard, 1886; Moniez, 1889; Pateff, 1926; Wetzel, 1929; Brunetti, 1933; Griepenburg, 1934, 1939; Lanza, 1949; Tosco, 1956; Varga, 1959; Varga and Takats, 1960; Doroszewski, 1960).

The Protista of subterranean waters belong to many groups and genera.
Mastigophora: *Chilomonas, Copromonas, Cercomonas, Oikomonas, Anisonema, Peranema, Bodo, Cercobodo,* etc.
Amoebina: *Amoeba, Astramoeba, Pelomyxa.*
Testacea: *Arcella, Difflugia, Euglypha, Trinema, Cyphoderia, Centropyxis,* etc.
Heliozoa: *Actinophrys, Actinosphaerium.*
Ciliophora: *Paramecium, Colpoda, Glaucoma, Stentor, Spirostomum, Euplotes, Stylonichia, Oxytricha, Carchesium, Vorticella, Cothurnia,* etc.

All these are common genera and are often found in surface waters. None of them are specifically cavernicoles. Moniez (1889) found in subterranean waters at Lille, specimens of *Phacus longicauda* Duj., possessing neither chlorophyll nor stigma. However, it is known that the disappearance of pigments in *Euglena* placed in darkness is a common occurrence, well known to protistologists (Zumstein, Ternetz, Baker, Lwoff). These are environmental effects rather than genetically determined characteristics. Therefore the colourless forms observed by Moniez should not be considered to be representatives of a cavernicolous race.

Aquatic subterranean Protista undoubtedly come from surface waters. Those individuals among them which are incapable of adapting themselves to the hypogeous environment soon disappear. However, many can live there and even multiply. Some of them encyst when conditions become unfavourable (Doroszewski, 1960).

The Protista which live in subterranean water feed upon minute organic particles, bacteria, or more rarely they are predatory species which feed upon other Protista.

Although not a single free-living aquatic protist can be considered as a true cavernicole, unicells constitute one of the regularly occurring elements of the subterranean aquatic fauna. They assuredly play a very important role, because they are the food of many cavernicolous Metazoa.

One of the most extraordinary discoveries in the field of subterranean protozoology is undoubtedly that reported by A.L.Brodsky (cited from Birstein and Boroutzky, 1950). This naturalist discovered an abundant population of Foraminifera in some wells in the Kara-Kum desert in the Trans-Caspian Province. The wells were about 20 metres deep and were fed by slightly brackish ground water. The foraminiferous fauna was composed of 10 species belonging to the genera *Spiroloculina*, *Biloculina*, *Triloculina*, *Lagena*, *Nodosaria*, *Discorbina*, *Globigerina* and *Textularia*. They are much smaller than the marine species of the same genera and the shell is thinner. They are typical representatives of the interstitial fauna.

B. PROTISTA OF THE CLAY DEPOSITS IN CAVES

Although the existence of aquatic Protista in caves has been well known for some time we know very little about the terrestrial forms.

Research work by B.Brunetti (1933) has shown that the soil of caves contains various Protista: Mastigophora, Sarcodina, and a few Ciliata. This fauna is poor in species and even more so in individuals, even when the medium is rich in organic matter. The forms found are commonplace species which have arrived from the surface.

Observations carried out by V.Caumartin and L.Bonnet in Pyrenean caves have confirmed the presence of Amoebina (both Nuda and Testacea) in cave clay. The *Amoebae* belong to the group *terricola-verrucosa*, with short pseudopodia. The encysted stages of *Amoebae* are also frequently found in cave clay. These Sarcodina certainly feed on the bacteria present in this medium, and it seems probable that they, in turn, form part of the diet of cavernicolous limnivores.

It is most desirable that a systematic study of these animals should be carried out to supplement and extend this brief information.

BIBLIOGRAPHY

BIRSTEIN, Y. A. and BOROUTZKY, E. V. (1950) La vie dans les eaux souterraines. *In La vie des eaux douces de l'URSS.* Tome III; Chapitre 28. Moscow-Leningrad.

BRUNETTI, B. (1933) Ricerche sui Protozoi dell' terreno. 3. I Protozoi del terreno delle grotte delle Buca Nova e del Castello (Monti di oltre Serchio). *Atti Soc. Toscana Sc. Nat. Pisa (Proc. Verb.)* **XLII.**

DOROSZEWSKI, M. (1960) Quelques remarques sur l'apparition des Infusoires dans les cavernes (*Biospeologica Polonica*, **IV**). *Speleologia*, **II.**

EHRENBERG, C. G. (1860) Über die mit dem *Proteus anguineus* zusammenlebenden mikroskopischen Lebensformen in den Bassins der Magdalenengrotte in Krain. *Monatsber. d. k. preuß. Akad. Wiss. Berlin.*

EHRENBERG, C. G. (1862) Zweite Mitteilung über die mikroskopischen Lebensformen als Nahrung des Höhlensalamanders. *Monatsber. d. k. preuß. Akad. Wiss. Berlin.*

GRIEPENBURG, W. (1933) Die Protozoenfauna einiger westfälischer Höhlen. *S. B. d. Gesell. Naturf. Freunde Berlin.*

GRIEPENBURG, W. (1934) Die Berghäuser Höhle bei Schwelm i. W. *Mitteil. Höhl. Karstf.*

GRIEPENBURG, W. (1939) Die Tierwelt der Höhlen bei Kallenhardt. *Mitteil. Höhl. Karstf.*

JOSEPH, G. (1879 a) Über Grotten-Infusorien. *Zool. Anz.* **II.**

JOSEPH, G. (1879 b) Weitere Mitteilungen aus dem Gebiete der Grottenfauna. *Zool. Anz.* **II.**

LANZA, B. (1949) Speleofauna toscana. I. Cenni storici ed elenco raggionato dei Protozoi, dei Vermi, dei Molluschi, dei Crostacei, dei Miriapodi e degli Aracnidi (Acari esclusi) cavernicoli delle Toscana. *Attual. Zool.* **VI.**

MONIEZ, R. (1889) Faune des eaux souterraines du département du Nord et, en particulier de la ville de Lille. *Rev. biol. Nord. France* **I.**

PACKARD, A. S. (1886) The Cave Fauna of North America, with remarks on the anatomy of the brain and origin of the blind species. *Mem. Nat. Acad. Sci. Wash.* **IV.**

PATEFF, P. (1926) Süßwasser-Rhizopoden aus der Höhle Salzlöcher (Schlesien). *Mitteil. Höhl. Karstf.*

SCHNEIDER, R. (1886) Amphibisches Leben in den Rhizomorphen bei Burgk. *Math. Nat. Naturw. Mitteil. Sitzb. Preuß. Akad. Wiss.* **VII.**

TOSCO, U. (1956) Rudimenti di Spelaeoprotozoologia. *Studia spelaeol.*

VARGA, L. (1959) Beiträge zur Kenntnis der aquatilen Mikrofauna der Baradla-Höhle bei Aggtelek. (*Biospeologica hungarica*, **III**). *Acta Zool. Acad. Sci. Hung.* **IV.**

VARGA, L. and TAKATS, T. (1960) Mikrobiologische Untersuchungen des Schlammes eines wasserlosen Teiches der Aggteleker Baradla-Höhle. *Acta Zool. Acad. Sci. Hung.* **VI.**

VEJDOVSKY, FR. (1880) Über die Rhizopoden der Brunnenwässer Prags. *S. B. k. böhm. Gesell. Wiss.*

WETZEL, A. (1929) Die Protozoen der Schneeberger Erzbergwerke. *Zool. Jahrb. Abt. System* **LVI.**

CHAPTER VII

THE CAVERNICOLOUS INVERTEBRATES (EXCLUDING ARTHROPODA)

INTRODUCTION —

CAVERNICOLOUS METAZOA

Many groups of animals which have a majority of terrestrial or fresh-water forms possess representative species in the subterranean environment. In the case of the lower Metazoa it is often difficult to decide whether species captured underground are true cavernicoles or merely stray specimens from the surface. However, those Metazoan groups which are almost exclusively marine are represented underground by very few species.

A list of Metazoan groups which possess no members in the subterranean fauna is given below:

Ctenophora	Amphineura	Echinodermata
Endoprocta	Scaphopoda	Pterobranchiata
Bryozoa†	Opisthobranchiata	Enteropneusta
Brachiopoda	Cephalopoda	Tunicata
Phoronidea		Cephalochorda
Pogonophora	Merostomata	
Sipunculoidea	Cirripedia	Cyclostoma
Chaetognatha	Leptostraca	Selachii
	Cumacea	Crossopterygii
	Tanaidacea	Dipnoi
	Euphausiacea	Anura
	Stomatopoda	

Sponges

Sponges are found only exceptionally in caves. Specimens of *Ephydatia mülleri* (= *Spongilla stygia* Joseph) were reported by Arndt (1933) in the caves of Karst and Herzegovina. Spicules of *Spongilla fragilis* were reported by Kofoid (1900) from Mammoth Cave. Both species are surface forms.

† Remy (1937) reported the presence of *Plumatella* in the entrance of the Ponor Crnulja (Herzegovina) in a zone still exposed to daylight. On the other hand, the Bryozoa are not at all uncommon in water pipes (Arndt, 1933).

Coelenterata

The presence of *Hydra* in caves has been reported from time to time. It seems sure that this creature is present there accidentally rather than normally.

However, *Hydra* is frequently found in ground water. P.A.Chappuis (1944, 1946) has shown that the green hydra *(Hydra viridissima)* is a normal inhabitant of ground water in Transylvania. However, this occurrence seems only to be found in the southern Carpathians and has not been discovered elsewhere. *Hydra vulgaris* occurs rather rarely in ground water of the River Save (Meštrov, 1960). Husmann (1956), observed only two specimens of *Pelmatohydra oligactis* in the ground water of the Weser basin.

The presence of *Hydra* in interstitial media may appear surprising at first glance because in ponds and lakes hydras are found fixed to plants and floating free in the water. But today it is known that the marine littoral fauna contains numerous hydroid types perfectly adapted to interstitial life: *Protohydra, Halammohydra, Psammohydra, Armohydra, Otohydra* (Delamare-Deboutteville, 1960).

Turbellaria

(a) Rhabdocoela. Many species belonging to the Rhabdocoela have been found in wells, springs and ground water. They are members of the interstitial fauna *(Mesostoma hallezianum, Prorhynchus putealis, Protomonotresis centrophora).*

On the other hand, only common species which have strayed in the subterranean medium have been taken in caves. However, there are two exceptions. *Krumbachia subterranea* has been captured in several caves in Germany and must be regarded as truly cavernicolous (Reisinger, 1933; Griepenburg, 1934, 1939). As for the white, blind species *Vortex cavicolus* captured in Carter Cave (Kentucky) it is probably a true cavernicole, but its systematic position is problematical (Packard, 1883).

(b) Tricladida. The order Tricladida may be divided into three suborders, Maricola, Paludicola and Terricola. Only the Paludicola possess cavernicolous species.†

The tricladid paludicoles (commonly called planarians) contain numerous subterranean species. They are white and depigmented. Many have no eyes but this is not a general feature, since certain subterranean species may possess either normally developed or reduced eyes (see p. 415). It will

---

† No terricolous cavernicolous planarians exist. Komarek (1919) errected the genus *Geopaludicola* for the species *absoloni*, which he regarded as a terrestrial planarian. This species is, in fact, a typical paludicole (Kenk, de Beauchamp) which must be renamed *Planaria absoloni*.

be noted later that cavernicolous planarians differ from surface forms quite sharply in both behaviour and reproduction.

Our knowledge of the light-avoiding planarians remained rudimentary for a long time as can be seen by reference to the review by de Beauchamp (1920). But it has been advancing more speedily during the past 40 years, mainly due to important work by four zoologists: P. de Beauchamp, Miss L. Hyman, R. Kenk and J. Komarek.

Ecology – The ecology of the hypogeous planarians is imperfectly known. Certain species, such as *Dendrocoelum boettgeri*, and some species of *Fonticola*, populate ground water and belong to the interstitial fauna. However, most of the light-avoiding planarians have been collected in caves.

Geographical Distribution – The Paludicola are confined to the palaearctic region with the exception of *Curtisia foremani* which has reached South America. The geographical distribution of the hypogean planarians is less widespread. They have been reported from four regions:

(1) The whole of Europe, from Spain to Caucasia, excluding the more northerly regions (British Isles and Scandinavia). In Europe hypogeous species are indeed more numerous than epigeous ones.

(2) *Acromyadenium maroccanum* can, at least provisionally, be placed in the light-avoiding planarians, although it has been captured in mountain torrents. It is indigenous to the Middle Atlas Mountains. In Africa and Madagascar the only planarian collected in caves is the common surface species *Dugesia gonocephala*.

(3) Asia, a centre of remarkable turbellarian evolution, certainly possesses light-avoiding planarians, but little is known of them. They are common in the Batu caves, in Malaya, but have not been exactly identified. The Japanese species are also poorly known (Kawakatsu, 1960). *Planaria papillifera* Ijima and Kaburaki, which according to Professor Okugawa, should be placed in the genus *Phagocata*, has been found in a well at Tokyo. Its grey colour and the presence of eyes indicate that this planarian is a troglophile.

(4) North America possesses 14 species of hypogeous planarians (Hyman, 1960). They are found all over the United States from Oregon to Pennsylvania, and from Iowa to Texas and Florida.

Classification – The light-avoiding planarians are spread amongst three families.

(1) Dendrocoelidae Hallez. The members of this family are characterised by a network of longitudinal and circular muscles in the internal musculature of the pharynx. They possess anteriorly a simple or complex adhesive organ, sometimes constituting a sucker *(Dendrocoelopsis spinosopenis)*, which serves as a locomotory organ, and also sometimes for capturing

prey. These planarians are large, white and deprived of pigment, even in the surface forms (a remarkable example of preadaptation). The majority of light-avoiding planarians belong to this family, accounting at the present for some 40 species. North America possesses only two species, *Procotyla typhlops* and *Sorocelis americana*.

(2) Kenkiidae Hyman. The longitudinal and circular muscles are separated in the pharynx (as in the Planariidae). The adhesive organ is superficially similar to those found in the Dendrocoelidae but constructed in a different manner. The species are white, depigmented and have no eyes. The family is indigenous to North America and contains ten species. The most notable is *Sphalloplana percaeca* discovered by A.S. Packard in 1874, in Mammoth Cave.

(3) Planariidae Stimpson. The circular and longitudinal muscles of the internal musculature of the pharynx are separated. There are only a few hypogeous representatives. The species of the genus *Fonticola* are found in springs and ground water but are not true subterranean forms (de Beauchamp, 1939). There are only five true cavernicoles in this family. They are *Planaria absoloni*, *Atrioplanaria racovitzai* and *Crenobia anophthalma* (Europe), and *Phagocata subterranea* and *P. cavernicola* (North America).

De Beauchamp (1955) has described another species of the Planariidae of which both the systematic and the ecological positions remain uncertain. This species, *Polycellis benazzii*, was captured in the Tana di Spettari, at Toirano, in the province of Savona (Italy). The surface layers of this planarian are without pigment, but the pharynx contains chromatophores. The eyes are reduced in size, but 9 to 10 pairs are present at some distance from the margin.

Nemertea

A species of this group has been captured in ground water. It is *Prostoma clepsinoides* Dugès var. *putealis* de Beauchamp, which differs from the type commonly found in surface waters by the complete absence of pigment and of eye spots, and by the presence of tactile setae along the anterior border (de Beauchamp, 1932). The light-avoiding form was captured in a well near Dijon, and in ground water of the Rhine near Strasbourg. It was later found by L. Botosaneanu in a temporary resurgent in the Bains d'Hercule in Rumania (Motas, 1961).

Another cavernicolous species was captured in Vjetrenica, near Zavala, and again in a cave near Bileca in Herzegovina. This species has been described by Tarman (1961), under the name of *Prostoma hercegovinense*. This nemertine is without pigment or eyes.

Nematoda

Free- living nematodes have frequently been reported from ground water, caves and also mines (Hnatenwytsch, 1929; Altherr, 1938). An excellent review of these nematodes has been published by Andrassy (1959).

Some are terrestrial, found in the clay or earth deposits in caves, others are aquatic, but almost all are more or less amphibious. They feed upon organic matter or bacteria, but *Dorylaimus bokori* sucks the cell sap of

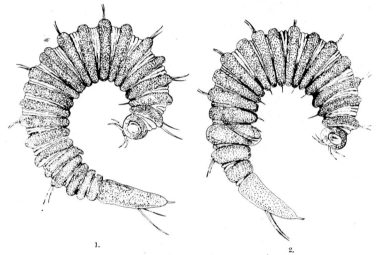

Fig. 5. *Desmoscolex aquaedulcis* (after Stammer).
1. male; 2. female

fungi. The species of the genus *Mylonchulus* are carnivorous. The majority of these nematodes can survive unfavourable periods as resistant encysted larvae.

Most of the forms found underground belong to genera occurring widely in the epigeous habitat (*Rhabditis, Diplogaster, Cephalobius, Plectus, Mono-hystera, Trilobus, Dorylaimus*, etc.). There can be little doubt that the vast majority of these nematodes are surface species which have been carried down to the subterranean environment, and which are able to exist there because of their remarkable powers of adaptation.

Must it be concluded that there are no truly cavernicolous nematodes? It is difficult to answer this question because many free-living nematodes are unpigmented and lack eyes. Not a single morphological characteristic allows the separation of possible true cavernicoles from surface forms. Besides, there is the ever present possibility that species which have at present only been found in caves will at some future date be found on the surface. Andrassy proposed, with some reservations, that *Mylonchulus cavensis* should be considered a true cavernicole.

On the other hand special mention must be made of three remarkable species *(Desmoscolex aquaedulcis, Halaimus stammeri* and *Thalassolaimus aquaedulcis)* described by Stammer (1935) and by Schneider (1960) (Fig. 5). Specimens of these three species have come from the caves of Karst, Jugoslavia. A new species of *Desmoscolex*, still to be described, has been collected by A. Chodorowski, in two caves in the Pyrénées ariégeoises. The interest of these nematodes is that they belong to the family Desmoscolecidae, of which all other members are entirely marine, and belong to the littoral fauna (E. Schulz, S. A. Gerlach). No representatives of this family were known to inhabit fresh water before the discovery of these species in fresh subterranean water. In Stammer's view these species are a relict fauna of the Tertiary era, and of marine origin. Have they passed through a stage living in fresh water on the surface of the earth, before penetrating the subterranean world or did they come directly from the sea? In the present state of our knowledge it is impossible to answer this question.

Rotifera and Gastrotricha

Rotifers and more rarely Gastrotricha, have been reported from caves. It can be stated that they are always surface forms, which have strayed into the subterranean environment but are unable to exist there for long.

This is, however, not true for the ground water. The banks of rivers and lakes contain numerous psammic Rotifera, the study of which has been advanced in particular by Wisniewsky in Poland, and by Pennak in the United States. However, the study of lacustrine or river sand banks does not belong to biospeology but to hydrobiology.

ANNELIDA

Polychaeta

The polychaetes are the most typical and most abundant group of the annelids, and are essentially marine. No freshwater forms are known.

However, the first cavernicolous polychaete, *Troglochaetus beranecki* was described in 1921. A related species discovered in the Akiyoshi-do in Japan has been reported by Uéno (1957) under the name of *Troglochaetus*, in fact it belongs to a new genus, *Speochaetes*, which is yet to be described.

A third representative of the cavernicolous polychaetes, belonging to a different group, was described in 1930 (although discovered in 1913). It has been named *Marifugia cavatica*.

(a) Troglochaetus beranecki. This species was discovered on the 6th December 1919 by Théodore Delachaux in the Grotte du Ver which opens into the gorges of the Areuse (Canton of Neuchatel, Switzerland). The

Swiss zoologist was most surprised when it was proved that the new animal was a polychaete. It was named *Troglochaetus beranecki* (Delachaux, 1921) (Fig. 6).

Affinities – Delachaux, who was not an annelid specialist, associated *Troglochaetus* with the Nereidiformes and in particular the Eunicidae. Remane (1928) showed that this association was spurious. *Troglochaetus*

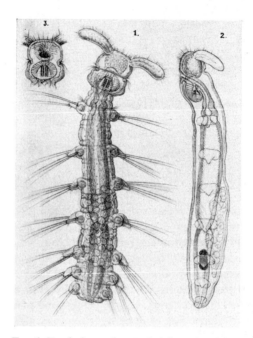

FIG. 6. *Troglochaetus beranecki* (after Delachaux).

was found to belong to the Nerillidae†. Representatives of this family comprise a characteristic element of the interstitial fauna. These poly-chaetes are of very small size (0·3–2 mm) and have many regressive char-acteristics.

Troglochaetus is allied to *Nerillidium mediterraneum* which inhabits the sands of the Naples shore together with *Amphioxus* (Remane, 1928). It is even more closely related to *Thalassochaetus palpifoliaceus* from the Gulf of Kiel (Ax, 1954). *Troglochaetus* is distinguished from marine forms by the absence of cirri on the anal segment. It is worthy of note, however, that the anal cirri are short in *Nerillidium australis* and very short in *N. stygicola*.

Origin – *Troglochaetus* would seem to be a marine relic which reached

† The Nerillidae have formerly been classed in the Archiannelida, an essentially hetero-geneous collection of neotonous and larviform species (Remane, Beklemishev).

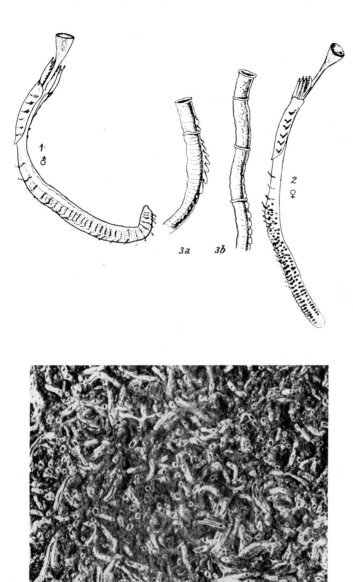

FIG. 7. *Marifugia cavatica* (after Absolon and Hrabe).
1. male; *2.* female; *3a, b.* tubes.

Central Europe by ascending the arm of the sea which once occupied the Rhône Valley (Remane, 1928)†.

Ecology – This species is found both in caves and in ground water, but is more common in the latter. It is a true member of the interstitial fauna (Stammer, 1937; Husmann, 1956; Picard, 1962). Even in caves it appears to search for the interstitial medium (Plesa, 1957). The interstitial mode of life which it follows is a reason for its wide distribution.

Geographical Distribution – Since the date of its discovery this annelid has been found in many parts of central and eastern Europe: Switzerland (Delachaux); Lorraine (Picard); Alsace (Hertzog); Germany (Noll; Ankel; Haine; Noll and Stammer; Husmann; Schulz and Uhlmann); Silesia (Stammer; Pax), Czechoslovakia (Kulhavy); Hungary (Andrassy); and in Rumania (Plesa).

(b) Marifugia Cavatica. This polychaete was discovered on 27th August, 1913 by the Moravian naturalist Karl Absolon in the Ponor Crnulja which opens into Popovo Polje, in Herzegovina. But it was not until 1927, during a meeting of the Association française pour l'Avancement des Sciences held at Constantine, that the discovery was made public (Absolon, 1927). Three years later a description of this species was made (Absolon and Hrabe, 1930).

It is a sedentary annelid belonging to the Serpulidae, and seems to have affinities with the genera *Placostegus* and *Ditrupa*. It is transparent, without pigment or eyes, and lives in a more or less calcified tube (Fig. 7). These tubes are produced by millions of specimens in the Ponor Crnulja, where they form immense colonies covering an area of many tens of square metres to a depth of nearly a metre. Professor Remy visited the site of the discovery in 1936 and said afterwards:

> Cette région de la grotte m'a offert un des plus beaux spectacles que j'ai jamais vus. Le manteau de Vers recouvre des gros blocs du plancher, et en certains endroits particulièrement bien garnis, l'on hésite à passer pour ne pas écraser le tapis; au plafond, les tubes forment un revêtement bosselé, moutonné, auquel de place en place sont fixés des pendentifs, des lustres pouvant atteindre 50 cm de large; les parois latérales sont ornées; de corniches, de consoles, de draperies ondulées qui se raccordent aux "nuages" du plafond et se déploient jusqu'au sol.

The colonies are found under water during the winter, because the autumn rain transforms the Popovo Polje into a lake. During the summer the colonies are above water but are kept damp by water which condenses on them. Thus they remain in a state of reduced metabolism during part of the year (Hawes, 1938).

Marifugia Cavatica is not restricted to the Popovo Polje, it is widely distributed throughout the Adriatic area. It has been recorded from Istria,

† Stammer (1937), following Remane, believes that *Troglochaetus* is a Tertiary relic, but considers that it was a freshwater species widespread in Europe during the Miocene. The present day geographical distribution of this species is more in accord with Remane's interpretation than with Stammer's.

Croatia, Slovenia, Dalmatia and Herzegovina (Stammer, 1933, 1935; Remy, 1937). This extensive distribution and the fact that this species rapidly dies in sea water, does not support Absolon's interpretation that this species is a recent immigrant from the marine environment which has become adapted to life in fresh water in a similar way to *Merceriella enigmatica*. It is, indeed, quite clearly an ancient and possibly a tertiary, relict species and the possibility that it became adapted to fresh water before penetrating the subterranean environment, must not be lightly dismissed.

Oligochaeta

The cavernicolous oligochaetes have been the subject of many studies by L. Černosvitov, L. Cognetti de Martiis, S. Hrabe, I. Malevitch, A. Michaelsen, S. Ohfuchi, S. Omodeo, I. Sciacchitano, J. Stephensen and Fr. Vejdovsky. The reader is referred to the excellent work of Michaelsen (1933) and of Černosvitov (1939). However, many questions relating to the cavernicolous oligochaetes are far from being clearly answered.

The oligochaetes may be systematically divided into the Terricolae, or terrestrial forms, and the Limicolae or aquatic forms. However, it may be mentioned that such an essentially ecological division is often difficult to apply to cavernicoles (cf. Chapter XVII); this is certainly the case with the subterranean oligochaetes.

(a) Terricolae. Many terricolous oligochaetes have been reported from caves but most of them are common species also found in the superficial layers of the earth's surface. The existence of truly cavernicolous Terricolae will be difficult to recognise because the surface-living members of this group are in any case lacking pigmentation or only lightly pigmented, and have no eyes (Michaelsen, 1933). Černosvitov (1939) justly pointed out that: "It is necessary to consider that the conditions of life that all Terricolae normally lead differ but little from those which normally exist in caves. It would hardly be expected therefore, that the species confined exclusively to caves (if they exist) will be numerous, or that they will have developed peculiarities due to this mode of life".

However, it may be reported that several species of terricolous oligochaetes have been captured in caves in Assam and Japan. They belong to the genera *Enchytraeus, Dravida, Megascolides, Glyphidrilus* and *Pheretima* and appear to be true cavernicoles. These species are distinguished from surface forms by the total depigmentation of their surface layers and also by the absence of differential coloration between their dorsal and ventral surfaces (Stephensen, 1924; Ohfuchi, 1941).

(b) Limicolae. The limicoles are more interesting to the biospeologist because a certain number of them are truly cavernicolous, whilst others may be regarded as living fossils. Most of them belong to the family Lumbriculidae.

B 4

The genus *Trichodrilus*, of which all the species are hypogeous (apart from *T. allobrogum* which is also found in superficial waters), may be mentioned (Černosvitov, Hrabe). The hypogeous biotopes may be very varied. For example the present author dredged from the bottom of the Lac de Saint-Point (Jura français) specimens of *Trichodrilus claparedei*, a species which has also been found in a well at Skoplje in Macedonia, and in a cave in Herzegovina (Hrabe, 1938).

The genus *Dorydrilus* similarly contains only subterranean species: *D. michaelaeni* first discovered at the bottom of Lake Léman, and then in caves in the Jura français (Juget, 1959); *D. wiardi* and *D. mirabilis* which have been collected in caves, wells and quarries in Europe.

Pelodrilus bureschi, of some Bulgarian caves and *P. leruthi* of the Pyrénées ariègoises, are ancient relict species (Michaelsen, 1926; Hrabe, 1958). The second species is also found in the hypotelminorheic environment, which medium probably corresponds to its primitive habitat.

The interstitial medium contains oligochaetes which are not represented in the cavernicolous fauna. This is the case in the Naididae (Botea, 1960).

Hirudinea

Leeches are from time to time discovered in caves (de Beauchamp; Harant and Vernières; Percy Moore; Moniez; Remy; Sciacchitano), but they belong to common species and do not differ at all from surface forms.

Only two cases have been reported at the time of writing of cave- dwelling leeches possessing a modified anatomy.†

Herpobdella octoculata var. *pallida* has a transparent body, but the eyes are normal.

Herpobdella (Dina) absoloni is more interesting. This species is very close to the epigeous form *H. (Dina) lineata* (Komarek, 1953); but it differs from the latter by the disappearance of pigments in the surface layers of the body, and of the eyes (Mrazek, 1913). This cavernicolous leech was discovered by K. Absolon in the caves of Popovo Polje, in Herzegovina and has been described by Johannson (1913). It was also found in Montenegro by Mrazek (1913). It is a relic form which must have inhabited the great Pleistocene lakes which then covered the Balkans (Sket, 1961). A related form has been discovered in the Koutais caves in Transcaucasia.

MOLLUSCA

It was mentioned earlier that many Molluscan groups are exclusively marine, and therefore do not contain cavernicolous species. These groups are the Amphineura, Scaphopoda, Opisthobranchiata and the Cephalopoda.

† A depigmented leech has been discovered in a cave in Japan, but it has not yet been described (Uéno, 1957).

Lamellibranchiata

Pisidium is sometimes found in subterranean water. *Pisidium nitidum* is a common species found in such habitats as the Grotte de Sainte-Croix near Saint-Girons (Ariège). It is, in fact, really a hypotelminorheic species which has been carried into the caves.

However, Boettger (1939) proposed that some species of *Pisidium* are true cavernicoles. Two examples are *P. subterraneum* from the caves of Rion near Kutais in Transcaucasia (Shadin, 1932), and *P. cavernicum*, from a cave in the Ryu-Kyu Islands (Japan) (Mori, 1938).

Another lamellibranch which is certainly cavernicolous and also an old Pontain relict, belongs to the genus *Congeria* which is related to *Dreissensia*. This is *C. kusceri* discovered by Kuščer in 1934 (Karaman, 1935) in a spring in Karst and more recently in the Popovo Polje in Herzegovina (Bole, 1961; 1962).

Gasteropoda

The cavernicolous gasteropods are a large group both in numbers and in species. Wagner (1935) reported the presence in Planina Cave, in Karst, of deposits containing millions of shells of cavernicolous species, mainly *Valvata subpiscinalis* and *Frauenfeldia lacheineri*.

However, our knowledge of the cavernicolous gasteropods remains fragmentary. It is often difficult clearly to separate the truly cavernicolous species from those which frequent caves without being completely restricted to them. The anatomy of these molluscs as well as their ecology is very badly known. Their biology remains totally unknown, and we are forced in order to obtain some idea, to consider a blind form which lives in crevices in the soil, and which is therefore an endogean and not a true cavernicole. It has the advantage however, of being the subject of an excellent monograph by Wächtler (1929). This species is *Caecilioides acicula*. It is widely distributed in Europe and related to the genus *Hohenwartia*, the members of which live under stones in the Mediterranean region.

Cavernicolous gasteropods may be recognised by the following characteristics (Wagner, 1914): their surface layers are depigmented, and therefore, small forms are hyaline. The eyes are reduced or absent. The shells are thin and often transparent, but when opaque, of a whitish colour. The size of many cavernicolous molluscs is reduced, but this is not a general rule.

Terrestrial Gasteropoda. Basommatophora. Auriculidae. The first cavernicolous mollusc to be described came from the Adelsberg cave. It was reported by Rossmässler in 1839 under the name of *Carychium spelaeum*. In 1860, Bourguignat erected the genus *Zospeum* for this species (Fig. 8). Today about 15 species belonging to this genus are known from the north-east of Italy and Carniola (Conci, 1956). They are typical cavernicoles, blind and depigmented.

Carychium stygium, described by Call in 1897, was discovered in Mammoth Cave and later in other caves in Kentucky (Buzzard's Cave; White's Cave).

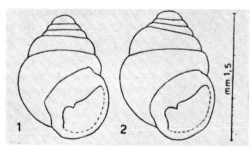

FIG. 8. *Zospeum globosum* (after Conci).

Stylommatophora. Testacellidae. The members of this family are typically endogeans which are found deeply buried in the soil. Two species can be called cavernicoles. These are *Testacella barbei*, a completely depigmented form found in the caves of the Central Pyrenees (Lucas, 1958) and *Daudebardia cavicola* which is found in "Baradla", a cave near Aggtelek in Hungary (Soos, 1927).

Limacidae. Amalia cavicola belongs to this family. It is a pigmented but mottled form which comes from the caves of Dalmatia, and is perhaps a troglophile (Simroth, 1916).

Zonitidae. This is the family which contains the majority of cavernicolous species (and individuals) in the Mollusca (Boettger, 1939). Some are troglophiles but others are truly cavernicolous Mollusca. According to Wagner (1914) out of 35 cavernicolous molluscs found in Southern Dalmatia and Herzegovina, 15 belonged to the family Zonitidae, of which 8 could be regarded as troglobia. Out of 12 cavernicolous molluscs obtained in Transcaucasia 5 belonged to the Zonitidae (Birstein, 1950).

The Zonitidae have been particularly well investigated by Riedel (1957, 1959 a and b). The principal genus is *Oxychilus*, which contains both troglophiles and troglobia. The latter are distributed throughout the eastern Mediterranean from Dalmatia to Turkey and Caucasia.

The carnivorous habit of *Oxychilus* has been known for a long time (Hele, 1886; Wagner, 1914; Zimmermann, 1926; Boettger, 1935; Riedel, 1957; Tercafs, 1960, 1961). The members of this genus are, in fact, polyphagous and feed just as well on plant debris as on animal matter (Riedel).

We will now mention a few other troglobia belonging to the family Zonitidae: *Meledella werneri* which lives in caves in the island of Mljet (Riedel, 1960a); the species of the genus *Spelaeopatula* from the caves of

Herzegovina and Albania; members of the genus *Lindbergia* from caves in Bulgaria, Greece and Crete (Riedel, 1959a and c, 1960b). The representatives of this latter genus are depigmented, and their eyes are reduced.

Enidae: Spelaeoconcha pagenetri comes from the caves of Curzola and Brazza.

Valloniidae: Spelaeodiscus hauffeni is found in the caves of Slovenia.

In tropical Asia the terrestrial cavernicolous gasteropods are represented essentially by the "Achatinelles" belonging to the family Subulinidae and the genus *Opeas. Opeas cavernicola* lives in the Siju Cave in the Garo Hills in Assam. This species resembles the epigeous form *O. gracilis*, but lacks pigmentation. The majority of individuals (about 93%) possess reduced and degenerate eyes but still with pigment; while the others have the eyes without pigment or crystallin (Annandale and Chopra, 1924). The Batu Caves in the State of Selangor in Malaya, possess two species of *Opeas*, *O. doveri* and *O. dimorpha*. Their eyes are reduced in size but are still pigmented.

Aquatic Gasteropoda. The composition of the fauna of subterranean waters is quite different from that of surface waters. While the Pulmonata flourish in rivers and lakes they are very rare in underground waters. Only a single species can be named which deserves the title of a cavernicole; this is *Ancylus sandbergeri*, discovered by Wiedersheim in 1873 in a cave at Württemberg.

All other aquatic cavernicolous molluscs belong to the prosobranchs, and to the family Hydrobiidae (Bythinellidae), except if certain Valvatidae *(Valvata erythropomatia, V. spelaea, V. moquini)*.

The Cavernicolous Hybrobiidae: The genera *Bythinella* and *Belgrandia* are normally found in springs. It is only exceptionally that they find their way into underground waters, for example, *Bythinella padiraci* inhabits the Padirac river (Lot, France) and *Belgrandia kusceri* has been found in caves around Trieste.

The genus *Paulia* Bourguignat (= *Avenionia* Nicolas) is closely related to *Bythinella*, but its species have a depigmented body, which in life appears pinkish-white and transparent. The eyes, however, are normal and well pigmented. Two species are known: *P. berenguieri*, which has been collected in springs around Avignon, and *P. bourguignati* which has been reported in France from the department of Aube, and in Belgium from around Liège. These two species are often recovered from ground water but it is rare to find the living animals, usually only their shells are discovered in alluvium. The same may be said for the species of *Paladilhia*.

Another large genus in this family is *Lartetia*. It is not related to *Bythinella* but to *Hydrobia*, a genus inhabiting brackish water (Boettger, 1939). *Lartetia* contains many species in which conchylogists have interested themselves (Clessin, Locard, Geyer, Lais, Bolling). This genus seems to have originated in Karst, but it has now spread into a large part of Central

Europe. For a long time the shells of these animals were found in alluvium, but the habitat of the living animal was unknown. It was Quenstedt (1863) and Meinert (1868) who established that these molluscs lived in sub-terranean waters, principally in flooded crevices in calcareous rock systems.† Several species have been collected from springs and wells, but this is not their natural biotope and they do not reproduce there. The

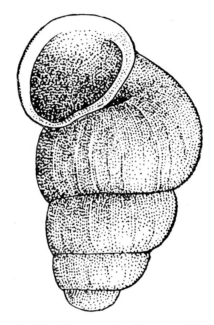

FIG. 9. *Akiyoshia uenoi* (after Kurada and Habe).

bodies of these molluscs are only slightly pigmented. Wiedersheim (1873) showed that their eyes are much reduced. However, the anatomy and biology of these hypogeous forms is still little known.

Many other genera of this family *(Pseudamnicola, Frauenfeldia, Belgrandia, Lanzaia, Iglica, Hauffenia, Hadziella)*, contain cavernicolous species which are particularly abundant in those areas of Karst which border the Adriatic (Bole, 1960, 1961a and b). It is necessary to investigate the internal anatomy of these species in order to determine their systematic position, because their shells are often very variable structures.

This review may be concluded by reporting that cavernicolous Hydrobiidae are equally well known outside Europe.

In the United States *Amnicola proserpina* comes from caves in Missouri, and *Horatia micra* has been taken in a subterranean river in Texas.

† *Lartetia rhenana* is frequently found in the interstitial environment.

In Japan, they are represented by two blind species with long tentacles belonging to the genus *Akiyoshia* (Fig. 9) (Kuroda and Habe, 1954; Habe, 1961).

In the ground waters of New Zealand *Potamopyrgos spelaeus* may be found (Chilton, 1894).

BIBLIOGRAPHY

Sponges

ARNDT, W. (1933) *Ephydatia mülleri* (Liebk.) als Bewohnerin von Höhlen der Herzegovina und des Karstes. *Mitt. Höhl. Karstf.*

JOSEPH, G. (1881) Mitteilungen über einen Grottenschwamm (*Spongilla stygia* n. sp.), einen Grottenpolypen (*Hydra pellucida* n. sp.). *Jahresb. schles. Gesell.* **LIX.**

KOFOID, CH. A. (1900) The plankton of Echo River, Mammoth Cave. *Trans. Amer. Microsc. Soc.* **XXI.**

Coelenterata

CHAPPUIS, P. A. (1944) Die Grundwasserfauna der Körös und des Szamos. *Mat. Termes. Kozlem.* **XL.**

CHAPPUIS, P. A. (1946) Un nouveau biotope de la faune souterraine aquatique. *Bull. Sect. Sci. Acad. Roumaine* **XXIX.**

DELAMARE-DEBOUTTEVILLE, CL. (1960) *Biologie des eaux souterraines littorales et continentales.* Paris, Hermann.

HUSMANN, S. (1956) Untersuchungen über die Grundwasserfauna zwischen Harz und Weser. *Archiv f. Hydrobiol.* **LII.**

MEŠTROV, M. (1960) Faunistisch-ökologische und biozönologische Untersuchungen unterirdischer Gewässer des Savetales. *Biošloki Glasnik.* **XIII.**

Rhabdocoela

GRIEPENBURG, W. (1934) Die Berghauser Höhle bei Schwelm i. W.*Mitteil. Höhl. Karstf.*

GRIEPENBURG, W. (1939) Die Tierwelt der beiden Hüllöcher im Sauerland. *Mitteil. Höhl. Karstf.*

PACKARD, A. S. (1883) A cave-inhabiting flat-worm. *Amer. Natural.* **XVII.**

REISINGER, E. (1933) Neues zur vitalen Nervenfärbung (gleichzeitig ein Beitrag zur Kenntnis der Protoplanelliden). *Verhandl. Deutsch. Zool. Gesell.* **XXXV.**

Tricladida

BEAUCHAMP, P. DE (1920) Turbellariés et Hirudinées (1ère série). *Biospeologica*, **XLIII.** *Archiv. Zool. exper. géner.* **LX.**

BEAUCHAMP, P. DE in BOLIVAR, C. and JEANNEL, R. (1931) Campagne spéléologique dans l'Amérique du Nord en 1928. *Biospeologica*, **LVI.** 2. Turbellariés Triclades. *Arch. Zool. exper. génér.* **LXXI.**

BEAUCHAMP, P. DE (1932) Turbellariés, Hirudinées, Branchiodellides (2ème série). *Biospeologica*, **LVIII.** *Archiv. Zool. expér. gener.* **LXXIII.**

BEAUCHAMP, P. DE (1939) La systématique et l'éthologie des *Fonticola* (Turb. Triclades). *Vestnik Cs. Zool. Spolec. Praze* **VI–VII.**

BEAUCHAMP, P. DE (1949) Turbellariés (3ème série). *Biospeologica*, **LXIX.** *Archiv. Zool. expér. gener.* **LXXXVI.** N. et. R.

BEAUCHAMP, P. DE (1955) Nouvelles diagnoses de Triclades obscuricoles. X. *Polycelis benazii* n. sp., dans une grotte de Ligurie. *Bull. Soc. Zool. France* **LXXX.**

HUSMANN, S. (1956) Untersuchungen über die Grundwasserfauna zwischen Harz und Weser. *Archiv f. Hydrobiol.* **LII.**

HYMAN, L.H. (1937) Studies on the morphology, taxonomy and distribution of North American Triclad Turbellaria. VIII. Some cave planarians of the United States. *Trans. Amer. Microsc. Soc.* **LVI.**

HYMAN, L.H. (1939) North American Triclad Turbellaria. X. Additional species of Cave Planarians. *Ecology,* **LVIII.**

HYMAN, L.H. (1945) North American Triclad Turbellaria. xi. New chiefly cavernicolous planarians. *Amer. Midl. Nat.* **XXXIV.**

HYMAN, L.H. (1954) North American Triclad Turbellaria. xiii. Three new cave planarians. *Proc. U.S. Nat. Mus.* **CIII.**

HYMAN, L.H. (1960) Cave planarians in the United States. *In* Symposium: Speciation and raciation in cavernicoles. *Amer. Midl. Nat.* **LXIV.**

KAWAKATSU, M. (1960) Notes on the freshwater planarians found in the subterranean waters of the Akiyoshi District. *Japan. J. Zool.* **XII.**

KENK, R. (1923–24) *Dendrocoelides spelaea* n. sp., eine neue höhlenbewohnende Tricladenart aus Slovenien. *Izviesc. Jugosl. Akad. Zagreb.*

KENK, R. (1935a) Studies on Virginian Triclads. *J. Elisha Mitchell Sci. Soc.* **LI.**

KENK, R. (1935b) Die Höhlentrikladen Sloveniens. *Verhandl. Intern. Ver. Theor. Angew. Limnologie* **VII.**

KOMAREK, J. (1919) O temnostnich Tricladach (Vermes, Turbellaria) z krasu balkanskych na zàklade sběru Dra. Karla Absolon. *Casopis Morav. Mus. Zem.* **XVI.**

Nemertea

BEAUCHAMP, P. DE (1932) Sur une Némerte obscuricole. *Bull. Soc. Zool. France* **LVII.**

MOTAS, C. (1961) La Speologia in Romania. *Rassegna Speleol. Ital.* **XIII.**

TARMAN, K. (1961) *Prostoma hercegovinense* n. sp., jamski Nemertin iz Hercegovinskih jam. *Drugi Jugoslav. Speleol. Kongress. Zagreb.*

Nematoda

ALTHERR, E. (1938) La faune des mines de Bex, avec étude spéciale des Nématodes. *Rev. suisse Zool.* **XLV.**

ANDRASSY, I. (1959) Nematoden aus der Tropfsteinhöhle "Baradla" bei Aggtelek (Ungarn), nebst einer Übersicht der bisher aus Höhlen bekannten freilebenden Nematoden-Arten. *Acta Zool. Acad. Sci. Hung.* **IV.**

HNATENWYTSCH, B. (1929) Die Fauna der Erzgruben von Schneeberg in Erzgebirge. *Zool. Jahrb. Abt. System.* **LVI.**

SCHNEIDER, W. (1960) Neue freilebende Nematoden aus Höhlen und Brunnen. I. Nematoden aus jugoslawischen Höhlen. *Zool. Anz.* **CXXXI.**

STAMMER, H.J. (1935) *Desmoscolex aquaedulcis* n. sp., der erste süßwasserbewohnende Desmoscolecide aus einer slowenischen Höhle (Nemat.). *Zool. Anz.* **CIX.**

Troglochaetus

AX, P. (1954) *Thalassochaetus palpifoliaceus* nov. gen., nov. sp. *(Archiannelida, Nerillidae),* ein neuer mariner Verwandter von *Troglochaetus beranecki* Delachaux. *Zool. Anz.* **CLIII.**

DELACHAUX, TH. (1921) Un Polychète d'eau douce cavernicole, *Troglochaetus beranecki* nov. gen. n. sp. (Note préliminaire). *Bull. Soc. Neuchâtel. Sc. Nat.* **XLV.**

HUSMANN, S. (1956) Untersuchungen über die Grundwasserfauna zwischen Harz und Weser. *Archiv f. Hydrobiol.* **LIII.**

PICARD, J. Y. (1962) Contribution à la connaissance de la faune psammique de Lorraine. *Vie et Milieu,* **XIII.**

PLESA, C. (1957) Un animal cavernicol pentru fauna Republicii Populare Romine: *Troglochaetus beranecki* Delachaux (Archiannelida). *Comm. Acad. Republ. Popul. Romine.* **VII.**

REMANE, A. (1928) *Nerillidium mediterraneum* n. sp. und seine tiergeographische Bedeutung. *Zool. Anz.* **LXXVII.**
STAMMER, H. J. (1937) Der Höhlenarchannelide, *Troglochaetus beranecki*, in Schlesien. *Zool. Anz.* **CXVIII.**

Marifugia cavatica

ABSOLON, K. (1927) Les grandes Amphipodes aveugles dans les grottes balkaniques. *Assoc. franç. Avanc. Sc. C. R. 51ème Session*, Constantine.
ABSOLON, K. and HRABE, S. (1930) Über einen neuen Süßwasser-Polychaeten aus den Höhlengewässern der Herzegowina. *Zool. Anz.* **LXXXVIII.**
HAWES, R. S. (1938) Effect on organisms of summer drought in caves. *Nature, Lond.* **CXLI.**
REMY, P. (1937) Sur *Marifugia cavatica* Absolon et Hrabe, Serpulide des eaux douces souterraines du Karst adriatique. *Bull. Mus. Hist. Nat. Paris* (2) **IX.**
STAMMER, H. J. (1933) Einige seltene oder neue Höhlentiere. *Zool. Anz.* 6 Suppl. Bd.
STAMMER, H. J. (1935) Untersuchungen über die Tierwelt der Karsthöhlengewässer. *Verhandl. Intern. Ver. Theor. Ang. Limnologie* **VII.**

Oligochaeta

BOTEA, F. (1960) Despre unele Naididae (Oligochaeta) gasite infreaticul vaii rîuli Doftana. *Communic. Acad. Popul. romĭne.* **X.**
ČERNOSVITOV, L. (1939) Etudes biospéologiques. X. Catalogue des Oligochètes hypogés. *Bull. Mus. R. Hist. Nat. Belgique* **XV.**
HRABE, S. (1958) *Trichodrilus moravicus* und *claparedei*, neue Lumbriculiden. *Zool. Anz.* **CXXI.**
HRABE, S. (1958) A new species of the Oligochaeta from the Southwest of France. *Notes biospéologiques* **XIII.**
JUGET, J. (1959) Recherches sur la faune aquatique de deux grottes du Jura méridional français: la grotte de la Balme (Isère) et la grotte de Corveissiat (Ain). *Annal. Spéleol.* **XIV.**
MALEVITCH, I. I. (1947) Les Oligochètes des cavernes du Caucase. *Biospeologica Sovietica*, **IX.** *Bull. Soc. Nat. Moscou, Sect. Biol.* **LII.**
MICHAELSEN, W. (1926) *Pelodrilus Bureschi*, ein Süßwasser-Höhlenoligochät aus Bulgarien. *Arb. d. Bulg. Naturf. Gesell.* **XII.**
MICHAELSEN, W. (1933) Über Höhlen-Oligochaeten. *Mitteil. Höhl. Karstf.*
OHFUCHI, S. (1941) The cavernicolous Oligochaeta of Japan. I. *Sc. Rep. Tohoku Univ.* 4. *Biol.* **XVI.**
STEPHENSEN, J. (1924) Oligochaeta of the Siju Caves, Garo Hills, Assam. *Rec. Indian Mus.* **XXVI.**

Hirudinea

JOHANNSON, L. (1914) Über eine neue von Dr. K. Absolon in der Herzegovina entdeckte höhlenbewohnende Herpobdellide. *Zool. Anz.* **XLII.**
KOMAREK, J. (1953) Herkunft der Süßwasser-Endemiten der dinarischen Gebirge; Revision der Arten, Artenentstehung bei Höhlentieren. *Archiv f. Hydrobiol.* **XLVIII.**
MRAZEK, A. (1913) Einige Bemerkungen über *Dina absoloni* Joh. *Zool. Anz.* **XLIII.**
SKET, B. (1961) Najdbe Pijavk (Hirudinea) v Podzemeljskem Okolju. *Drugi Jugosl. Speleol. Kongress*. Zagreb.
UÉNO, S. (1957) Blind aquatic beetles of Japan, with some accounts of the fauna of Japanese subterranean waters. *Arch. f. Hydrobiol.* **LIII.**

Lamellibranchiata

BOETTGER, C. R. (1939) Etudes Biospéologiques. VI. Die subterrane Molluskenfauna Belgiens. *Mém. Mus. R. Hist. Nat. Belgique* No. 88.

BOLE, J. (1961) Über einige Forschungsprobleme bezüglich der unterirdischen Mollusken-fauna. *Drugi Jugosl. Speleol. Kongres.* Zagreb.

BOLE, J. (1962) *Congeria kusceri* sp. n. (Bivalvia; Dreissenidae) *Biol. Vestnik.* **X.**

KARAMAN, ST. (1935) Die Fauna der unterirdischen Gewässer Jugoslawiens. *Verhandl. Intern. Ver. f. Theor. Angew. Limnologie* **VII,** Beograd.

MORI, S. (1938) Classification of Japanese *Pisidium. Mem. Coll. Sc. Kyoto Imper. Univ. Ser. B.* **XIV.**

SHADIN, W.J. (1932) Die Süßwassermollusken aus der Rion-Höhle bei Kutais (Trans-kaukasien, Georgien). *Archiv. f. Molluskenkunde,* **LXIV.**

Gasteropoda

ANNANDALE, N. and CHOPRA, B. (1924) Molluscs of the Siju Cave, Garo Hills, Assam. *Rec. Indian Mus.* **XXVI.**

BIRSTEIN, J. A. (1950) Peščernaja fauna zapadnogo Zakavkazja (Cave fauna of the western Transcaucasus). *Zool. Žurn.* Moskwa. **XXIX.**

BOETTGER, C.R. (1935) Étude biologique des cavernes de la Belgique et du Limbourg hollandais. **XXII.** Mollusca. *Mitteil. Höhl. Karstf.*

BOETTGER, C.R. (1939) Études biospéologiques. VI. Die subterrane Molluskenfauna Bel-giens. *Mém. Mus. R. Hist. Nat. Belgique.* No. 88.

BOLE, J. (1960) Zur Problematik der Gattung *Lanzaia* Brusina (Gastropoda). *Biol. Vestnik.* **VII.**

BOLE, J. (1961a) O. Nekaterih Problemih Proucevanja subterane Malakofaune. *Drugi Jugosl. Speleol. Kongres.* Zagreb.

BOLE, J. (1961b) Nove Hidrobide (Gastropoda) iz Podzemeljskih voda zahodnega balkana. *Biol. Vestnik.* **IX.**

CHILTON, CH. (1894) The subterranean Crustacea of New Zealand, with some remarks on the fauna of caves and wells. *Trans. Linn. Soc. London.* 2nd Ser. *Zool.* **VI.**

CONCI, (1956) Nuovi rinvenimento di Molluschi troglobi del genere *Zospeum* in caverne delle Prealpe Trentine e Venete (Italia settentrionale). *Comunic.* 1ère *Congr. Speleol. Intern.* Paris **III.**

GHOSH, E. (1929) Fauna of the Batu Caves, Selangor. VI. Mollusca. *J. Fed. Malay States Mus.* **XIV.**

HABE, T. (1961) Two new subterranean aquatic snails. *Venus,* **XXI.**

HELE, F. M. (1886) *Zonites drapalnardi* in captivity. *Journal of Conchology* **V.**

KURODA, T. and HABE, T. (1954) New aquatic gastropods from Japan. *Venus* **XVIII.**

LUCAS, A. (1958) Diagnose d'une Testacelle cavernicole des Pyrénées. *Notes biospéolo-giques.* **XIII.**

MEINERT, Fr., mentioned in WIEDERSHEIM, R. (1873) Beiträge zur Kenntnis der würtem-bergischen Höhlenfauna. *Verhandl. phys. med. Gesell. Würzburg.* N. F. **IV.**

QUENSTEDT, F.A. (1864) Geologische Ausflüge in Schwaben.

RIEDEL, A. (1957) Revision der Zonitiden Polens. *Annales Zoologici.* Warszawa. **XVI.**

RIEDEL, A. (1959a) Die von Dr. K.Lindberg in Griechenland gesammelten *Zonitidae* (Gastropoda), *Annales Zoologici.* Warszawa. **XVIII.**

RIEDEL, A. (1959b) Über drei Zonitiden-Arten (Gastropoda) aus den Höhlen der Türkei. *Annales Zoologici.* Warszawa. **XVIII.**

RIEDEL, A. (1959c) Sur une nouvelle espèce de Mollusque cavernicole en Bulgarie. *Speleo-logia.* Warszawa. **I.**

RIEDEL, A. (1960a) Über *Meledella werneri* Sturany 1908 (Gastropoda; Zonitidae). *Frag-menta balcania Mus. Maced. Sc. Nat.* **III.**

RIEDEL, A. (1960b) Die Gattung *Lindbergia* Riedel (Gastropoda, Zonitidae) nebst An-gaben über *Vitrea illyrica* (A.J.Wagner). *Annales Zoologici.* Warszawa. **XVIII.**

SIMROTH, H. (1916) Über einige von Herrn Dr. Absolon in der Herzegowina erbeutete höhlenbewohnende Nacktschnecken. *Nachrichtenbl. deutsch. Malacozool. Gesell.* **XLVIII.**

Soos, L. (1927) Contribution to the knowledge of the mollusc fauna of some Hungarian caves. *Allattani Közlem.* **XXIV.**

TERCAFS, R.R. (1960) *Oxychilus cellarius* Müll., un Mollusque cavernicole se nourrissant de Lépidoptères vivants. *Rassegna Speleol. ital.* **XII.**

TERCAFS, R.R. (1961) Comparaison entre les individus épigés et cavernicoles d'un Gastéropode troglophile, *Oxychilus cellarius* Müll. *Ann. Soc. R. Zool. Belgique.* **XCI.**

WÄCHTLER, W. (1929a) Zur Lebensweise der *Caecilioides acicula* Müll. *Archiv f. Molluskunde* **LXI.**

WÄCHTLER, W. (1929b) Anatomie und Biologie der augenlosen Landlungenschnecke *Caecilioides acicula* Müll. *Z. Morphol. Ökol. Tiere.* **XIII.**

WAGNER, A.J. (1914) Höhlenschnecken aus Süddalmatien und der Herzegowina. *S.B. K. Akad. Wiss. Math. Naturw. Kl. Wien* **CXXIII.**

WAGNER, H. (1935) Über die Mollusken-Fauna der Planina-Höhle. *Mitteil. Höhl. Karstf.*

WIEDERSHEIM, R. (1873) Beiträge zur Kenntnis der württembergischen Höhlenfauna. *Verhandl. phys. Med. Gesell. Würzburg* N.F. **IV.**

ZIMMERMANN, K. (1926) Landschnecken im Kampf. *Der Naturforscher* **III.**

CHAPTER VIII

THE ARACHNIDS

A. ARTHROPODS

The great majority of subterranean animals are arthropods. Their predominance in nature is such that many biospeologists tend to neglect species which do not belong to this phylum. This in turn accentuates the predominance of arthropods in biospeological collections.

The arthropods may be divided into three groups, the Chelicerates, the Crustacea and the Tracheates, which probably have independent evolutionary origins, their resemblances being due to convergence.†

B. CHELICERATES

The chelicerates may be divided into two groups: the aquatic Merostomata and the terrestrial Arachnida. A small group with uncertain affinities are included with the chelicerates: the Pycnogonidea. The Merostomata and the Pycnogonidea are exclusively marine, and do not possess any cavernicolous members. Our study is therefore limited to the Arachnida.

C. ARACHNIDA

Petrunkewitch (1949) in his detailed study of palaeozoic arachnids recognised sixteen orders, five of which are known only as fossils. The eleven remaining orders all possess cavernicolous representatives, with the single exception of the Solifuga, which belong to warm, dry regions of the globe.

D. SCORPIONIDEA

No scorpion leads a truly cavernicolous life. *Euscorpius carpathicus* and *E. italicus* have been reported from caves in southern Europe; *Chaeri-*

---

† The position of the Tardigrades remains in doubt. Zoologists hesitate to place them in the arthropods rather than in the annelids. As the occurrence of the tardigrades in caves is certainly accidental (Joseph, Tosco) we can shelve this problem as it is not relevant to the subject matter of the book.

bus cavernicola from caves in Sumatra; and *Centrurus yucatanus* from a cavern in Mexico. But none of these species may be considered as true cavernicoles.

Eugène Simon (1879) in Volume VII of his extensive work *Les Arachnides de France*, mentioned a scorpion which he described as blind and for which reason he named *Belisarius* (*B. xambeui* Simon). This name is doubly incorrect, first of all because Bélisaire did not have his eyes put out, as the medieval legend leads one to believe, and secondly because this scorpion is not blind anyway. The median eyes are missing but the lateral ones persist, although they are reduced (Vachon, 1944).

This scorpion has been discovered in two caves at La Preste (Pyrénées-Orientales). They may be found regularly in the Grotte de Montbolo above Arles-sur-Tech. But they are, in fact, endogeans, which are more usually found under stones and in fallen material in forest or humid valleys in the mountainous regions of French and Spanish Catalonia (Auber, 1959).

E. PSEUDOSCORPIONIDEA (CHERNETES; CHELONETHIDA)

The pseudoscorpions are an order which is well represented in the subterranean fauna. About a sixth of all known pseudoscorpions (293 out of 1857) are cavernicoles (Chamberlin and Malcolm, 1960). Our knowledge of the pseudoscorpions is relatively well advanced owing to a great deal of work by M. Beier, J. C. Chamberlin, E. Ellingsen, J. Hadži, I. Lapchov, K. Morikawa, E. Simon and M. Vachon.

Distribution of Cavernicolous Pseudoscorpions in the Systematic Classification

The Pseudoscorpionidea are divided into three sub-orders: Chthoniinea, Neobisiinea and Cheliferina. Now, Beier (1940, 1949) and Chamberlin and Malcolm (1960) have shown that the majority of cavernicolous pseudoscorpions belong to the first two sub-orders (92% after Chamberlin), whilst the Cheliferina which is very rich in species contains few hypogeous representatives. This difference certainly relates to the mode of life of the three sub-orders. The Chthoniinea and the Neobisiinea live under stones or in masses of dead leaves, that is, in dark and humid places. They are thus "preadapted" to cavernicolous life. The Cheliferina are, on the other hand, corticicoles or pholeophiles, that is inhabitants of relatively dry media. Chamberlin and Malcolm are doubtful of the existence of true cavernicoles in the Cheliferina. For them, the presence of Cheliferina in caves must have resulted from phoresy. These arachnids may fix themselves on to bats when they may be transported into caves where they would become acclimatized. These animals would then be "chiropterophiles" rather than troglophiles.

Geographical Distribution of Cavernicolous Pseudoscorpions

Our knowledge of the geographical distribution of cavernicolous pseudo-scorpions is relatively complete with respect to Europe, North America and Japan. These conditions are the most interesting because the forms in tropical caves are much less modified than those species which populate caves in temperate or Mediterranean regions. Several species *(Negroroncus)* have been captured in caves in tropical and eastern Africa. A cavernicolous pseudoscorpion *(Chthoniella cavernicola)* was found in a cave in Table Mountain, Cape Province. Finally, *Stygiochelifer cavernae* comes from the Gua Lava Cave, in Java.

Whilst it must be admitted that there are great gaps in our knowledge of the tropical cavernicolous pseudoscorpions, it is, however, true to say that the mediterranean zone, and particularly those regions of Karst which border the Adriatic, are the richest areas of the world for hypogeous pseudo-scorpions. Beier (1939), in an important work devoted to pseudoscorpions collected by K. Absolon, enumerated about a hundred species or sub-species of cavernicolous pseudoscorpions, found in the Balkan States.

The Phylogenetic Age of Cavernicolous Pseudoscorpions

We are indebted to Beier (1940, 1949) for some interesting data on this question. The great majority of the European cavernicolous species belong to three genera. These are *Chthonius, Neobisium* (the cavernicolous forms are classed in the sub-genus *Blothrus*) and *Roncus* (the cavernicolous forms are classed in the sub-genus *Parablothrus*). It appears that most species of *Chthonius*, with the exception of some specialised forms, are of recent origin. Their cavernicolous characters fluctuate and show extensive indi-vidual variation.

Blothrus and *Parablothrus* comprise many evolutionary lines, some little specialised while others are highly modified. But the origin of almost all species may be easily found in epigeous types still living in Europe.

This is, however, not the case with certain genera belonging to the family Syarinidae, which possess no epigeous species in Europe†; examples are *Troglochthonius* (Dalmatia), *Troglobisium* (Catalonia), *Pseudoblothrus* (Alps and Jura Mountains, Crimea), and *Haboblothrus* (Southern Italy). On the contrary they are related to circum-tropical forms such as *Tyranno-chthonius*. The cavernicolous genera of this family in Europe may be re-garded as "living fossils", relict species of the ancient tropical fauna which formerly occupied the Mediterranean regions.

There is a similar contrast in North America (Chamberlin and Malcolm, 1960). Whilst most of the cavernicolous pseudoscorpions are related to

† This family has many epigeous forms in North America.

surface forms, this is not true for the Vachoniidae and the Hyidae. These two families are represented in America by highly specialised caverni- coles, inhabiting the Mexican caves, but no epigeous species have been collected in the New World. The epigeous Vachoniidae are known only in South Africa, while the epigeous Hyidae populate the tropical oriental regions (Philippines, Java, Sumatra, Micronesia). Thus the Mexican cavernicoles represent tropical relicts.

The Characteristics of Cavernicolous Pseudoscorpions (Lapchov, 1940; Beier, 1940, 1949; Chamberlin and Malcolm, 1960).

All lines of the Pseudoscorpionidea offer examples of parallel evolution showing a progressive specialisation with the following characteristics:

(a) Disappearance of cuticular pigment.
(b) Reduction and disappearance of the eyes. It is the posterior pair which are first reduced, then — because they are older in phylogeny — the anterior pair.
(c) Lengthening of the appendages, and in particular of the prehensile appendages which acquire considerable size but at the same time become very slender (Figs. 10 and 11).
(d) Cavernicoles are either dwarfs related to epigeous forms (sub-genus *Neochthonius*), or giant form.

F. OPILIONIDS

Formerly the cavernicolous opilionids were studied mainly by E. Simon in Europe and A. S. Packard in America, but in recent years extensive re- search has been carried out by many arachnologists including K. Absolon, Ed. Dresco, C. J. and M. L. Goodnight, J. Hadži, J. Kratochvil, and C. Fr. Roewer.

The opilionids may be divided into three sub-orders, the Cypho- phthalmes, the Laniatores and the Palpatores, which are so different from each other that they may conveniently be treated separately.

(a) Sub-order Cyphophthalmes

These very primitive and minute opilionids (2–3 mm) resemble the Aca- rina. There is only one family in this group, the Sironidae. They are endo- geous forms living in the soil; during heavy rain they approach the surface of the soil. They have a wide geographical distribution and have been collected in all the continents. But today they are most frequently collected from areas which once formed part of the ancient continent of Gondwana-

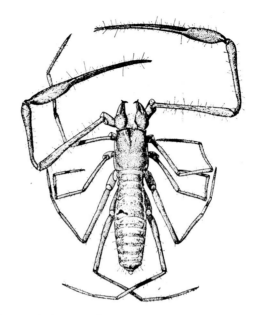

FIG. 10. A cavernicolous pseudoscorpion, *Neobisium tuzeti* (after Vachon).

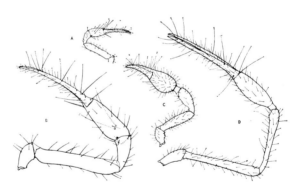

FIG. 11. Pedipalps in the Chthoniidae, (A) an epigeous, and (B) a cavernicolous type; those of the Neobisiidae, (C) an epigean, (D) a cavernicolous type (after Chamberlin and Malcolm).

land (Kratochvil, 1958). The European forms represent the remains of a subtropical and tropical fauna.

Although the majority of Sironidae are endogeans several species are exclusively cavernicolous. Six species from Jugoslavia of the genus *Siro* are cavernicolous (Kratochvil, 1937). *S. duricornis*, described by Joseph, is sometimes epigeous and sometimes cavernicolous. A remarkable species, *Trauteeva paradoxa* has recently been discovered in a cave near Gradesnica in Bulgaria (Kratochvil, 1958a). Finally, *Speleosiro argasiformis* has been taken in a cave near Cape Town, in South Africa.

These primitive opilionids which lead in the main an endogeous life, are little modified. All, with the exception of the genus *Stylocellus*, lack eyes. In *Siro gjordjevici* the legs are elongated and this elongation is more marked in *Trauteeva paradoxa*.

(b) Sub-order Laniatores

The Laniatores are characterised by their strong palpi which terminate in a prehensile claw. They are essentially tropical or sub-tropical forms, particularly abundant in tropical rain forest. The species which are found in North America and Europe are relics of an ancient fauna of warmer times, going back to the Tertiary.

The cavernicolous Laniatores can be divided into two groups:
(1) Three families of this sub-order have cavernicolous species: the Para-lodidae *(Paralola bureši)*, from a cave in Bulgaria, the Phalangodidae and the Assamiidae (*Sudaria jacobsoni* from a cave in Sumatra).

The most important of these families is the Phalangodidae which has many members in the warmer regions of the world. Several species are found in Europe *(Scotolemon* and *Querilhacia)*, in North America *(Phalangodes* and *Bishopella)*, and in Japan *(Strisilvea cavicola)*. The forms which inhabit the Northern Hemisphere are found in mosses, under stones and also in caves. The strictly caverni-colous species are few in number, and are not highly modified anatomically. Depigmentation of the integument and a slight re-duction of the eyes are all that is found. They can be considered to be recent cavernicoles.
(2) More interesting to the biospeologist are three other families of the Laniatores which Kratochvil (1958a) has united in a superfamily, the Travunoidea; they are the Triaenonychidae, the Synthetonychidae and the Travuniidae.

The representatives of the first family are distributed across those south-ern countries which once formed the continent of Gondwanaland, that is: South America, South Africa, Madagascar, Australia, Tasmania, New Zealand and New Caledonia. Only one South American genus,

Sclerobunus, is represented in North America by three species, one of which is cavernicolous (*S. cavicolens*, Morrison's Cave, Montana).

The family Synthetonychidae was erected recently by Forster (1954) and is indigenous to New Zealand.

Regarding the third family, the Travuniidae, this was created in 1932 by Absolon and Kratochvil. This is not to say that its species were not

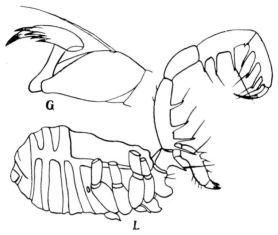

FIG. 12. *Travunia anophthalma* (after Absolon and Kratochvil). G, last tarsal segment of the fourth leg with the peltonychium.

known until that date, for Lucas (1860) had already described a species *(leprieuri)*, but at this time he placed it in the genus *Scotolemon* and in the family Phalangodidae. Our knowledge of the Travuniidae has been excellently reviewed by Roewer (1935), and Kratochvil (1958a). The family contains eight genera and eighteen species. Its representatives are all native to Europe, and inhabit mountainous regions from the Pyrenees to Montenegro. All are cavernicoles with the single exception of *Peltonychia helvetica*, which inhabits the mountainous regions of Ticino.

It cannot be doubted that the Travunoidea are an extremely ancient group which have been driven back to the present day remnant of Gondwanaland on the one hand, and to European caves on the other. The antiquity of this group can once more be demonstrated by reference to the morphology of the Travuniidae. They are more highly modified than the hypogeous members of the Phalangodidae. The Travuniidae are depigmented and pink or white in colour. The eyes are reduced or lacking. The tuber oculorum may itself be absent; in two species (*Arbasus caecus* of the massif d'Arbas in the Central Pyrenees, and *Travunia anophthalma* (Fig.12), from various caves in South Dalmatia), it has disappeared without trace. It cannot be doubted that the Travuniidae are cavernicoles of ancient origin, in fact true "living fossils".

(c) Sub-order Palpatores

The Palpatores may be distinguished from the Laniatores by the fact
that the palpi are slender and do not terminate in a claw as in the former
group. Hansen and Sörensen have divided the Palpatores into two tribes,
the Eupnoi and the Dyspnoi.

The Eupnoi contain a single family, the Phalangiidae, which includes a
considerable number of species. The family contains no true cavernicoles,
but only forms which are occasionally found in caves. The species most
specialised for hypogeous life is *Nelima troglodytes*, native to the Jugoslavian Karst. However, Hadži (1935) who has observed the behaviour
of this species reports that this opilionid is found only in caves during
the winter; in the summer it lives outside. Caves are thus a wintering place
for it.

Only two of the four families comprising the Dyspnoi contain cavernicoles.

Fig. 13. *Buresiolla karamani* (after Hadži).

The Nemastomidae has a wide geographical distribution in the Northern
Hemisphere (Europe, Caucasia, Asia Minor, North Africa and North
America). Most of the species of this family are epigeans in humid forests
and mountains in the neighbourhood of streams. The genera *Crosbycus*,
Histricostoma, *Nemastoma* and *Mitostoma* contain some cavernicolous
species, but they are generally little modified and still possess eyes. However, *Nemastoma inops* which is native to caves in Kentucky is completely
anophthalmic. The more specialised species are grouped into the genus
Buresiolla, established by Kratochvil (1958). This genus contains three
species all of which are cavernicolous: *B. bureschi* (Bulgaria), *B. karamani*

(Bosnia), and *B. tunetorum* (Tunisia). All three species are anophthalmic, but *B. karamani* has no ocular tubercule, and its appendages are extraordinarily elongated (Fig. 13). The second pair of legs is eight times the length of the body (Hadži, 1940).

The Ischyropsalidae is remarkable for the extreme development of the chelicerae, which may be longer than the body. There are two European genera in this family: *Ischyropsalis* and *Sabacon*. These opilionids inhabit the mountainous areas of Europe; they are found in humid woodland under mosses and bark. Some species are found in caves, but their black pigmentation and their morphology are not perceptibly altered from those of related epigeous forms. This is the case with *I. luteipes*, a species found in mountainous areas, which lives amongst mosses and avoids light, but is also found in caves and can be called a troglophile.

Several species are strictly cavernicolous. Examples are *I. pyrenea* (Central Pyrenees), *I. navarrensis* (Navarre), *I. corcyraea* (Corfu), *I. pentelica* (Attica) and *I. troglodytes* (Crete). In these species the eyes are depigmented. Finally in *I. strandi* (= *I. ruffoi*), a species collected in the cave of Monte Baldo, near Verona (Italy), the eyes have completely degenerated (Kratochvil, 1936).

G. PALPIGRADA

The Palpigrada are microscopic arachnids which are less than 2 mm in length. They may easily be recognised by the long multi-articulate flagella which extend from the body, their locomotory pedipalps and their first pair of legs which are transformed into tactile organs. There are thirty-five species known to science at the present time, which are grouped into four genera. The Palpigrada are found in all the warmer regions of the globe. In Europe they are localised in the Mediterranean region.

The great majority of Palpigrada are endogeans, and live under stones. Ten cavernicolous species have been described, and a list of these has been drawn up by Janetschek (1957). They are all European with the exception of one species captured in Cuba. They are found in the caves of the Mediterranean regions of Europe and in the caves of the Alps and the Carpathian Mountains. All the evidence suggests that they represent the relics of a warmth-loving fauna, which have taken refuge in the hypogeous habitat.

As all palpigrades are depigmented and anophthalmic it is difficult to distinguish the cavernicolous from the endogeous species. At most, it has been pointed out that the legs of the cavernicoles are more slender than those of the endogeans (Hansen, 1926).

H. PEDIPALPIA

The old order Pedipalpia is heterogeneous and should be cancelled. However, it is convenient to retain it in a biospeological work because the different groups of the Pedipalpia behave in an analogous fashion with respect to the subterranean environment. All avoid light, are hydrophilic and are localised in the warmer regions of the world. No member of the Pedipalpia has succeeded in reaching the temperate regions, or at least in maintaining itself in these lands. The several species which have penetrated caves are only slightly modified, as is the rule for tropical cavernicoles. Some are found as commonly outside the caves as they are in them, others are perhaps strictly cavernicolous, but these are little modified. In most species the eyes are normal; however, in the family Schizomidae the ocular regression is the rule. In addition, in *Charinus diblemma* the integument is depigmented and the lateral eyes are atrophied.

There appears below a list of examples of cavernicolous Pedipalpia.

ORDER UROPYGES

Thelyphonidae

Hypoctonus wood-mansoni
H. formosus } Farm — Cave (Burma)

Schizomidae (Tartaridae)

Schizomus cavernicolens	Caves in Yucatan
S. machadoi	Caves in the Congo
S. cavernicolus	Farm Cave (Burma)
S. sijuensis	Siju Cave (Assam)
Trithyreus cavernicola	Caves in E. Africa and Zanzibar
T. schoudeteni	Caves in the Congo
T. parvus	A cave in Gabon

ORDER AMBLYPYGES

Charontidae

Charinus: many species in African caves
Sarax sarawakensis	Batu Cave (Malaya)
S. brachydactylus	Caves in the Philippines

Categaeus pusillus
Stygophrynus cavernicola } Caves in Malaya
S. cerberus

Tarantulidae (Phryinidae)

Damon diadema
D. medius
Phyrnichus alluaudi } Caves in equatorial Africa
P. telekii

Tarantula fuscimana   Caves in Mexico

I. ARANEIDA

The araneids together with the acarines constitute the most important group of the Arachnida. More than 20,000 species are known at the present time. Several hundred species of the Araneida have been collected in caves. Some species are found only occasionally; others are partly modified for cavernicolous life, and finally strictly cavernicolous araneids are known which differ markedly from the epigeans. There is a considerable bibliography relating to the cavernicolous Araneida. The reader should consult two excellent reviews by Kästner (1926) and Fage (1931 b). Very recently, Yaginuma (1962) has given a list of cave spiders reported in Japan. This list includes 40 species: 23 troglobious, 3 troglophilous and 14 trogloxenous.

The Geographical Distribution of the Cavernicolous Araneida

This group is found in all parts of the world. However, unquestionably the great majority of them are concentrated around the Mediterranean. The cavernicolous Araneida are more poorly represented in North America than in Europe (Berland, 1931).

The Classification of the Araneida

Although it has been criticised and modified we will retain the old classification proposed by Eugène Simon, because it is not the purpose of this book to argue the merits of various classifications but to provide the biospeologist with a single one which is convenient to use. This classification is outlined below:

Orthognathes {Liphistiomorphes
              {Mygalomorphes

Labidognathes {Cribellates
              {Ecribellates   {Haplogynes
                              {Entelegynes

The Systematic Division of the Cavernicolous Araneida

Liphistiomorphes. These very primitive spiders belong to the Far East (S. Japan, Indo-China, Burma, Malaya, Java and Sumatra). Only one species, *Liphistius batuensis* from the Batu Caves, Malaya, is cavernicolous. Its habits have been studied by Bristowe (1952). The spider constructs an earthen cell, closed at each end by a valve. The structure is lined with silk (Fig. 14). From the upper opening proceed 6 or 7 silken "fishing lines" which serve to capture prey, which consist mainly of Orthoptera *(Paradiestrammena)*.

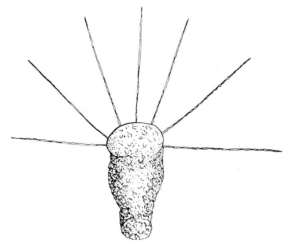

FIG. 14. An earthen cell built by *Liphistius batuensis* (after Bristowe).

Mygalomorphes. No cavernicolous representatives of this group are known from temperate countries. Two species have been found in the tropics. *Accola caeca* was discovered by Eugène Simon in the Antipolo Cave on the Isle of Luzon in the Philippines. This species is blind, as its name indicates. Other representatives of this genus are found in Venezuela, where they live in humid and dimly lit forests under fallen tree trunks. Another blind member of the Mygalomorphes has been captured by F. Silvestri in the Grotta Bellamar in Cuba, and described by L. Fage under the name *Troglothele caeca.*

Cribellates. This group contains no true cavernicoles in temperate regions. Some are, however, found in warmer countries. They belong to the families Filistatidae, Amaurobiidae, Eresiidae and Uloboridae.

Haplogynes. Almost all the families belonging to this group contain cavernicoles: Dysderidae, Leptonetidae, Telemidae, Oonopidae, Sicariidae and the Ochyroceradtidae (Fage, 1912). The first four families are of special interest to biospeologists.

The Dysderidae contains the oldest known cavernicolous spider, *Stalita taenaria* described by Schiödte (1848), from the caves the of Adelsberg area. Today many species of *Stalita* are known from Julian Venetia to Montenegro; a species is also found in Crete. All are cavernicoles, most are blind and several others have reduced eyes. *Stalitochaera kabiliana*, which belongs to the same family, has been found in caves in Algeria.

The Leptonetidae, which have been revised by Fage (1913) are minute spiders, between 1 and 3 mm in length. They are indigenous around the Mediterranean: Northern Spain and Italy, the South of France, Corsica, Dalmatia, Herzegovina and Algeria. A single species *(Leptoneta caeca)* is found in caves in Texas. A questionable species exists in Japan. Several *Leptoneta* species are found in forests, under stones and in moss. However, most of them live in caves. All possess more or less reduced eyes. They weave their webs in cracks in the walls of the caves and between stalactites.

The Telemidae have sometimes been classified with the Leptonetidae. Today they have been separated because these spiders are without lungs; they belong to the Apneumones (as do the representatives of the related family, the Caponidae). The four stigmata, two anterior and two posterior, open into trachael systems.

Telema tenella is a microscopic spider from 1 to 1·5 mm in length. It is blind and without pigment, and has been collected from some caves in the Pyrénées-Orientales. Some *Telema* species which have not been specifically identified, have recently been discovered in the Gabon and Congo caves (Lawrence, 1958; Strinati, 1960). On the other hand, Fage (1921a) reported another member of this family which has a respiratory system of the same type. It is of similar size (1·4 mm) but partly pigmented and possessing normal eyes. These specimens were collected in the Kulumuzi Cave in East Africa. Fage named this species *Apneumonella oculata*. Another species of the Telemidae, *Cangoderces lewisi* has been collected from a cave in South Africa.

However the Telemidae are not strictly localised in caves. De Barros Machado (1956) has shown that they are widespread in the tropical forests of Angola and the Congo. There can be little doubt that the microscopic Pyrenean cavernicolous spiders are a relic fauna of the tropical sylvicolous population which occupied Europe during the Tertiary.

Some cavernicolous and anophthalmic spiders, for instance *Dysderoides typhlops*, captured by E. A. Glennie, in a cave at Simla, in the Himalayan mountains, belong to Oonopidae.

Entelegynes. Many families belonging to this group contain cavernicoles which are little specialised. Such families are the Pholcidae *(Pholcus)* and the Theriidae. The four families Tetragnathidae, Erigonidae, Linyphiidae and Agelenidae ought to hold the attention of biospeologists for a long time.

The Tetragnathidae contains troglophiles belonging to the genera *Meta* and *Nesticus*. However, certain species of the latter genus are slightly modified troglobia. This is the case with *Nesticus borutzkii*, *N. zaitzevi*, *N. caucasicus* and *N. birsteini*, which are natives of the caves of Trans-caucasia (Kharitonov, 1947); and also with *Typhlonesticus parvus*, from a cave in Herzegovina.

The Erigonidae (= Micryphantidae) contains a genus which is of great interest to biospeologists: *Porhomma* (Fage, 1931a). Unlike many caverni-coles which are relicts of a tropical fauna, *Porhomma* is probably a "glacial relic". Indeed while the species of this genus live above ground in the North of Europe, they are localised in the interiors of caves in more souther-ly regions. This distribution may be considered as a consequence of the retreat of the great glaciers at the end of the Ice Age and the subsequent warming of the climate.

The Linyphiidae contains the genera *Centromerus* and *Leptyphantes*, the different species of which live either in caves or on the surface. The in-habitants of the caves are generally little specialised. The species of the genus *Troglohyphantes* are similar in this respect (Fage, 1919), although certain of them are strict, anophthalmic cavernicoles, e.g. *T. caecus*, of the Grotte de Betharram, in the Pyrenees, and *T. cantabricus anophthalmus* from the Grotte de Santian near Santander. In America species of the genera *Anthrobia*, *Phanetta* and *Bathyphantes* have been found in caves in Kentucky and Indiana.

The family Agelenidae is composed of shade-loving *(Tegenaria, Chori-zomma* and *Phanotea)* as well as strictly anophthalmic cavernicoles: *Iberina*, of the Cantabrian Mountains and the Pyrenees, and *Hadites* of the Jugos-lavian Karst.

The Origins of the Cavernicolous Spiders (Fage, 1931 b)

It is perhaps in this group of cavernicolous animals that the gradation from epigeous life to that in the reclusion of subterranean cavities proceeds by the least perceptible steps. The genera *Robertus, Leptoneta, Troglo-hyphantes, Leptyphantes, Centromerus* and *Porhomma* contain some species which live in woods amongst mosses or dead leaves, whilst others frequent the entrances of caves or may be occasional visitors. Finally there are those which are firmly established in the subterranean medium. The caverni-colous species are more or less depigmented, and their ocular apparatus is reduced or in extreme cases, totally absent; their appendages may become elongated. These modifications already manifest themselves in certain surface forms. This proves that a life hidden in mosses or in dead leaves and in a shaded and humid environment is a "preparation" for the caverni-colous habitat.

This is the reason for the unequal distribution of cavernicolous types amongst the various families of the Araneida. Only those spiders which already lead a humicolous or muscicolous life are eligible to evolve into cavernicoles. The hunting spiders, which pursue their prey with the aid of their eyes, such as the Lycosidae and Salticidae, have never given rise to cavernicolous species. The same applies to the Thomisidae, which live in full sunlight on flowers.

J. RICINULIDA

This order contains only a few species, the individuals of which are very rare. Two species have been collected from caves in Mexico, *Cryptocellus pearsi*, from various locations in Yucatan, and *C. boneti*, from the famous Cacahuamilpa Cave. These two species may be regarded as troglophiles or guanobia rather than true cavernicoles.

K. ACARINA

The Acarina are a large order of arachnids with regressed organisation. They may be classified into three ecological groups: parasites, terrestrial forms and aquatic forms. The first of these will be dealt with in a later chapter.

L. TERRESTRIAL ACARINA

The terrestrial mites which inhabit caves have been the subject of important studies by A. Bonnet, J. Cooreman, G. Lombardini, I. Trägårdh, H. Vitzthum and C. Willmann. Although there have been many studies on the taxonomy of the Acarina there has been little attempt to investigate their ecology. Also, it is difficult to distinguish between occasional guests and true cavernicoles in this group (Willmann, 1938).

The Acarina most frequently found in caves belong to the Gamasides of which some are creophages and feed on the dead bodies of insects, while others are coprophages or guanobia. The Oribatei are also found underground amongst humus and detritus.

Several species have been found only in the subterranean environment, but it is not possible at the moment to state whether or not they are true troglobia.

Parasitidae — Two species of *Eugamasus (E. loricatus* and *E. magnus)* are widespread in European caves.
Thrombiidae — *Typhlothrombium aelleni:* Swiss Jura (Cooreman, 1954).
Spelaeothrombium congoensis and *S. leleupi:* The Congo (André, 1957).
S. caecum: Dalmatia (Willmann, 1940).
Anomalothrombium madagascariense: Madagascar (André, 1938).

Rhagiidae. Rhagidia longipes, a totally depigmented form, with fragile elongated legs is common in European caves. It is related to the epigeous form *R. terricola* (= *gigas*); it has never been found on the surface and can be regarded as a true cavernicole (Vitzthum, 1925; Cooreman, 1959). Another species of this genus, *R. cavicola*, has been reported from the Mammoth Cave (Bailey, 1933).

Eupodidae. A number of species belonging to this family are troglophiles. *Linopodes notatorius* L. is widespread in Europe in the subterranean environment.

Oribatidae. Belba langsdorfi is generally held to be a true cavernicole (Griepenburg, 1939).

M. AMPHIBIOUS ACARINA

During the past thirty years evidence has accumulated which supports the idea of the existence of hypogeous mites which, while they belong to the terrestrial group, nevertheless lead an aquatic life, or one which is more or less amphibious.

The genus *Schwiebea* belongs to the sub-order Sarcoptiformes and the family Tyroglyphidae. *S. cavernicola* is an amphibious species. It is found in moist soil and in decayed wood; but is captured with equal fre-

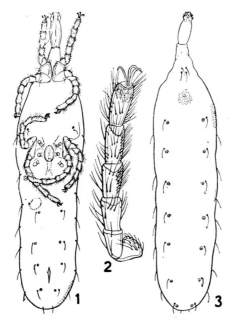

FIG. 15. *Charonothrombium racovitzai* (after Motas and Tanasachi).
1. ventral view; 2. first leg; 3. dorsal view

quency in the subterranean water in caves (Karst) and in ground water (Vitzthum, 1932; Stammer, 1936; Noll and Stammer, 1953; Husmann, 1956). Another species of the same genus has been collected in the Grotte de l'Eglise, at Bas Nistos in the central Pyrenees (Cooreman, 1959).

On the other hand one must place alongside the Thrombiidae, which are so widespread both on the surface and inside of the soil, those Acarina which populate subterranean nappes. The latter constitute the family Stygothrombiidae and include the genera *Stygothrombium*, *Cerberothrombium* and *Charonothrombium* (Viets, 1932, 1934, 1955; Motas and Tanasachi, 1946; Walter, 1947; André, 1949). These forms are widespread in Europe. A species of *Stygothrombium* is known from Japan (Imamura, 1956).

Stygothrombium chappuisi is an exclusively aquatic mite and cannot live in air despite Husmann's affirmation (Schwoerbel, 1961).

These mites are confined to ground water, but *Cerberothrombium vermiforme*, described by Sokolow and first named *Baicalacarus vermiformis*, lives in the littoral zone of Lake Baikal. The Stygothrombiidae are remarkable for their elongated bodies which are often vermiform (Fig. 15) and of a translucent appearance. They possess no eyes. They can be considered the ultimate limit of specialised evolution.

N. AQUATIC ACARINA

Hydracarina belonging to epigeous species which have most certainly been brought in from the surface, have often been found in subterranean waters. The presence of these specimens led to the belief, which was held for many years, that subterranean waters were without true cavernicolous Hydracarina. It is known today that aquatic hypogeous Acarina are to be found underground but they are confined to a biotope which has been explored only in the last few years.

P.A.Chappuis was the first to find a subterranean aquatic mite. It was a porohalacarid, taken in a well at Bale and described in 1917 by Walter under the name *Soldanellonyx chappuisi*. This species was found later in a cave in Indiana in the United States. This species is, however, not localised to the subterranean environment. It has also been dredged from the bottom of Swiss lakes, found in a spring in Harz, and at high altitudes (2000 m.) in the Grimsel and Gothard Mountains.

Another species of the same genus, *S. monardi*, although found in caves in the Pyrenees, Crete and the United States, has also been taken on the edge of Swiss lakes and in marshes in Denmark, Holland and Germany. These Acarina cannot be considered cavernicoles since they are ubiquitous.

The true aquatic, hypogeous Acarina were discovered in 1931 by Stanko Karaman around Skoplje in the Yugoslavian part of Macedonia. They were described by Viets in 1932. Since then other aquatic hypogeous species have been discovered in Europe. Numerous publications have been devoted to these arachnids. Prominent workers in this field have been E. Angelier, C. Motas and his associates, L. Szalay, C. Walter, and above all K. Viets, who has produced some most fundamental work in this field.

Other species have been discovered in the United States (described by Walter, 1931) and in subterranean waters in Japan. The latter have been described by T. Imamura and T. Uchida. One hundred and ninety species of aquatic, hypogeous Acarina are known today (Schwoerbel, 1961).

The Distribution of Aquatic and Hypogeous Acarina in the Systematic Classification

The aquatic Acarina which inhabit either surface or subterranean waters are divided into two groups.

(a) Hydrachnellae whose representatives may be found in either stagnant or running surface waters. Numerous forms have been captured in subterranean waters although many of them cannot be taken as exclusively hypogeous (Szalay, 1949). The subterranean forms have been divided into several families (Viets, 1935, Imamura, 1959a). Lists of species have been published: for Europe by Motas and Tanasachi (1946), Motas, Tanasachi and Orghidan (1947, 1957, 1958), Szalay (1949) and Viets (1935); and for Japan by Imamura (1959b).

(b) The great majority of Halacaridae are marine forms, although a few inhabit fresh water. The latter are separated from the marine forms and placed in the family Porohalacaridae. All Porohalacaridae are found in the hyporheic medium (Schwoerbel, 1961).

Subterranean forms of Porohalacaridae are much less numerous than those in the Hydrachnellae. However, many genera have been established for the subterranean porohalacarids, which Viets (1933) has divided into four sub-families.

The Ecology of the Aquatic Hypogeous Acarina

Aquatic acarina are rarely found in caves. They are phreotobia which populate the hyporheic medium.

Such forms are pale and depigmented. The eyes are reduced, depigmented or sometimes completely absent (Stygohydracarus, Acherontacarus, Stygohalacarus, Parasoldanellonyx). Hairs used for swimming are absent on the legs. The phreaticolous forms do not swim (Chappuis, 1944) but crawl between the grains of sand. Many species of Wandesia which are known in Europe possess elongated, vermiform bodies, and thus somewhat resemble Stygothrombium.

The development of aquatic hypogeous Acarina differs from that of epigeous forms in that they are at no stage parasitic on insects, as is frequently the case amongst the surface Hydracarina (Schwoerbel, 1961 b).

The Origin of the Aquatic Hypogeous Acarina

Migration of surface forms into the subterranean waters is still occurring (Schwoerbel, 1961 a). That is to say, a certain number of Hydracarina which belong to the hyporheic medium are recent hypogeans. Other types can be considered as relics. They show no affinities with the surface forms from their countries of origin. They are, however, similar to species found in aquatic mosses in tropical countries (Motas, Tanasachi and Orghidan, 1947). Their subterranean evolution is thus a consequence of the climatic changes which have taken place in the more northerly regions.

BIBLIOGRAPHY

Arachnida

PETRUNKEVITCH, Al. (1949) A Study of Palaeozoic Arachnida. *Trans. Connecticut Acad. Arts. Sc.* **XXXVII.**

Scorpionidea

AUBER, M. (1959) Observations sur le biotope et la biologie du Scorpion aveugle: *Belisarius xambeui* Simon. *Vie et Milieu*, **X.**
SIMON, E. (1879) *Les Arachnides de France.* Tome **VII,** Paris.
VACHON, M. (1944) Remarques sur le Scorpion aveugle du Roussillon, *Belisarius xambeui* E. S. *Bull. Mus. Hist. Nat. Paris* (2) **XVI.**

Pseudoscorpionidea

BEIER, M. (1939) Die Höhlenpseudoskorpione der Balkanhalbinsel. Eine auf dem Material "Biospeologica balcanica" basierende Synopsis. *Stud. allg. Karstf. Biol. Ser.* No. 10.
BEIER, M. (1940) Phylogenie der troglobionten Pseudoscorpione. *Sixth Intern. Congr. Entomol. Trans.* **II.**
BEIER, M. (1949) Türkische Pseudoscorpione. *Istanbul Univ. Fen. Fak. Mec. Ser. B.* **XIV.**
CHAMBERLIN, J.C. and MALCOLM, D.R. (1960) The occurrence of false scorpions in caves with special reference to cavernicolous adaptation and to cave species in the North-American fauna (Arachnida-Chelonethida). *In Symposium: Speciation and Raciation in Cavernicoles. Amer. Midl Nat.* **LXIV.**
LAPCHOV, I.I. (1940) Die Höhlen-Pseudoscorpiones Transkaukasiens. *Biospeologica Sovietica*, V. *Bull. Soc. Nat. Moscou. Sect. Biol.* **XLIX.**

Opilionids

ABSOLON, K. and KRATOCHVIL, J. (1932) Zur Kenntnis der höhlenbewohnenden Araneae der illyrischen Karstgebiete. *Mitteil. Höhl. Karstf.*
HADŽI, J. (1935) Ein eigentümlicher neuer Höhlen-Opilionid aus Nord-Amerika, *Cladonychium corii* g.n.sp.n. *Biol. general.* **XI.**
HADŽI, J. (1940) Zwei interessante neue Opilionenarten der Gattung *Nemastoma*. *Glasnik, Bull. Soc. Sc. Skolpje.* **XXII.**
KRATOCHVIL, J. (1936) *Ischyropsalis strandi* n. sp., un Opilion cavernicole nouveau d'Italie. *Festschr. 60. Geburtstag v. Prof. Embrik Strand.* **I.** Riga.
KRATOCHVIL, J. (1937) Essai d'une nouvelle classification du genre *Siro*. *Veštn. česko. Zool. Společ. Praze.* **V.**

KRATOCHVIL, J. (1958a) Höhlenweberknechte Bulgariens (Cyphophthalmi und Lania-tores). *Práce brn. zakl. C.S.A.V.* **XXX.**
KRATOCHVIL, J. (1958b) Höhlenweberknechte Bulgariens (Palpatores-Nemastomatidae). *Práce brn. zakl. C.S.A.V.* **XXX.**
ROEWER, C. F. (1935) Opiliones (Fünfte Serie). Zugleich eine Revision aller bisher bekann-ter europäischen Laniatores. *Biospeologica,* **LXII.** *Archiv. Zool. expér. géner.* **LXXVIII.**

Palpigrada

HANSEN, H. J. (1926) Palpigradi (deuxième Série). *Biospeologica,* **LIII.** *Archiv. Zool. expér. géner.* **LXV.**
JANETSCHEK, H. (1957) Das seltsamste Tier Tirols. Palpenläufer (Arachn., Palpigradida): Stellung, Verbreitung, Arten, Bibliographie. *Festschr. 50jähr. Bes. Kufstein. Mittelsch. 1907–1957. Kufstein. Buch,* **III.**

Pedipalpia

ANNANDALE, N., BROWN, J.C., GRAVELY, F.H. (1914) The limestone caves of Burma and the Malay Peninsula. *J. Proc. Asiatic Soc. Bengal.* N.S. **IX.**
BRISTOWE, W.S. (1952) The arachnid fauna of the Batu Caves in Malaya. *Ann. Mag. Nat. Hist.* (12) **V.**
CHAMBERLIN, R.V. and IVIE, W. (1938) Arachnida of the orders Pedipalpia, Scorpionida and Ricinulida. *In* PEARSE, A.S., Fauna of the Caves of Yucatan. *Carnegie Inst. Washington, Public.* No. 491.
FAGE, L. (1929) Fauna of the Batu Caves, Selangor. X. Arachnida: Pedipalpi (Part) and Araneae. *J. feder. Malay States Mus.* **XIV.**
FAGE, L., SIMON, E. (1936) Arachnida. **III.** Pedipalpi; Scorpiones; Solifuga et Araneae (1ère partie). *Miss. Sc. OMO* **III,** fas. 30.
GRAVELY, F.H. (1924) Tartarides from the Siju cave, Garo Hills, Assam. *Rec. Indian Mus.* **XXVI.**
HANSEN, H.J. (1926) *Trithyreus cavernicola* n.sp., a new form of the Tribe Tartarides (the order Pedipalpi) from tropical East-Africa and Zanzibar. *Biospeologica,* **LII.** *Archiv. Zool. expér. géner.* **LXV.**
LAWRENCE, R.F. (1958) A collection of cavernicolous Arachnida from French Equatorial Africa. *Rev. suisse Zool.* **LXV.**

Araneida

BARROS MACHADO, A. DE (1956) Captures d'Araignées *Telemidae* au Congo belge et paléogéographie de cette famille. *Folia sc. Afric. Centralis.* **II.**
BERLAND, L. (1931) Arachnides Araneides. *In* BOLIVAR, C. et JEANNEL, R., Campagne Spéléo-logique dans l'Amérique du Nord (1928) No. 7. *Biospeleologica,* **LVI.** *Archiv. Zool. expér. géner.* **LXXI.**
BRISTOWE, W.S. (1952) The arachnid fauna of the Batu Caves in Malaya. *Ann. Mag. Nat. Hist.* (12) **V.**
FAGE, L. (1912) Études sur les Araignées cavernicoles. I. Revision des *Ochyroceratidae* (n. fam.). *Biospeologica,* **XXV.** *Archiv. Zool. expér. géner.* (5) **X.**
FAGE, L. (1913) Études sur les Araignées cavernicoles. II. Revision des *Leptonetidae. Biospeologica,* **XXIX.** *Archiv. Zool. expér. géner.* **L.**
FAGE, L. (1919) Études sur les Araignées cavernicoles. III. Le genre *Troglohyphantes. Biospeologica,* **XL.** *Archiv. Zool. expér. géner.* **LVIII.**
FAGE, L. (1921) Sur quelques Araignées Apneumones. *Compt. rend. Acad. Sci. Paris,* **CLXXII.**
FAGE, L. (1931a) Remarques sur la distribution géographique actuelle des Araignées du genre *Porhomma. Compt. rend. Soc. Biogéogr. Paris.* No. 68.
FAGE, L. (1931b) Araneae (5ème Série), précédée d'un essai sur l'évolution souterraine et son déterminisme. *Biospeologica,* **LV.** *Archiv. Zool. expér. géner.* **LXXI.**

KÄSTNER, A. (1926–27) Überblick über die in den letzten 20 Jahren bekannt gewordenen Höhlenspinnen. *Mitteil. Höhl. Karstf.*

KHARITONOV, D. E. (1947) Spiders and harvestspiders from the caves of the Black Sea Coast of the Caucasus. *Biospeologica Sovietica,* **VIII.** *Bull. Soc. Nat. Moscou. Sect. Biol.* **LIII.**

LAWRENCE, R. F. (1958) A collection of cavernicolous Arachnida from French Equatorial Africa. *Rev. suisse Zool.* **LXV.**

SCHIÖDTE, J. C. (1849) Specimen Faunae subterraneae. Bidrag til den underjordiske Fauna. Kjöbenhavn.

STRINATI, P. (1960) La Faune actuelle de trois grottes d'Afrique Équatoriale française. *Annal. Speleol.* **XV.**

YAGINUMA, T (1962) Cave spiders in Japan. *Bull. Osaka Mus. Nat. Hist.* No. 15.

Ricinulida

BOLIVAR Y PIELTAIN, C. (1941) Estudio de un Ricinulideo de la Caverna de Cacahuamilpa, Guerrero, Mex. *Revista d. Soc. Mexic. Hist. Nat.* **II.**

CHAMBERLAIN, R. V. and IVIE, W. (1938) Arachnida of the orders Pedipalpa, Scorpionida and Ricinulida. *In* Fauna of the Caves of Yucatan by A. S. Pearse. **VII.** *Carnegie Inst. Washington, Public.* No. 491.

Terrestrial Acarina

ANDRÉ, M. (1938) Un Thrombidion cavernicole de Madagascar. *Bull. Soc. Zool. France* **LXIII.**

ANDRÉ, M. (1957) Contribution à l'étude des Thrombidions du Congo belge. *Rev. Zool. Bot. Afric.* **LVI.**

BAILEY, V. (1933) Cave life of Kentucky, mainly in the Mammoth Cave region. *Amer. Midl. Nat.* **XIV.**

COOREMAN, J. (1954) Notes sur quelques Acariens de la faune cavernicole. *Bull. Mus. Hist. Nat. Belgique* **XXX,** No. 34.

COOREMAN, J. (1959) Notes sur quelques Acariens de la faune cavernicole (2ème Série). *Bull. Inst. R. Sci. Nat. Belgique* **XXXV.**

GRIEPENBURG, W. (1939) Die Tierwelt der Höhlen bei Kallenhardt. *Mitteil. Höhl. Karstf.*

VITZTHUM, M. (1925) Die unterirdische Acarofauna. *Gen. Zeit. Naturw.* **LXII.**

WILLMANN, C. (1938) Die Acarofauna der Höhlen des Fränkischen Jura und einiger anderer Höhlen. *Mitteil. Höhl. Karstf.*

WILLMANN, C. (1940) Neue Milben aus Höhlen der Balkanhalbinsel, gesammelt von Prof. Dr. K. Absolon, Brünn. *Zool. Anz.* **CXXIX.**

Amphibious Acarina

Schwiebea

COOREMAN, J. (1959) Notes sur quelques Acariens de la faune cavernicole (2ème série). *Bull. Inst. R. Sci. Nat. Belgique* **XXXV,** No. 34.

HUSMANN, S. (1956) Untersuchungen über die Grundwasserfauna zwischen Harz und Weser. *Archiv. f. Hydrobiol.* **LII.**

NOLL, W. and STAMMER, H. J. (1953) Die Grundwasserfauna des Untermaingebietes von Hanau bis Würzburg, mit Einschluß des Spessarts. *Mitteil. Naturw. Mus. Aschaffenburg.* N. E. **VI.**

STAMMER, H. J. (1936) Die Höhlenfauna des Glatzer Schneeberges. 8. Die Wasserfauna der Schneeberghöhlen. *Beitr. Biol. Glatz. Schneeb.* 2. Heft.

VITZTHUM, H. GRAF (1932) Acarinen aus dem Karst (Excl. Oribatei). *Zool. Jahrb. Abt. System.* **LXIII.**

Stygothrombiidae

ANDRÉ, M. (1949a) Une nouvelle espèce de Thrombidion (Stygothrombiidae) recueillie en France dans un cours d'eau phréatique. *Bull. Mus. Hist. Nat. Paris* (2) **XXI.**

ANDRÉ, M. (1949 b) Les *Stygothrombium* (Acariens) de la faune française. *Bull. Mus. Hist. Nat. Paris.* (2) **XXI.**

IMAMURA, T. (1956) Some subterranean water-mites from Hyogo Prefecture, Japan. *Communic. Premier Congr. Intern. Spéléol. Paris.* **III.** *Biol.*

MOTAS, C. and TANASACHI, J. (1946) Acariens phréaticoles de Transylvanie. *Notat. Biol.* Bucarest. **IV.**

SCHWOERBEL, J. (1961) Subterrane Wassermilben (Acari: Hydrachnellae, Porohalacaridae und Stygothrombiidae), ihre Ökologie und Bedeutung für die Abgrenzung eines aquatischen Lebensraumes zwischen Oberfläche und Grundwasser. *Archiv. f. Hydrobiol. Suppl.* **XXV.**

VIETS, K. (1932) Weitere Milben aus unterirdischen Gewässern. *Zool. Anz.* **C.**

VIETS, K. (1934) Siebente Mitteilung über Wassermilben aus unterirdischen Gewässern. *Zool. Anz.* **CVI.**

VIETS, K. (1955) In subterranen Gewässern Deutschlands lebende Wassermilben (Hydrachnellae, Porohalacaridae und Stygothrombiidae). *Archiv. f. Hydrobiol.* **L.**

WALTER, C. (1947) Neue Acari (Hydrachnellae, Porohalacaridae, Thrombiidae) aus subterranen Gewässern der Schweiz und Rumäniens. *Verhandl. Naturf. Gesell. Basel.* **LVIII.**

Aquatic Acarina

ANGELIER, E. (1953) Recherches écologiques et biogéographiques sur la faune des sables submergés. *Archiv. Zool. expér. géner.* **XC.**

ANGELIER, E. (1960) Acariens psammiques (Hydrachnellae et Porohalacaridae), in ANGELIER, E., Hydrobiologie de la Corse. *Suppl. No. 8, Vie et Milieu.*

CHAPPUIS, P. A. (1944) Die Grundwasserfauna der Körös und des Szamos. *Mat. Termes. Kozlem.* **XL.**

IMAMURA, T. (1959 a) Water-mites (Hydrachnellae) of subterranean waters in Kantô District, Japan. *Acarologia,* **I.**

IMAMURA, T. (1959 b) Check-list of the troglobiontic Trombiidae, Porohalacaridae and Hydrachnellae of Japan. *Bull. Biogeogr. Soc. Japan.* **XXI.**

MOTAS, C. and TANASACHI, J. (1946) Acariens phréaticoles de Transylvanie. *Notat. Biol.* Bucarest. **IV.**

MOTAS, C., TANASACHI, J. and ORGHIDAN, TR. (1947) Hydracariens phréaticoles de Roumanie. *Notat. Biol.* Bucarest. **V.**

MOTAS, C., TANASACHI, J. and ORGHIDAN, TR. (1957) Über einige neue phreatische Hydrachnellae aus Rumänien und über Phreatobiologie, ein neues Kapitel der Limnologie. *Abh. Naturw. Bremen.* **XXXV.**

MOTAS, C., TANASACHI, J. et ORGHIDAN, TR. (1958) Hydrachnelles phréaticoles de la R. P. Roumaine. *Vešt. Česk. Zool. Spel.* **XXII.**

SCHWOERBEL, J. (1961 a) Subterrane Wassermilben (Acari: Hydrachnellae, Porohalacaridae und Stygothrombiidae), ihre Ökologie und Bedeutung für die Abgrenzung eines aquatischen Lebensraumes zwischen Oberfläche und Grundwasser. *Archiv. f. Hydrobiol. Supplement* **XXV.**

SCHWOERBEL, J. (1961 b) Entstehung von Grundwasserarten bei Süßwassermilben (Hydrachellnae) und die Bedeutung der parasitischen Larvenphase. *Naturwiss.* **XLVIII.**

SZALAY, L. (1949) Über die Hydracarinen der unterirdischen Gewässer. *Hydrobiologia* **II.**

VIETS, K. (1933) Vierte Mitteilung über Wassermilben aus unterirdischen Gewässern (Hydrachnellae und Halacaridae, Acari). *Zool. Anz.* **CII.**

VIETS, K. (1935) Wassermilben aus unterirdischen Gewässern Jugoslawiens (Achte Mitteilung). *Verh. Intern. Ver. theor. angew. Limnologie* **VII.**

WALTER, C. (1931) Arachnides Halacarides. *In* Campagne Spéléologique de C. Bolivar et R. Jeannel dans l'Amérique du Nord (1928); No. 6. *Biospeologica,* **LVI.** *Archiv. Zool. expér. géner.* **LXXI.**

CHAPTER IX

THE CRUSTACEA

A. INTRODUCTION

Until 1800 the Crustacea were classified with the Insecta. In that year Cuvier's *Leçons d'Anatomie Comparée* appeared. In this work the autonomy of the Crustacea was established. Today zoologists are inclined to separate the Crustacea from the Tracheata completely and recognise a separate origin for both. The Crustacea differ from Tracheata; (1) by their aquatic mode of life (Tracheata have, from the origin, a terrestrial mode of life); (2) by their two pairs of antennae (one pair in tracheates); (3) by their biramous appendages (always uniramous in tracheates).

The Crustacea may be divided into a lower group, the Entomostraca, and a higher group, the Malacostraca. It must be borne in mind, however, that although the Malacostraca are a well defined group, the Entomostraca are not.

The latter are made up of a number of rather unrelated groups, four of which are exclusively marine (Cephalocarida, Mystacocarida, Ascothoracia and Cirripedia), while the remaining three (Branchiopoda, Copepoda and Ostracoda), have freshwater representatives. Only the last three groups contain cavernicolous species.

B. BRANCHIOPODA

This group is divided into four sub-classes: Anostraca, Notostraca, Conchostraca and Cladocera. No Notostraca are known from subterranean water. The anostracan, *Branchipus pellucidus* and the conchostracan, *Estheria caeca* described by Joseph (1882), must be relegated to the mythical world.

Our knowledge of the cavernicolous Cladocera is summarised in an excellent paper by Rammes (1933). The few species which have been found are, in fact, epigeous species. They are starving specimens because the algae and flagellates upon which they feed are absent in underground waters. Thus they cannot subsist in this environment for long, and still less breed there.

C. COPEPODA

The copepods are mainly a marine group, but there are a number of freshwater species. Their marine origin cannot be doubted, however, since the latter forms are closely related to marine species.

The Copepoda are classified into two sub-orders, the Gymnoplea and the Podoplea, and the latter group is again sub-divided into the Cyclopoidea and the Harpacticoidea.

(a) Gymnoplea

This sub-order is commonly represented in the marine plankton. Its members in the freshwater plankton belong to the Centropagidae and to the genera *Diaptomus*, *Heterocope* and *Eurytemora*.

With regard to the cave-dwelling Centropagidae, *Anadiaptomus poseidon spelaea* has been collected from caves in the Namoroka Reserve in Madagascar, but this is simply a hypogeous variety of an epigeous species (Brehm, 1953; Paulian and Grjebine, 1953).

However, Tafall (1942) discovered a species of *Diaptomus* in the Cueva la Chica and the Cueva de los Sabinos near Valles, Mexico. The species was clearly a true cavernicole since it lacked pigmentation and was anophthalmic. Tafall described it under the name *D. (Microdiaptomus) cokeri*. It is the smallest species of the genus known, being only 0·75 mm in length, and the antennules are clothed in long flattened sensile hairs.

(b) Cyclopoidea

It was not until 1907 that the first cavernicolous representatives of this group were discovered by Steinmann and Graeter. The hypogeous members of this group have been studied principally by E. Graeter and P. A. Chappuis and more recently by F. Kiefer and K. Lindberg.

The cavernicolous Cyclopodea can be divided into three ecological groups:

(1) The first group includes ubiquitous species which are abundant in both surface and subterranean waters. *Eucyclops serrulatus* and *Tropocyclops prasinus* may be placed in this category.

The specimens of these species captured underground do not differ appreciably from the surface forms although hypogeous specimens of *E. serrulatus* may be more or less depigmented (Graeter, 1910).

(2) The second group comprises troglophiles which are regular inhabitants of the hypogeous medium although they are sometimes captured in surface water. Examples are *Megacyclops viridis* and *Paracyclops fimbriatus*.

Certain genera belonging to the Cyclopoidea contain a majority of troglophilic or almost rigorously troglobious species. *Diacyclops*, with the troglophilic species *D. bicuspidatus*, *D. bisetosus* and *D. languidoides*, and especially *Acanthocyclops*, which has a majority of almost troglobious species *(A. phreaticus, A. kieferi, A. reductus, A. venustus* and *A. sensitivus)*, may be included in this section.

Within this group certain species are found in caves and others in ground waters. An example of the latter type is *Acanthocyclops sensitivus* which is an interstitial form.

These hypogeous species exhibit certain morphological changes. Examples found in subterranean water are usually more or less depigmented. The subterranean specimens of *Eucyclops graeteri* Chappuis (= *C. macrurus* var. *subterranea* Graeter) appear to be anophthalmic, and have thus been described by Graeter (1910). However, if this subterranean form is cultured in light the ocular apparatus is "reacquired" (Chappuis, 1920). In fact, the eye has not disappeared, but, in the hypogeous specimens, the pigment vanishes. The pigment reappears when the animal is returned to the light. There is a similar change in *Eucyclops teras* and *Diacyclops languidoides zschokhei*.

This second group corresponds to the troglophiles which are in a state of evolutive unstableness which may be referred to as cavernicoles "*en puissance*".

(3) The third group is composed of very minute forms, rarely exceeding 0·5 mm, which have undergone morphological simplification and a considerable reduction in the numbers of segments in the appendages and of the bristles on them.

Carcinologists have established special genera for the species of this group (Kiefer, 1937; Lindberg, 1954, 1956). The genus *Graeteriella* was erected by Brehm to contain the most ancient species of this group, namely *G. unisetiger*, which was originally described by Graeter in 1908. This species is recognised by its furca of which each branch bears only one bristle. The eye is present, but reduced. This microscopic copepod does not swim; it is a crawling form and has a similar type of locomotion to the harpacticoids. It was discovered in a cave in the Swiss Jura, and has since been found over a large part of Europe from the north of Spain to Georgia (Lindberg, 1954; Kiefer, 1957).

The genus *Speocyclops* (Kiefer, 1937) contains fourteen species which are all cavernicolous. They are distributed along the area to the north of the Mediterranean from the Pyrenees to Caucasia.

Menzel (1926) and Chappuis (1927) first recognised the affinities of the cavernicolous copepods. Their findings have been verified by Kiefer (1937) and more recently by Lindberg (1954, 1956). The troglobious species resemble those typical forms which live in humid mosses or in the minute aquaria provided by the sheath-like leaves of bromeliaceous epiphytes.

The latter forms belong to the genera *Bryocyclops* (Africa, Madagascar, India and the Malay Archipelago) and *Muscocyclops* from South America.

Carcinologists have suggested that the troglobious cyclopoids are derived from forms which made up part of the tropical fauna of the Ancient Continent during the first part of the Tertiary. At the start of the Quarternary the tropical forms were destroyed in Europe following the increased severity of the climate there. Only those few species which succeeded in adapting themselves to life in subterranean waters survived, thus escaping the rigours of the Ice Age. It is these relict species of an originally tropical fauna that are found in European caves today.

(c) Harpacticoidea

The harpacticoids are small, often microscopic copepods. They swim poorly, or not at all. Under the binocular microscope they can be seen crawling in the debris in which they live.

Although cavernicolous harpacticoids are numerous our knowledge of them is recent and still very incomplete. The systematic classification of these microscopic Crustacea has developed mainly through the studies of P.A.Chappuis, who from 1914 to 1960 produced more than one hundred publications relating to the copepods. The majority of these concern the harpacticoids. Chappuis not only described numerous new species, but created several new genera and thus brought clarity to a group whose study had previously been in a state of great confusion.

The important monograph by K.Lang (1948) must also be mentioned here. This included a list of all known Harpacticoidea and thus rendered a great service to carcinologists and biospeologists alike.

The Cavernicolous Harpacticoidea. These may be divided into several families: the Canthocamptidae, Ameiridae, Parastenocaridae, Ectinosomidae, Viguierellidae, etc. (Chappuis, 1944). There is no place in this book to review the cavernicolous harpacticoids; that is a matter for specialists, instead, an ecological classification will be given. Such a classification is essentially provisional owing to the fragmentary state of our knowledge of the ecology of these animals.

Apart from this consideration, it is often difficult to decide whether to regard a newly discovered species as a cavernicole, simply because no criterion exists which allows the separation of strictly subterranean forms from surface species. The genus *Viguierella* contains two species, of which one *V. caeca (= Phyllognathopus viguieri)* is widespread in the surface waters of the tropics and is also occasionally found in the subterranean waters of temperate regions. However, all specimens of the species whether epigeans or cavernicoles are anophthalmic.†

† *Epactophanes richardi* which lives in mosses is also anophthalmic.

The Origins of the Cavernicolous Harpacticoidea. The cavernicolous harpacticoids are directly related to surface species, but there are a few genera which are entirely cavernicolous. Such genera are *Antrocamptus, Chappuisius, Ceuthonectes,* and *Spelaeocamptus.* Sufficient is known of the phyletic relationships of the cavernicolous harpacticoids for it to be stated that they have been derived from several distinct epigeous lines.

(1) The Ameiridae contains only the two genera *Nitocra* and *Nitocrella. Nitocra* contains marine or brackish-water species while the majority of the species of *Nitrocrella* are freshwater cavernicoles. It is probable that the transition from the marine to the subterranean environment has been a direct one, as is indicated by the fact that *Nitocra affinis* has been captured in both the Suez Canal and in "Zinzulusa" cave in Otrante Land (southern Italy).

(2) Several cavernicolous harpacticoids are directly related to epigeous, muscicolous species (i.e. those species which live in very wet mosses such as those which grow in bogs, marshes and on the banks of streams). This is the case with the species of *Bryocamptus,* a genus widespread in the Holartic region. The majority of these species are epigeous, but a few are adapted for life underground. Biospeologists who have visited the Grotte de Sainte-Marie, near La Preste (Pyrénées-Orientales) have been impressed by the large numbers of two species, *B. zschokkei* and *B. pygmaeus,* which crawl on the humid bark of the branches which accumulate in the terminal chamber. But these two species are commonplace muscicoles, which live in the film of water which often covers surface plants. However, certain species of *Bryocamptus* seem to be strict cavernicoles: *B. caucasicus, B. dentatus, B. pyrenaicus, B. typhlops* and *B. unisetosus.*

(3) The genus *Moraria* contains forms which are very elongated. Some of them live in mud at the edges of lakes or in mosses. Some species which live in decaying wood or dead leaves (e.g. *M. varica* and *M. arboricola*) may be regarded as truly amphibious. Lastly a number of species are strictly cavernicolous *(M. catalana, M. denticulata, M. scotenophila* and *M. subterranea).*

(4) Certain cavernicolous harpacticoids seem to have been derived from species which formerly cosmopolitan (and some species are today so distributed), but are now restricted to tropical areas with the exception of a few temperate species which have taken refuge in caves. This is the case with the two genera *Attheyella* and *Elaphoidella* which may be regarded as the two most primitive representatives of the Canthocamptidae. These two genera are almost cosmopolitan.

The genus *Elaphoidella* includes 81 species or sub-species (Chappuis, 1958b) found in all parts of the world, but principally in the tropics. *E. bidens* (Roy, 1932) a species with parthenogenetic reproduction, is cosmopolitan. While the tropical species of this genus live in lakes, marshes and mosses, the 34 species found in Europe are all cavernicoles with the exception of *E. gracilis*. Therefore it appears that the European species represent the relict of an ancient tropical fauna living in the Tertiary (Chappuis, 1933, 1937, 1958b).

(5) *Parastenocaris* contains 70 species (Chappuis, 1958a, 1959), with an area of distribution from the Arctic to the tropics. They are forms characteristic of the interstitial medium. This environment may contain salt,

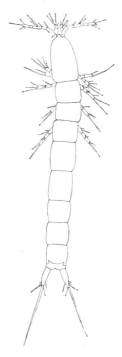

FIG. 16. *Parastenocaris* sp. (original figure by R. Rouch).

brackish, or fresh water. The evolution of freshwater from the saltwater forms has clearly taken place (Noodt, 1962). The euryhalinity of these forms facilitates their evolution towards the freshwater environment. Their bodies, which are extremely elongated and vermiform (Fig. 16) allow them to slip easily between the particles of sand.

The interstitial medium is probably the original one for this genus. However, several species have become muscicoles in tropical regions (Indonesia, Guiana), whilst others have become cavernicoles in temperate regions. This is the case with *Parastenocaris cantabrica*, *P. dentulatus*, *P. micheli* and *P. stammeri*, from the western Pyrenees and the Cantabrian Mountains, and with *P. proserpina* from southern Italy.

D. OSTRACODA

The hypogeous ostracods are rarely found in caves and thus the great majority of them are phreatobia.

The systematics of the hypogeous Ostracoda are essentially the work of one man, W. Klie. The distribution of hypogeous species in the systematic classification is narrowly limited. With the exception of *Cypridopsis subterranea* (which is widespread in Western Europe) most hypogeous species belong to the genus *Candona* and to the groups *rostrata* (Klie, 1935, 1938) and *Cryptocandona*, also a few to the groups *compressa* and *mixta*.

Hypogeous ostracods are known from Europe, North America and Japan. The ostracod *Darwinula protacta* has been discovered leading a troglobious life in a cave in Irumi, in the Congo (Rome, 1953).

The hypogeous Ostracoda differ from the surface forms in that their carapaces are transparent, there is a greater development of the sensory organs at the apices of the second antennae (Klie, 1931) and they are anopthhalmic. However, we must remark that the eye of *Candona eremita* present in the young and not lost until the adult stage is attained (Vejdovsky, 1882).

E. MALACOSTRACA

The Malacostraca are characterised by the constancy of the number of segments which make up the body (19) and of the appendages.

The Malacostraca may be divided into fifteen sub-orders, seven of which possess cavernicolous species. These are:

Super-Orders	Orders
Syncarida	Bathynellacea
	Thermosbaenacea
	Spelaeogriphacea
Peracarida	Mysidacea
	Isopoda
	Amphipoda
Eucarida	Decapoda

F. SYNCARIDA

This group has the peculiarity of having been described originally from fossil forms (in 1847, by Jordan). In 1886 Packard gave them the name Syncarida. The syncarids of the Lower Carboniferous era were marine. In the Upper Carboniferous and the Permian they invaded freshwater, and were widespread in Europe, North America and Brazil.

Characters of the Cavernicolous Oniscoids. It would be entirely artificial

The first living representative of this super-order, *Anaspides tasmaniae*, was discovered in the mountainous regions of Tasmania by G. M. Thomson in 1893. At first it was classified as a schizopod, but was eventually placed in its rightful position amongst the Syncarida by Calman (1896). Three other species were found afterwards: *Paranaspides lacustris* and *Micraspides calmani* in Tasmania, and *Koonunga cursor* near Melbourne.

However, in 1880 Vejdovsky discovered in some wells in Prague two specimens of a microscopic crustacean which he described in his great work on the fauna of wells of the city of Prague (1882), under the name *Bathynella natans*. But he did not recognise its affinities and placed it in the "incertae sedis".

Calman in 1899 recognised the true systematic position of these small Crustacea and placed them in the Syncarida.

Thus, the representatives of the three orders which constitute the Syncarida were discovered: the Gampsonychidae which are exclusively fossils, and the Anaspidacea and the Bathynellacea which have living representatives.

The Bathynellacea, although showing simplification and reduction of many parts of the body, are more primitive than the Anaspidacea notably by reason of their free first-thoracic segment and of their furca (Nicholls, 1946; Siewing, 1958). The members of the Bathynellacea are thus the closest living relatives of the primitive Syncarida. On the other hand the Anaspidacea are related to the higher Malacostraca. The burrowing forms of the Anaspidacea, for example *Koonunga*, with its reduced eyes, and *Micraspides*, which is anophthalmic and depigmented, are two examples of "preadaptation" to the hypogeous way of life. However, they should not be regarded as intermediate between the Anaspidacea and the Bathynellacea as has occasionally, but falsely, been upheld.

The history of the Syncarida shows that the Bathynellacea represent extremely ancient relics of the freshwater fauna of the Carboniferous.

The Bathynellacea — These species are very small (1–2 mm) and possess elongated, rather vermiform bodies. They possess many regressive characters: the absence of a pigmented integument, anophthalmia, the reduction of the antennary flagella and that of exopodites on the pereiopods.

Their development is direct (Bartok, 1944; Chappuis, 1948; Miura and Morimoto, 1953; Jakobi, 1954).

During the past few years the Bathynellacea have been the subject of numerous publications including those by J. M. Braga, A. Capart, P. A. Chappuis, Cl. Delamare-Deboutteville, H. Jakobi, St. Karaman, Y. Morimoto, Y. Miura, R. Siewing, P. Torok, and M. Uéno. The reader will find an excellent and comprehensive review of our knowledge of the Bathynellacea by Botosaneanu (1959).

The Bathynellacea contain some 60 species distributed in five genera.

B 5a

These are *Allobathynella, Austrobathynella, Bathynella* (Fig. 17), *Parabathynella* (Fig. 18), and *Thermobathynella* (Delamare-Deboutteville, 1960).

The Bathynellacea are only rarely found in caves. They are one of the characteristic inhabitants of ground water and the interstitial environment (Hertzog, 1930; Nicholls, 1946; Jakobi, 1954; Husmann, 1956; Spooner, 1961).

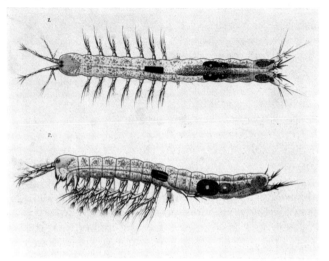

FIG. 17. *Bathynella natans* (after Chappuis).

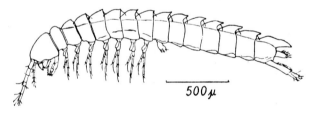

FIG. 18. *Parabathynella carinata* (after M. Uéno).

They do, however, populate other media. *Thermobathynella amyxi* has been collected in brackish water in the estuary of the Amazon and another congeneric species, *T. adami*, has been found in a hot spring (55 °C) in the Congo.

The Bathynellacea are found in all parts of the world, Europe, Equatorial Africa (the Congo, and in Lake Tanganyika and Lake Bangweulo), Madagascar, Lake Baikal, Malaya, Brazil and Patagonia, so that this order may be considered to be cosmopolitan. The bathynellacean fauna is well represented in Japan where at least a dozen species of these microscopic Crustacea have been described.

G. THERMOSBAENACEA

This is the most primitive order of the Peracarida.† It includes two genera, *Thermosbaena* and *Monodella*.

Thermosbaena. This genus was established in 1924 by Theodore Monod, in a study made on the only known species *T. mirabilis*. In 1927 the same author created the order Thermosbaenacea. This crustacean is depigmented; the eyes are rudimentary (Siewing, 1958). It shows many regressive

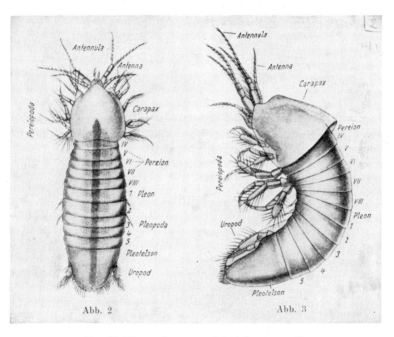

FIG. 19. *Thermosbaena mirabilis* (after Monod).

modifications (Monod, 1927) (Fig. 19). The development is direct and takes place in the dorsal incubation pocket, between the body and the carapace (Barker, 1956, 1962).

One of the most characteristic features of this species is, of course, its habitat. It was discovered in 1923 by L. G. Seurat, in the baths fed by the warm springs of El Hamma, near Gabès (Tunisia). The waters of these springs contain a high concentration of chlorides and sulphates. The temperature is 47 °C. *Thermosbaena* is certainly perfectly adapted to this high

† The majority of workers have not followed Siewing (1958) in his attempt to exclude the Thermosbaenacea from the Peracarida and erect for them a new division of the Malacostraca (the Pancarida).

temperature because it becomes moribund at 35 °C and dies at 30 °C (Barker, 1959).

Now this species would not have been mentioned here if it had not, on certain occasions, been found in the interstitial habitat (Barker, 1959). It is probable that it is of marine origin. According to Barker (1959), they were restricted to chotts during the transformation of the Mesogean Sea into the Mediterranean. Finally they gained access to the baths of El Hamma, the only known locality for this species.

Monodella. The second representative of the Thermosbaenacea is the genus *Monodella*, created in 1949 by Ruffo for a species related to, but more primitive than, the genus *Thermosbaena*. The species of *Monodella* measure 3–4 mm in length. They are anophthalmic and depigmented (Fig. 20). The brood pouch of the gravid female is dorsal as in *Thermosbaena*. It is formed between the dorsal wall of the body and the carapace and resembles the brood pouch of the Cladocera rather than that of the Peracarida. The development is direct (Stella, 1955).

There are four species in the genus: *Monodella halophila* (Fig. 20) has been collected by St. Karamann (1953) in more or less salty, littoral, subterranean water, around Dubrovnik (Raguse).

M. relicta has recently been discovered in a mineralised, limnocrene spring on the shores of the Dead Sea (Israel) (Dov Por, 1962).

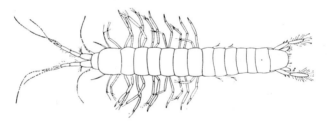

FIG. 20. *Monodella halophila* (after Karaman).

M. stygicola was the first species of the genus to be discovered. It was found by S. Ruffo (1949), in "L'Abisso", a cave near Otrante in the southeast of Italy. The waters of this cave are slightly brackish since it opens only some 200 metres from the sea.

The fourth species *M. argentarii* from the "Punta degli Stretti", a cave which opens on the Monte Argentario, a region which is almost an island connected to the Tyrrhenian coast of Tuscany by a narrow isthmus. The waters in this cave are quite fresh, and this species is planktonic (Stella, 1951, 1953).

These four species are the successful end-products of ecological evolution. The archetypal *Monodella* was probably originally a marine, littoral and interstitial form, which has now become adapted to brackishwater, then to freshwater and lastly to pelagic ways of life.

H. SPELAEOGRIPHACEA

Some South African speleologists in 1956 discovered in "Bats Caves" in Table Mountain above Cape Town a peculiar eyeless crustacean which was first investigated by K.H.Barnand and then studied by Miss Isabella Gordon (1957, 1960) and described under the name *Spelaeogriphus lepidops* (Fig. 21).

These Crustacea have a small carapace. Since the females possess oostegites on their pereiopods, they should be placed in the Peracarida. How-

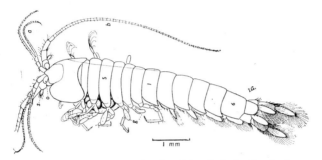

FIG. 21. *Spelaeogriphus lepidops* (after Isabella Gordon).

ever, they do not fit into any of the existing orders of this group although they show certain affinities with the Thermosbaenacea and the Tanaidacea, and even with the Syncarida. Miss Gordon has established a new order in the Peracarida to accommodate this species, the Spelaeogriphacea. *Spelaeogriphus* is most certainly the last relict of a primitive peracarid type.

I. MYSIDACEA

Until the two preceding orders were discovered, the Mysidacea were considered to be the most primitive of the Peracarida. Since the Carboniferous this order has been represented by varied and numerous genera.

The mysids are mainly marine Crustacea, but there are a minority of species which inhabit freshwater. The most ancient of these forms known to science is *Mysis relicta* found in Scandinavia, Russia, Germany, Ireland and North America. Some twenty other freshwater species are known today (Stammer, 1936); they are particularly well represented in the Ponto-Caspian area.

The Cavernicolous Mysidacea.

In the present state of our knowledge there are nine cavernicolous mysids divided into three genera.

Mysidae: Heteromysis cotti Calman 1932. Jameo de Agua, in the island of Lanzarote, Eastern Canary Islands, in salt water. It is thus a marine cavernicole.

Troglomysis vjetrenicensis Stammer 1936. Vjetrenica Cave, near Zavala (Herzegovina), in freshwater.

Antromysis cenotensis Creaser 1936. Caves and cenotes in Yucatan, Mexico, in freshwater.

Antromysis sp. Bolivar 1944. Cueva del Quintanal, at Alquizar, Cuba.

Lepidopodae: Lepidops servatus (Fage, 1925). Found in Lake Machumwi-Ndogo, Zanzibar, which is at the bottom of a swallow-hole caused by the "fall-in" of an ancient cave. The water of the lake is slightly brackish.

Spelaeomysis bottazzii Caroli 1924. Found in caves in Otranteland, south-eastern Italy, Zinzulusa, l'Abisso, Buco dei Diavoli, and a spring near Bari (Ruffo, 1949 b). Fresh or brackishwater.

Typhlolepidomysis quinterensis Villalobos 1951. Grutas de Quintera, Tamaulipas, Mexico. Freshwater.

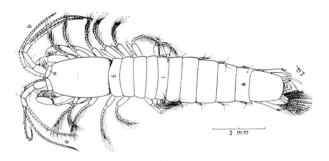

FIG. 22. *Stygiomysis holthuisi* (after Isabella Gordon).

Stygiomysidae: Stygiomysis hydruntina Caroli 1937. Found in caves in Otranteland, in south-eastern Italy, l'Abisso, Buco dei Diavoli. In fresh or slightly brackish water.

Stygiomysis holthuisi Gordon 1960. Devil's Hole, Isle of Saint Martin, Lesser Antilles. In brackish water (Fig. 22).

The Origin of the Cavernicolous Mycidacea. There is no doubt of the marine origin of this group. In fact about half of them have been found in brackish water and one other *(Heteromysis cotti)*, in water of which the composition was very nearly that of sea water.

The Morphological Characters of the Cavernicolous Mysids. The species of

Stygiomysis, with their vermiform bodies and reduced carapaces (Fig. 22) are the most modified of the cavernicolous Mysidacea.

Lepidops servatus and *Heteromysis cotti* retain ocular rudiments in the form of a few omatidia at the end of the ocular peduncle. The ocular rudiments of *H. cotti* contain pigment. The other cavernicolous mysids are anophthalmic. The structures observed by Stammer (1936) in *Troglomysis vjetrenicensis*, and interpreted by him as degenerate eyes, correspond in fact to part of the endocrine system, the "X organ" (Hanström, 1948).

The Mysidae have a statocyst at the base of the uropod, a sense-organ which the two other families lack.

J. ISOPODA

The isopods are one of the best-represented orders of Crustacea in the subterranean world. They are divided into 10 sub-orders, 6 of which contain cavernicolous forms. These are the Cymothoidea, Sphaeromidea, Asellota, Anthuridea, Phreatoicidea and Oniscoidea. Only the last sub-order includes terrestrial forms, the other five being aquatic.

(a) Cymothoidea

This group is divided into six families of which the Cirolanidae is the most primitive, and contains subterranean species. The hypogeous Cirolanidae are depigmented and anophthalmic. They may be found either in subterranean cavities or in ground water. This group is so close to marine forms that there can be little doubt that they were derived from them directly. But it is questionable whether or not they have passed through an epigeous freshwater stage in this evolution. Our knowledge is not sufficient for us to be able to answer this question at the present time.

From the systematic point of view the hypogeous Cirolanidae may be divided into three groups:

(1) The greatest number of this group fall into the sub-family Cirolaninae, erected by Racovitza in 1912, which contains the genera related to *Cirolana*.†

The genus *Cirolana* includes many cavernicolous species from Central America: *Cirolana (Speocirolana) pelaezi*, *C. (Speocirolana) bolivari* (Mexico) and *C. (Troglocirolana) cubensis* (Cuba). One must also include here *Cirolanides texensis* (caves and wells in Texas).

The genus *Typhlocirolana* contains numerous species found in ground waters in North Africa and the Sahara. The systematics of this genus are

† *Saharolana seurati*, from a stream in southern Tunisia, and *Anina lacustris*, from pools in Zanzibar, but which also inhabits the caves of Shimoni, East Africa, are not true cavernicoles.

complex and remain uncertain despite some excellent work by Gurney (1908), Racovitza (1912), Monod (1930) and Nourisson (1956). The principal species is *T. fontis* which inhabits most parts of northern Africa. *T. rifana* comes from phreatic waters of Melilla. Lastly *T. moraguesi* (Fig. 23), from the Cueva del Drach in the Island Majorca (Balearics) is related to African species. This species was found by Racovitza (1905) and it was this discovery which determined his vocation as a biospeologist.

Regarding *Conilera stygia*, which was described by A. S. Packard from

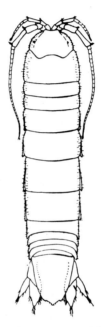

FIG. 23. *Typhlocirolana moraguesi* (after Racovitza).

a well near Monterey (Mexico), its systematic position remains to be determined.

(2) The sub-family Bathynominae, is represented by the genus *Bathynomus* with two giant species: *B. giganteus* (which reaches a length of 226 mm) and *B. döderleini* (124 mm). These two species are abyssal. The first species has been found in two distinct races, of which one has been dredged up in the Gulf of Mexico and the other from the Indian Ocean. The second species comes from the seas around Japan.

Now surprising as it may seem, Racovitza has shown that the cavernicolous isopod *Sphaeromides raymondi* discovered by P. Raymond in 1897, in the underground river of the Dragonière, in Ardèche, must belong to this sub-family. This species has been found in a number of other caves and resurgents in the French departments of Ardèche and Hérault.

Another species, *S. virei*, formerly placed in the genus *Troglaega* by Brian, should be classified after Racovitza, in the genus *Sphaeromides*. This species is indigenous to the caves of Istria, but a sub-species (*montenegrina*, Sket) has recently been discovered in Obodoska pečina, near Rijeka Crnojeviča, in the neighbourhood of Skadarsko Jez (Lake Scutari).

Hence, *Sphaeromides*, a freshwater cavernicole is related to abyssal species. This is, however, not a unique case. Examples of this situation will be given later when the galatheids are considered. *Sphaeromides* and *Bathynomus* are relicts — which have taken refuge in subterranean cavities or marine abysses — of a larger group, whose members were widespread in the Cretaceous, and whose remains are called genus *Palaega*.

(3) The sub-family Faucheriinae is notable for the fact that its representatives possess the ability to roll themselves into balls (volvation). This peculiarity is widespread in the isopods, but in the Cirolanidae it is only found in the Faucheriinae.

Faucheria faucheri is a very small isopod (3·5 mm in length), which is known in the departments of Gard, Hérault and Aude, in the South of France.

Creaseriella anops is native to the cenotes and caves of Yucatan, in Mexico.

There remains the last species of the cavernicolous Cirolanidae, the systematic position of which has not yet been determined. This species is *Anopsilana poissoni*, which has been found in Mitoho Cave, Madagascar. It feeds on the blood of the blind fish, *Typhleotris madagascariensis*.

(b) Sphaeromidea

This sub-order is rich in both genera and species. The bodies of the members of this group are very generally strongly convex, so that they can roll themselves up into a ball (volvation), to which the name "Spherome" has been given.

A number of species of this sub-order inhabit subterranean cavities. They are all indigenous to Europe, and found nowhere else.

The systematics of this group have been the subject of numerous studies by A. Arcangeli, A. Brian, A. Dollfus, R. Fabiani, G. Feruglio, E. Hubault, St. Karaman, E. G. Racovitza, B. Sket, H. J. Stammer, H. Strouhal and A. Viré.

The cavernicolous representatives of this sub-order occur (according to the classification of H. J. Hansen, 1905), in the family Sphaeromidae, the group Platybranchiatae and the section Monolistrini. The Monolistrini contains three genera *Caecosphaeroma*, *Monolistra* and *Microlistra* (Sket, 1961).

The genus *Caecosphaeroma* (Fig. 24), is native to Western Europe, or more exactly to France. *C. virei* inhabits subterranean cavities in the Jura.

C. burgundum is widespread in eastern France (departments Moselle, Meurthe-et-Moselle, Haute-Saône, Côte d'Or and Yonne). A sub-species *C. burgundum rupis-fucaldi*, has been found in the departments Charente and Dordogne.

They are probably recent cavernicoles; in fact, when they are placed in the light they quickly become pigmented (Paris, 1925). In addition, the population of *C. virei* which inhabits the Poncin Cave (Ain) is made up of individuals which are more or less pigmented (Remy, 1948).

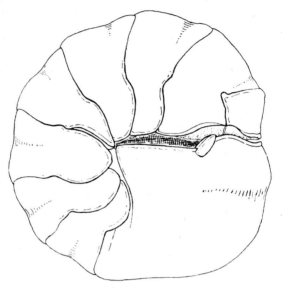

FIG. 24. *Caecosphaeroma virei* (after Racovitza).

FIG. 25. *Monolistra spinosa* (after Racovitza).

The genera *Monolistra* (Fig. 25) and *Microlistra* possess respectively 9 and 3 species or sub-species (Sket, 1961). The geographical distribution of these two genera is quite different from that of *Caecosphaeroma*. They are found in the Southern Alps, and their southward extension, the Dinaric Alps (Lombardy, Venetia, Friuli, Istria, Slovenia, and Herzegovina).

The Sphaeromidea populate underground rivers and water-courses. Their geographical distribution to which we shall have occasion to return later (p. 274), proves that the cavernicolous members of this group have originated from marine forms which became adapted to brackish water, then to freshwater, and finally to life underground.

(c) Asellota

The representatives of this group show both primitive structures and highly specialised characters, especially with regard to the pleon and the pleopods. Hansen (1904), divided the group into three families, the Parasellidae, the Stenetriidae and the Asellidae.

(1) Parasellidae. This is morphologically the most varied family in the Asellota and at the same time the richest in species. In 1916 Hansen divided the Parasellidae into 12 groups which are today regarded as sub-families, or even families by some carcinologists.

The immense majority of parasellids are marine and are particularly abundant in cold waters, and profundal regions between 1000 and 3000 metres.

During the last 30 years phreatobious forms have been described. They closely resemble marine forms. Birstein (1961) associates them with the genera *Thambena* and *Microthambena* which are abyssal forms. The first of these genera comes from the North Atlantic and the second from the North-West Pacific. There can be no doubt that the hypogeous parasellids gain access to the subterranean environment by way of the littoral interstitial medium. This is why biospeologists who study the Parasellidae (and the Crustacea in general), must always bear the marine fauna in mind.

The Parasellidae should be regarded as the relicts of an ancient fauna which for the most part have taken refuge in either the marine depths, or underground. Analogous conditions are found in other Crustacea (Cirolanidae, Amphipoda, Galatheidae) and also in the fishes (Brotulidae).

The discovery of hypogeous parasellids is a comparatively recent event due to the Jugoslavian biospeologist Stanko Karaman. In 1933 this zoologist discovered two species of the Asellota, which he described as *Microparasellus puteanus* and *M. stygius*, in the ground water at Skoplje, in the Jugoslavian part of Macedonia. In 1934 he discovered a new phreatobious species for which he created the genus *Microcharon*. Finally Chappuis and Delamare-Deboutteville (1952) erected the new genus *Angeliera*. Karaman created the family Microparasellidae in which to include the new genera which he had described. However, it appears more in conformity with the systematic hierarchy to consider this group as a sub-family of the Parasellidae.†

The genus *Microparasellus* contains three freshwater phreatobious species, two native to Macedonia and the other to the Lebanon. The genus *Microcharon* is made up of freshwater and marine species, distributed throughout the Mediterranean region *sensu lato;* two species (*M. teissieri*

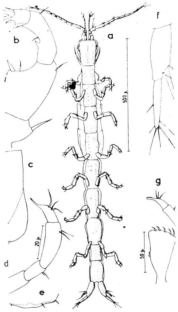

FIG. 26. *Angeliera phreaticola* (after Delamare-Deboutteville and Chappuis).

a, general view; b, fifth pereiopod male; c, first pleopod male; d, palp of the maxilliped; e, posterior edge of the pleotelson; f, uropod; g, maxilliped.

and *M. hatrisi*), have been found in Manche (Spooner, 1959). Lastly, the genus *Angeliera* is represented by only a single species which lives in both fresh and salt water, and which has an extremely wide geographical distribution, since specimens have been captured in Europe, India, and Madagascar.

The Microparasellidae are typical inhabitants of the interstitial medium. They are found in either sandy marine beaches or in ground water in fluvial valleys. They have slender and extremely elongated bodies (Fig. 26), they are depigmented, anophthalmic and have the ability to move backwards or forwards with equal facility. All these characteristics indicate that their adaptation to the interstitial environment is longstanding.

† Bocquet and Levi (1955) have rejected the name Microparasellidae and replaced it with that of Microjanirinae.

Two other genera occupy a separate place. The first of these is *Proto-charon* with two species. *P. arenicola* has been found in a resurgent on Reunion Island, and *P. antarctica* on Amsterdam Island. The other genus is *Mackinia*, with the single species *M. japonica*, which has been found in a well near Tokyo. With regard to the Tasmanian species *Pseudasellus nichollsi*, its systematic position within the Parasellidae is in some doubt in the present state of our knowledge.

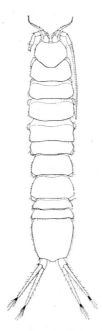

FIG. 27. *Stenasellus virei* (after Racovitza).

(2) Stenetriidae. Hansen (1904) established this family for the single genus *Stenetrium* of which all the species are marine. This genus may be compared with two others, the systematic positions of which remain to be made more precise. These are *Synasellus* which is indigenous to Portugal and has six species (Braga, 1959); and *Stygasellus* with the single species *S. phreaticus*, discovered in ground water in Szamos, in Transylvania (Chappuis, 1943, 1944, 1948).

(3) Asellidae. This family contains many genera of which the most primitive is, without doubt, *Stenasellus*. This genus is sufficiently distinct to have merited the creation of a sub-family for it, the Stenasellinae, erected by Arcangeli (1958) but raised by this author to the rank of family, a position which seems rather exaggerated. Hansen (1905) and later Racovitza (1924, 1950) have given details of the morphology of this genus, and in particular of the development of the first two pleonites whose structure is normal, in

contrast with other members of the Asellidae, where they are regressed. Numerous species of *Stenasellus* have recently been described. All (Fig.27), of them lead a hypogeous life. In Eurasia the area of distribution stretches from Portugal to Transcaspia. But other species recently discovered have shown the vast geographical distribution of *Stenasellus* which includes Eurasia and large part of Africa. Several species have been found in Portuguese Guinea, in the Ivory Coast and the Congo. The vast geographical distribution together with the consistent hypogeous habit of this genus point to an ancient origin. *Stenasellus* is a relict, and more exactly a thermophilous relict, strong evidence for which assumption is provided by its distriaction. It is found in tropical Africa and *Stenasellus thermalis hungaricus* is found in a warm spring (18–24°C) at Podused, near Zagreb, Jugoslavia: (Meštrov, 1960).

S. *virei*, the most ancient species known, inhabits the hypotelminorheic medium, and it is from this habitat that it finds its way into subterranean streams and gours. The primitive origin accounts for their burrowing habits which have often been observed (Meštrov, 1962).

The true Asellidae includes the genera *Asellus* (= *Caecidotea* †), *Johannella, Mancasellus, Lirceus,* and *Anneckella.*

Anneckella contains a single species which *A. ficki* has been found in a spring on the steep slopes of Drakensberg, in South Africa. This species is a member of the phreatic fauna. *Johannella purpurea* has been found in a sulphur spring at El Hamma, Algeria. *Mancasellus* and *Lirceus* contain no truly hypogeous species.

Asellus is a large genus with 60–70 species or sub-species. Its geographical distribution embraces all the holarctic zone: Europe, North Africa, North and West Asia, the Japanese Archipelago and also North America. Many workers (Birstein, Dudich, Karaman, Stammer) have often tried to separate the species of *Asellus* into distinct sub-genera. These attempts have met with little general satisfaction and much criticism (Chappuis, 1953, 1955). At the very most, one can try to distinguish the evolutionary lines within the genus and group the species accordingly.

The Asellidae are very numerous in caves and wells of Japan. They belong to the genera or sub-genera *Asellus, Mackinia, Nipponasellus, Uenasellus* and *Phreatoasellus* (Matsumoto, 1956, 1958, 1960, 1961, 1962).

Numerous cavernicolous species of *Asellus* have been described. However, the distinction between surface and subterranean forms is often delicate. For instance, in the Postojna Jama, one may find intermediates between pigmented individuals and totally depigmented forms of *Asellus aquaticus* (Kosswig, 1940). A similar case is that of *A. strinatii* found in Inonu Cave, Turkey (Chappuis, 1955). These species can be taken as

† Miller (1923) and Chappuis (1950, 1957) have established the synonymity of *Asellus* and *Caecidotea*.

cavernicoles "en puissance", that is to say they have recently settled in underground waters.

Other species of *Asellus* appear to have had a more ancient origin. The truth of this statement rests on the fact that the "nuptial crochet" of the 4th pereiopod of the male is not yet differentiated in certain hypogeous species (Racovitza, 1920; Chappuis, 1949, 1953). Examples of this may be seen in *A. cavaticus*, a species found in many European caves, and *A. stygius* from subterranean waters in North America.

Despite the numerous publications which have been devoted to the epigeous and hypogeous members of *Asellus*, new studies are necessary in order to find the ancestral forms with which to link the various evolutionary lines that comprise the cavernicolous species of this genus.

(d) Anthuridea

The Anthuridea have a rather peculiar morphology which separates them from other Isopoda. There are two families, the Anthuridae and the Microcerberidae.

The Anthuridae contains two species which have been found in springs and ground waters: *Cruregens fontinalis* comes from New Zealand, and *Cyathura milloti* from Reunion Island.

The Microcerberidae are species which live a hypogeous existence. They are blind, depigmented and of microscopic size (0·8–1·6 mm in length).

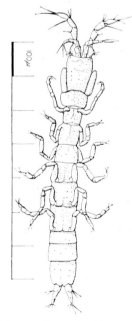

Fig. 28. *Microcerberus pauliani* (after Delamare-Deboutteville).

Their morphology is rather different from that of the Anthuridae (Fig. 28). *Microcerberus* is the only genus in this family, and was discovered by Karaman (1932) in some wells in the Skoplje region. In 1940 Karaman created the family Microcerberidae and recognised its relationship with the Anthuridae. Lang (1961) instituted for these isopods the sub-order Microcerberidea. These microscopic isopods inhabit the interstitial medium; some are marine, but others freshwater forms, the last-mentioned owing their origin to the first, without doubt. *Microcerberus* is cosmopolitan, having been found in the Mediterranean areas of Europe, in North Africa, Angola, Madagascar, Reunion Island, India, the Bahamas, Mexico and Brazil. This immensely widespread distribution implies a very ancient origin.

(e) Phreatoicoidea

This group has fairly recently been monographed by Nicholls (1943–1944). They are aberrant isopods, laterally compressed as are the amphipods, with which, perhaps, they are distantly related in phylogeny (Fig. 29). The Phreatoicoidea are an ancient group, which have been perfectly differenti-

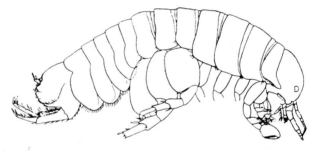

FIG. 29. *Phreatomerus latipes* (after G. E. Nicholls).

ated from the Triassic. They are indigenous to those countries which once formed Gondwanaland (South Africa, India, Australia, Tasmania and New Zealand). They are typically freshwater forms although some spend an almost terrestrial life.

Several hypogeous species of the Phreatoicoidea have been collected in shafts and ground waters: *Phreatomerus latipes, Phreatoicus typicus* and *P. orarii,* and *Neophreatoicus assimilis* are all from New Zealand, whilst *Nichollsia kashiense* and *N. menoni* are from India.

(f) Oniscoidea

With 300 genera and over 2000 species, the Oniscoidea, or terrestrial Isopoda, is one of the largest groups in the order.

to try to establish within the Oniscoidea rigid limits separating the endo-geans from the cavernicoles. Not only are there numerous intermediates connecting these two categories, but also many cavernicolous species are occasionally found leading an endogeous life. The example, amongst many others, of *Mesoniscus alpicola* may be mentioned. This species is endogeous in countries with a cold climate and also in Alpine regions, but becomes cavernicolous at lower elevations (Pesta, 1925; Chappuis, 1944; Strou-hal 1947).

The troglobia are entirely depigmented. The eyes are more or less reduced, although rudiments may be found by modern anatomical methods (de Lattin, 1939).

Depigmentation and reduction of the eyes occurs well before the fully cavernicolous life is reached. Cases of partial or total depigmentation in epigeous oniscoids are very frequent. Besides, only some of the primitive Oniscoidea related to marine forms, for example *Ligia*, *Scyphax* and *Ac-toecia*, possess large eyes. These are composed of numerous ommatidia which are directly joined to one another, and thus may be regarded as compound eyes. In all other Oniscoidea there are far fewer ommatidia. Moreover these are separated from each other, and are no longer covered by a continuous corneal layer. In these cases the eyes are of the diss-ociated type. The tendency towards ocular reduction which may be obser-ved in all epigeous Oniscoidea is further advanced in the humicoles, hygro-philes, endogeans, myrmecophiles and termitophiles. The cavernicoles are the last stage of a regressive evolutionary line which began long before the Oniscoidea penetrated into the subterranean world.

No matter what section of biospeology interests the naturalist, the study of troglophiles should not be neglected in favour of one concerning only the troglobia. In the case of the Oniscoidea this recommendation becomes an obligation. A list of cavernicolous oniscoids (i.e. those found in caves) drawn up with these considerations in mind, with only accidental trog-loxenes excluded, includes 230 species.

The Origin and the Geographical Distribution of the Cavernicolous Oni-scoidea. As is the rule, the cavernicolous oniscoids are the relics of an ancient fauna once widely distributed on the surface of the earth. The reasons for this change in habitat must be sought in the climatic changes which occurred in those areas which were originally inhabited by these isopods. Only those species which took refuge in the subterranean envi-ronment and which were able to survive there, have been perpetuated to this day.

During the Quaternary the holarctic region was partly covered by glaciers and was subjected to a very hard climate. This is the reason why the great majority of the cavernicolous oniscoids are found in Europe and North America; they are the relics of a pre-glacial period.

The tropical regions of the globe were only slightly affected by the

glaciers, and here there are few cavernicoles. Of the 230 species of cavernicolous oniscoids, only 22 have been found in the tropics (that is to say 1/10
of the total). Moreover, these tropical species are often only slightly modified troglophiles. Only the five following species are completely depigmented
and anophthalmic and can be regarded as true troglobia:

> *Troglophiloscia silvestrii* Brian – Cuba
> *Setaphora caeca* (Budde-Lund) – Burma
> *Setaphora kempi* (Collinge) – Assam
> *Trogleubelum tenebrarum* (Van Name) – the Congo
> *Neosanfilippia venezuelana* Brian – Venezuela.

*Distribution of the Cavernicolous Oniscoidea in the Systematic Classifica-
tion.* The distribution of the cavernicolous species between the 21 families
is most unequal. The cavernicolous state is related to the two following
conditions, one an ecological factor and the other a phyletic one.

(1) It is the rule that cavernicoles most often originate from hygrophilic
forms. The action of this factor may be clearly traced in the case of the
Oniscoidea. The Trichoniscidae, which are essentially hygrophilic, humicolous or endogeous forms, constitute the majority of cavernicolous oniscoids. One hundred and forty-one species out of the total of two hundred
and thirty cavernicolous oniscoids (that is to say 61 %) belong to this family.
If they are joined with the other members of the trichoniscoid complex,
that is with the members of the families Styloniscidae and Buddelundiel-
lidae, the proportion is raised to 152 out of 230 (about two thirds of the
total). It may be added that the family Ligiidae, the members of which
have still not evolved complete independence from the aquatic environment, contains eight cavernicolous species.

On the other hand, the species adapted to dry climates have only very
rarely given rise to hypogeous forms. The immense group of quinque-
tracleate porcellionids possess only four species which have been collected
in caves, and only one of these can be considered as a true cavernicole:
Nagurus cerrutii from Sardinia.

(2) While the cavernicoles represent relicts of an ancient fauna, the
recently evolved species which are well adapted to life in the world as it is
today, possess few hypogeous representatives. This is why the European
and Mediterranean genera *Porcellio* and *Armadillidium*, which contain
so many species, possess very few cavernicoles.

The family Armadillidiidae may be divided into two phyletic series, the
elumean series and the armadillidian series. The former group is characterised by a primitive cephalic structure, whilst in the latter group the cephalic
morphology is complex. But, of the thirteen cavernicolous Armadillidiidae,
only two belong to the armadillidian series *(Armadillidium pruvoti* and
A. cavernarum), and even these must be considered endogeous rather than
true cavernicoles.

K. AMPHIPODA

The amphipods together with the isopods are one of the best-represented groups of the Crustacea found in subterranean waters. W.T.Calman, in his classic work, divided the Amphipoda into four sub-orders: the Gammaridea, the Hyperiidea, the Caprellidea and the Ingolfiellidea. The second and third sub-orders contain no subterranean forms.

(a) Ingolfiellidea

This group contains the single family Ingolfiellidae, with the single genus *Ingolfiella*.[†] These amphipods are an ancient group which show some characteristics of isopods associated with an amphipod organisation. The claw of the gnathopod is constructed upon a peculiar plan, in that

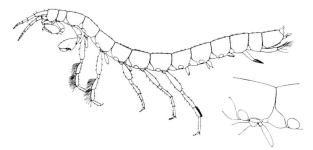

FIG. 30. *Ingolfiella leleupi* (after Ruffo).
Right, the fourth and the fifth segments, enlarged.

the dactylopodite and the propodite close down against the enlarged carpos (Fig. 30).

The Species of the Genus Ingolfiella: The family was established by H.J.Hansen (1903) to include two marine species. The first, *I. abyssi*, was dredged from a depth of 3500 metres in the Davis Straits, between Canada and Greenland. The second species, *I. littoralis*, was found at a depth of only 2 metres on the coast of Siam.

In 1933 Karaman collected in ground waters around Skoplje an unusual form of amphipod which he named *Balcanella acherontis*. Hertzog (1935) showed in 1935 that *Balcanella* was a synonym of *Ingolfiella* and stated that Karaman's species should be renamed *Ingolfiella acherontis*. In 1957, Karaman found another species of *Ingolfiella* in ground waters of Vardar, which form he named *I. petrovskii*. In 1959, he found a third species, *I. macedonica*. More recently, N. Coineau (1963) collected *I. catalanensis* in ground waters in Catalonia.

[†] Karaman (1959) wished to establish a genus for each of the species he described, but this position is manifestly an exaggerated one.

Besides these species, Ruffo (1950) described a new species, *I. leleupi*, discovered by Leleup and Dartevelle in Kakontwee Cave, near Jadotville in Katanga. This species was found again in Mwana Kussu Cave near Kasongo in Maniema in the Congo (Leleup, 1955), and later, in underground waters in Northern Rhodesia (Ingle, 1961). This species is huge compared with other members of its genus, measuring 12·5–14·5 mm, whilst the size of other species is between 1·5 and 2·5 mm. According to Leleup this species lives in warm water (26–27 °C). It is guanobious and occasionally carnivorous.

Finally we may mention the last few species to be discovered: *I. ruffoi*, which has been found in the marine littoral interstitial environment along the coast of Peru (Siewing, 1958); *I. chilensis* in the sea-shore of Chile (Noodt, 1959); and *I. britannica*, in the ground waters of the shore, at Plymouth (Spooner, 1959, 1960).

The Distribution of Ingolfiella *in Various Habitats:* This genus is found in many habitats. The diversity of the media which the different species inhabit, which at first may appear bewildering, is in fact easily accounted for. Probably the original habitat of *Ingolfiella* was the marine littoral interstitial one. *I. littoralis*, *I. ruffoi*, *I. chilensis* and *I. britannica* have remained in the primitive environment. In cold waters the genus is represented by only a single relict in the abyssal zone *(I. abyssi)*. Certain evolutionary lines have departed from the marine environment and have colonised the continental freshwater interstitial medium, for example *I. acherontis*, *I. petrovskii I. macedonia* and *I. catalanensis*. Finally, *I. leleupi* has colonized caves. These last four species may be regarded as hypogeous relicts.

(b) Gammaridae

The great majority of amphipod species belong to this sub-order. There are numerous families, but only three of them possess representatives leading hypogeous lives. They are the Talitridae, the Calliopidae and the Gammaridae.

The Hypogeous Talitrids. Two species of the genus *Hyale*, *H. jeanneli* and *H. incerta* have been found in springs in Zanzibar. They are depigmented and anophthalmic (Chevreux, 1913). The first species is also found in lakes, and thus cannot be a true cavernicole.

Hyalella anophthalma is depigmented and anophthalmic; it was collected in the Cueva de Rio Gueque, Venezuela (Ruffo, 1957).

The hypogeous calliopids. Paraleptamphopus subterraneus, which is blind and depigmented, has been collected in subterranean waters in New Zealand (Chilton, 1894, 1909, 1912). This species has very probably been derived from the epigeous species *P. coeruleus*, which inhabits the waters of mountainous areas in the same region. Since *P. subterraneus* is found not

only in ground water, but also in surface streams and lakes, it may therefore be considered to be merely a facultative hypogean.

The Hypogeous Gammarids

The Gammaridae contain the great majority of subterranean Amphipoda. This group has been studied by many workers, some of whom have
become eminent in this field: U. d'Ancona, J. Balazuc, J. A. Birstein,
Ed. Chevreux, H. Gauthier, L. Hertzog, St. Karaman, S. Ruffo, K. Schäferna, A. Schellenberg, C. R. Shoemaker, B. Sket, C. Spence Bate, Fr. Vejdovsky, A. Wrzesniowski. For a bibliography of older publications the
reader may wish to refer to a once most complete review by Alois Humbert (1876).

The Distribution of Hypogeous Gammaridae in the Systematic Classification.
The hypogeous gammarids are indeed a heterogeneous assembly. They are
composed of many quite distinct phyletic lines. Using the excellent classification proposed by Ruffo (1956), the hypogeous amphipods may be
divided into nine groups, although here it is perhaps worthwhile to add
a tenth group in which to place the Caucasian genera *Lyurella*, *Zenke-
vitschia* and *Anopogammarus*, which seem to be very isolated.

(1) Group Bogidiella *(Bogidiellidae)*. The first group contains only the
genus *Bogidiella* (= *Jugocrangonyx*). This genus is widespread in Europe
(Spain, France, Italy and Jugoslavia), in Israel, North Africa, Mexico and
Brazil.

(2) Group Hadzia *(Hadziidae): Hadzia, Metaniphargus, Quadrivisio.*
According to Schellenberg (1937) the genus *Hadzia* is related to that of
Quadrivisio, which is widespread in tropical seas. Two species of the genus
Quadrivisio have reduced eyes; one of them has been found in a cave in
East Africa. *Hadzia* contains three species which are native to Yugoslavia
and the extreme south-east of Italy, near Otranto.

One must associate *Hadzia* with the genus *Weckelia*, which has been
created by Shoemaker (1942) to include "*Gammarus caecus*", an anophthalmic species discovered by Eigenmann in the Grotte Modesta, in Cuba
(Weckel, 1907) and also another anophthalmic species, *Paraweckelia silvai*,
from the caves of Caguanes, Cuba (Shoemaker, 1959).

(3) Group Niphargus. This group contains the genera *Eriopisa*, *Eriopisella*, *Niphargus*, *Niphargopsis*, and *Niphargellus*.

Niphargus is one of the most important genera of hypogeous gammarids.
It contains about sixty species and almost as many sub-species. *Niphargus*
species vary in size between 5 mm in *N. foreli thienemanni* and 33 mm in
N. orcinus virei.

The observably marked differences in size between the different species
of this genus gives some support to the ideas of Brehm (1955). The hypothesis of this author is that the evolution of the species of *Niphargus* has

been the result of a process of partial neotony, a phenomenon to which Garstang's term paedomorphosis may be more exactly applied. This is to say that certain structures have been maintained in the adults of some species, which in other species are juvenile and thus transient ones.

The genus *Niphargus* is indigenous to Central Europe but is absent from the North and extreme South (Spain, Sicily and the Peloponnesian Peninsula) of that continent (Ruffo, 1956).

Niphargopsis has only a single, Alpine, species *N. casparyi*. *Niphargellus* is represented by three species found in Germany and in the South of England.

(4) Group Crangonyx. This group which was recognised by Schellenberg (1936) contains numerous genera. The genus *Crangonyx* contains about ten species, of which some have eyes, some are microphthalmic and some anophthalmic. They are of variable size (1·7–18 mm). The geographical distribution of this genus is a wide one, covering Eurasia, South Africa and North America.

Metacrangonyx has four species, three of which have been found in Morocco and one in the Balearics. *Microniphargus* is related to the marine genus *Eriopisella* (Schellenberg, 1934) and its species inhabit caves in Belgium and the Rhineland. *Pseudoniphargus* is represented by a single species, *P. africanus*, which has been collected in North Africa, the Iberian Peninsula, the South of France, Corsica, Dalmatia and Madeira.

Synurella is a genus which is joined to *Crangonyx*. The former genus contains several species which are epigeous and possess eyes. Their geographical distribution includes Central Europe, Siberia and Alaska. There is a tendency in these species for ocular reduction to occur. According to Stankovič (1960) they have been the origin of many cavernicolous forms. The best-known species is *S. ambulans* which is widespread in Central Europe, including North Italy, Germany, Moravia, Poland and the Balkans (Ruffo, 1950).

One must include in this fourth group the following genera which are native to North America. These are *Eucrangonyx*, *Allocrangonyx*, *Stygobromus*, *Apocrangonyx*, *Stygonectes* and *Sympleonia*. *Stygobromus pusillus* has been collected in the Lake Teletsk (Altay), a fact which establishes the ancient origin of this genus. To the above genera we must link the following, *Eocrangonyx* (Asia), *Pseudocrangonyx* (Japan), *Protocrangonyx* (Australia), *Paracrangonyx* (New Zealand), and also probably *Austroniphargus bryophilus*, an anaphthalmic species found in humid mosses at an altitude of 2600 metres in Madagascar. However, the latter species is probably a phreatobian.

(5) Group Salentinella. This group contains only the very isolated genus *Salentinella*. One of the species, *S. gracillima* inhabits brackish water in Dalmatia and Puglia in Southern Italy. The five other species are found in fresh water. They are *S. angelieri* (Corsica, Italy, and Peloponnesia), *S. fran-*

ciscoloi (Alpes Maritimes), *S. gineti* (Pyrénées ariègeoises), *S. delamaeri* (Catalonia) and a species which has not yet been fully described, from the Balearics.

(6) Group Neoniphargus. Group six contains a single genus, *Neoniphargus*, which is made up of about thirty epigeous species which inhabit freshwater lakes, springs and rivers in Australia and Tasmania. In addition, there are two hypogeous species. These are *N. indicus* which has been found in a coal mine in Bengal (Chilton, 1923; Stephensen, 1931), and *N. kojimai* which was discovered in the sand filters of the waterworks at Tokyo (Uéno, 1955).

(7) Group Phreatogammarus. The genus *Phreatogammarus* is localised in New Zealand (Chilton, 1918).

(8) Group Typhlogammarus. *Typhlogammarus* is an isolated genus, and represents an ancient relict. There is a single species *T. mrazeki* which is found in subterranean cavities in Herzegovina and Montenegro. It is depigmented and anophthalmic.

(9) Group Gammarus. It is now known that *Gammarus*, which was once regarded by early biospeologists as the ancestor of *Niphargus* belongs to a totally different group. However, it has given rise to several hypogeous forms. A race of *Gammarus pulex*, called *G. pulex subterraneus*, has frequently been observed in the Harz mines in Germany (Schneider, 1885; Mühlmann, 1938; Anders, 1956). This race is completely depigmented but the eyes are quite normal and they are rather longer than the surface form. If they are cultured in light they become pigmented olive green, brown or red. According to Anders, the three colours of pigment, would be under the control of three allelomorphic genes.

Gammarus (Neogammarus) rhipidiophorus is another species of interest to biospeologists. It is found in brackish water in the interstitial environment. It is microphthalmic and may be taken to be a type intermediate between epigeous and hypogeous forms. This species has been collected in France, Italy, and Tunisia.

Rivulogammarus komareki imeretinus and *R. k. caucasicus* have been collected from caves in Transcaucasia; the first sub-species has reduced eyes (Birstein, 1933).

Finally troglobious species of *Gammarus* are known which are depigmented and anophthalmic. *Gammarus (Metohia) carinata*, which was discovered by Absolon in subterranean water in the Gacko region of Herzegovina, has been described by Karaman (1953). It completely lacks pigment and eyes.

The Marine Origin of the Hypogeous Gammaridae. There can be no doubt that the hypogeous gammarids are derived from marine forms (with the exception of a few *Gammarus* species). However, we may go further and state that all hypogeous amphipods have been directly evolved from marine types. This took place via the interstitial medium. Species inhabiting this medium are characteristically euryhaline and this property allows them to migrate from marine into brackish waters and finally into fresh subterranean water (Ruffo, 1956). The euryhalinity of *Niphargus* has been established by precise physiological experiments (p. 345).

We will now trace the successive evolutionary stages of this transference. Here are some examples:

(1) Group Bogidiella. Most of the species of *Bogidiella* live in fresh water; however, some do inhabit brackish water *(B. albertimagni dalmatica)*, and a few occupy the marine interstitial medium. Thus there can be little doubt of the marine origin of this genus. It is, in fact, a relict of a marine population which once occupied the ancient Sea of Tethys (Ruffo).

(2) Group Hadzia. This genus has probably been derived from the marine genus *Quadrivisio*, as has been mentioned previously. *Hadzia fragilis* which lives in freshwater in the Vjeternica cave (Popovo-Polje) has also been found in slightly brackish water at Dubrovnik and at Cavtat (Karaman, 1953).

(3) Group Niphargus. The marine origin of *Niphargus* has been recognised for a long time (Chevreux, 1920; Schellenberg, 1933). As we have noted, *Niphargus* resembles *Eriopisa*, which is represented by one marine species *E. elongata*. This species is remarkable for the fact that it has no eyes, and is found in coastal regions. These two genera are so closely related that Chilton (1920), an experienced carcinologist. described an anophthalmic phreatobian under the name *Niphargus phillipensis*, and also a microphthalmic species from brackish waters in Bengal and Malaya, under the name *N. chilkensis* (Chilton, 1921), when these species in fact belong to *Eriopisa* (Schellenberg, 1931, 1933; Gauthier, 1936).

(4) Group Crangonyx. The marine origin of *Pseudoniphargus africanus* can hardly be doubted, as this species lives with apparently equal facility in marine, brackish and fresh waters (Balazuc and Angelier, 1952). Migration from the sea to fresh water has taken place via the interstitial medium.

(5) Group Salentinella. *Salentinella gracillima* lives in brackish water while the other species inhabit fresh water.

(6) Group Gammarus. *Gammarus rhipidiophorus* is a brackish-water form, thus filling the intermediate evolutionary position between the sea and freshwater species.

Regressive Evolution in the Hypogeous Gammaridae. Depigmentation and the reduction of the eyes, and finally their disappearance, are typical manifestations of regressive and orthogenetic evolution. The hypogeous amphipods provide us with good examples, of which two will be given.

(1) There are about ten species of *Crangonyx*, some of which are epigeous and have normal eyes. An example is *C. gracilis*, which is widespread in North America. Other species are microphthalmic, for instance, *C. packardi* which also comes from North America. Finally there are a number of

anophthalmic species, for example, *C. subterraneus* and *C. paxi* from Europe, *C. chlebnikovi* and *C. arsenjevi* from Asia, *C. robertsi* from South Africa and *C. mucronatus* from North America.

(2) The genus *Synurella* contains many species and sub-species. The following species which have been studied by Karaman (1931) provide an excellent example of regressive and orthogenetic evolution. *S. ambulans* is an epigeous form which inhabits cold waters in Central Europe. From this generalised type several more specialised and localised species have been evolved. For example, *S. jugoslavica kolombatovici* is a sub-species which lives in springs and has well developed eyes, while *S. jugoslavica jugoslavica* inhabits pools fed by springs, but is depigmented and the eyes are reduced. This series may be ended with two cavernicolous forms, *S. jugoslavica subterranea* which may be anophthalmic or with the eyes reduced to spots of pigment, and *S. tenebrarum* in which the eyes have totally disappeared.

The Habitats of the Hypogeous Gammaridae. The hypogeous gammarids occupy a variety of habitats and they may pass freely from one of them to another. For example, *Niphargus foreli* is found in (1) the littoral zone of alpine lakes, (2) the profundal zone of sub-alpine lakes, (3) in

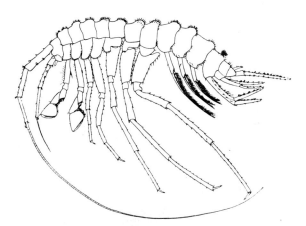

FIG. 31. *Niphargus balcanicus* (after Karaman)

springs, (4) in caves. The forms of the species which inhabit these various media differ very little from each other and at most they could be regarded as sub-species (Schellenberg, 1934). Later (p. 281) it will be shown that *Niphargus* species, and in particular *N. aquilex*, inhabits surface waters—forest ponds, marshes, etc. at certain times of the year. Usually, however, each species has a preference for a particular biotope. Karaman (1932) distinguishes six types of habitat:

(1) *Niphargus* is abundant in the hypotelminorheic medium (Meštrov, 1962). This habitat may well account for the burrowing habits of this genus.

(2) The interstitial medium constituted by ground water. This biotope is occupied by many species including *Niphargus fontanis, N. nolli, N. skopljensis, Microniphargus leruthi, Bogidiella albertimagni,* and *Gammarus rhipidiophorus.*

(3) Subterranean streams *(Niphargus pancini)*.

(4) Underground lakes and rivers. These are populated by the truly cavernicolous amphipods. The species which inhabit this biotope are generally of comparatively large size. *Niphargus balcanicus* (Fig. 31) which according to Absolon attains a length of 50 mm, the species of *Niphargus* belonging to the *orcinus* group *(Orniphargus), Typhlogammarus mrazeki* and *Synurella jugoslavica,* may be included here.

(5) Springs are regularly inhabited by certain hypogeous species which are not, however, rigorously localised in such an environment: for instance *Niphargus illidzensis* and *N. tatrensis,* etc. *Bogidiella albertimagni glacialis* have been found in mountain springs at a height of 1900 metres in Jugoslavia.

(6) In the profundal zones of the great sub-Alpine lakes. *Niphargus foreli* and *N. foreli ohridanus* may be found here.

L. DECAPODA

The decapods may be classified according to a system first proposed by Boas (1880) and then revised by W. T. Calman in his *Treatise upon Crustacea* which appeared in 1909.

Sub-order Natantia
Sub-order Reptantia

Section 1. Palinura
Section 2. Astacura
Section 3. Anomura
Section 4. Brachyura.

Only the section Palinura contains no cavernicolous species.

(1) Natantia

A most complete review of the cavernicolous Natantia has been written by L. B. Holthuis (1956). One can do no better than produce a summary of it and add the new data which have been acquired since 1956.

The hypogeous Natantia can be divided into three families, the Atyidae, the Palaemonidae and the Hippolytidae.

(a) Atyidae. The most primitive forms of this family *(Xiphocaris)* are related to the marine family Acanthephyridae. However, all the species belonging to the Atyidae are found in fresh water. This adaptation to fresh water is certainly ancient because the remains of *Caridina* have been collected in France in Oligocene beds corresponding to freshwater deposits.

Further proof of their antiquity is provided by their geographical distribution. This family is represented in nearly all regions of the globe with the exception of the Arctic zones (Bouvier, 1925).

It is therefore not surprising that a number of the Atyidae are found in subterranean waters. About twenty hypogeous species belonging to this family are known today.

In Europe nearly all the cavernicolous forms belong to the genus *Troglocaris*, which, with the genus *Atyaephyra* (an epigeous, freshwater genus) belongs to the paratyian series. The most ancient species known is *Troglocaris anophthalmus* (= *T. schmidti*) which is widespread in the area between Venetia Juliana and Herzegovina. Three other species of this genus are *T. inermis* (Cévennes méridionales, in the south of France), *T. hercegovinensis* (Herzegovina) and *T. kutaissiana* (western Transcaucasia). Matjašič (1956) has created the genus *Spelaeocaris* for the species *S. pretneri* which has been found in caves in Herzegovina. The larval development of *Troglocaris anophthalmus* has been recorded by Stammer (1933, 1935) and Matjašič (1958).

The North American hypogeous species also belong to the same paratyian series: *Palaemonias ganteri*, from Mammoth Cave, *P. alabamae*, from Shelta Cave, Alabama, and species of the genus *Typhlatya*: *T. pearsi* (Yucatan), *T. garciai* (Cuba), and *T. nana* (Porto Rico).

An atyid from caves on the island of Shikoku (Japan) belongs to the caridian series. It is *Caridina japonica sikokuensis*, a troglobious sub-species closely related to the epigeous type (Torii, 1953; Uéno, 1957).

In Africa there are two known hypogeous atyids. The first belongs to the caridian series: *Caridina lovoensis* (from a cave near Thysville, in the lower Congo); the second to the caridellian series: *Caridinopsis brevinaris* (Guinea). The troglobious species from Madagascar should probably be placed in the caridellian series. These include *Typhopatsa pauliani, Parisia microphthalma, P. edentata* and *P. macrophthalma* (Holthuis, 1956).

Lastly, two cavernicolous Atyidae have been collected from wells in Western Australia (North West Cape). They are *Stygiocaris lancifera* and *S. stylifera* (Holthuis, 1960; Mees, 1962). Another species, *Antecaridina lauensis*, has been found in a cave on Fiji.

In all these cavernicolous atyids depigmentation has occurred, in most species the eyes are reduced, with exceptions of *Parisia macrophthalma* (eyes normal) and *Antecaridina lauensis* (eyes black-pigmented but facets regressed).

(b) Palaemonidae. This family is composed mainly of marine forms, though a few species inhabit brackish or fresh waters *(Palaemonetes).*

Three cavernicolous species found in North America belong to the genus *Palaemonetes* or to a closely related genus. They are *P. cummingi* (Florida) and *P. antrorum* (artesian well in San Marcos, Texas) and *Creaseria morleyi* (cenotes and caves in Yucatan).

Four species which inhabit the caves in Cuba have been grouped into the same genus by Holthuis: *Troglocubanus calcis, T. eigenmanni, T. gibarensis* and *T. inermis.*

Macrobrachium cavernicola is indigenous to the caves of Assam, in particular the Siju Caves, and in a cave near Tcherrapoundji (Khasi Mountains) (Lindberg, 1960).

The Mediterranean region possesses three cavernicolous species belonging to this family. They belong to the genus *Typhlocaris. T. galilea* has been found in a pool fed by a spring near the Sea of Galilee; *T. lethaea* comes from the Grotte del Lete, near Benghazi, in Libya and *T. salentina* inhabits various caves in south-eastern Italy (Zinzulusa, l'Abisso, Grotta dei Diavoli).

All of these species are depigmented and have reduced eyes. However, *Macrobrachium cavernicola* has red chromatophores in its integument; the eyes, although having their facets reduced, still retain their sensory cells and black pigmentation.

(c) Hippolytidae. This family contains only one hypogeous species, *Barbouria cubensis,* and it is not yet certain that it is truly cavernicolous. The species lives in Cuba in a cave filled with brackish water near the sea. This crustacean is purple in colour and the eyes are well developed and pigmented.

(2) Reptantia

(a) Astacura. The only subterranean family is the Astacidae (= Potamobiidae). They are of Asian origin. They gradually spread into Europe where they appeared in the Neocomian (Lower Cretaceous). They have also found their way into America and have been collected in Mexico from early Tertiary deposits. They occupied the United States in the Oligocene. The first hypogeous forms made their appearance in the Pliocene (Rhoades, 1962).

Crayfish have frequently been captured in subterranean rivers in Europe, but they all belong to the common species *Potamobius astacus,* and do not differ from the surface form. By contrast almost a score of species of cavernicolous crayfish are known from North America. It is a peculiar fact, the reason for which eludes us, that while the evolution of cavernicolous crayfish and cavernicolous fishes has taken place in North America, no comparable evolution of these forms has occurred in Europe.

Hobbs (1942, 1952), Hobbs and Barr (1960) and Rhoades (1962) have revised the cavernicolous crayfish of North America and recognise eighteen species. They are divided in the genera *Procambarus* (6 spp.); *Troglocambarus* (1 spp.); *Orconectes* (5 spp.) and *Cambarus* (6 spp.).

The evolution of the North American species and sub-species of *Cambarus* is of some interest particularly on account of the gradual series of stages of transition between the surface and cavernicolous species.

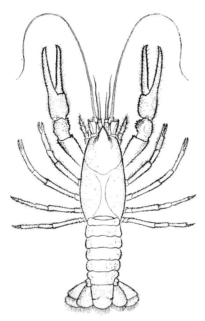

Fig. 32. *Cambarus setosus* (after Garman).

Such an evolution has taken place on several occasions in different phyletic lines, since the present-day assembly of hypogeous forms is a polyphyletic complex. Certain hypogeous sub-species, for example *Cambarus bartoni tenebrosus* (Mammoth Cave) and *C. bartoni laevis* (caves in Indiana), are closely related to the epigeous *C. bartoni bartoni*. These hypogeous sub-species differ only in their paler coloration, a slight reduction in the size of the eye, and a certain lengthening of the body (Garman, 1921; Eberley, 1960). *C. cahni* has a depigmented integument but the eyes still retain their pigmentation. The other cavernicolous species of *Cambarus* are totally depigmented and their eyes are more or less reduced (Fig. 32).

(b) Anomura. The Anomura are represented underground by the Galatheidae. They have been captured in a subterranean lake called the Jameo de Agua† (Lanzarote, eastern Canary Isles). This lake communicates with

† Often confused in error with the Cueva de las Verdes; cf. Fage and Monod (1936).

the sea, as proved by its corresponding rise and fall with the tides, and is of salt water. This cave contains two remarkable cavernicolous crustaceans, a mysid, *Heteromysis cotti* (p. 118), and an aonmuran, *Munidopsis polymorpha*. These are almost unique examples of purely marine cavernicoles.

Munidopsis polymorpha (Fig. 33) is pale pinky-yellow in colour and its eyes are regressed and depigmented. Fuchs (1894) and Calman (1904) have remarked that almost all other species of *Munidopsis* are abyssal forms. They concluded that the marine cavernicoles have been derived from abyssal forms. This interpretation is certainly erroneous. Both the cavernicolous

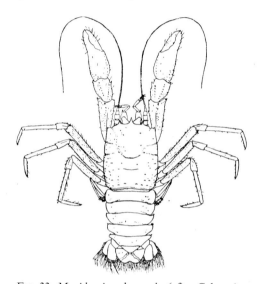

FIG. 33. *Munidopsis polymorpha* (after Calman).

and the abyssal forms are relicts of an ancient fauna. Both have been derived from littoral forms. They have followed separate lines of evolution, although their morphology may be analogous.

(c) Brachyura. Up until quite recently all crabs which had been found in caves were trogloxenes or at most troglophiles. In this ecological category we must include the following species: *Paratelphusa convexa* and *Sesarma jacobsoni* from caves in Java (Ihle, 1912); *Paratelphusa falcidigitis*, from the Siju Caves in Assam; and *Epilobocera cubensis* from some caves in Cuba. *Geotelphusa dehaani* from some caves in Japan may also be included here, although those individuals which live underground have lost their ocular pigment (Torii, 1953; Uéno, 1957).

Recently, however, a truly cavernicolous crab has been discovered (Rioja, 1953). This species, *Typhlopseudotelphusa mocinoi*, which is totally depigmented and anophthalmic has been discovered in the Cueva del Tio Ticho, at Comitàn, Mexico.

BIBLIOGRAPHY
Branchiopoda

RAMMES, W. (1933) Cladoceren der Adelsberger Grotten. *Mitteil. Höhl. Karstf.*

Gymnoplea

BREHM, V. (1953) Cladocères et Copépodes Calanoïdes de Madagascar. *Naturaliste Malgache.* **V.**

PAULIAN, R. and GRJEBINE, A. (1953) Une campagne spéléologique dans la réserve naturelle de Namoroka. *Naturaliste Malgache.* **V.**

TAFALL, B.F.O. (1942) *Diaptomus (Microdiaptomus) cokeri*, nuevo subgenero y especie de Diaptomido de las cuevas de la region de Valles (San Luis Potosi, Mexico). *Ciencia* **III.**

Cyclopoidea

CHAPPUIS, P.A. (1920) Die Fauna der unterirdischen Gewässer der Umgebung von Basel. *Archiv. f. Hydrobiol.* **XIV.**

CHAPPUIS, P.A. (1928) Neue Harpacticiden aus Java. *Treubia*, **X.**

GRAETER, E. (1910) Die Copepoden der unterirdischen Gewässer. *Archiv. f. Hydrobiol.* **VI.**

ITO, T. (1952) Four new copepods from subterranean waters of Japan. *Rep. Fac. Fish. Prefect Univ. Mie.* **I.**

ITO, T. (1953) Subterranean-water copepods from Japan. I–III. *Suido Kyokai Zasshi*, Tokyo. **CCXXI, CCXXIV, CCXXVIII.**

ITO, T. (1954) Cyclopoid copepods of Japanese subterranean waters. *Rep. Fac. Fish. Prefect Univ. Mie.* **I.**

ITO, T. (1956) On the copepod fauna of subterranean waters of Japan. *Bull. Biogeogr. Soc. Japan.* **XVI–XIX.**

ITO, T. (1957) Groundwater copepods from south-western Japan. *Hydrobiologia* **XI.**

KIEFER, F. (1937) Über Systematik und geographische Verbreitung einiger Gruppen stark verkümmerter Cyclopiden (Crustacea, Copepoda). *Zool. Jahrb. Abt. System.* **LXX.**

KIEFER, FR. (1957) *Graeteriella unisetiger* (E.Graeter), ein für Italien neuer Cyclopide (Crust. Cop.) aus dem Grundwasser der Etsch. *Mem. Mus. Civ. Stor. Nat. Verona.* **VI.**

LINDBERG, K. (1954) Un Cyclopide (Crust. Copépode) troglobie de Madagascar, avec remarques sur un groupe de Cyclopides trés évolués, cavernicoles et muscicoles. *Hydrobiologia* **VI.**

LINDBERG, K. (1956) Les Cyclopides (Crust. Copépodes) trés évolués en tant qu'habitants des eaux souterraines. Revue des travaux récents concernant les *Bryocyclops* Kiefer et *Specocyclops* Kiefer. *Commun. Premier Congrès Intern. Spéléol. Paris* **III.**

MENZEL, R. (1926) Zum Vorkommen der Harpacticidengattung *Viguierella* Maupas im malayischen Archipel. *Zool. Anz.* **LXV.**

Harpacticoidea

CHAPPUIS, P.A. (1933) Copépodes (1ère Série), avec l'énumération de tous les Copépodes cavernicoles connus en 1931. *Biospeologica*, **LIX.** *Archiv. Zool. expér. géner.* **LXXVI.**

CHAPPUIS, P.A. (1937) Über die Systematik und geographische Verbreitung einiger Harpacticoiden-Gattungen (Crust. Copepoda). *Intern. Rev. gesamt. Hydrobiol. Hydrogr.* **XXXV.**

CHAPPUIS, P.A. (1944) Die Harpacticoiden Copepoden der europäischen Binnengewässer. *Archiv f. Naturg.* **XII.**

CHAPPUIS, P.A. (1958a) Le genre *Parastenocaris. Vie et Milieu* **VIII.**

CHAPPUIS, P. A. (1958 b) Sur la validité du genre *Elaphoidella* Chappuis. *Notes biospéol.* **XIII.**

CHAPPUIS, P. A. (1959) Biogéographie du genre *Parastenocaris. Vie et Milieu* **IX.**

LANG, K. (1948) *Monographie der Harpacticiden.* Häkan Ohlssons Boktryckeri. 2 Vol. Lund.

MIURA, Y. (1962) Subterranean harpacticoid copepods of the Amami Group of the Ryukyu Islands. *Annot. Zool. Japon.* **XXXV.**

NOODJ, J. (1962) Limnisch-subterrane Copepoden der Gattung *Parastenocaris* Kessler aus Mittelamerika. *Beitr. neotr. Fauna. II.*

ROY, J. (1932) Copépodes et Cladocères de l'Ouest de la France. Thèse, Paris. Gap.

Ostracoda

KLIE, W. (1931) Zwei neue Arten der Ostracoden-Gattung *Candona* aus unterirdischen Gewässern im südöstlichen Europa. *Zool. Anz.* **XCVI.**

KLIE, W. (1935) Drei neue Höhlenostracoden aus der Umgebung von Laibach. *Zool. Anz.* **CXI.**

KLIE, W. (1938) Ostracoden aus dem Grundwasser der oberrheinischen Tiefebene. *Archiv. f. Naturg.* N. F. **VII.**

ROME, R. (1953) Ostracodes cavernicoles de la grotte de Tsebahu, Mont Hyo (Irumu, Congo Belge). *Rev. Zool. Bot. Afric.* **XLVII.**

VEJDOVSKY, F. (1882) *Tierische Organismen der Brunnengewässer von Prag.* Prag.

Syncarida

BARTOK, P. A. (1944) A *Bathynella chappuisi* fejlödés – Morphologiàja. – *Acta Sc. Math. Nat.* Koloszvar **XXI.**

BOTOSANEANU, L. (1959) Fauna Republicii Populare Romine. *Crustacea*, **IV,** 5, Bathynellacea.

CALMAN, W. T. (1899) On the characters of Crustacean Genus *Bathynella* Vejd. *J. Linn, Soc. Zool.* **XXVII.**

CHAPPUIS, P. A. (1948) Le développement larvaire de *Bathynella. Bull. Soc. Sc. Cluj* **X.**

DELAMARE-DEBOUTTEVILLE, CL. (1960) Présence d'un Syncaride d'un genre nouveau dans les eaux interstitielles de la Patagonie andine, et remarques biogéographiques. *Compt. rend. Acad. Sci. Paris* **CCLI.**

HERTZOG, L. (1930) Notes sur quelques Crustacés nouveaux pour la plaine d'Alsace (Bas-Rhin). *Bull. Assoc. phil. Alsace-Lorraine* **VII.**

HUSMANN, S. (1956) Untersuchungen über die Grundwasserfauna zwischen Harz und Weser. *Archiv f. Hydrobiol.* **LII.**

JAKOBI, H (1954) Biologie, Entwicklungsgeschichte und Systematik von *Bathynella natans* Vejd. *Zool. Jahr. Abt. System.* **LXXXIII.**

MIURA, Y. and MORIMOTO, Y. (1953) Larval development of *Bathynella morimotoi* Uéno. *Annot. Zool. Japon.* **XXVI.**

NICHOLLS, A. G. (1946) Syncarida in relation to the interstitial habitat. *Nature, Lond.* **CLVIII.**

SIEWING, R. (1958) Syncarida. *Bronns Kl. Ordn. Tierr.* **V,** Abt. 1; 4. Buch, Teil II.

SPOONER, G. M. (1961) *Bathynella* and other interstitial Crustacea in Southern England. *Nature, Lond.* **CXC.**

VEJDOVSKY, F. (1882) *Tierische Organismen der Brunnengewässer von Prag.* Prag.

Thermosbaenacea

Thermosbaena

BARKER, D. (1956) The morphology, reproduction and behaviour of *Thermosbaena mirabilis* Monod. *Proc. XIV. Intern. Zool. Congr.* Copenhagen.

BARKER, D. (1959) The distribution and systematic position of the *Thermosbaenacea*. *Hydrobiologia* **XIII.**

BARKER, D. (1962) A study of *Thermosbaena mirabilis* (Malacostraca, Peracarida) and its reproduction. *Quart. J. Microsc. Sc.* **CIII.**

BRUNN, A. F. (1940) Observations on *Thermosbaena mirabilis* Monod from the hot springs of Hamma, Tunisia. *Vidensk. Medd. Dansk. Naturh. Foren.* **CIII.**

MONOD, TH. (1927) *Thermosbaena mirabilis* Monod. Remarques sur sa morphologie et sa position systématique. *Faune des Colonies françaises* **I.**

MONOD, TH. (1940) Thermosbaenacea. *Bronns Kl. Ordn. Tierr.* **V,** Abt. 4, 4. Buch, Bd. I.

SIEWING, R. (1958) Anatomie und Histologie von *Thermosbaena mirabilis.* Ein Beitrag zur Phylogenie der Reihe Pancarida (Thermosbaenacea). *Akad. Wiss. Lit. Mainz. Abhandl. math. -nat. Kl.* No. 7.

Monodella

DOV POR, FR. (1962) Un nouveau Thermosbénacé, *Monodella relicta*, n. sp. dans la dépression de la Mer Morte. *Crustaceana* **III.**

KARAMAN, ST. (1953) Über einen Vertreter der Ordnung Thermosbaenacea (Crust. Peracarida) aus Yugoslavien, *Monodella halophila* n. sp. *Acta Adriatica* **V.**

RUFFO, S. (1949 a) *Monodella stygicola* n. g. n. sp., nuevo Crostaceo Thermosbenaceo delle acque sotteranee della Peninsola Salentina. *Archiv. Zool. Ital.* **XXXIV.**

RUFFO, S. (1949 b) Sur *Monodella stygicola* Ruffo des eaux souterraines de l'Italie méridionale, deuxième espèce connue de l'ordre des Thermosbenacés. *Hydrobiologia* **II.**

STELLA, E. (1951) *Monodella argentarii* n. sp. di Thermosbaenacea (Crustacea Peracarida), limnotroglobio di Monte Argentario. *Archiv. Zool. Ital.* **XXXVI.**

STELLA, E. (1953) Sur *Monodella argentarii* Stella, espèce de Crustacé Thermosbenacé des eaux d'une grotte de l'Italie centrale (Monte Argentario, Toscana). *Hydrobiologia* **V.**

STELLA, E. (1955) Behaviour and development of *Monodella argentarii* Stella, a thermosbaenacean from an Italian cave. *Verhandl. Intern. Ver. Limnol.* **XII.**

STELLA, E. (1959) Ulteriori osservazioni sulla riproduzione e lo sviluppo di *Monodella argentarii* (Pancarida, Thermosbaenacea). *Riv. Biol. Ital.* **LI.**

Spelaeogriphacea

GORDON, I. (1957) On *Spelaeogriphus*, a new cavernicolous crustacean from South Africa. *Bull. Brit. Mus. Zool.* **V,** No. 2.

GORDON, I. (1960) On *Stygiomysis* from the West Indies, with a note on *Spelaeogriphus* (Crustacea, Peracarida). *Bull. Brit. Mus. Zool.* **VI,** No. 5.

Mysidacea

BOLIVAR Y PIELTAIN, C. (1944) Exploracion biologica de algunas cavernas de Cuba. *Ciencia* **IV.**

CALMAN, W. T. (1932) A cave dwelling crustacean of the family Mysidacae from the Island of Lanzarote. *Ann. Mag. Nat. Hist.* (10), **X.**

CAROLI, E. (1924) Su di un Misidaceo cavernicolo (*Spelaeomysis bottazzii* n. g. n. sp.) di terra d'Otrante. *Atti Accad. Naz. Lincei. Cl. Sc. Fis. Mat. Nat.* **XXXIII.**

CAROLI, E. (1937) *Stygiomysis hydruntina* n. g. n. sp. Misidaceo cavernicolo di Terra d'Otrante, rappresentante di una nuova famiglia. Nota preliminare. *Boll. Zool. Torino.* **VIII.**

CREASER, E. P. (1936) Crustaceans from Yucatan. *In* The cenotes of Yucatan by A. S. Pearse, E. P. Creaser and F. G. Hall. *Carnegie Inst. Washington, Public.* No. 457.

CREASER, E. P. (1938) Larger cave Crustacea of the Yucatan Peninsula. *In* Fauna of the caves of Yucatan. A. S. Pearse, *Carnegie Inst. Washington, Public.* No. 491.

FAGE, L. (1925) *Lepidophthalmus servatus* Fage. Type nouveau de Mysidacé des eaux souterraines de Zanzibar. *Biospeologica,* **LI.** *Archiv. Zool. expér. géner.* **LXIII.**

GORDON, I. (1960) On a *Stygiomysis* from the West Indies, with a note on *Spelaeogriphus* (Crustacea, Peracarida). *Bull. Brit. Mus. Zool.* **VI.**

HANSTRÖM, B. (1948) The brain, the sense organs and the excretory organs of the head in the Crustacea Malacostraca. *Bull. biol. France. Belgique* **XXXIII.**

STAMMER, H.J. (1936) Ein neuer Höhlenschizopode, *Troglomysis vjetrenicensis* n.g.n.sp. Zugleich eine Übersicht der bisher aus dem Brack- und Süßwasser bekannten Schizopoden, ihrer geographischen Verbreitung und ihrer ökologische Einteilung, - sowie eine Zusammenstellung der blinden Schizopoden. *Zool. Jahr. Abt. System.* **LXVIII.**

VILLALOBOS, A. (1951) Un nuevo Misidaceo de las Grutas de Quintero en el Estado de Tamaulipas. *Annal. Inst. Biol. Mexico.* **XXII.**

Cymothoidea

The old bibliography (before 1912) is mentioned in the publication of Racovitza. It is not reproduced below.

BAKER, J.K. (1960) New range extensions for the cave isopod, *Cirolanides texensis* Benedict. *Nat. Speleol. Soc. News* **XVIII.**

BOLIVAR Y PIELTAIN, C. (1950) Estudio de una *Cirolana* cavernicola nueva de la region de Valles, San Luis Potosi, Mexico. *Ciencia* **X.**

BRIAN, A. (1923) Descrizione di un rarissimo Isopodo cavernicolo: *Troglaega virei* Valle. *Annal. Mus. Stor. Nat. Genova* **LI.**

BRIAN, A. (1924) Nuove osservazione sulla *Troglaega virei* Valle e notizie sulle localita di rinvenimento. *Annal. Mus. Stor. Nat. Genova* **LI.**

CREASER, E.P. (1936) Crustaceans from Yucatan. *Carnegie Inst. Washington Public.* No. 457.

CREASER, E.P. (1938) Larger cave Crustacea of Yucatan Peninsula. *In* A.S.Pearse, Fauna of the caves of Yucatan. *Carnegie Inst. Washington Public.* No. 491.

MONOD, TH. (1930) Contribution à l'étude des Cirolanides. *Annal. Sci. Nat. Zool.* (10) **XIII.**

NOURISSON, M. (1956) Étude morphologique comparative et critique des *Typhlocirolana* (Crustacés Isopodes Cirolanidae) du Maroc et d'Algérie. *Bull. Soc. Sci. Nat. Maroc.* **XXXVI.**

PAULIAN, R. and DELAMARE-DEBOUTTEVILLE, CL. (1956) Un Cirolanide cavernicole à Madagascar (Isopode). *Mém. Inst. Sci. Madagascar. Sér. A.* **XI.**

RACOVITZA, E.G. (1912) Cirolanides (Première Série). *Biospeologica* **XXVII.** *Archiv. Zool. expér. génér.* (5) **X.**

REMY, P. (1951) Stations de Crustacés obscuricoles. *Biospeologica,* **LXXII,** Appendice. *Archiv. Zool. expér. génér.* **LXXXVIII.**

RIOJA, E. (1953) Estudios carcinologicos, XXX. Observaciones sobre los Cirolanidos cavernicolos de Mexico (Crustaceos, Isopodos). *Anal. Inst. Biologia* **XXIV.**

RIOJA, E. (1957) Estudios Carcinologicos. XXXV, Datos sobre algunos Isopodos cavernicolos de la Isla de Cuba. *Anal. Inst. Biologia* **XXVII.**

Sphaeromidea

The old bibliography (before 1910) is mentioned in the publication of Racovitza. It is not reproduced below.

ARCANGELI, A. (1935) Isopodi del Museo Civico di Storia Naturale di Milano. *Atti. Soc. Ital. Sci. Nat.* **LXXIV.**

ARCANGELI, A. (1942a) *Monolistra (Typhlosphaeroma) pavani*, nuova specie di Isopodo Sferomide cavernicolo. *Boll. Mus. Zool. Anat. Comp. Torino* **XLIX.**

ARCANGELI, A. (1942b) Note sopra alcuni Sferomidi cavernicoli italiani (Crostacei Isopodi aquatici). *Boll. Mus. Zool. Anat. Comp. Torino* **XLIX.**

BRIAN, A.(1931) Determinazione di un nuovo materiale di Isopodi cavernicoli raccolti dal Rag. L. Boldori sulle Alpi. *Mem. Soc. Entomol. Ital.* **X.**

HUBAULT, E. (1934) Étude faunistique d'eaux souterraines à la lisière septentrionale du bassin d'Aquitaine. *Bull. biol. France. Belgique* **LXVIII.**

HUBAULT, E. (1937) *Monolistra hercegoviniensis* Absolon, Sphéromien cavernicole d'Herzégovine et *Sphaeromicola stammeri* Klie, son commensal. *Archiv. Zool. expér. génér.* **LXXVIII.**

KARAMAN, ST. (1954) Über die jugoslavischen Arten des Genus *Monolistra* (Isopoda). *Acta Mus. Maced. Sci. Nat.* **II.**

PARIS, P. (1925) La faune cavernicole de la Côte d'Or. *Assoc. franç. Avanc. Sci. Compt. rendu* 49ème *Sess.* Grenoble.

RACOVITZA, E. G. (1910) Sphéromiens (Première Série) et Revision des *Monolistrini* (Isopodes sphéromiens). *Biospeologica* **XIII.** *Archiv. Zool. expér. géner.* (5) **IV.**

RACOVITZA, E. G. (1929a) *Microlistra spinosa* n.g.n.sp., Isopode Sphéromien cavernicole nouveau de Slovénie. *Bull. Soc. Stiin. Cluj.* **IV.**

RACOVITZA, E. G. (1929b) *Microlistra spinosissima* n.sp., Isopode Sphéromien cavernicole nouveau de Slovénie. *Bul. Soc. Stiin. Cluj.* **IV.**

REMY, P. (1943) Nouvelle station du Sphéromien troglobie, *Caecosphaeroma (Vireia) burgundum* Dollfus var. *rupis-fucaldi* Hubault et de l'Ostracode commensal, *Sphaeromicola topsenti* Paris. *Bull. Soc. Zool. France* **LXVIII.**

REMY, P. (1948) Sur la distribution géographique des Sphaeromidae *Monolistra* et des *Spheromicola*, leurs Ostracodes commensaux. *Bull. Mus. Hist. Nat. Marseille* **VIII.**

RUFFO, S. (1938) Studio sulla fauna cavernicola della Ragione Veronese. *Boll. Ist. Entomol. R. Univ. Bologna* **X.**

SKET, B. (1960) Einige neue Formen der Malacostraca aus Jugoslavien. *Bull. Sci.* **V.**

SKET, B. (1961) Über unsere Monolistrini (Isopoda). *Drugi Jugosl. Speleol. Kongress.* Zagreb.

STAMMER, H.J. (1930) Eine neue Höhlensphäromide aus dem Karst, *Monolistra (Typhlosphaeroma) schottlaenderi* und die Verbreitung des Genus *Monolistra. Zool. Anz.* **LXXXVIII.**

STROUHAL, H. (1928) Eine neue Höhlen-Sphaeromide (Isop.). *Zool. Anz.* **LXXVII.**

Asellota

Parasellidae

The bibliography before 1960 is to be found in the publication of Chappuis and Delamare-Deboutteville. It is not reproduced below.

BIRSTEIN, J.A. (1961) *Microthambema tenuis* n.g.n.sp. (Isopoda Asellota) and the relations of some asellote isopods. *Crustaceana* **II.**

CHAPPUIS, P.A. and DELAMARE-DEBOUTTEVILLE, CL. (1960) État de nos connaissances sur une famille et une sous-famille: les Microparasellides et les *Microcerberinae* (Isopodes) des eaux souterraines. *In* DELAMARE-DEBOUTTEVILLE, CL., *Biologie des eaux souterrains littorales et continentales*, Paris.

KARAMAN, ST. (1959) Über eine neue *Microcharon*-Art (Crust. Isopoda) aus dem Karstgebiete der Herzegowina. *Acta Zool. Acad. Sci. Hungar.* **IV.**

MEŠTROV, M. (1960) Faunističko-ekološka i biocenološka istraživanja podzemnik voda savske nizine. *Bioloski Glasnik.* **XIII.**

SIEWING, R. (1959) *Angeliera xarifae*, ein neuer Isopode aus dem Küstengrundwasser der Insel Abd-el-Kuri (Golf von Aden). *Zool. Anz.* **CLXIII.**

SPOONER, G.M. (1959) The occurrence of *Microcharon* in the Plymouth offshore bottom fauna, with description of a new species. *J. Marine Assoc. U. K.* **XXXVIII.**

Stenetriidae

BRAGA, J.M. (1959) Le genre *Synasellus* et ses espèces. *Anaïs. Facult. Cienc. Porto.* **XLI.**

BRAGA, J.M. (1960) *Synasellus albicastrensis*, un nouveau Asellide troglobie du Portugal. *Anaïs. Facult. Cienc. Porto.* **XLIII.**

CHAPPUIS, P. A. (1943 *a*) *Stygonectes phreaticus* n. g., n. sp., ein neuer Isopode aus dem Grundwasser der Körös bei Baràtka (Bihar). *Fragm. Fauna Hungarica.* **VI.**
CHAPPUIS, P. A. (1943 *b*) *Stygasellus* nom. nov. für *Stygonectes* Chappuis 1943. *Fragm. Fauna Hungarica.* **VI.**
CHAPPUIS, P. A. (1944) Die Grundwasserfauna der Körös und des Szamos. *Matem. Termes. Közlem.* **XL.**
CHAPPUIS, P. A. (1948) Sur deux genres d'Asellides aberrants: *Stygasellus* et *Synasellus* (Crust. Isop.). *Bull. Soc. Sc. Cluj.* **X.**

Stenasellus

The bibliography before 1950 is compiled in the publication of Racovitza. It is not reproduced below.

BIRSTEIN, A. and STAROSTIN, I. V. (1949) Nouvelle race d'Aselle (*Stenasellus*) en U.R.S.S., originaire de Turkménie et sa signification pour la zoogéographie de l'Asie centrale. *Dokl. Akad. Nauk. SSSR* **LXIX.**
BRAGA, J. M. (1950) Sur deux *Stenasellus* (Crust. Isopoda) de la Guinée portugaise. *Anaïs Faculd. Cien. Porto.* **XXXV.**
BRAGA, J. M. (1962) Sur la distribution géographique des *Stenasellus* de la péninsule ibérique et description d'une espèce nouvelle de ce genre. *Anaïs Faculd, Cienc. Porto.* **XLIV.**
CHAPPUIS, P. A. (1951) Isopodes et Copépodes cavernicoles. *Rev. Zool. Bot. Afric.* **XLIV.**
CHAPPUIS, P. A. (1952) Un nouveau *Stenasellus* du Congo belge. *Rev. Zool. Bot. Afric.* **XLV.**
KARAMAN, ST. (1952) Über die jugoslawischen *Stenasellus*-Arten. *Fragmenta balcanica.* **I.**
MEŠTROV, M. (1960) *Stenasellus hungaricus thermalis* ssp. n. (Crustacea, Isopoda). Nalaz predglacijalne vrste u toplim izvorima kod Zagreba. *Bioloski Glasnik,* **XIII.**
MEŠTROV, M. (1962) Un nouveau milieu aquatique souterrain: le biotope hypotelminorhéique. *Compt. rend. Acad. Sci. Paris* **CCLIV.**
RACOVITZA, E. G. (1950) Asellides, première série: *Stenasellus.* Biospeologica, **LXX.** *Archiv. Zool. expér. géner.* **LXXXVII.**
SKET, B. (1958) Ein interessanter Fund der Malacostraca (Crust.) aus der Herzegowina und Crna Gora. *Bull. Sci.* **IV.**

Asellus

The reader will find the bilbiography, as far as 1951, in the treatise of Birstein. Some publications of importance which have appeared since, are mentioned below.

BIRSTEIN, J. A. (1951) *Faune de l'U.R.S.S.* VII, 5, Asellota. Moscou.
BRAGA, J. M. (1958) Un *Asellus* remarquable des eaux souterraines du Portugal (*Asellus Pauloae* n. sp.). *Mem. Est. Mus. Zool. Univers. Coimbra.* No. 254.
BRESSON, J. (1955) Aselles de sources et de grottes d'Eurasie et d'Amérique du Nord. *Archiv. Zool. expér. géner. N and R.* **XCII.**
CHAPPUIS, P. A. (1949) Les Asellides d'Europe et pays limitrophes. *Archiv. Zool. expér. géner.* **LXXXVI,** *N* and *R.*
CHAPPUIS, P. A. (1950) Campagne spéléologique de C. Bolivar at R. Jeannel dans l'Amérique du Nord. 13, Asellides. Biospeologica, **LXXI.** *Archiv. Zool. expér. géner.* **LXXXVII.**
CHAPPUIS, P. A. (1953) Sur la systématique du genre *Asellus. Notes biospéol.* **VIII.**
CHAPPUIS, P. A. (1955) Remarques générales sur le genre *Asellus* et description de quatre espèces nouvelles. *Notes biospéol.* **X.**
CHAPPUIS, P. A. (1957) Un Asellide nouveau de l'Amérique du Nord. *Notes biospéol.* **XII.**
KARAMAN, ST. (1950) Über zwei *Asellus*-Arten aus dem herzegowinisch-dalmatischen Karst. *Acad. Serbe Sci. Monographies* **CLXIII.**
KARAMAN, ST. (1952) Über einen neuen *Asellus* aus dem Grundwasser Südwest-Deutschlands. *Nach. Naturw. Museum Stadt Aschaffenburg* No. 34.

RACOVITZA, E. G. (1950) Campagne spéléologique de C. Bolivar et R. Jeannel dans l'Amérique du Nord (1928). 12, *Asellus stygius* (Packard, 1871). *Biospeologica*, **LXXI** *Archiv. Zool. expér. géner.* **LXXXVII.**

SKET, B. (1958) Einige interessante Funde der Malacostraca (Crust.) aus der Herzogewina und Crna Gora. *Bull. Sci.* **IV.**

SKET, B. (1960) Einige neue Formen der Malacostraca aus Jugoslawien. III. *Bull. Sci.* **V.**

Anneckella

CHAPPUIS, P. A. and DELAMARE-DEBOUTTEVILLE, CL. (1957) Un nouvel Asellide de l'Afrique du Sud. *Notes biospéol.* **XII.**

Johannella

MONOD, TH. (1924) Sur quelques Asellides nouveaux des eaux douces de l'Afrique du Nord. *Bull. Soc. Hist. Nat. Afrique Nord.* **XV.**

Asellids of Japan

MATSUMOTO, K. (1956) On the two subterranean water isopods, *Mackinia japonica* gen. et sp. nov. and *Asellus hubrichti* sp. nov. *Jap. Soc. Sci. Fish.* **XXI.**

MATSUMOTO, K. (1958) A new subterranean-water isopod, *Asellus kagaensis* sp. nov., from the wells of Japan. *Bull. Biogeogr. Soc. Jap.* **XX.**

MATSUMOTO, K. (1960a) Subterranean isopods of the Shikoku District, with the description of three new species. *Bull. Biogeogr. Soc. Jap.* **XXII.**

MATSUMOTO, K. (1960b) The subterranean-water isopods of the Akiyoshi limestone Area. *Jap. J. Zool.* **XII.**

MATSUMOTO, K. (1961a) The subterranean isopods of Honshû, with the description of four new species. *Bull. Biogeogr. Soc. Jap.* **XXII.**

MATSUMOTO, K. (1961b) The subterranean isopods from the Amami Group (Ryukyu Islands), with a description of a new species. *Annot. Zool. Jap.* **XXXIV.**

MATSUMOTO, K. (1962a) A new subterranean asellid from Hokkaido. *Jap. J. Zool.* **XIII.**

MATSUMOTO, K. (1962b) Two new genera and a new subgenus of the family Asellidae of Japan. *Annot. Zool. Jap.* **XXXV.**

Anthuridea

The bibliography of the Microcerberidae is mentioned in the publication of Chappuis and Delamare-Deboutteville (1960). Some recent papers are pointed out below.

CHAPPUIS, P. A., DELAMARE-DEBOUTTEVILLE, CL. and PAULIAN, R. (1956) Crustacés des eaux souterraines littorales d'une résurgénce d'eau douce a la Réunion. *Mém. Inst. Sci. Madagascar. Ser. A.* **XI.**

CHAPPUIS, P. A. and DELAMARE-DEBOUTTEVILLE, CL. (1959) Un Microcerberinae nouveau de Roumanie. *Vie et Milieu.* **IX.**

CHAPPUIS, P. A. and DELAMARE-DEBOUTTEVILLE, CL. (1960) État de nos connaissances sur une famille et une sous-famille: les Microparasellides et les Microcerberinae (Isopodes) *in* DELAMARE-DEBOUTTEVILLE, CL., *Biologie des eaux souterraines et littorales.* Paris.

DELAMARE-DEBOUTTEVILLE, CL. and CHAPPUIS, P. A. (1956) Compléments à la diagnose de quelques *Microcerberus. Vie et Milieu* **VII.**

LANG, K. (1961) Contributions to the knowledge of the genus *Microcerberus* Karaman (Crustacea, Isopoda) with a description of a new species from the central Californian coast. *Arkiv f. Zool.* **XIII.**

Phreatoicoidea

The bibliography of the Phreatoicoidea up to 1944 is mentioned by Nicholls. Amongst the recent papers, the following publications should be pointed out.

CHOPRA, B. and TIWARI, K. K. (1950) On a new genus of phreatoicid isopod from wells in Banaras. *Rec. Ind. Mus.* **XLVII.**

DAHL, E. (1955) Some aspects of the ontogeny of *Mesamphisopus capensis* (Barnard) and the affinities of the Isopoda Phreatoicidea. *K. fysiogr. Sällsk. Lund Förh.* **XXIV.**

NICHOLLS, G.E. (1943–1944) The Phreatoicoidea. *Pap. Proc. Roy. Soc. Tasmania.*

TIWARI, K.K. (1952) The morphology of *Nichollsia kashiense* Chopra and Tiwari (Crust. Isopoda Phreatoicidea). I. The head and endophragmal system. *Proc. Indian. Acad. Sci. Sect. B.* **XXXV.**

TIWARI, K.K. (1958) Another new species of *Nichollsia* (Crustacea: Isopoda; Phreatoicoidea). *Rec. Indian. Mus.* **LIII.**

Oniscoidea

The bibliography of the cavernicolous Oniscoidea is very large. It cannot be reproduced here. The reader will find an extensive bibliography in the second volume of the "Isopodes terrestres" in the *Faune de France*.

ARCANGELI, A., (1945) Considerazioni sopra l'origine delle forme cavernicole degli Isopodi terrestri. *Attualitá Zoologiche* **VI.** *Supplem. al Archiv. Zool. ital.* **XXXIV.**

CHAPPUIS, P. A. (1944) Die Gattung *Mesoniscus* Carl (Crust. Isop.). *Rev. suisse Zool.* **LI.**

LATTIN, G. DE (1939) Untersuchungen an Isopodenaugen (Unter besonderer Berücksichtigung der blinden Arten). *Zool. Jahrb. Abt. Anat.* **LXV.**

PESTA, O. (1925) Zur Kenntnis von *Mesoniscus alpicola* (Heller). *Speläol. Jahrb.* **V/VI.**

STROUHAL, H. (1947) Der troglophile *Mesoniscus alpicola* (Heller). *Anz. österr. Akad. Wiss. Sitz. math.-naturw. Kl.* **LXXXIV.**

VANDEL, A. (1960–1962) Isopodes terrestres. *Faune de France,* Paris. Première Partie No. 64. Seconde Partie, No. 66.

Ingolfiellidea

The bibliography of the genus *Ingolfiella* is included in the publication of Karaman. The papers published since 1959 are mentioned below.

COINEAU, N. (1963) Présence du sous-ordre des Ingolfiellidae Reibisch (Crustacea Amphipoda) dans les eaux souterraines continentales de France. *Comt. rend. Acad. Sci. Paris* **CCLVI.**

INGLE, R.W. (1961) The occurrence of *Ingolfiella leleupi* Ruffo (Amphipoda: Ingolfiellidae) in the Lusaka ground-water of Northern Rhodesia. *Ann. Mag. Nat. Hist.* (13) **IV,** No. 43.

KARAMAN, ST. (1959) Über die Ingolfielliden Jugoslawiens. *Biol. Glasn.* **XII.**

LELEUP, N. (1955) A propos de l'archaïsme et de l'écologie de l'*Ingolfiella leleupi* Ruffo (Crust. Ingolf.). *Notes biospéol.* **X.**

NOODT, W. (1959) Estudios sobre Crustaceos chilenos de aguas subterraneas. 1. *Ingolfiella chilensis* n. sp. de la playa marina de Chile central (Crustacea Amphipoda). *Invest. zool. chil.* **V.**

RUFFO, S. (1950) Considerazioni sulla posizione sistematica e sulla distribuzione geografica degli Ingolfielli. *Boll. d. Zool.* **XVII.**

SIEWING, R. (1958) *Ingolfiella ruffoi,* n. sp., eine neue Ingolfiellida aus dem Grundwasser der peruanischen Küste. *Kieler Meeresforschung* **XIV.**

SPOONER, G.M. (1959) New members of the British marine bottom fauna. *Nature, Lond.* **CLXXXIII.**

SPOONER, G. M. (1960) The occurence of *Ingolfiella* in the Eddystone shell gravel, with description of a new species *J. mar. Assoc. U. K.* **XXXIX.**

Gammaridae

The bibliography of the hypogeous Gammaridae includes about 400 publications. Only some of them are mentioned here.

ABSOLON, CH. (1927) Les grands Amphipodes aveugles dans les grottes balkaniques. *Assoc. franç. Avanc. Sci.* Constantine.

ANCONA, U. D' (1942) I *Niphargus* italiani. Tentativo di valutazione critica delle minori unita sistematiche. *Mem. Ist. Ital. Spel. Ser. Biol.* **IV.**

ANDERS, F. (1956) Über Ausbildung und Vererbung von Körperfarbe bei *Gammarus pulex subterraneus* (Schneider), einer normalerweise pigmentlosen Höhlenform des gemeinen Bachflohkrebses. *Zeit. indukt. Abstamm. Vererb.* **LXXXVII.**

BALAZUC, J. and ANGELIER, E. (1952) Sur la capture à Banyuls-sur-mer (P.O.) de *Pseudoniphargus africanus* Chevreux 1901 (Amphipode, Gammaridae). *Bull. Soc. Zool. France* **LXXVI.**

BALAZUC, J. (1954) Les Amphipodes troglobies et phréatobies de la faune gallo-rhénane. *Biospeologica,* **LXXIV.** *Archiv. Zool. expér. géner.* **XCI.**

BIRSTEIN, J.A. (1933) Malacostraca der Kutais-Höhlen am Rion (Transkaukasus, Georgien). *Zool. Anz.* **CIV.**

BIRSTEIN, J.A. (1961) Biospeologica Sovietica. XIV. Les Amphipodes souterrains de Crimée. *Bull. Soc. Natural. Moscou. Sect. Biol.* **LXVI.**

BREHM, V. (1955) *Niphargus*-Probleme. *Sitzb. Österr. Akad. Wiss.* Abt. I, No. 164.

CHEVREUX, ED. (1901) Amphipodes des eaux souterraines de France et d'Algérie. *Bull. Soc. Zool. France* **XXVI.**

CHEVREUX, ED. (1910) Les Amphipodes d'Algérie et de Tunisie. *Mém. Soc. Zool. France* **XXIII.**

CHEVREUX, ED. (1913) Crustacés. II. Amphipoda. *In Result. sc. Voyage. Ch. Alluaud et R.Jeannel en Afrique Orientale (1911-1912).* Paris.

CHEVREUX, ED. (1920) Sur quelques Amphipodes nouveaux ou peu connus des côtes de Bretagne. *Bull. Soc. Zool. France* **XLV.**

CHILTON, CH. (1894) The subterranean Crustacea of New Zealand; with some general remarks on the fauna of caves and wells. *Trans. Linn. Soc. London* **VI.**

CHILTON, CH. (1918) Some New Zealand Amphipoda belonging to the genus *Phreatogammarus. J. Zool. Research.* **III.**

CHILTON, CH. (1920) *Niphargus philippensis,* a new species of amphipod from the underground waters of the Philippine Islands. *Phillipine J. Sc.* Manille. **XVII.**

CHILTON, CH. (1921) Amphipoda. Fauna of the Chilka Lake. *Mem. Indian Mus.* **V.**

CHILTON, CH. (1923) A blind amphipod from a mine in Bengal. *Rec. Indian. Mus.* **XXV.**

COINEAU, N. (1962) *Salentinella delamarei,* nouvel Amphipode Gammaridae des eaux phréatiques du Tech (Pyrénées Orientales). *Vie et Milieu.* **XIII.**

GAUTHIER, H. (1936) *Eriopisa seurati,* nouvel Amphipode du Sud-tunisien. *Bull. Soc. Hist. Nat. Afric. Nord.* **XXVII.**

HUMBERT, A. (1876) Description du *Niphargus puteanus* var. *Forelii. In* Matériaux pour servir l'étude de la Faune profonde du Lac Léman. Article XXXIX. *Bull. Soc. Vaudoise Sci. Nat.* **XIV,** No. 76.

KARAMAN, ST. (1931) Über die Synurellen Jugoslawiens. *Prirod. Razprav.* **I.**

KARAMAN, ST. (1932) 5. Beitrag zur Kenntnis der Süßwasser-Amphipoden. *Prirod. Razprav.* **II.**

KARAMAN, ST. (1950) *Neogammarus rhipidiophorus* Catta aus unterirdischen Gewässern der Adriatischen Küste. *Posebna Izd. Nauk. Prir. Mat. N.S.* No. 2.

KARAMAN, ST. (1953) Über subterrane Amphipoden und Isopoden des Karstes von Dubrovnik und seines Hinterlandes. *Acta Mus. Macedon. Sc. Nat.* **I,** No. 7.

KARAMAN, S.L. (1960) Weitere Beiträge zur Kenntnis der jugoslawischen Niphargiden. *Glasn. Prirod. Muz. Beogradu. Ser. B.*

LERUTH, R. (1938) Notes d'Hydrobiologie souterraine. IV. Remarques écologiques sur le genre *Niphargus. Bull. Soc. R. Sci. Liège.*

MEŠTROV, M. (1962) Un nouveau milieu aquatique souterrain: le biotope hypotelminorhéique. *Compt. rend. Acad. Sci. Paris* **CCLIV.**

MÜHLMANN, H. (1938) Variationsstatistische Untersuchungen und Beobachtungen an unter- und oberirdischen Populationen von *Gammarus pulex* (L.). *Zool. Anz.* **CXXII.**

RUFFO, S. (1948) *Hadzia minuta* n. sp. *(Hadziidae)* e *Salentinella gracillima* n. g. n. sp. (Gammaridae), nuovi Anfipodi troglobi dell'Italia meridionale. *Boll. Soc. Natural. Napoli* **LVI.**

RUFFO, S. (1950) Studi sui Crostacei Anfipodi. **XXI.** Nuove osservazione sulla distribuzione di *Synurella ambulans* (F. Müller) in Italia. *Atti Accad. Agricolt. Sc. Lett. Verona.* (5) **XXXV.**

RUFFO, S. (1951) Rinvenimento di *Gammarus (Neogammarus) rhipidiophorus* Catta nelle acque sotterranee della Liguria. *Doriana*, **I,** No. 18.

RUFFO, S. (1953) Studi sui Crostacei Anfipodi. **XXXV.** Nuove osservazioni sul genere *Salentinella* Ruffo (Amphipoda, Gammaridae). *Boll. Soc. Entomol. Ital.* **LXXXIII.**

RUFFO, S. (1956) Lo stato attuale delle conoscenze sulla distribuzione geografica degli Anfipodi delle acque sotterranee a dei paesi mediterranei. *Communic. 1er Congr. Intern. Speoleol.* Paris. **III.**

RUFFO, S. (1957) Una nuova specie trogobie di *Hyalella* di Venezuela (Amphipoda, Talitridae). *Annali. Mus. Civ. Stor. Nat. Genova* **LXIX.**

SCHÄFERNA, K. (1922) Amphipoda balcanica. *Česk. Spol. Nauk. Vestnik.* Prag. **II.**

SCHELLENBERG, A. (1931) Amphipoden der Sunda-Expeditionen Thienemann und Rensch. *Archiv f. Hydrobiol. Suppl. Bd.* **VIII.**

SCHELLENBERG, A. (1932) Deutsche subterrane Amphipoden. *Zool. Anz.* **XCIX.**

SCHELLENBERG, A. (1933) *Niphargus*-Probleme. *Mitt. Zool. Mus. Berlin* **XIX.**

SCHELLENBERG, A. (1934a). Amphipoden aus Quellen, Seen und Höhlen. *Zool. Anz.* **CVI.**

SCHELLENBERG, A. (1934b) Eine neue Amphipoden-Gattung aus einer belgischen Höhle, nebst Bemerkungen über die Gattung *Crangonyx. Zool. Anz.* **CVI.**

SCHELLENBERG, A. (1935) Schlüssel der Amphipodengattung *Niphargus* mit Fundortangaben und mehreren neuen Formen. *Zool. Anz.* **CXI.**

SCHELLENBERG, A. (1936a) Bemerkungen zu meinen *Niphargus*-Schlüssel und zur Verbreitung und Variabilität der Arten nebst Beschreibung neuer *Niphargus*-Formen. *Mitteil. Zool. Mus. Berlin* **XXII.**

SCHELLENBERG, A. (1936b) Die Amphipodengattungen um *Crangonyx*, ihre Verbreitung und ihre Arten. *Mitteil. Zool. Mus. Berlin* **XXII.**

SCHELLENBERG, A. (1937) Litorale Amphipoden des Tropischen Pazifiks. *K. Sv. Vet. Handl.* (3) **XVI.**

SCHNEIDER, R. (1885) Der unterirdische *Gammarus* von Clausthal *(G. pulex* var. *subterraneus). Sitzber. Akad. Berlin.*

SHOEMAKER, C.R. (1942) Notes on some American fresh-water amphipod crustaceans and descriptions of a new genus and two new species. *Smiths. Misc. Coll.* **CI.**

SHOEMAKER, C.R. (1959) Three new cave amphipods from the West Indies. *J. Wash. Acad. Sci.* **XLIX.**

SKET, B. (1958) Über Zoogeographie und Phylogenie der Niphargiden. *Verh. deutsch. Zool. Gesell.*

STANKOVIČ, S. (1960) The Balkan Lake Ohrid and its living world. *Monographiae biologicae* **IX.**

STEPHENSEN, K. (1931) *Neoniphargus indicus* Chilton, an Indian freshwater amphipod. *Rec. Indian Mus. Calcutta* **XXXIII.**

UÉNO, M. (1955) Occurrence of a freshwater gammarid (Amphipoda) of the *Niphargus* group in Japan. *Bull. Biogeogr. Soc. Japan* **XVI–XIX.**

VILLALOBOS, F.A. (1961) Un Anfipodo cavernicolo nuevo de Mexico, *Bogidiella tabascensis*, n.sp. *An. Inst. Biol. Mexico* **XXXI.**

WECKEL, A.L. (1907) The freshwater Amphipoda of North America. *Proc. U.S. Nat. Mus.* **XXXII.**

Natantia

The full bibliography of hypogeous Natantia has been compiled in the publication of Holthuis (1956). Only the recent papers are mentioned below.

HOLTHUIS, L.B. (1956) An enumeration of the Crustacea Decapoda Natantia inhabiting subterranean waters. *Vie et Milieu* **VII.**

HOLTHUIS, L.B. (1956) The troglobic Atyidae of Madagascar (Crustacea Decapoda Natantia). *Mem. Inst. Sci. Madagascar* Ser. A. **XI.**

HOLTHUIS, L.B. (1960) Two new atyid shrimps from subterranean waters of N.W. Australia (Decapoda Natantia). *Crustaceana* **I.**

LINDBERG, K. (1960) Revue des recherches biospéologiques en Asie moyenne et dans le sud du Continent asiatique. *Rass. Speleol. Ital.* **XII.**

MATJAŠIČ, J. (1956) Ein neuer Höhlendecapode aus Jugoslawien. *Zool. Anz.* **CLVII.**

MATJAŠIČ, J. (1958) Postembrionali razvoj jamske koziu *Troglocaris. Biol. Vest. Jugosl.* **VI.**

MATJAŠIČ, J. (1960) O vjetrenishik troglokarisih. *Biol. Vest. Jugosl.* **VII.**

MEES, G.E. (1962) The subterranean freshwater fauna of Yardie Creek station, North West Cape, Western Australia. *J. R. Soc. West Austral.* **XLV.**

SMALLEY, A.E. (1961) A new cave shrimp from south-eastern United States (Decapoda, Atyidae). *Crustaceana* **III.**

STAMMER, H.J. (1933) Einige seltene oder neue Höhlentiere. *Zool. Anz. 6 Suppl. Bd.*

STAMMER, H.J. (1935) Untersuchungen über die Tierwelt der Karsthöhlengewässer. *Verh. Intern. Ver. theor. angew. Limnologie.* **VII.**

TORII, H. (1953) Fauna der Ryugado Sinterhöhle in Kochi Präfektur (Die Berichte der speläobiologischen Expeditionen. VI). *Annot. Zool. Japon.* **XXVI.**

UÉNO, S.I. (1957) Blind aquatic beetles of Japan, with some accounts of the fauna of Japanese subterranean waters. *Archiv. f. Hydrobiol.* **LIII.**

Reptantia
Astacura
EBERLEY, W.R. (1960) Competition and evolution in cave crayfishes of southern Indiana. *System. Zool.* **IX.**

GARMAN, H. (1921) Two interesting crustaceans from Kentucky. *Trans. Kentucky Acad. Sci.* **I.**

HOBBS, H.H. (1942) A generic revision of the crayfishes of the subfamily *Cambarinae* (Decapoda, Astacinae) with the description of a new genus and species. *Amer. Midl. Nat.* **XXVIII.**

HOBBS, H.H. Jr. (1952) A new albinotic crayfish of the genus *Cambarus* from the southern Missouri with a key to the albinistic species of the genus (Decapoda, Astacidae), *Amer. Midl. Nat.* **XLVIII.**

HOBBS, H.H. and BARR, TH.C. (1960) The origins and affinities of the troglobitic crayfishes of North America (Decap. Astacidae). I. The Genus *Cambarus. Amer. Midl. Nat.* **LXIV.**

RHOADES, R. (1962) The evolution of crayfishes of the genus *Orconectes* section *limosus.* (Crustacea, Decapoda). *Ohio. J. Sci.* **LXII.**

Anomura
CALMAN, W.T. (1904) On *Munidopsis polymorpha* Koelbel, a cave-dwelling marine crustacean from the Canary Islands. *Ann. Mag. Nat. Hist.* (7) **XIV.**

FAGE, L., and MONOD, TH. (1936) La faune marine du Jameo de Agua, lac souterrain de l'île de Lanzarote. *Biospeologica,* **LXIII.** *Archiv. Zool. expér. génér.* **LXXVIII.**

FUCHS, TH. (1894) Über Tiefseetiere in Höhlen. *Ann. K. K. Naturhist. Hofmuseum. Wien.* **IX,** *Notizer.*

Brachyura
IHLE, J.E.W. (1912) Über eine kleine Brachyurensammlung aus unterirdischen Flüssen von Java. *Notes Leyden Museum* **XXXIV.**

RIOJA, E. (1953) Estudios carcinologicos. **XXVIII.** Descripcion de un nuevo genero de Potamonidos cavernicolos y ciegos de la Cueva del Tio Ticho, Comitàn, Chis. *Annal. Inst. Biol. Mexico* **XXIII.**

TORII, H. (1953) Fauna der Ryugado Sinterhöhle in Kochi Präfektur (Die Berichte der speläobiologischen Expeditionen VI). *Annot. Zool. Japon.* **XXVI.**

UÉNO, S.I. (1957) Blind aquatic beetles of Japan with some accounts of the fauna of Japanese subterranean waters. *Archiv. f. Hydrobiol.* **LIII.**

CHAPTER X

ONYCHOPHORA AND MYRIAPODA

A. TRACHEATA

There is a present-day tendency for zoologists to regard the tracheates as an entirely independent group of the arthropods with a separate origin from the Crustacea and the chelicerates. There are good reasons to believe that the tracheates have been derived from some annelid type which was already adapted to terrestrial life. The Onychophora provide us with a clear image of such transitional types.

The following classification of the tracheates is the one according to Tiegs and Snodgrass:

Super-Class	Class
Monognatha	Onychophora
Dignatha	{ Diplopoda { Pauropoda
Trignatha	⌈ Chilopoda ⎨ Symphyla ⎝ Hexapoda (= Insecta).

The Diplopoda, Pauropoda, Chilopoda and Symphyla are frequently, but artificially, united in the heterogeneous group Myriapoda.

The Pauropoda are found only rarely in European caves (Remy and Husson, 1938; Griepenburg, 1939) and those of North America, but according to Uéno (1957) there are cavernicolous forms in Japan.

The Symphyla are endogeans which are sometimes found at the entrances of caves *(Scutigerella immaculuta, Hanseniella livea)*, but no species can be considered to be truly cavernicolous. Scheller (1961) has described a new species of *Symphyella, S. major*, which is known from only two specimens captured in a cave in the Canton of Vaud (Switzerland). But there are no overriding reasons for regarding this species as truly cavernicolous.

However, the absence of Symphyla in caves may appear surprising, since they are without exception depigmented and anophthalmic. But it is not sufficient for an animal to be merely depigmented and anophthalmic to allow it to be a valid candidate for cavernicolous life. The Symphyla have been unable to penetrate the subterranean world simply because their diet, which consists entirely of green plants, does not allow them to do so.

B. ONYCHOPHORA

These forms avoid light and are nocturnal and extremely hygrophilic. These characteristics would seem to predispose them to cavernicolous life. In fact, out of the 60–70 species known, only one is a cavernicole. This is *Peripatopsis alba* (Lawrence, 1931). It was found in a cave in the sandstone of Table Mountain above Cape Town, South Africa. It is completely white, blind and depigmented. The legs are slightly longer than those of the related silvicolous species *P. balfouri*.

C. DIPLOPODA

The Studies concerning the Cavernicolous Diplopods

The diplopods are for the most part extremely hygrophilic Tracheata, and are frequently found in caves. A large number of hypogeous species have been described. The cavernicolous diplopods of France, Italy and the Balkans are particularly well known. The first have for the most part been described by H. W. Brölemann and H. Ribaut. The Italian forms have been studied by P. Manfredi; while the subterranean forms frequenting the Balkans have been studied by J. Lang, K. Strasser, I. Tabacaru and above all by K. W. Verhoeff. A revision has been given in the posthumous publication of C. Attem's work (1959).

The cavernicolous Diplopoda of Morocco have been listed by Schubart (1960).

The cavernicolous Diplopoda of North America have been the subject of numerous publications by C. H. Bollman; N. B. Causey; R. V. Chamberlin; E. D. Cope; K. Dearolf; R. L. Hoffman; H. F. Loomis; J. MacNeill; A. S. Packard and J. A. Ryder. The cavernicolous Diplopoda of America have been listed by Chamberlin and Hoffman (1958) and by Nicholas (1960). Their number is certainly lower than that for Europe because North America has only some 90 species and sub-species of cavernicolous diplopods (Causey, 1960).

The cavernicolous Diplopoda of Japan and Korea have been described by A. Haga, H. Takashima and K. W. Verhoeff, and those of China by I. Loksa.

Cavernicolous diplopods are less common in the tropics. However, many species have been collected in the caves of India, Assam, Burma, Siam, Java, East Africa, the Congo, and Cuba.

The Distribution of Cavernicolous Diplopods in the Systematic Classification

The cavernicolous diplopods are so numerous that there is no question of considering here the details of so complex a systematic classification. Such an undertaking could only be the work of a highly qualified specialist. The following discussion will be limited to considerations of a very general nature.

The Pselaphognatha (Polyxenidae) are represented underground only by accidental specimens belonging to species common on the surface *(Polyxenus lagurus* and *P. lucidus).*

The Colobognatha do not seem to possess any representatives in the subterranean world.

On the other hand, the four other sub-classes of diplopods have given rise to cavernicolous types.

The Oniscomorpha are lapidicoles or humicoles, or sometimes troglophiles (e.g. *Gervaisia*). Many genera are subterranean (e.g. *Spelaeoglomeris, Stygioglomeris* and *Rhyparomeris*, etc.) and Ribaut (1954, 1955) has produced dichotomous keys to these genera. Tabacaru (1958, 1960) described some cavernicolous species of *Gervaisia* from Rumania.

The Polydesmoidea are represented in Europe by some cavernicolous species belonging to the genera: *Brachydesmus, Polydesmus, Devillea,* etc. In North America the Polydesmoidea comprise numerous cavernicolous species belonging to the genera *Antriadesmus, Brachydesmus, Scytonotus, Speodesmus* and *Speorthus.* In Japan, the cavernicolous Polydesmoidea belong to the genus *Epanerchodus* (Myoshi, 1951) and more exactly to three sub-genera established by Verhoeff (1941 c): *Riuerchodus, Stygoerchodus* and *Antrochodus.*

The Nematomorpha includes many orders containing cavernicolous types. The Chordeumoidea are represented in Europe by a large number of species of which some are humicoles but other cavernicoles (*Chordeuma, Anthogona, Cranogona, Scutogona, Acherosoma, Brölemanneuma, Attractasoma,* etc.). In North America the Chordeumoidea make up about 80% of the cavernicolous forms known. The two genera *Pseudotremia* (Cleidogonidae) and *Scoterpes* (Conotylidae) have, in particular, a large number of species.

Another order of the Nematomorpha, the Lysiopetaloidea also contains cavernicolous forms. In Europe they are localised in the Balkan regions *(Apfelbeckia, Antropetalum, Rhopalopetalum* and *Lysiopetalum).* These genera can be described more exactly as troglophiles than cavernicoles (Verhoeff, 1941 b). In North America the Lysiopetalidae are represented by three species of the genus *Tetracion.*

However, cavernicolous species are particularly abundant in the sub-class Iuliformes and the order Iuloidea. The family Blaniulidae is one of the most important in the sub-class, containing 45% of cavernicolous species (Brölemann, 1923). Numerous species from Europe and North Africa have been described and placed in the genera *Archichoneiulus, Archiboreiulus, Typhloblaniulus,* etc. However, the Blaniulidae are not restricted to Europe. Verhoeff (1938, 1939, 1941a) has described many blaniulids from caves in Japan and Korea, for which species he has erected several new genera: *Antrokoreana, Scleroprotopus, Nipponiosoma, Lavabates.* Similarly, the related family Iulidae contains numerous cavernicoles included in many genera (*Cylindroiulus, Typhloiulus, Trogloiulus,* etc.).

The order Spirostreptomorpha is represented in the subterranean environment by the family Cambalidae. Several species belonging to this family have been discovered in North America (*Cambala, Eclomus, Troglocambala*). A number of species belonging to the Cambalopsidae have been recorded from caves in Assam, Burma and Java.

Cavernicoles of Recent and of Ancient Origin

Several cavernicolous Diplopoda are closely related to surface forms and may thus be regarded as recent cavernicoles. This is the case with many of the species of *Brachydesmus* and *Polydesmus.*

On the other hand, many diplopods are ancient cavernicoles and represent the relics of an earlier fauna which no longer possesses surface forms at least in their country of origin.

The Oniscomorpha contains a number of species of this ancient type which have taken refuge underground (*Doderoa, Adenomeris, Spelaeoglomeris, Stygioglomeris*).

Within the Polydesmoidea, we may mention the genus *Devillea* with the two species: *D. doderoi* from caves in Sardinia and *D. tuberculata* from caves in the Alpes Maritimes. This European genus is quite isolated from its relatives. It belongs to the Leptodesmidae, most of whose members are American. The Trichopolydesmidae also belongs to the Polydesmoidea. It is essentially a New World group, particularly well represented in the northern parts of South America. However, there are two European species of this family, *Trichopolydesmus eremitus* (from a cave near Herkulesbad, Rumania), and *Serradium hirsutipes* (from a cave near Rovereto, Trentino, Italy, and another cave in the province of Verona). It cannot be doubted that these two cavernicoles are the ancient relics of a tertiary fauna (Verhoeff, 1941 d).

The European Blaniulidae are directly related to the North American family Paraiulidae (Brölemann, 1923). The study of the hypogeous Iuloidea as well as that of the cavernicolous Polydesmoidea makes it clear that an ancient Atlantic fauna once existed. This fauna expanded and became widespread in both Old and New Worlds at a time when the Atlantic was not fully formed.

The Characters of Cavernicolous Diplopoda

(1) The Size and Number of Segments. One of the most remarkable characteristics exhibited by cavernicolous diplopods has been reported by H.C.Brölemann (1923) in his study of the hypogeous Blaniulidae. This French myriapodologist discovered that the cavernicolous forms are very much larger than those of the surface. They possess more segments and thus undergo more moults. In consequence they take longer to reach maturity and their length of life is probably extended. We will give later an interpretation of these characteristics (p. 369).

(2) The Ocular Apparatus. All present-day diplopods have the eyes reduced. This was not the case in ancestral forms. The diplopods of the Carboniferous had two large compound eyes composed of 60–80 ommatidia, and those of *Glomeropsis ovalis* had as many as 600 elements.

The eyes of living diplopods are made up of scattered ommatidia, representing a reduced condition similar to that found in oniscoids. Although the end-point of regressive evolution of the ocular apparatus is anophthalmia, this state is not confined to cavernicoles. *Blaniulus guttulatus* which feeds on the strawberries in our gardens is quite without eyes, as are all the species of the tribe Blaniulini.†

As all the species of *Blaniulus* are lucicoles (except *B. lichensteini* from caves in Hérault, France) it is not surprising that the cavernicolous *Typhloblaniulus*, which are closely related to them, are also anophthalmic. It may be justly concluded that the ancestors of *Typhloblaniulus* were certainly anophthalmic before they entered into caves.

(3) The Antennae. In addition to depigmentation and anophthalmia, regressive manifestations common to all hypogeans, some cavernicolous diplopods show remarkable modifications to the structure of their antennae. While in the chilopods the length of these organs and the number of antennary segments are variable, in the diplopods they are remarkably constant. The antenna of the diplopods is a conservative structure. This condition is present in many other organs of these archaic tracheates.

In all diplopods the antennae are composed of seven segments. In the large majority the seventh segment is wider than its length and shorter than the two segments preceding it. Verhoeff has observed that in four genera of cavernicolous diplopods there is a noticeable lengthening of the seventh antennary segment which may become as long as wide, and even longer than its width. Similarly the fifth and sixth segments may elongate. Finally there is an increase in the number of sensile hairs. In epigeous types they number 5–8 but there are 18–24 in *Typhloiulus* (Fig. 34).

† Except *Typhloblaniulus troglodites eulophus*, an endogeous form living in Sardinia.

Verhoeff regards these sensile hairs (or the similar minute cup-like structures present in *Epanerchodus*) which are found on the seventh antennary segment as hygroscopic organs. He believes them to supply the animal with information concerning the humidity of the air. Although this interpretation is plausible, since hair-like hygroscopic organs are known

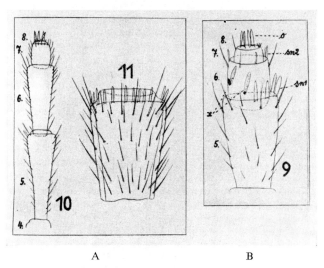

A   B

Fig. 34. The four terminal segments of the antennae in some Diplopoda. A, *Cylindroiulus zinalensis*, an epigeous form. B, *Typhloiulus maximus*, a cavernicolous form. *Typhloiulus maximus;* extremity of the fifth segment (after Verhoeff).

in some insects, it should be regarded as purely hypothetical when applied to these diplopods.

These modifications of the antennae are present in four genera, although the latter are systematically very different from each other:

POLYDESMOIDEA   Polydesmidae: *Epanerchodus* (Verhoeff, 1941c)
                Trichopolydesmidae: *Serradium* (Verhoeff, 1941d)

IULOIDEA   Iulidae: *Typhloiulus* (Verhoeff, 1930)
                Blaniulidae: *Antrokoreana* (Verhoeff, 1938).

D. CHILOPODA

The chilopods are not so interesting to biospeologists as are the diplopods. Their cavernicolous representatives are less numerous and less differentiated from surface forms, many of them barely so.

The Chilopoda may be divided into four sub-classes: Geophilomorpha, Scolopendromorpha, Lithobiomorpha and Scutigeromorpha.

Although the Geophilomorpha are depigmented and anophthalmic and thus, one might believe, pre-adapted to cavernicolous life, no members of this sub-class can be regarded as strict cavernicoles (Verhoeff, 1937).

In the Scolopendromorpha, the Cryptopidae are all anophthalmic. A number of the species of *Cryptops* are cavernicolous (Verhoeff, 1933), but they are rather similar to surface forms. However, *C. (Trigonocryptops) longicornis*, from the Cueva de la Pileta, near Ronda (Andalusia), has remarkably long antennae and legs (Ribaut, 1915). *C. jeanneli* from caves in

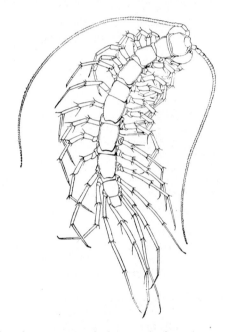

FIG. 35. *Lithobius drescoi*, a cavernicolous species from the caves of the province of Santander (Spain) (after Demange).

the Alpes Maritimes may perhaps also be truly cavernicolous (Matič, 1960).

The Scutigeromorpha possess no cavernicolous representatives in Europe (Verhoeff, 1937). *Scutigera coleoptrata* has been found in cellars, but does not penetrate caves. Some species of this group have been reported from tropical caves (Verhoeff, 1939). They belong to the genera *Therenonema* (Assam), *Therenopoda* (Burma and Riu-Kiu Islands of Japan) and *Ballonema* (East Africa). These genera are not truly cavernicolous, however, but simply lucifugous and hygrophilic forms which occasionally frequent caves. Verhoeff calls them "large-eyed nocturnal animals", recalling the state of affairs in owls and other nocturnal hunters.

Only the Lithobiomorpha have given rise to true cavernicoles. *Lithobius*, the most important genus in the sub-class, contains troglophiles and also well-defined cavernicoles as is evident from the studies of W. K. Verhoeff, H. Ribaut, J. M. Demange, and Z. Matič devoted to them. Troglobious species have been found mainly in the Pyrenees, the Cantabrian Mountains and the Adriatic areas of Karst (Matič, 1957). However, a troglobious species has been discovered in the cave of Daia Chiker, near Taza, Morocco, and has been described by Verhoeff (1936) under the name *L. chikerensis*.

These cavernicolous forms differ from those of the surface by a number of remarkable characteristics. The number and size of the ocelli are reduced and in several species the ocular apparatus has completely disappeared. Matič (1960) has prepared a list and a key to the anophthalmic species. On the other hand Tömösvary's organ increases in size in cavernicolous forms.

The antennae become elongated and may equal the length of the body. At the same time there is a corresponding increase in the number of the antennary segments. There are between 20 and 50 in the majority of *Lithobius* species but over a hundred in certain cavernicolous species: 106 in *L. matulici* (Herzegovina); 100–103 in *L. drescoi* (Cantabrian Mountains).

Finally the legs elongate, so giving some cavernicolous *Lithobius* the shape of *Scutigera* (Fig. 35).

BIBLIOGRAPHY

Pauropoda and *Symphyla*

GRIEPENBURG, W. (1939) Die Tierwelt der beiden Hüllöcher im Sauerland. *Mitteil. Höhl. Karstf.*

REMY, P. and HUSSON, R. (1938) Les Pauropodes des Galeries de Mines et des Cavernes naturelles. *Compt. rend. Premier Congr. Lorrain. Soc. Sav. Est. France.*

SCHELLER, U. (1961) Cave Symphyla from Switzerland. *Rev. Suisse Zool.* **LXVIII.**

UÉNO, S. I. (1957) Blind aquatic beetles of Japan with some accounts of the fauna of Japanese subterranean waters. *Archiv f. Hydrobiol.* **LIII.**

Onychophora

GRINDBERG, J. R. (1957) The white Peripatus, *Peripatopsis alba* Lawrence. *Bull. South Afric. Speleol. Assoc. Cape Section.*

LAWRENCE, R. F. (1931) A new *Peripatopsis* from the Table Mountain Caves. *Annal. South Afric. Mus.* **XXX.**

LAWRENCE, R. F. (1953) *The biology of the cryptic fauna of forests.* Cape Town and Amsterdam.

Diplopoda

ATTEMS, C. (1959) Die Myriapoden der Höhlen der Balkanhalbinsel. Nach dem Material der "Biospeologica balcanica". *Annal. Naturhist. Mus. Wien.* **LXIII.**

BRÖLEMANN, H.W. (1910) Symphyles, Psélaphognathes et Lysiopétalides (Myriapodes) (1ère Série). *Biospeologica*, **XVII.** *Archiv. Zool. expér. géner.* (5) **V.**

BRÖLEMANN, H.W. (1913) Gloméridés (Myriapodes) (1ère Série). *Biospeologica*, **XXXI.** *Archiv. Zool. expér. géner.* **LII.**

BRÖLEMANN, H.W. (1923) Blaniulidae (Myriapodes) (1ère Série). *Biospeologica*, **XLVIII.** *Archiv. zool. expér. géner.* **LXI.**

CAUSEY, N.B. (1960) Speciation in North American cave millipeds. *Amer. Midl. Nat.* **LXIV.**

CHAMBERLIN, R.V. and HOFFMAN, R.L. (1958) Check-List of the millipeds of North America. *U.S. Nat. Mus. Bull.* **CCXII.**

MANFREDI, P. (1932a) I Miriapodi cavernicoli italiani. *Le Grotte d'Italia* **VI.**

MANFREDI, P. (1932b) Contributo alla conoscenza della fauna cavernicola italiana. *Natura,* Milano **XXIII.**

MANFREDI, P. (1936) Secondo elenco di Miriapodi cavernicoli italiani. *Le Grotte d'Italia* (2). **I.**

MYOSHI, Y. (1951) Beiträge zur Kenntnis japanischer Myriopoden. II. Neber drei neue Arten von Epanerchodus (Polydesmidae). *Zool. Magaz.* **LX.**

NICHOLAS, BROTHER G. (1960) Check-list of macroscopic troglobitic organisms of the United States. *Amer. Midl. Nat.* **LXIV.**

RIBAUT, H. (1913) Ascospermophora (Myriapodes) (1ère Série). *Biospeologica,* **XXVIII.** *Archiv. Zool. expér. géner.* **L.**

RIBAUT, H. (1954) Un nouveau genre de Gloméride cavernicole. *Notes biospéol.* **IX.**

RIBAUT, H. (1955) *Rhyparomeris,* nouveau genre de Gloméride cavernicole (Diplopoda, Adenomeridae). *Bull. Soc. Hist. Nat. Toulouse* **XC.**

SCHUBART, O. (1960) Eine neue cavernicole Stylodesmide aus Marokko (Diplopoda, Proterospermophora). *Bull. Soc. Sci. Nat. Phys. Maroc.* **XL.**

TABACARU, I. (1958) Beiträge zur Kenntnis der cavernicolen *Gervaisia*–Arten. *Gervaisia orghidani* n. sp. und *G. jonescui* (Bröl.) *Zool. Anz.* **CLXI.**

TABACARU, I. (1960) Neue cavernicole *Gervaisia*–Arten: *G. racovitzai* n. sp., und *G. spelaea* n. sp. und *G. dobrogica* n. sp. *Zool. Anz.* **CLXV**

VERHOEFF, K.W. (1930) Arthropoden aus südostalpinen Höhlen, gesammelt von Karl Strasser. 5. Aufsatz. *Mitteil. Höhl. Karstf.*

VERHOEFF, K.W. (1938) Ostasiatische Höhlendiplopoden (148. Diplopoden-Aufsatz). *Mitteil. Höhl. Karstf.*

VERHOEFF, K.W. (1939) *Scleroprotopus* in japanischen Höhlen. *Mitteil. Höhl. Karstf.*

VERHOEFF, K.W. (1941a) Ostasiatische Diplopoden aus Höhlen; 3.Aufsatz. *Zeit. Karst. Höhlenk.*

VERHOEFF, K.W. (1941b) Über eine neue, cavernicole *Lysiopetalum*–Art. *Zeit. Karst. Höhlenk.*

VERHOEFF, K.W. (1941c) Ostasiatische Diplopoden aus Höhlen. 4.Aufsatz. *Zeit. Karst. Höhlenk.*

VERHOEFF, K.W. (1941d) Höhlen-Diplopoden aus dem Trentino. *Zeit. Karst. Höhlenk.*

Chilopoda

DEMANGE, J.M. (1956) Étude sur *Lithobius troglodytes* et ses variétés, et description de deux formes nouvelles. *Notes biospéol.* **XI.**

DEMANGE, J.M. (1958) Contribution à la connaissance de la faune cavernicole de l'Espagne (Myriapodes, Chilopodes, Lithobioidea). *Speleon.* **IX.**

DEMANGE, J.M. (1959) Contribution à la connaissance de la faune cavernicole de l'Espagne (Myriapodes) (2ème Note). *Speleon.* **X.**

DEMANGE, J.M. (1961) Un nouveau *Lithobius* cavernicole de Roumanie [Bulgarie]. *Annal. Spéléol.* **XVI.**

MATIČ, Z. (1957) Description d'un nouveau *Lithobius* cavernicole des Pyrénées espagnoles (Myriapoda, Chilopoda). *Notes biospéol,* **XII.**

MATIČ, Z. (1958) Contribution à la connaissance des Lithobiides cavernicoles de France (Collection *Biospeologica* VIIème et VIIIème Séries). *Notes biospéol.* **XIII.**

MATIČ, Z. (1959) Contribution à la connaissance des Lithobiides cavernicoles de la péninsule ibérique (Collection *Biospeologica* VIIème Série). *Biospeologica,* **LXXXIX.** *Archiv. Zool. expér. géner. N & R.* **XCVIII.**

Matič, Z. (1959) Noi contributii la cunoasterea Lithobiidelor cavernicole din Peninsula Iberica. *Stud. Cer. Biol. Cluj.* **X**.

Matič, Z. (1960) Beiträge zur Kenntnis der blinden *Lithobius*-Arten (Chilopoda-Myriapoda) Europas. *Zool. Anz.* **CLXIV**.

Matič, Z. (1960) Die Cryptodiden (Myriapoda, Chilopoda) der Sammlung des Speologischen Institutes "E.Gh.Racovitza" aus Cluj. *Zool. Anz.* **CLXV**.

Ribaut, H. (1915) Notostigmophora, Scolopendromorpha, Geophilomorpha (Myriapodes) (1ère Série). *Biospeologica*, **XXXVI**. *Archiv. Zool. expér. géner.* **LV**.

Verhoeff, K.W. (1930) Arthropoden aus südostalpinen Höhlen, gesammelt von Karl Strasser. 5.Aufsatz. *Mitteil. Höhl. Karstf.*

Verhoeff, K.W. (1933) Arthropoden aus südostalpinen Höhlen, gesammelt von Karl Strasser. 7.Aufsatz. *Mitteil. Höhl. Karstf.*

Verhoeff, K.W. (1936) Über einige Myriopoden und einen Isopoden aus mediterranen Höhlen. *Mitteil. Höhl. Karstf.*

Verhoeff, K.W. (1937) Chilopoden und Diplopoden aus jugoslawischen Höhlen, gesammelt von Dr. St.Karaman, Skoplje. *Mitteil. Höhl. Karstf.*

Verhoeff, K.W. (1939) Eine Höhlen-Scutigeride der Riu Kiu Insel Okinawa. *Mitteil. Höhl. Karstf.*

CHAPTER XI

THE APTERYGOTE INSECTS

A. THE INSECTA OR HEXAPODA

The insects or hexapods are probably derived from a line related to the Myriapoda, and in particular, the Symphyla. They have most likely arisen as a result of the phenomenon of paedomorphosis, involving the transformation of a hexapod larva into the sexual stage. The class Insecta is divided into two sub-classes, the Apterygota and the Pterygota.

B. APTERYGOTA

If the Apterygota and the Pterygota have a common origin it is certainly an ancient one, probably before the Devonian. We know today that apterism in the Apterygota represents a primitive character, not a secondary one. The development of certain Apterygota shows very primitive features (e. g. the anamorphosis of Protura). On the other hand the persistence of moulting throughout life may be regarded as primitive.

Paclt (1956) who has produced a recent monograph on the Apterygota, estimates that the group is monophyletic.

The Apterygota are divided into four orders:

Super Order Entotropha   (Collembola; Protura; Diplura)
Super Order Ectotropha   (Thysanura).

Cavernicolous Protura are not known. *Acerentulus catalanus* has been found at the entrance to the Grotte d'En Brixot in the Pyrénées Orientales, but this insect is not an inhabitant of the cave itself.

C. COLLEMBOLA

Introduction: The Collembola are insects of very small size. They measure in general 1–2 mm, although the extreme size-limits are 0·25 *(Neela)* and 8 mm.

The abdomen is made up of only six segments, the lowest number known in the Tracheata. This feature may be considered in addition to that of small size, as a paedomorphic character.

An extensive bibliography on the Collembola has been given by J.T. Salmon (1951) and J. Paclt (1956).

The Cavernicolous Collembola

The Collembola are very hygrophilic insects, preadapted to live in the subterranean world. Thus it is not surprising that they are abundant in caves.

We possess precise information on cavernicolous Collembola from only three regions of the world.

The first region, and that which has been best studied, comprises Europe and the Mediterranean region (South Europe, North Africa and Asia Minor). Numerous publications have been devoted to the cavernicolous Collembola of this region. The leading workers in this field include: K. Absolon, F. Bonet, K. Börner, P. Cassagnau, Cl. Delamare-Deboutteville, J.R. Denis, H. Gisin, E. Handschin, C.N. Jonesco, H. Kseneman, R. Moniez, J. Stach, I. Tarsia In Curia, H. Wankel and V. Willem.

The most recent work by H. Gisin, *Collembolen Europas* (1960) allows the zoologist to orientate himself rapidly in the systematics and distribution of European Collembola. Of the 842 species listed by Gisin, 147 or 17% have been found in caves. Of course, all these species are not true troglobia; numerous species are facultative or occasional guests.

The second region of the world in which the cavernicolous Collembola are well known is the Japanese Archipelago. The essentials of our knowledge of these insects may be found in the detailed work of R. Yosii, consisting of two monographs (Yosii, 1954, 1956).

The third region to be considered is North America. The cavernicolous Collembola of the United States are imperfectly known. The most recent paper on the subject is by K. Christiansen (1960), who has listed 30 species of Collembola found in caves in the United States. However, he regards this number as being only a very small fraction of the fauna which must populate the underground cavities of that country. The cavernicolous Collembola of Mexico constitute a varied and original fauna. It has been studied by F. Bonet, but much work is still needed.

Finally we possess a number of isolated observations on the Collembola collected in the subterranean cavities of Liban, India, Assam, Malaya, Cuba and Trinidad.

The studies which have been mentioned are devoted in particular to the systematics of the cavernicolous Collembola. Little is known of their ecology, but the reader will find an excellent summary in a paper by F. Bonet (1931), unfortunately already old.

The Distribution of Cavernicolous Collembola in the Systematic Classification

In general 13 families of Collembola are recognised, although some ento-mologists increase this number by elevating some tribes to the family status. Other workers reduce this number by relegating certain families to sub-families. Gisin, for example, has divided the Collembola into only five families. These questions will not be dealt with here; but it is advisable to bring into the light the very unequal distribution of cavernicoles among the families of the Collembola. This distribution is, moreover, different in different regions of the world.

Firstly let us examine the sub-order Arthropleona (all the segments of the body distinct). Two families, albeit poor in species, the Poduridae and the Actaletidae, contain no cavernicolous forms.

The Isotomidae do not seem to have given rise to true cavernicoles in Europe or Japan, but only to troglophiles and occasional cavernicoles. Nevertheless the American species *Folsomia cavernicola* is probably a true troglobian.

The Anuridae (or Neanuridae) are represented in the caves of Europe by only one troglophilic species. On the other hand, Japanese caves are much richer in the representatives of this family. Yosii has enumerated 18 species, many of which are troglophiles or even trogloxenes. However, some are completely depigmented and anophthalmic and certainly troglo-bia. This worker places such species in the genus *Caecoloba* which now contains five species native to Japan. *Caecoloba* is related to the genus *Lobella* which is epigeous or troglophilic, and has eyes.

The Hypogastruridae and the Onychiuridae are sometimes extremely abundant in caves. But they are more correctly regarded as guanobia and detriticoles than true cavernicoles. Numerous species are found on the surface as well as in caves. They may be considered to be recent caverni-coles, derived from humicoles found on the surface of the soil and living the primitive way of life of the Collembola. However, certain species of the Hypogastruridae belonging to the genera *Schaefferia* and *Mesochorutes* are true troglobia. They are very different from the surface forms and their origin is certainly an ancient one.

Most troglobia belong to four closely-related families, which are occa-sionally regarded as a single family. These are the Entomobryidae *(Si-nella, Pseudosinella, Metasinella, Verhoeffiella)*, the Tomoceridae *(Tomo-cerus* and *Tritomurus)*, the Paronellidae *(Troglopedetes)* and the Cypho-deridae *(Oncopodura)*. These are very agile forms found on concretions or upon the walls of caves covered with stalagmites. The Tomoceridae are so abundant in Japanese caves, that Yosii has called them the "Tomoceridae-caves".

The sub-order Symphypleona (in which the thoracic segments and the first abdominal segments are fused) contains a number of cavernicoles. The troglobious Sminthuridae belonging to the genera *Arrhopalites* and *Pararrhopalites* live on the humid surfaces of stalactites but it is particularly easy to collect these small jumping forms on the surface of gours. The Dicyrtomidae contain only a single cavernicolous form in Europe. Finally the microscopic Neelidae contain no cavernicoles in Europe and only a single such species in America.

Characters of Cavernicolous Collembola

1. The cavernicolous forms are distinguished from those of the surface by a partial or total depigmentation of the body.

2. The eyes of hypogeous forms are reduced or lack pigment *(Arrhopalites caecus* and *A. acanthophthalmus)*. It must be mentioned, however, that depigmentation of the ocular apparatus has also been observed in forms which are not, or which are only occasionally, cavernicoles *(Pseudosinella immaculata)*.

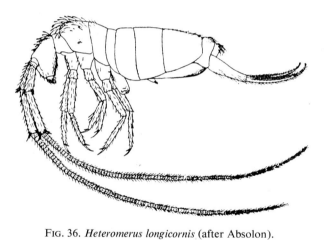

FIG. 36. *Heteromerus longicornis* (after Absolon).

In the troglobia the eyes have completely disappeared. This regressive manifestation is not confined to cavernicolous types. All the representatives of the families Onychiuridae, Cyphoderidae and Neelidae, together with numerous species of the genus *Willemia* (Hypogastruridae) are anophthalmic, irrespective of their habits. Anophthalmia is a phyletic manifestation and not an ecological one.

3. The antennary and post-antennary sense organs are sometimes better developed in cavernicoles than in epigeans, but the very precise studies of J. Stach (1934) on the genus *Onychiurus* have shown that such observed differences are always weak. The post-antennary apparatus probably plays a hygroscopic role.

The sense organs inserted on the third segment of the antennae in the Entomobryidae are particularly well developed in the cavernicolous forms (Christiansen, 1961).

4. The claws which terminate the legs are clearly more elongated in cavernicoles than in epigeans.

5. Certain cavernicolous species of the Entomobryidae are remarkable for the length of their antennae, legs, and furca. This is the case in *Heteromerus longicornis* from caves in Herzegovina and the Dinaric regions of

FIG. 37. *Metasinella acrobates* (after Denis).

Karst (Absolon, 1900), (Fig. 36); and also in *Metasinella acrobates* from the Cueva de Bellamar, Cuba (Denis, 1929) (Fig. 37). Such hypertely represents the end of a long orthogenetic evolution. A remarkable fact is that in the genus *Metasinella*, this extraordinary lengthening of appendages is found only in a single species, *M. acrobates*, and not in the other two species, *M. topotypica*, from caves of Cuba and *M. falcifera*, from caves in Yucatan (Bonet, 1944). It may be added that the instances of this extraordinary lengthening are found only in the family Entomobryidae, whose morphology is remarkable by reason of their slenderness. No modifications of this type have been found in other families of the Collembola. Later on (p. 484) some important consequences will be derived from these facts, which it is hoped will aid the understanding of the evolution of the cavernicoles.

Role of Collembola in the Economy of the Cavernicolous Association

The abundance of Collembola in the subterranean world is related to their adaptation to a polyphagous type of nutrition. They play a special role in the living hypogeous world by constituting an inexhaustible reserve of food, constantly used by carnivores, in particular the carabid Coleoptera.

D. DIPLURA

The Diplura are divided into four families; the Procampodeidae, the Campodeidae, the Projapygidae, and the Japygidae. The first family contains only two species, neither of which is cavernicolous. No cavernicolous Projapygidae are known.

(a) Campodeidae

This is the apterygote group which, with the Collembola, contains the majority of cavernicolous forms. The hypogeous Campodeidae are known mainly through the studies of J . R . Denis, R . Moniez, J . Stach, A . Viré and P . Wygodzinsky and above all B . Condé and F . Silvestri. Condé (1955) has given an excellent monograph on this family, and a complete bibliography has been provided by Paclt (1956).

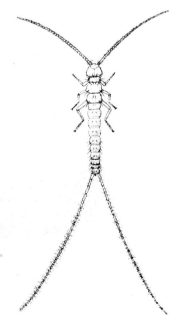

FIG. 38. *Campodea procera* (after Condé).

All the Campodeidae lead a subterranean life. Some are endogeans which live in crevices in the ground, or under stones, while others are cavernicoles.

The genus *Campodea* contains mainly endogeous species, but several of them can be regarded as troglophiles or even troglobia. The genus *Plusiocampa* contains the immense majority of cavernicolous forms from the Mediterranean region and North America. Finally a number of generally monospecific genera are known which are strictly cavernicoles. These in-

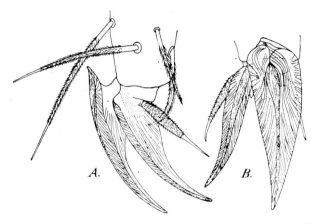

FIG. 39. The distal extremity of the third leg of *Hystrichocampa pelletieri*. *A*, anterior aspect; *B*, tergal aspect (after Condé).

clude: *Hystrichocampa* (Jura, France), *Patrizicampa* (Sardinia), *Tachycampa* and *Jeannelicampa* (North Africa), *Vandelicampa* (Lebanon), *Juxtlacampa*, *Parallocampa*, *Paratachycampa* (Mexico), *Simlacampa* (India), and *Antrocampa* (South Africa). The cavernicolous campodeids are similar to the endogeous forms; all are white, depigmented and anophthalmic irrespective of habitat. The cavernicoles differ in a very general way from the endogeans by their rather greater size, longer antennae and cerci (Fig. 38), an increased number of antennal segments, and claws furnished with well developed, striated, latero-tergal crests (Fig. 39). One of the most highly modified forms is *Anisocampa leleupi* from the Bats Caves in Table Mountains (Cape Province). This species measures 6·5 mm in length not including the appendages. The antennae are one and a half times the length of the body and consist of 52 segments. The cerci are as long as the body (Condé, 1962).

It will be shown elsewhere (p. 485) that the origin of these characters is not related to the cavernicolous mode of life.

(b) Japygidae

The cavernicolous Japygidae are few. Our knowledge of them is exclusively due to the researches of two entomologists, one French and the other Italian; J.Pagès and F.Silvestri.

The Mediterranean region contains a number of cavernicolous forms of this family: *Japyx gallicus* and *J. doderi* (Pyrénées); *Metajapyx moroderi* (Spain); *M. moroderi patrizianus* (Sardinia); *Dipljapyx fagniezi* (Alpes Maritimes), and *D. nexus* (Italy). A cavernicolous species is known from California *(Japyx kofoidi)* and two more from Africa: *Catajapyx jeanneli* (East Africa), and *Austrjapyx leleupi* (the Congo).

All the Japygidae are decoloured and anophthalmic, and the cavernicolous forms differ very little from the endogeous members of the family.

E. THYSANURA

The Thysanura are very poorly represented in caves. This order has been divided by Börner into two sub-orders, the Archeognatha (= Machiloidea) and the Zygentoma.

The Machilidae are represented by only a few "regular trogloxenes" in European caves; but they are seasonal, the subterranean environment is merely a place of hibernation for them (Tercafs, 1960). Wygodzinsky (1958) has given a list of machilids found in caves in the south east of Europe, but gives no information on their ecology.

In America the Machilidae are represented in the subterranean world by *Machilis cavernicola* from Mammoth Cave, described by Tellkampf in 1844; a specimen of this species has been found in the same cave by A.S.Packard. This cavernicolous form would be decoloured. According to Packard, another machilid has been captured in Wyandotte Cave.

Among the Zygentoma only the Nicoletiidae will be considered. All the members of this family are without pigment and are anophthalmic. The majority live in ant hills or termite nests. A few species of *Nicoletia* have been reported from caves in the Mediterranean region. Silvestri (1938) has described *Nicoletia jeanneli* from a cave in the department of Var, in Southern France. Joseph has discovered and described two species from caves in Karst; these are *Nicoletia cavicola* and *Troglodromicus cavicola*. In America, Ulrich has described *Nicoletia texensis*, from caves in Texas, and Bilimek has discovered *Lepisma anophthalma* in the famous Cacahuamilpa Cave, near to Mexico. The latter thysanuran is probably better regarded as a species of *Nicoletia*.

BIBLIOGRAPHY

Collembola

ABSOLON, K. (1900) Über zwei neue Collembolen aus den Höhlen des österreichischen Occupationsgebietes. *Zool. Anz.* **XXIII.**

ABSOLON, K. and KSENEMANN, H. (1942) *Troglopedetini.* Vergleichende Studie über eine altertümliche höhlenbewohnende Kollembolengruppe aus den dinarischen Karstgebieten. *Stud. Karstf. (biol. Ser.)* No. 16.

BONET, F. (1929) Colembolos cavernicolas de España. *Eos.* **V.**

BONET, F. (1931) Estudios sobre Colembolos cavernicolas con especial referencia a los de la fauna española *Mem. Soc. Esp. Hist. Nat.* **XIV.**

BONET, F. (1944) Sobre el genero *Metasinella* Denis y algunos otres Colembolos cavernicolas de Cuba. *Ciencia* **V.**

CARPENTER, G.H. (1897) The Collembola of Mitchelstown Cave. *Irish Naturalist* **VI.**

CASSAGNAU, P. and DELAMARE-DEBOUTTEVILLE, CL. (1953) Les *Arrhopalites* et *Pararrhopalites* d'Europe (Colemboles, Symphypleones cavernicoles). *Notes biospéol.* **VIII.**

CHRISTIANSEN, K. (1960a) The genus *Pseudosinella* (Collembola, Entomobryidae) in caves of the United States. *Psyche* **LXVII.**

CHRISTIANSEN, K. (1960b) The genus *Sinella* Brook (Collembola, Entomobryidae) in Nearctic Caves. *Ann. Entomol. Soc. America* **LIII.**

CHRISTIANSEN, K. (1960c) A preliminary survey of the knowledge of North American cave Collembola. *Amer. Midl. Nat.* **LXIV.**

CHRISTIANSEN, K. (1961) Convergence and parallelism in cave Entomobryidae. *Evolution* **XV.**

DENIS, J.R. (1929) Notes sur les Collemboles récoltés dans ses voyages par le Prof. F. Silvestri. *Boll. Labor. Zool. Portici.* **XXII.**

GISIN, H. (1960) *Collembolenfauna Europas.* Genève.

HANDSCHIN, E. (1926) Subterrane Collembolengesellschaften. *Archiv f. Naturg.* **XCI,** Abt. A.

PACLT, J. (1956) *Biologie der primär flügellosen Insekten.* Jena.

SALMON, J.T. (1951) Keys and Bibliography to the Collembola. *Public. Victoria Univ. Coll. Zool.* No. 8.

STACH, J. (1934) Die in den Höhlen Europas vorkommenden Arten der Gattung *Onychiurus* Gervais. *Ann. Mus. Polon.* **X.**

YOSII, R. (1954) Höhlencollembolen Japans. I. *Kontyu.* **XX.** II. *Jap. Jour. Zool.* **XI.**

YOSII, R. (1956) Monographie zur Höhlencollembolen Japans. *Contrib. Biol. Lab. Kyoto Univ.* No. 3.

Campodeidae

CONDÉ, B. (1955) Matériaux pour une Monographie des Diploures Campodéidés. *Mém. Mus. Hist. Nat. Paris N.S. Ser. A. Zool.* **XII.**

CONDÉ, B. (1961, 1962) Découverte d'un Campodéide troglobie en Afrique Australe. *Compt. rend. Troisieme Congr. Intern. Spéleologie, Vienne.*

PACLT, J. (1956) *Biologie der primär flügellosen Insekten.* Jena.

Japygidae

A most complete review has been given by Paclt:

PACLT, J. (1956) *Biologie der primär flügellosen Insekten.* Jena, 1956.

Thysanura

SILVESTRI, F. (1938) Descrizione di una nuova specie di *Nicoletia* vivente in una grotta della Francia (Insecta, Thysanura). *Rev. franç. Entomol.* **V.**

TERCAFS, R.R. (1960) Répartition géographique et remarques écologiques sur les Machilidae (Apterygota, Thysanura) cavernicoles de Belgique. *Les Naturalistes belges* **XLI.**

WYGODZINSKY, P. (1958) Notes et descriptions de *Machilida* et *Thysanura* palaearctiques. *Rev. franç. Entomol.* **XXV.**

CHAPTER XII

THE PTERYGOTE INSECTS
(EXCLUDING COLEOPTERA)

A. THE PTERYGOTA

The Pterygota is much the best represented group in the world today, for it contains more than a million species.

The cavernicolous Pterygota are numerous, in both species and individuals, but their distribution in the systematic classification is quite different from that which may be observed in the other subdivisions of the Arthropoda. While most of the orders of the Arachnida, Crustacea, Myriapoda and Apterygota contain cavernicolous species, this is not the case within the Pterygota. The true troglobia are found only in two tribes of the Coleoptera. The other pterygote orders contain no true cavernicoles or only rare forms which may uncommonly be categorised as specialised troglophiles. Thus, contrary to that which is commonly believed, the insects are very refractory to cavernicolous life.

This perhaps surprising conclusion may be explained by the following two considerations:

(1) Firstly there is the question of nutrition. The strictly phytophagous species cannot live in caves by reason of the lack of food-stuffs which make up their normal diet. But the majority of the Pterygota are phytophagous species.

Only omnivores and carnivores can live in caves, and it is thus certainly not by chance that the only two cavernicolous tribes of the Coleoptera correspond exactly, one to a carnivorous type and the other to an omnivorous one.

(2) The cavernicoles belong essentially to relict groups wihch have taken refuge in the depths of the earth and are unadapted to conditions which reign in the world at the present time. But this is certainly not the case with the Pterygota, which are perfectly in accord with our time. They are a freely expanding group. The Lepidoptera, Diptera, Hymenoptera and Coleoptera are by far the most numerous specimens of the present terrestrial fauna. Expansion, and refuge in subterranean retreats, are incompatible conditions. This is why the pterygote groups which contain true troglobia are exceptions.

B. CLASSIFICATION OF THE INSECTS

The thirty orders which today make up the class Insecta have been grouped by Handlirsch and Martynov into eight super-orders: Palaeoptera, Blattoidea, Orthopteroidea, Coleopteroidea, Psocoidea, Neuropteroidea, Hymenopteroidea and Panorpoidea (= Mecopteroidea).

The Palaeoptera (Ephemeroptera and Odonata) do not contain cavernicolous species.

C. BLATTOIDEA

This super-order has been divided into three orders:

Order Dictyoptera $\begin{cases} \text{sub-order} & \text{Blattodea} \\ \text{sub-order} & \text{Mantodea} \end{cases}$

Order Zoraptera

Order Isoptera.

Only the order Dictyoptera, or more exactly the sub-order Blattodea contains cavernicoles. No Mantodea live underground.

The basis of our knowledge of the cavernicolous cockroachs rests on the work of Lucien Chopard who has produced nearly 50 publications devoted to the study of cavernicolous Blattoidea and Orthopteroidea.

At present 33 species of Blattodea have been collected in caves (Chopard, 1954). However, of these 33 species, only a few can be regarded as true cavernicoles. The cavernicolous Blattodea are in general tropical forms, as are the Dictyoptera as a whole.

Excluding the trogloxenes, the Blattodea found underground may be divided into three categories, the troglophiles, the guanobia and the troglobia.

Troglophilic Blattodea

The majority of the Blattodea found in caves belong to this category. These forms are but little modified.

Guanobious Blattodea

Numerous Blattodea are associated with guano. They are generally well pigmented and possess eyes. We may mention as examples; *Pycnoscelus niger* (Tongking); *P. striatus* (Batu Caves, Burma); *P. surinamensis* (Siju Caves, Assam); *Dyscologamia pilosa* (Malaya); *Gyna vetula* and *G. kazungulana* (East Africa); *Apotrogia* deplanata* and *A. trochaini* (Congo), and *Arenivaga erratica* (Texas).

---

* The name *Apotrogia* must be substituted for *Acanthogyna* (Chopard, 1952).

Transitory Forms between Troglophiles and Troglobia

A certain number of Blattodea can be regarded as intermediate between troglophiles and troglobia. This is the case with *Ischnoperta cavernicola* from Sarawak and *I. remyi* from Tongking and *Typhloblattodes madecassus* from Madagascar. Two genera are particularly interesting from this point of view. These are *Alluaudellina* and *Nocticola*, upon which a few details will be given.

The Genus Alluaudellina. *A. cavernicola* was originally regarded as a troglobian but it is, in fact, a troglophile which is linked with some termitophilous species. The first example of this insect was discovered by Ch. Alluaud in 1909 in the Kulumuzi Caves in the district of Tanga (Tanga-

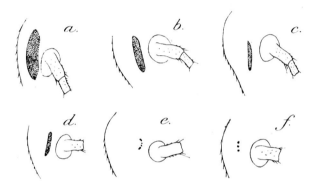

FIG. 40. Variations in the structure of the eye of males and females of *Alluaudellina cavernicola* (after Chopard).
a, macropterous and macrophthalmous male; *b*, macropterous and microphthalmous male; *c*, micropterous and microphthalmous male; *d*, female with eye present; *e* and *f*, females with eyes very reduced.

nyika), and was described by Shelford in 1910. Other specimens were discovered in 1912 by Alluaud and Jeannel from the same caves and also in caves in the neighbourhood of Shimoni (Chopard, 1932).

This species has also been found under rocks in bamboo forest on Mount Kinangop. These individuals are very similar to the cavernicolous forms (Chopard, 1936). It must be concluded that this blattid is a troglophile, and an endogean, but not a troglobian. In addition, another species of this genus, *A. himalayensis* has been collected on the surface in the region of Darjeeling (north-east India).

This blattid is yellowish white in colour and the legs and cerci are very long. The antennae are three times the length of the body. The elytra of the male are well developed and exceed the abdomen in length, but the wings are shorter than the elytra. However, there is polymorphism in the males and both macropterous and micropterous types are present. The latter has shorter elytra and the wings are reduced to small lobes. The females possess neither wings nor elytra.

The variations of the ocular system of this blattid are particularly interesting. Certain males possess relatively well developed eyes, whilst others have the eyes reduced to three ommatidia, but intermediates exist between the two extreme types (Fig. 40). In the female the eyes are always reduced, and this reduction is manifest in various stages of regression. One can, by uniting specimens from the two sexes, obtain a graded series, starting from an almost normal eye and ending in a completely regressed one. It would seem, however, that no completely anophthalmic forms exist.

It is a remarkable fact that the regression of the wings and eyes occurs in a parallel way. Ocular regression precedes the reduction of wings so that one can observe a series of individuals showing the following gradation:

macropterous and macrophthalmous types
macropterous and microphthalmous types
micropterous and microphthalmous types.

It will be shown later (p. 435) that similar relationships are found in the Coleoptera.

Thus, *A. cavernicola* may be regarded as a species in a state of morphological instability. This is a very clear example of the march of regressive evolution.

The Genus Nocticola. This genus appears to be related to the last one and also to epigeous genera such as *Cardacus* and *Cardacopus* (Chopard, 1951). *Nocticola* contains half a dozen species which are cavernicolous, or termitophilic *(N. termitophila* and *T. sinensis)* or endogeous *(N. bolivari* and *N. remyi)*.

The cavernicoles are represented by three species; two of them *(N. caeca* and *N. simoni)* were discovered by Eugène Simon in caves on the island of Luzon (Philippines) and the third, *N. decaryi*, has been captured in caves hollowed from the interior of Cretaceous calcareous strata near Majunga (Madagascar).

The antennae and legs are long but the elytra are shortened. The wings are reduced to stumps in the males and are absent in the females. The eyes are also reduced, and made up of an incurved band composed of a few ommatidia. The development of the eyes is delayed. The larvae possess only three ommatidia, or even no trace of the ocular apparatus at all.

Troglobia

The blattids which one can regard as troglobia, that is to say those which are completely anophthalmic and which bear extremely long appendages (antennae, legs and cerci), include only two species. Both belong to the same genus, *Spelaeoblatta*, which would seem to be related to *Nocticola*

(Chopard, 1951). These two species, *S. caeca* and *S. gestroi* have been collected in the same region of tropical Asia (Burma).

Spelaeoblatta is the most modified cavernicolous blattid known. It is depigmented and without eyes and ocelli. It is completely apterous, the wings and elytra are wanting. The appendages are remarkably elongated (Bolivar, 1897; Chopard, 1921).

D. ORTHOPTEROIDEA

The super-order Orthopteroidea comprises four orders, the Plecoptera, the Embioptera, the Orthoptera and the Dermaptera. Only the last two orders contain some cavernicolous forms.

(a) Orthoptera

The Orthoptera contains 107 cavernicolous species (Chopard, 1954); but these are not distributed evenly in the different subdivisions of this order. They belong to three sub-orders only: the Gryllacridoidea, the Gryllodea and the Notoptera (= Grylloblattidae). On the other hand there are no known cavernicoles in the sub-orders Phasmida, Tettigonidea (locusts) and Acridiidae (crickets). The reason for this must be sought in the fact that these sub-orders are phytophagus, and thus cannot find the food they require underground.

Cavernicolous Orthoptera have been the subject of numerous studies by B. Baccetti; F. Capra; L. Chopard; A. Griffini; T. Hubbell; F. M. Hutton; H. H. Karny; B. Lanza; A. M. Richards and S. H. Scudder.

(1) Gryllacridoidea. This sub-order contains the majority of cavernicolous Orthoptera, in fact, 69 species from a total of 107 (Chopard, 1954). All hypogeous types belong to a single family, the Rhaphidophoridae.

The Family Rhaphidophoridae: It is not by chance that this family contains the majority of cavernicolous Orthoptera. Their morphology and their physiology predispose them for hypogeous life. In fact, they are all apterous forms, with long, slender limbs. They are hygrophilic and omnivoruos insects; they are "scavengers" which feed on animal or vegetable remains. It is a diet which permits them to exist in caves while vegetarians would die there of starvation. They are well qualified to utilise efficiently the food resources which are offered to them in the subterranean world. E. Jacobson records the case of *Rhaphidophora picea* feeding on the bodies of bats in the caves of Java (Chopard, 1924). More curious, however, are the feeding habits of *R. oophaga*, which have been reported by Lord Medway. This species, which inhabits a cave on Sarawak (Borneo) normally feeds on insect debris and the bodies of birds and bats, but they also invade the nests of salanganes *(Callocalia maxima* and *C. fuciphaga)* to feed on their eggs and attack their young (Chopard, 1959). The

B 7a

Rhaphidophoridae of New Zealand, leave the caves during the night to search mosses, liverworts and the green plants which grow around the entrances (Richards, 1962).

The Rhaphidophoridae contain numerous cavernicolous species of which the majority have been described by Lucien Chopard. They are particularly common in tropical countries, in particular in Asia where they are represented by the genera *Rhaphidophora, Diestrammena, Paradiestrammena*

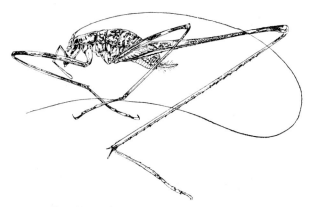

FIG. 41. *Speleotettix tindalei* (after Chopard).

and *Tachycines*. Cavernicolous species are also known from New Zealand *(Pachyrhamma, Pleioplectron, Neonetus, Weta); Speleotettix* comes from Australia and New Zealand, and *Speleiacris* from South Africa.

The cavernicolous Rhaphipophoridae of New Zealand are particularly well known, thanks to the detailed studies which A.M.Richards has devoted to their anatomy (1955), systematics (1960) and biology (1961, 1962).

All these forms are troglophiles which occasionally leave the caves during the night in the manner that E.Jacobson has observed for *Rhaphidophora picea.*

Their limbs may be greatly elongated. This is the case in *Diestrammena,* and especially *Speleotettix* (Fig. 41), but other genera such as *Rhaphidophora* have normal appendages.

The eyes may be normal or reduced. The ocular apparatus of *Tachycines cuenoti* shows various stages of regression which depends on the individual. This is an example of evolutionary instability (Chopard, 1929).

The only two anophthalmic and truly troglobious Orthoptera, are *Diestrammena caeca* (Lakadong Cave, Assam) and *D. cassini* (a cave in Tongking). In the first species the eyes are reduced to a hardly visible depigmented spot, and in the second species the eye has completely failed to appear.

The males of certain genera of the Rhaphidophoridae *(Troglophilus, Ceutophilus, Hadenoecus)* possess exsertile, glandular organs (the glands of Hancock) inserted on the abdomen. They serve to facilitate the bringing together of the sexes. Seliškar (1923) who carried out a detailed study of these organs in *Troglophilus*, believed that these organs are particularly well-developed in cavernicolous species because they aid the meeting of the sexes which has to take place in complete darkness. However, the exsertile organs are completely absent in the Rhaphidophoridae of New Zealand (Richards, 1961).

The Holarctic Rhaphidophora: The Rhaphidophora of the present day are localised in tropical regions and in southern areas of the world. But at one time they populated the Northern Hemisphere, as has been proved by the discovery of authentic rhaphidophorids in the amber of eastern Prussia, namely *Prorhaphidophora antiqua* and *P. zeuneri* (Chopard, 1936). This discovery shows that at the beginning of the Oligocene the climate of Europe was warm and humid, analogous to that of present-day tropical regions.

At the present time, the holarctic region contains four rhaphidophorid genera which are cavernicoles or endogeans. These are *Dolichopoda* and

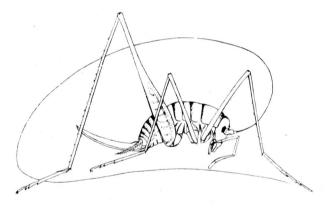

FIG. 42. *Dolichopoda bormansi* (after Chopard).

Troglophilus in Europe, and *Hadenoecus* and *Ceutophilus* in North America. *Dolichopoda* and *Hadenoecus* have elongated appendages and are certainly related to each other. They are placed in the tribe Dolichopodini. By contrast *Troglophilus* and *Ceutophilus* have shorter and more robust appendages and are classified in the Ceutophilini (Chopard, 1931).

The origin of this group must go back to the Tertiary because, with the exception of the cavernicoles, all the Rhaphidophoridae are tropical or southern forms. The European and American cavernicolous forms are a relict of an ancient tropical fauna.

The genus *Dolichopoda* is remarkable because of its depigmented integument, its reduced eyes and the absence of wings or elytra (Fig. 42). The palps and legs are elongated and the antennae are five times longer than the body (75–80 mm for a body length of 16 mm). The genus is widespread in the caves of the Mediterranean region, from Spain to the Caucasus Mountains and Asia Minor (Chopard, 1932). This genus is particularly well-represented in Italy where it is so polymorphic that entomologists have had to refer to the chromosomal patterns in order to distinguish the species. The males of the sub-genus *Dolichopoda* possess a diploid number equal to 31, and in those of the sub-genus *Chopardina* it equals 35 (Baccetti and Capra, 1959).

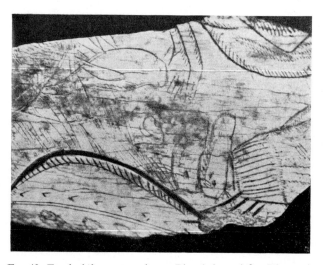

Fig. 43. *Troglophilus* engraved on a Bison's bone (after Bégouen).

The genus *Troglophilus* is distinguished from the preceding genus by a more compact and robust body and shorter appendages. The species of this genus are today widespread from Italy to Asia Minor. But the remarkable discovery of a prehistoric engraving (Fig. 43) depicting *Troglophilus*, shows that the geographical distribution of this genus was at one time more extended. This engraving was discovered by Count Bégouen while excavating in the Grotte des Trois Frères in the Pyrénées ariègeoises (Chopard, 1928; H. and L. Bégouen, 1928; H. Bégouen, 1929).

The American genus *Ceutophilus* has been the subject of a very complete monograph by Hubbell (1936).

The holarctic rhaphidophorids cannot be regarded as troglobia. They are really troglophiles. In addition their way of life is very varied. *Hadenoecus puteanus* and *Ceutophilus brevipes* are only exceptionally found in caves. *Hadenoecus subterraneus* leaves the caves after twilight, in order to seek its food, returning inside before dawn (Nicholas, 1962). Nevertheless, *Ceutophilus stygius* does not perform such migrations.

Likewise, it has been known for a long time that species of *Dolichopoda* come out from their retreats at night to hunt outside the caves (Chopard, 1917, 1932; Lanza, 1954).

Regarding *Troglophilus*, those which do live in caves are frequently found near the entrances (Waldner, 1940). They may also be found in woods, under rocks and bark (Remy, 1931). They are carnivores which feed on dead insects and some vegetable matter (Remy, 1931).

(2) Gryllodea. The Gryllodea, which are more or less omnivorous and hygrophilic insects, have given rise to cavernicolous forms. But again, the cavernicolous forms are not distributed uniformly between the dozen families which constitute this sub-order. If one excludes a guanobious mole-cricket *(Gryllotalpa nigripennis)* found in the Batu Caves (Selangor State), a few European gryllids which are troglophilic *(Gryllomorpha, Pelatoptila* and *Aneuroptila),* a nemobiid from Assam *(Speonemobius decoloratus)* and a pentacentrid from the caves of Yucatan, Mexico *(Tohila atelomma),* all the other cavernicolous Gryllodea belong to the family Phalangopsidae.

The phalangopsids are hygrophilic and inhabit woods; they are widespread in the tropical regions. They have given rise to a number of cavernicoles which entomologists have placed in several genera:

Tropical Africa: *Homeogryllus, Pholeogryllus, Phaeophilacris, Speluncaris*

Madagascar: *Malgasia* (= *Voeltzkowia*)

Malaya and Indonesia: *Parendacustes, Arachnomimus*

Central America: *Amphiacusta, Parcophus*

Jamaica: *Uvarioviella*

Uruguay: *Dyscophogryllus.*

The majority of cavernicolous Gryllodea are troglophiles. *Amphiacusta azteca* has been captured in Mexico both on the prairies and in a cave.

The cavernicolous Gryllodea are generally but slightly modified. The colour of the body is pale; the eyes and ocelli are reduced. The wings are regressed or absent; the elytra short, or even absent; they may, however, retain a virtually unmodified anatomy; in the male a functional stridulatory mirror is sometimes present. This is the case in *Amphiacusta bolivari* which Candido Bolivar has heard stridulating in the Cueva de Atoyac in the province of Vera Cruz, Mexico (Chopard, 1947).

The most modified species which tends towards the troglobious type is *Arachnomimus microphthalmus* from the Batu Caves (Selangor, Malaya). It is a depigmented form with reduced eyes and with greatly elongated appendages.

(3) Notoptera. This sub-order is composed of a single family the Grylloblattidae. It is a very small group which comprises only two genera, *Grylloblatta* (North America) and *Galloisiana* (Japan).

They are apterous thysanuriform insects with multiarticulate cerci which resemble antennae. These primitive Orthoptera are very ancient relicts which resemble Protorthoptera more than present day forms. They are preadapted for cavernicolous life, depigmented and yellowish in colour. The eyes are reduced, and the ocelli absent. The species of *Galloisiana* live in the mountainous forests of Japan. But one species, *G. chujoi* has been collected in Japanese caves (Gurney, 1961).

(b) Dermaptera

The Indo-Malayan region contains about half a dozen Dermaptera which are occasionally troglobia and which hardly differ from surface forms. They are of mediocre interest to biospeologists. However, J. Millot, and then Herniaux and Villiers (1955) have captured a forficulid, *Diplatys milloti* (Chopard, 1940) in some caves in Guinea, which has some very definite cavernicolous characteristics. These include depigmentation of the integument, and elongation of the antennae and the legs, but the eyes are hardly smaller than those of lucicoles. This can, therefore, only be regarded as a troglophile.

E. PSOCOIDEA

Three orders of unequal importance have been united in the super-order Psocoidea. They are the Psocoptera, the Thysanoptera and the Hemiptera. The Thysanoptera contains no cavernicolous members.

(a) Psocoptera

Very little is known about the cavernicolous Psocoptera. Several cavernicolous species have been reported, from Europe and Algeria (Enderlein, Badonnel), the United States (Packard, Call, Bailey), and from the Congo (Badonnel). However, it is difficult to assess the value of these observations. This is for the following reasons. *Psyllipsocus troglodytes* which has been described by Enderlein as one of the most modified cavernicolous Psocoptera, is only, according to Badonnel (1943), an environmentally induced variety of the common form *P. ramburi*. In fact, the culture of the larvae of this species in a warm and humid environment brings about the loss of pigment and the disappearance of the eyes.

(b) Rhynchota (= Hemiptera)

This group contains no true cavernicoles. However, certain reduviid bugs are frequently found in the entrances of tropical caves. They will be referred to later in a chapter dealing with the parietal fauna.

Synave (1952, 1953) and Paulian and Grjebine (1953) have considered *Typhlobrixia namorokensis* (Cixiidae, Homoptera), which has been found in subterranean galleries in the Namoroka Reserve, Madagascar, to be a troglobian. This insect is partially depigmented. Its eyes are reduced but still contain red pigment. The ocelli are also strongly reduced and the frontal ocellus has entirely disappeared. However, this insect is not a troglobian but an endogean. They feed on roots which hang down in the roof of the cave and it is this habit which causes them to stray accidentally into the cave itself.

F. NEUROPTEROIDEA

Neuroptera

A number of ant-lions (Myrmeleonidae) which have been captured in caves are mentioned in the entomological bibliography. They include *Neglurus vitripennis* (Batu caves, Malaya), the larvae of which live about 300 metres from the entrance. *Myrmecaefunis sp.* also lives in these caves. *Eremoleon longior*, which has been found in three caves in Yucatan, Mexico, should be regarded as a troglophile.

G. HYMENOPTEROIDEA

Hymenoptera

Ichneumonidae and Proctotrypidae are regularly found at the entrances of caves. They form part of the parietal fauna to which we shall refer later.

The so-called "cavernicolous ants" are simply endogeous species which have been recovered at the mouths of caves, or in guano (Wheeler, 1922, 1924, 1938). No ant can be considered to be a true cavernicole (Wilson, 1962).

H. MECOPTEROIDEA

The super-order Mecopteroidea contains four orders, the Mecoptera, the Trichoptera, the Lepidoptera and the Diptera. The first order contains no cavernicoles.

(a) Trichoptera

Many species of caddis-fly are to be found at the entrances of caves. They form part of the parietal association which will be dealt with later.

However, the life cycle of *Wormaldia subterranea* takes place entirely underground (Radovanovič, 1932, 1935). The larvae live in streams which traverse the Podpeška Jama and the Pokrito brezno, which are caves to the south of Ljubljana. As the larvae have eyes and the adults are not obviously modified for cavernicolous life we must not exclude the possibility that this species may also be found on the surface. They should not be accepted as true cavernicoles.

(b) Lepidoptera

The Lepidoptera, which are insects characteristically attracted to warmth and to flowers, are because of this not adapted to embark upon a cavernicolous life. However:

(1) Some Geometridae and Noctuidae, and also Orneodidae and Acrolepiidae, are regularly found at the entrances of caves. They form part of the parietal association and we shall deal with them later.

(2) Some Microlepidoptera should not be regarded as true cavernicoles. They are troglophiles or guanobia and their larvae feed upon bat droppings.

About a dozen species are known, belonging to the families Pyralidae (*Pyralis*), Tineidae *(Tinea, Crypsithyris)*, Gelechiidae *(Tricyananla)*, Cosmopterygidae *(Pyroderces)*, Schreckensteiniidae *(Oedematopoda)* and Oinophilidae *(Opogona, Wegneria)*.

In some species partial depigmentation has occurred *(Crypsithyris spelaea)*; in others the size of the eye has been reduced *(Wegneria cavernicola)*.

All the troglophilic Microlepidoptera are indigenous to the Indo-Malayan region (Assam, Burma, Siam, Malaya and Java). Species of *Opogona* have been found in caves in the Congo.

(c) Diptera

The Diptera are one of the largest insect orders. Several hundred species of Diptera have been found in caves. However, it is often difficult to find them a place in the existing ecological classification of cave fauna. It has often been stated that the Diptera contain no true cavernicoles but only occasional guests or at the most, troglophiles. This position is certainly too categorical (Bezzi, 1911), but it must be recognised that troglobia are very rare within the Diptera. In any case, they are much less modified than some other Diptera, for instance the termitophilic Diptera, or the Pupipara.

The Cavernicolous Diptera as Relicts. The cavernicolous Diptera form part of a group of apterous or sub-apterous flies which are ancient relicts and which may be found in areas which possess a rigorous climate (e.g. Antarctic islands, snow holes and caves), but which lack predators.

The Distribution of Cavernicolous Diptera Within the Systematic Classification. This is a most unequal distribution. Certain groups, particularly those within the higher Diptera, contain very few cavernicolous species.

This is the case in the Brachycera (with the exception of a few Empididae, belonging to the sub-family Atalantinae), and in the Calypteratae. Cavernicolous Diptera usually belong to ancient and primitive groups.

The Diptera found in caves belong to only a small number of families (Bezzi, 1911; Leruth, 1939). These are the four nematocerous families Tipulidae, Chironomidae, Sciaridae (Lycoriidae) and Mycetophilidae (= Fungivoridae); one family within the Aschiza, the Phoridae, and three families belonging to the Acalypteratae: the Borboridae (Cypselidae), the Helomyzidae and the Mormotomyidae.

The Ecological Classification of the Cavernicolous Diptera

The Parietal Association. A number of Diptera form part of this association, with which we shall deal later. These flies belong to those families which have just been mentioned and also to the Culicidae, and the Psychodidae.

Cold Caves and Snow-holes. Two species of the apterous tipulid *Chionea (C. alpina* and *C. lutescens)* have been found in cold caves in the Alps and the Balkans.

The snow-holes or "tesserefts" of the Djurdjura Mountains (Algeria) are the habitat of the chironomid *Cataliptus peyerimhoffi*. The wings are reduced to small lamellae, there are no halteres; the legs are very elongated, and give the insect the appearance of a spider (Bezzi, 1916).

The Guanobious Diptera. Many species of Diptera are always found in caves because their larvae feed on the guano deposited therein. They may thus be called guanobious Diptera.

One of the best known of these is the helomyzid *Heteromyiella atricornis*, which in France is called the "mouche du guano".

Two guanobious mycetophilids have larvae which possess the curious habit of spinning webs in the manner of spiders, but certainly for different reasons. These are *Macrocera fasciata*, the larvae of which have been studied by Enslin (1906), and *Speolepta leptogaster*, which has been investigated by Schmitz (1909, 1912).

The guanobious Diptera show little morphological modification with respect to surface forms. However, many guanobious Phoridae and Borboridae are unable to fly although their wings are apparently normal. They may be seen running over guano or on stalagmites without ever taking to the wing. In a few species the females have a strong tendency towards physogastry. This may be seen in the phorids *Triphleba aptina* (= *Phora aptina*) and *T. antricola* (Bezzi, 1911, 1914; Schmitz, 1918, 1929), and in the borborid *Limosina racovitzai*.

The Troglophilic Diptera. A number of Diptera which have had their origin in the parietal fauna have eventually invaded the interiors of caves and become exclusively cavernicolous.

Numerous species of *Sciara* inhabit caves; our knowledge of them is principally due to the work of Franz Lengensdorf. The majority live at the mouths of caves, but at least one species, *S. ofenkaulis*, is a true cavernicole which never leaves the subterranean environment. It has elongated legs and antennae. This species has been captured on numerous occasions in the caves in West Germany (Lengensdorf, 1924; Griepenburg, 1934) and more recently in France in caves in the department of Côte d'Or (Matile, 1960).

The very rare helomyzid, *Gymnomus troglodytes* provides us with an example of a new stage in the evolution of cavernicoles. It has been found in only a few caves in Jugoslavia (Bosnia, Herzegovina, Croatia and Slovenia). The eyes are reduced and the ocelli, although present, are very small.

A number of species of *Anopheles* and *Phlebotomus* are troglophiles, strictly confined underground.

Cavernicolous Species of Anopheles. In tropical Africa there are about six species of this genus whose life cycles are entirely completed in caves. The larvae develop in pools and gours in caves where they feed on bat droppings. The adults attack bats and suck their blood shortly after emergence, and in so doing transmit some Haemosporidea specific to bats (see Chapter XV) (Wanson and Lebied, 1945; Leleup and Lips, 1950, 1951; Abonnenc, 1954; Mattingley and Adam, 1955; Lips, 1960; Adam, 1962).

The cave-dwelling anophelines have normal wings and eyes, and are blackish in colour. They cannot therefore be regarded as true cavernicoles.

The Cavernicolous Species of Phlebotomus. These minute, biting Diptera belong to the family Psychodidae. A few species live in caves in equatorial Africa. These are: *Phlebotomus (Spelaeophlebotomus) gigas* Parrot and Schwetz, 1937, from caves in the Congo (Thysville and Matadi) and in Guinea (Kindia); *P. mirabilis* Parrot and Wanson, 1939, from caves in the Congo (Thysville); and *P. crypticola* Abonnenc, Adam and Bailly-Choumara, 1959, from caves in the Sudan.

These species have been found in totally dark parts of caves, often several hundred metres from the entrance. While the larvae have yet to be discovered it seems quite probable that the life cycle takes place entirely within caves. The adults suck the blood of Minioptera. The alary or auricular membranes are the most frequent points of attack (Wanson and Lebied, 1946). We may therefore regard these species of *Phlebotomus* as permanent cavernicoles. They also possess a number of morphological peculiarities characteristic of cavernicoles, for example, the body is but little pigmented and the eyes are reduced in size.

The Troglobious Diptera. None of the forms mentioned above should be regarded as true troglobia. Indeed there are but few Diptera which would seem to qualify for this title. However, we shall mention the three most characteristic examples of them. *Speomyia absoloni* (Borboridae), was discovered by Absolon in some caves in Herzegovina and described by Bezzi (1914). This fly is black in colour, but the wings and eyes are reduced and the ocelli have completely disappeared (Fig. 44). These characteristics are unquestionably morphological modifications typical of cavernicoles.

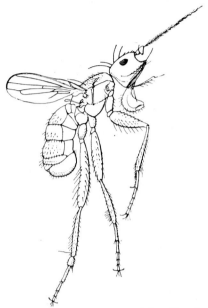

FIG. 44. *Speomyia absoloni* (after Séguy).

Our second example belongs to a new family, within the Acalypteratae, created by Austen in 1916 and called the Mormotomyidae. This new family is related to the Borboridae. The family was established to include a curious dipteran *Mormotomyia hirsuta* which was discovered in a cave in Kenya. This species is covered with dense hairs and this, together with its long legs gives it the appearance of a spider. The eyes are very small and the wings vestigial. The halteres are absent. This species is perhaps an ultra-specialised guanobian.

Our third example is a fly which is still more profoundly modified: the sciarid *Allopnyxia patrizii* (Freeman, 1952; Patrizi, 1956) (Fig. 45). This fly has been discovered in the deepest part of a cave in Latium, some 40 km from Rome.

There is extreme sexual dimorphism in this dipteran. The males have reduced wings with a simplified venation, but the thorax is, however, normal. The eyes are made up of eight depigmented ommatidia and the ocelli are present. (Fig. 45.2 and 4.)

In the female there is marked physogastry. The wings, halteres and even the scutellum are absent. The eyes have only four or five ommatidia and the ocelli are absent. (Fig. 45.1 and 3.)

This species is a true troglobian and its origin may be simply explained in the following terms. A number of sciarids are known from great humid

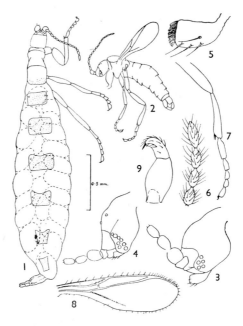

FIG. 45. *Allopnyxia patrizii* (after Freeman).
1, female; *2*, male; *3*, head of the female; *4*, head of the male; *5*, female palpus; *6*, segments 13–16 of female antenna; *7*, posterior leg of female; *8*, wing of male; *9*, clasper of male

forests. The males are winged, but the females are apterous and lack halteres, and also more or less physogastrous, though their eyes remain normal. Examples of such sciarids are *Peyerimhoffa terricola* (Algeria), *P. tanyae* (Madagascar), and *Campyloneura recondita*, which may be found by sifting beech-litter in Transylvania. It seems that such humicolous species similar to the forms from which cavernicolous types such as *Allopnyxia* have taken their origin. Thus apterism has been a characteristic acquired before such Diptera found their way into caves.

BIBLIOGRAPHY

Blattoidea

BOLIVAR, J. (1897) Viaggio di Leonardo Fea in Birmania e regioni vicini. 78. Nouvelle espèce cavernicole de la famille des Blattaires, *Spelaeoblatta gestroi. Ann. Mus. Stor. Nat. Genova* **XXXVIII.**

CHOPARD, L. (1921) On some cavernicolous Dermaptera and Orthoptera from Assam. *Rec. Indian Mus.* **XXII.**

CHOPARD, L. (1932) Un cas de microphthalmie liée à l'atrophie des ailes chez une Blatte cavernicole. *Soc. Entomol. France. Livre du Centenaire.* Paris.

CHOPARD, L. (1936) Orthoptères et Dermaptères (1ère Série). *Biospeologica,* **LXIV.** *Archiv. Zool. expér. géner.* **LXXVIII.**

CHOPARD, L. (1938) *La Biologie des Orthoptères,* Paris.

CHOPARD, L. (1951) Les Blattes cavernicoles du genre *Nocticola* Bol. *Eos, Tomo extraord.*

CHOPARD, L. (1952) Note sur quelques Orthoptèroides cavernicoles du Congo français. *Notes biospéol.* **VII.**

CHOPARD, L. (1954) Contribution à l'étude des Orthoptèroides cavernicoles. *Notes biospéol.* **IX.**

Orthopteroidea

BACCETTI, B. and CAPRA, F. (1959) Notulae orthopterologicae. XII. Revisione delle specie italiane del genre *Dolichopoda* Bol. (Orthopt. Rhaphidophoridae). *Redia.* **XLIV.**

BÉGOUEN, H. (1929) La Grotte des Trois-Frères. *Mitteil. Höhl. Karstf.*

BÉGOUEN, H. and L. (1928) Découvertes nouvelles dans la caverne des Trois-Frères, à Montesquieu-Avantès (Ariége). *Rev. anthropol.* **XXXVIII.**

CHOPARD, L. (1917) Note sur la biologie de *Dolichopoda palpata* Sulz. (Orth. Phasgonridae). *Bull. Soc. Entomol. France.*

CHOPARD, L. (1924) Note sur quelques Orthoptères cavernicoles de Sumatra et de Java. *Ann. Soc. Entomol. France* **XCIII.**

CHOPARD, L. (1928) Sur une gravure d'Insecte de l'époque magdalénienne. *Compt. rend. Soc. Biogeogr.* **V.**

CHOPARD, L. (1929) Note sur les Orthoptères cavernicoles du Tonkin. *Bull. Soc. Zool. France* **LIV.**

CHOPARD, L. (1931) Campagne spéléologique de C.Bolivar et R.Jeannel dans l'Amérique du Nord (1928). No. 8. Insectes Orthoptères. *Biospeologica,* **LVI.** *Archiv. Zool. expér. géner.* **LXXI.**

CHOPARD, L. (1932) Les Orthoptères cavernicoles de la faune paléarctique. *Biospeologica,* **LVII.** *Archiv. Zool. expér. géner.* **LXXIV.**

CHOPARD, L. (1936) Un remarquable genre d'Orthoptères de l'Ambre de la Baltique. *Livre jubilaire de M.Eugene-Louis Bouvier.* Paris.

CHOPARD, L. (1954) Contribution a l'étude des Orthoptèroides cavernicoles. *Notes biospéol.* **IX.**

CHOPARD, L. (1959) Sur les moeurs d'un *Rhaphidophora* cavernicole. *Annal. Spéléol.* **XIV.**

HUBBELL, T.H. (1936) A monographic revision of the genus *Ceutophilus. Univ. Florida Public.* **II.**

LANZA, B. (1954) Speleofauna toscana. III. Corologia degli Orttoteri cavernicoli toscani e note sistematiche sul genere *Dolichopoda. Monit. Zool. Ital.* **LXII.**

NICHOLAS, G. (1962) Nocturnal migration of *Hadenoecus subterraneus. Nat. Speleol. Soc. News.* **XX.**

REMY, P. (1931) Observations sur les moeurs de quelques Orthoptères cavernicoles. *Annal. Sc. Nat. Zool.* (10) **XIV.**

RICHARDS, A.M. (1955) The anatomy and morphology of the cave-orthopteran, *Macropathus filifer* Walker, 1869. *Trans. R. Soc. New Zealand* **LXXXIII.**

RICHARDS, A.M. (1960) Revision of the Rhaphidophoridae (Orthoptera) of New Zealand. Part VII. The Rhaphidophoridae of the Waipu Caves. *Trans. R. Soc. New Zealand* **LXXXVIII.**

RICHARDS, A.M. (1961) The life history of some species of Rhaphidophoridae (Orthoptera). *Trans. R. Soc. New Zealand* **LXXXVIII.**

RICHARDS, A.M. (1962) Feeding behaviour and enemies of Rhaphidophoridae (Orthoptera) from Waitomo Caves, New Zealand. *Trans. R. Soc. New Zeal. Zool.* **II.**

SELIŠKAR, A. (1923) Die männlichen Duftorgane der Höhlenheuschrecke *Troglophilus*. *Zool. Anz.* **LVII.**

WALDNER, F. (1940) Die Höhlenheuschrecke (*Troglophilus cavicola* Kollar) in Niederdonau. *Mitteil. Höhl. Karstf.*

Gryllodea

BACCETTI, B. (1960) Notulae orthopterologicae. XIV. Descrizione di un nuovo genere cavernicolo di Ortotteri scoperto in Sardegna (Orthopt. Gryllidae). *Studi Sassar. Sez.* **III.** *Agric.* **VII.**

CHOPARD, L. (1923) Description d'un Gryllide cavernicole de la Jamaïque. *Bull. Soc. Entomol. France.*

CHOPARD, L. (1924) Note sur quelques Orthoptères cavernicoles de Sumatra et de Java. *Ann. Soc. Entomol. France.* **XCIII.**

CHOPARD, L. (1924) On some cavernicolous Orthoptera and Dermaptera from Assam and Burma. *Rec. Indian Mus.* **XXVI.**

CHOPARD, L. (1929) Fauna of the Batu Caves, Selangor. XII. Orthoptera and Dermaptera. *J. Fed. Malay States Mus.* **XIV.**

CHOPARD, L. (1942) Trois Gryllides cavernicoles nouveaux du Congo belge (Orthop.). *Rev. franç. Entomol.* **IX.**

CHOPARD, L. (1946) Note sur quelques Orthoptères cavernicoles de Madagascar. *Rev. franç. Entomol.* **XII.**

CHOPARD, L. (1947) Note sur les Orthoptères cavernicoles du Mexique. *Ciencia.* **VIII.**

CHOPARD, L. (1949) Les Orthoptéroides cavernicoles de Madagascar. *Mem. Inst. Sci. Madagascar, Ser. A.* **III.**

CHOPARD, L. (1950) Orthoptéroïdes cavernicoles du Congo belge. *Rev. Zool. Bot. Afric.* **XLIII.**

CHOPARD, L. (1952) Note sur quelques Orthoptéroïdes cavernicoles du Congo français. *Notes biospéol.* **VII.**

CHOPARD, L. (1954) Contribution à l'étude des Orthoptéroides cavernicoles. *Notes biospéol.* **IX.**

CHOPARD, L. (1957) Contribution à la faune des Orthoptères des grottes du Congo belge. *Rev. Zool. Bot. Afric.* **LVI.**

CHOPARD, L. (1958) Contribution à la faune des Orthoptéroïdes des grottes du Congo belge. *Rev. Zool. Bot. Afric.* **LVIII.**

CHOPARD, L. and VILLIERS, A. (1957) Speleologica africana. *Gryllidae* des Grottes de Guinée. *Bull. Inst. franç. Afrique Noire.* **XIX.** Ser. A.

HUBBELL. T.H. (1938) New cave crickets from Yucatan with a review of the Pentacentrinae and studies on the genus *Amphiacusta* (Orthoptera, Gryllidae). *Carnegie Inst. Washington Public.* No. 491.

Notoptera

GURNEY, A.B. (1961) Further advances in the taxonomy and distribution of the Grylloblattidae (Orthoptera). *Proc. Biol. Soc. Washington* **LXXIV.**

Dermaptera

CHOPARD, L. (1940) Un remarquable Dermaptère cavernicole de l'Afrique occidentale, *Diplatys milloti* n. sp. *Bull. Soc. Zool. France* **LXV.**
HIERNAUX, C.R. and VILLIERS, A. (1955) Speologia africana. Étude préliminaire de six cavernes de Guinée. *Bull. Inst. franç. Afrique Noire. Ser. A.* **XVII.**

Psocoptera

BADONNEL, A. (1943) Psocoptères de Macédoine et Herzégovine (Voyages de P.Remy, R.Husson et A.Schweitzer). *Bull. Soc. Entomol. France* **XLVIII.**
BADONNEL, A. (1959) Un Psoque cavernicole du Moyen-Congo. *Rev. suisse Zool.* **LXVI.**
BAILEY, V. (1933) Cave life of Kentucky, mainly in the Mammoth Cave region. *Amer. Midl Nat.* **XIV.**
ENDERLEIN, G. (1909) Copeognatha (Erste Reihe). *Biospeologica,* **XI.** *Archiv. Zool. expér. géner.* (5), **I.**

Hemiptera

PAULIAN, R. and GRJEBINE, A. (1953) Une campagne spéléologique dans la Réserve naturelle de Namoroka. *Naturaliste Malgache.* **V.**
SYNAVE, H. (1952) Description de deux *Cixiidae* nouveaux appartenant au genre *Achaemenes* Stal et note préliminaire sur un Cixiide troglobie. *Bull. Inst. Sci. Nat. Belgique* **XXVIII.**
SYNAVE, H. (1953) Un Cixiide troglobie découvert dans les galeries souterraines du système de Namoroka (Hemiptera, Homoptera). *Naturaliste Malgache.* **V.**

Neuroptera

ANNANDALE, N. (1913) The limestone caves of Burma and the Malay Peninsula. *Proc. Asiatic Soc. Bengal* **IX.**
BANKS, N. (1929) Fauna of the Batu Caves, Selangor. XIII. Neuroptera. On the myrmeleonid, *Neglurus vitripennis* Navas. *J. Feder. Malay States. Mus.* **XIV.**
BANKS, N. (1938) A new myrmeleonid from Yucatan. *Carnegie Inst. Washington Public* No. 491.

Hymenoptera

WHEELER, W.M. (1922) The ants of Trinidad. *Amer. Mus. Novitates.* No. 45.
WHEELER, W.M. (1924) Hymenoptera of the Siju Cave, Garo Hills, Assam. I. *Triglyphothrix striatidens* Emery as a cave ant. *Rec. Indian Mus.* **XXVI.**
WHEELER, W.M. (1938) Ants from the caves of Yucatan. *Carnegie Inst. Washington Public.* No. 491.
WILSON, E. O. (1962) The Trinidad cave ant *Erebomyrma (Spelaeomyrmex) urichi.* (Wheeler) with a comment on cavernicolous ants in general. *Psyche* **LXIX.**

Trichoptera

RADOVANOVIČ, M. (1932) *Wormaldia subterranea* n. sp., eine neue, in den Höhlen Jugoslawiens aufgefundene Trichopteren-Art. *Zool. Anz.* **C.**
RADOVANOVIČ, M. (1935) Über die gegenwärtige Kenntnis der balkanischen Trichopteren. *Verh. Intern. Ver. Theor. angew. Limnologie.* **VII.**

Lepidoptera

ANNANDALE, N., BROWN, J.C. and GRAVELY, F.H. (1914) The limestone caves of Burma and the Malay Peninsula. *J. Proc. Asiatic Soc. Bengal. N.S.* **IX.**
DIAKONOFF, A. (1951) Notes on cave-dwelling Microlepidoptera with description of a new genus and species from East Java (Family *Oinophilidae*). *Zool. Mededel.* **XXXI.**
FLETCHER, T.B. (1924) Lepidoptera of the Siju Cave. I. *Pyralidae. Rec. Indian Mus.* **XXVI.**

MEYRICK, E. (1908) New Microlepidoptera from India and Burma. *Rec. Indian Mus.* **II.**
MEYRICK, E. (1924) Lepidoptera of the Siju Cave. II. *Tineidae. Rec. Indian Mus.* **XXVI.**
MEYRICK, E. (1929) Fauna of the Batu Caves, Selangor. XIV. Microlepidoptera. *J. Feder Malay States Mus.* **XIV.**

Diptera (general)

BEZZI, M. (1911) Diptères (1ère Série). *Biospeologica,* **XX.** *Archiv. Zool. expér. géner.* (5) **VIII.**
LERUTH, R. (1939) La Biologie du domaine souterrain et la faune cavernicole de Belgique. *Mém. Mus. Hist. Nat. Belgique* **LXXXVII.**
MATILE, L. (1959) Diptères. *In* Contribution à l'inventaire faunistique des cavités souterraines de l'ouest de la France. *Bull. Soc. Sci. Nat. Ouest France* **LV.**

Diptera from Cold Caves and Snow-holes

BEZZI, M. (1916) Sur un genre nouveau de Diptère subaptère des cavités souterraines du Djurdjura. *Bull. Soc. Hist. Nat. Afrique du Nord* **VII.**

Guanobious Diptera

BEZZI, M. (1911) Diptères (1ère série). *Biospeologica,* **XX.** *Archiv. Zool. expér. géner.* (5) **VIII.**
BEZZI, M. (1914) Ditteri cavernicoli dei Balcani raccolti dal Dott. K. Absolon (Brünn) (Seconda contribuzione). *Atti Soc. Ital. Sc. Nat.* **LIII.**
ENSLIN, E. (1906) Die Lebensweise der Larve von *Macrocerca fasciata* Meij. *Zeit. wiss. Insektenbiol.* **XI.**
SCHMITZ, H. (1909) Die Insektenfauna der Höhlen von Maastricht und Umgebung unter besonderer Berücksichtigung der Dipteren. *Tijd. v. Entomol.* **LII.**
SCHMITZ, H. (1912) Biologisch-anatomische Untersuchungen an einer höhlenbewohnenden Mycetophilidenlarve *Polylepta leptogaster* Winn. *Heerlen naturh. Genootsch.* Limburg.
SCHMITZ, H. (1918) Die Phoriden-Fauna von der Dr. Karl Absolon 1908–1918 besuchten mittel- und südosteuropäischen Höhlen. *Tijdschr. v. Entomol.* **LXI.**
SCHMITZ, H. (1929) *Revision der Phoriden nach forschungsgeschichtlichen, nomenklatorischen, systematischen und anatomischen, biologischen und faunistischen Gesichtspunkten.* Berlin und Bonn.

Troglophilous Diptera

GRIEPENBURG, W. (1934) Die Berghauser Höhle bei Schwelm i. W. *Mitteil Höhl. Karstf.*
LENGENSDORF, F.J. (1924–25) Beitrag zur Höhlenfauna des Siebengebirges unter besonderer Berücksichtigung der Dipteren. *Speläol. Jahrb.* **V/VI.**
LERUTH, R. (1941) Phoridae cavernicoles de Transylvanie (Diptera). *Études biospéologiques,* **XXVII.** *Mus. R. Hist. Nat. Belgique* **XVII.**
MATILE, L. (1960) Diptères cavernicoles de la Côte d'Or. *Sous le Plancher.*

Cavernicolous Anophelines

ABONNENC, E. (1954) Sur un Anophèle cavernicole de la Guinée: *Anopheles cavernicolus,* n. sp. *(Diptera, Culicidae).* Note préliminaire. *Bull. Inst. Franç. Afrique Noire. Ser. A.* **XVIII.**
ADAM, J.P. (1962) Un Anoplèle nouveau de la République du Congo (Brazzaville): *Anopheles (Neomyzomia) hamoni* n. sp. (Diptera, Cuicidae). *Bull. Soc. Pathol. exot.* **LV.**
LELEUP, N. (1950) Considérations sur l'éthologie et la dispersion actuelle des Anophèles cavernicoles du Congo Belge. *Rev. Zool. Bot. Afric.* **XLIII.**
LELEUP, N. and LIPS, M. (1950) Un Anophèle cavernicole nouveau du Katanga: *Anopheles rhodaini,* n. sp. *Rev. Zool. Bot. Afric.* **XLIII.**
LELEUP, N. and LIPS, M. (1951) Notes descriptives complémentaires sur *Anopheles rodhaini. Rev. Zool. Bot. Afric.* **XLIV.**
LIPS, M. (1960) Anophèles du Congo. 3, Faune des grottes et des anfractuosités. Référence. Récoltes. Répartition et importance médicale actuelle. *Riv. d. Parasitologia* **XXI.**

MATTINGLEY, P. F. and ADAM, J. P. (1955) A new species of cave-dwelling anopheline from the French Cameroon. *Ann. Trop. Med. Parasitol.* **XLVIII.**

WANSON, M. and LEBIED, B. (1945) Un nouvel Anophèle cavernicole du Congo, *Anopheles (Myzomyia) vanhoofi*, n. sp. *Rev. Zool. Bot. Afric.* **XXXIX.**

Cavernicolous Phlebotomines

ABONNENC, E., ADAM, J.P. and BAILLY-CHOUMARA, H. (1959) Sur trois Phlébotomes cavernicoles nouveaux de la région éthiopienne, *P. crypticola, P. balmicola, P. somaliense.* *Archiv. Inst. Pasteur Algérie* **XXXVII.**

ADAM, J. P., BAILLY-CHOUMARA, H. and ABONNENC, E. (1960) Notes écologiques sur quelques Phlébotomes cavernicoles de la région éthiopienne. *Archiv. Inst. Pasteur Algérie* **XXXVIII.**

PARROT, L. and SCHWETZ, J. (1937) Phlébotomes du Congo belge. VI. Trois espèces et une varieté nouvelles. *Rev. Zool. Bot. Afric.* **XXIX.**

PARROT, L. and WANSON, M. (1939) Phlébotomes du Congo belge. IX. *Phlebotomus (Prophlebotomus) mirabilis* n. sp. *Rev. Zool. Bot. Afric.* **XXXII.**

PARROT, L. and WANSON, M. (1946) Notes sur les Phlébotomes. LIII. Sur *Phlebotomus gigas* et sur *Phlebotomus mirabilis. Archiv. Inst. Pasteur Algérie* **XXIV.**

WANSON, M. and LEBIED, B. (1946) L'habitat des Phlébotomes cavernicoles de Thysville (Congo belge). *Archiv. Inst. Pasteur Algérie* **XXIV.**

Speomyia

BEZZI, M. (1914) *Speomyia absoloni* n.g.n.sp. (Dipt.), eine degenerierte Höhlenfliege aus dem herzegowinisch-montenegrischen Hochgebirge. *Zool. Anz.* **XLIV.**

Mormotomyia

AUSTEN, E.E. (1936) A remarkable semi-apterous fly (Diptera) found in a cave in East Africa, and representing a new family, genus and species. *Proc. Zool. Soc. Lond.*

Allopnyxia

FREEMAN, P. (1952) A new genus and species of Mycetophilidae (Diptera) allied to *Pnyxia* Johannsen from a cave in Italy. *Boll. Soc. Entomol. Ital.* **LXXXII.**

PATRIZI, S. (1956) Notes sur la faune cavernicole du Lazio et de la Sardaigne. *Communic. 1er Congr. Intern. Spéléol.* **III.**

CHAPTER XIII

THE COLEOPTERA

THE MAJORITY of cavernicolous insects belong to the Coleoptera but their distribution between the different sub-orders of this group is most unequal. This is shown in the table below.

The Caraboidea and Staphylinoidea contain nearly all the cavernicolous Coleoptera and at the same time the most specialised and highly modified forms.

The absence of cavernicoles in the large groups Lamellicornia and Phytophaga again shows that the phytophagous habit and cavernicolous life are incompatible. It is true that one may find members of certain genera of the Curculionidae (e.g. *Troglorhynchus, Alaeocyba*) in caves, but these have strayed from the endogeous environment. These beetles move down along the roots on which they feed and thus occasionally find their way into subterranean cavities. R.Jeannel called them "strayed endogeans".

The sub-orders and families of Coleoptera
containing cavernicoles

Sub-orders	Families
Caraboidea	Nebriidae Scaritidae Trechidae Harpalidae Pterostichidae Zuphiidae Brachynidae Dytiscidae
Staphylinoidea	Catopidae Liodidae Leptinidae Ptiliidae Staphylinidae Pselaphidae Histeridae

contd.

Sub-orders	Families
Cucujoidea	{ Dryopidae { Endomychidae
Heteromera	{ Tenebrionidae { Hylophilidae
Malacoderma	{ Cantharidae (Telephoridae) { Lampyridae
Lamellicornia	Null
Phytophaga	Null

A. CARABOIDEA

(a) Nebriidae

The three most interesting hypogeous species belonging to the Nebriidae are natives of North Africa. They are *Nebria exul, Spelaeonebria nudicollis* and *S. initialis.* These caraboids are found in the "tesserefts" and "anous" of the Jurjura Mountains, Algeria (Peyerimhoff, 1910, 1911, 1914). However, they are more correctly regarded as nivicoles, the inhabitants of snow-holes, than true cavernicoles.

Oreonebria ratzeri which lives a nivicolous life in the Jura Mountains has been captured in a cave in the Franc-Comptois Plateau, near Vercel (Doubs) (Colas, 1954).

(b) Scaritidae

The members of this group bore holes or live in burrows. The species of the genus *Reichia* are indigenous to the Mediterranean region. They are of very small size and lead an endogeous life. A dozen of colourless species of the genus *Antireichia* has been discovered by sifting mould in the mountain forests of the Kivu, the Ruanda and the Tanganyika (Basilewsky, 1951). They are cavernicole "en puissance".

The Scaritidae contains two cavernicolous species. The first one, *Spelaeodytes mirabilis,* has been found in a cave in Herzegovina but is only known from a single specimen (Miller, 1863). The second is *Italodytes stammeri* from Castellana Cave in the Murges (southern Italy) (Müller, 1939).

(c) Trechidae

This is one of the most important caraboid groups both for the number and the diversity of the species. The family may be divided into four sub-families: the Bembiinae, Pogoninae, Trechinae and Merizodinae.

The Bembiinae are essentially ripicolous and there are only a few cavernicolous forms (e.g. *Oreocys, Pseudolimnaeum*). The Pogoninae are for the most part halophiles. The Merizodinae are confined to Australasia; the genus, *Idacarabus*, is found in caves in Tasmania (Lea, 1910). Finally, the Trechinae contains very many species which may be lucicoles, endogeans or cavernicoles.

Sub-family Trechinae. This sub-family contains five tribes: Perileptini, Trechodini, Aepini, Homaloderini and Trechini. There are no cavernicolous species in the first three of these tribes. The Homaloderini, a group which had its origin in ancient Gondwanaland, contains a single caverni-colous species, *Iberotrechus bolivari* which was discovered in a cave in the province of Santander (western Spain). However, this species shows no particular morphological modifications.

The Trechini contain many species which have acquired the hypogeous way of life, that is they are either endogeans or cavernicoles. It may be recalled that the genus *Trechus* contains numerous species which have a microcavernicolous habit and frequent the nests and passages made by small rodents (Larson, 1939). One can say therefore that the Trechini are "pre-adapted" to cavernicolous life. The following discussion will be limited to the Trechini.

Research on the Trechinae. An excellent introduction to the complex systematics of this group is provided by the *Monographie des Trechinae* by René Jeannel. There are four volumes published from 1926 to 1930. Since then there has been a great deal of work on this group. The following discussion will contain references only to those publications which deal with the subterranean Trechinae and which have appeared since 1930.

H. Coiffait (1962) published a revision of the species of the cavernico-lous Trechini of the Pyrenees of which the number has increased consid-erably during the past thirty years. A study of the subterranean Trechini from Transcaucasia has been made by V. N. Kurnakov (1959). Regarding Japan, the work of A. Habu (1950), R. Jeannel (1953, 1962) and above all that of S. I. Uéno (1954–1960), must be mentioned. The Trechini of the United States have been studied by J. M. Valentine (1931–1952), R. Jeannel (1931), Jeannel and Henrot (1949) and more recently by C. H. Krekeler (1958) and T. C. Barr (1959–1960). A most interesting account of the hypo-geous Trechini of Mexico has been written by C. Bolivar (1943). Prelimi-nary studies of the cavernicolous Trechini of New Zealand have been undertaken by Britton (1958–1961).

Systematics of the Trechinae. Under this heading we will consider only the principal genera of the cavernicolous Trechini, and indicate their geographical distribution.

The classification established by R.Jeannel is based on the orientation of the copulatory apparatus. This organ may occupy a ventral position (isotope type) or a lateral one, and be thrown back on to the right side of the internal sac (anisotope type) (Fig. 46).

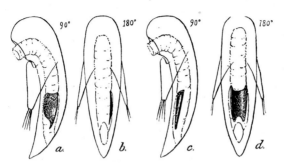

FIG. 46. The two types of genitalia in the *Trechini. a* and *b*, the anisotope type; *c* and *d* the isotope type (after Jeannel).

(1) The anisotope type is the primitive one in the Trechini. The original form of this organ in this tribe is found in the large genus *Trechus* which contains nearly 300 species which are distributed throughout the Northern Hemisphere.

Other forms, most of them cavernicolous and classified in numerous genera, are related to this primitive type. They will be dealt with according to their geographical distribution.

The Mediterranean region contains many Trechini of the anisotope type: *Speotrechus* (Cevennes and Causses), *Geotrechus* (Pyrenees) *Aphaenops* (Pyrenees), *Paraphaenops* (Catalonia), *Allegretia* (Italian Alps).

The adriatic, Dinaric and Balkan areas are particularly rich in cavernicolous Trechini. They include the genera *Aphaenopsis, Scotoplanes, Neotrechus, Orotrechus, Typhlotrechus, Nannotrechus* and *Pheggomistes.*

Pseudaphaenops is native to caves in the Crimea and *Jeannelius* and *Meganophthalmus* to those in the Caucase.

Similarly in North America there is a rich fauna of anisotope Trechini. The most important genus is *Pseudanophthalmus* which contains more than 80 species or sub-species. In addition the genera *Neaphaenops, Nelsonites* and *Darlingtonea* may be mentioned.

There are many cavernicolous Trechini in Japan. They belong to the genera *Trechiama, Nipponotrechus, Ryugadous* and *Kurasawatrechus.*

The genus *Duvaliomimus* is indigenous to New Zealand. It contains some cavernicolous species but these species are only weakly differentiated from the surface forms, which populate the mountainous regions of that country. However, *D. orpheus* possesses greatly reduced eyes, an elongated prono-tum and a pale colour. The genera *Scototrechus* and *Neanops* are closely related to *Duvaliomimus* (Britton, 1958–1961). The first genus possesses reduced eyes; the second is anophthalmous.

(2) The isotope type seems to be more recent than the anisotope type. The origin of this line is found in the genus *Duvalius* which contains numerous species distributed over a large part of Europe, Asia Minor and North Africa.

To this primitive type the following genera are related: *Apoduvalius* (Cantabrian Mountains; Algeria) (Jeannel, 1953a); *Trichaphaenops* (Alps and Jura); *Sardaphaenops* (Sardinia) (Cerruti and Henrot, 1956); *Anoph-thalmus* (Venetia Juliana; Istria; Slovenia; Croatia); *Aphaenopidius* (Slo-venia).

The isotopic type is represented in North America by the single species *Ameroduvalius jeanneli*, which has been found in a cave in Kentucky (Valentine, 1952).

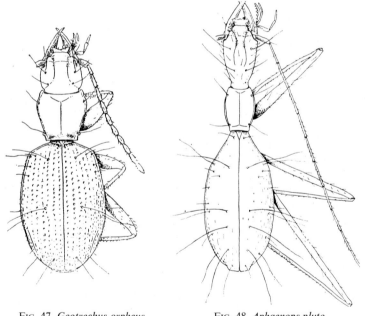

FIG. 47. *Geotrechus orpheus* (after Jeannel).

FIG. 48. *Aphaenops pluto* (after Jeannel).

Orthogenetic evolution. Phylogenetic classification concerns the system-atist. Biospeologists on the other hand are more interested in orthogenetic evolution which proceeds in a parallel manner, in different phyletic lines. Jeannel has recognised four morphological types of Trechini:

(1) The *primitive type* which is winged and pigmented; the eyes are present and the pronotum small.

(2) The *secondary type* which is derived from the first, and is pigmented and possesses eyes, but is apterous and has a large pronotum.

(3) The *anophthalmic type* resembles the secondary one but it is depigmented and the eyes are more or less reduced (Fig. 47).

(4) The *aphaenopsian type* which is depigmented and anophthalmic to the same degree as the type above, but differs from it in that the body and appendages are elongated, and a narrow pronotum is present (Fig. 48). Moreover, the integument becomes thin and pliable.

Evolution towards the aphaenopsian type has proceeded on many occasions on many different lines, but it has always ended with the establishment of similar morphological types. The aphaenopsian type is the last stage of orthogenetic evolution in the Trechinae.

In the Pyrenees, the aphaenopsian type is represented by the genus *Aphaenops*. This genus has a similar origin to *Geotrechus*, but its mode of life is entirely different.

The species of the genus *Neaphaenops*, from North America, also belong to the aphaenopsian type. They resemble *Pseudanophthalmus* which belongs to the anophthalmic type, the latter being related to the genus *Trechoblemus*, which is widespread in the Old World.

Mexaphaenops, from caves in Nuevo Leon, Mexico, also belongs to the aphaenopsian type. It is related to epigeous forms, belonging to the genus *Hydroduvalius*, which are depigmented, apterous, and microphthalmic, and found at high altitudes. *Mexaphaenops* is more distantly related to *Paratrechus*, the species of which inhabit mountainous regions, possess normal eyes, and may be winged or apterous (Bolivar, 1943).

The Evolution of the Trechinae Underground. Two different lines of evolution have taken place underground:

(1) The anophthalmic type leads an endogeous mode of life. Endogeans may be taken under large rocks especially on north facing slopes, in humid regions and also at the entrances of caves under fallen boulders. The endogeous mode of life is found in certain cavernicoles which have been derived from endogeans. This is the case with the genera *Duvalius* (isotope series) and *Geotrechus* (anisotope series), each of which contains species that are either endogeans or cavernicoles. The latter live in a similar manner to their endogean congeners, that is under rocks which rest on the clay deposits in caves and also under the stalagmitic floor.

(2) The aphaenopsian type, of which a typical European genus is *Aphaenops* and typical North American ones are *Nephaenops* and *Mexaphaenops*, has a different origin from the endogeans. They are probably derived from nivicoles. *Trechopsis lapiei* which lives in snow holes (tesserefts, anous) in the Jurjura Mountains (Algeria) and *Anophthalmus nivalis* (Slovenia) provide us with examples of the mode of life which the species

of *Aphaenops* must have led in the past. Furthermore, the members of the sub-genus *Arctaphaenops* of the genus *Trichaphaenops*, which live in ice caves, still exist in an environment similar to those which the primitive aphaenopsian lines once frequented. Examples of this situation are provided by *Arctaphaenops angulipennis* from caves of the Dachstein Mountains (Austria), *A. styriacus* from the "Bärenhöhle", and *A. gaudini* from the "chorouns" of Dévoluy (Jeannel, 1952).

The aphaenopsian types are found hunting over concretions, or upon the stalagmitic walls. They feed on Collembola, Diptera and Acarina.

What has just been said of this two-fold evolution shows that the qualification of it being "subterranean" is only partly correct. This is because the evolution described had its beginning long before caraboids found their way into caves, but its final stages have taken place under-ground, which is an essentially stable environment.

(d) Harpalidae

Until a short time ago the Harpalidae were unknown in the underground world. Recently, two species were described: *Notospeophonus castaneus*, captured in caves, near Portland, Victoria, Australia (Moore, 1962); and *Pholeodytes townsendi*, from a cave in New Zealand (Britton, 1962). The eyes of *Pholeodytes* are regressed, but they are surrounded by a circle of pigment.

(e) Pterostichidae

The Pterostichidae constitute a large family of Caraboidea. The members of this family are occasionally involved in subterranean evolution but less frequently than those of the Trechinae.

(1) There are typical endogeous members in this family which are de-pigmented, microphthalmic or even anophthalmic *(Lianoe, Tapinopterus, Speluncaris, Typhlochromus, Caecocaelus)*. Certain of these have given rise to cavernicolous species. The microphthalmic sub-genus *Lianoe* of the genus *Pterostichus* are typical endogeans, but two species *(microphthalmus* and *nadari)* have become cavernicoles and inhabit caves in the Pyrenees.

(2) In addition there are troglophiles and guanobia which, in Europe, are particularly frequent in the sub-family Sphodrinae *(Sphodropsis, Lae-mostenus, Ceutosphodrus, Antisphodrus)*. Certain species are depigmented and microphthalmic.

Troglophilic pterostichids have been reported in other parts of the world. In Africa the genera *Megalonyx* and *Speokokosia* belong to this group. In America the numerous representatives of the genus *Rhadine* which are related to *Agonum* (Anchomenini) are either lucicoles or troglophiles or troglobians (Barr, 1960). In Japan, the representatives of this tribe *(Jujiroa* and *Ja)* are very close to the north American *Rhadine* (Kéno, 1955).

Finally, a cavernicolous member of the Sphodrinae, *Prosphodrus waltoni*, has recently been discovered in New Zealand in caves in the Te Kuiti region (Britton, 1959). Its characters indicate that it is probably a troglophile.

(3) The Pterostichidae also contain a number of true troglobia which represent the final stages of subterranean evolution.

Some of these belong to the sub-family Pterostichinae, for example, *Troglorites*, which has two cavernicolous species: *T. breuili* which is native to Navarre (Jeannel, 1919), and *T. ochsi* from the Alpes Maritimes (Fagniez, 1921).

Others belong to the sub-family Molopinae, for instance *Zariqueya troglodytes* from Catalonia (Jeannel, 1924), *Henrotius jordai* and *H. henroti* from Majorca (Jeannel, 1950, 1953) and *Speomolops sardous* from Sardinia (Patrizi, 1955).

In America, *Comstockia subterranea*, from caves in Texas and New Mexico is a true troglobian which has an aphaenopsian appearance. According to Jeannel (1931) they are related to the troglophilic genus *Rhadine*. Similarly *Spelaeorhadine araizai*, which is another species with aphaenopsian facies, has been discovered more recently in a cave in the Sierra de Bustamente in Nuevo Leon, Mexico (Bolivar, 1944).

(f) Brachynidae

The bombardier beetles contain a number of species which have been found in caves in East Africa (Shimoni and Kulumuzi caves), and a cave in the Transvaal near Pretoria. They belong to the genera *Brachynillus*, *Crepidogaster*, *Plagiopyga* and *Pheropsophus*. While there is little information concerning their ecology they may be provisionally regarded as troglophiles. However, *Brachynillus varendorffi* which was described by Reitter (1904), is completely depigmented and has no eyes. It lives in the deeper parts of the Kulumuzi Cave, upon clay deposits, and according to Jeannel (1914), may well be a troglobian.

(g) Dytiscidae

Hypogeous dytiscids are known from Europe, Africa and Asia.

(1) Europe. The first cavernicolous dytiscid was discovered by Sietti (1904) in a well at Beausset (Var) in Southern France. It was described by Abeille de Perrin (1904) under the name *Siettitia balsetensis*. Many years later it was found by Colas (1946) in a well at Sablettes, near Toulon.

A species related to the last has been found in a well at Avignon. Only a single specimen of this species, *S. avenionensis* has so far been captured (Guignot, 1925).

These two species live in ground water. They are depigmented; the eyes are present but lack pigment and the ocelli are absent (Regimbart, 1905; Jeannel, 1906). The body is covered with long fine hair (Fig. 49).

Regimbart (1905) and Jeannel (1906) have shown that *Siettitia* is near the hydroporids of the genus *Graptodytes* and in particular to the light-avoiding species *G. aurasius*, which has been found in a stream near Tgout on the Saharan slopes of the Aurès Mountains, Algeria. However, this African species possesses normal eyes, which are well pigmented.

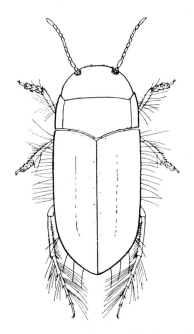

FIG. 49. *Siettitia balsetensis* (after Spandl).

Birstein and Boroutzky (1950) have reported that a blind dytiscid had been captured in the Riongès Cave near Kutaïs (Georgia), but this species does not seem to have been described.

(2) Africa. C. Alluaud and P. A. Chappuis have captured a completely blind and depigmented dytiscid, *Bidessus chappuisi* in a well at Banfora in the Upper Volta, West Africa (Peschet, 1932).

Two other African species have been found in caves. They are *Liodessus antrias* from Guinea and *Copelatus strinatii* from the Middle Congo, but they are morphologically similar to surface forms (Guignot, 1955, 1958).

(3) Asia. In 1951 Yoshinobu Morimoto found the first Japanese caverni-colous dytiscid. It is a hydroporid, which has been described by S. Uéno (1957) under the name of *Morimotoa phreatica*. This beetle is apterous, depigmented and anophthalmic and thus further regressed than *Siettitia*.

However, like the latter species, the body is covered with long fine hair. Larvae corresponding to several instars have been described.

Another archaic form, *Phreatodytes relictus*, which belongs to the Noterinae is also depigmented, anophthalmic and covered with long hair (Fig. 50). Its larvae have no ocelli (Fig. 51).

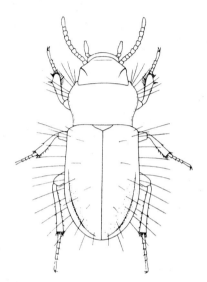

FIG. 50. *Phreatodytes relictus* (after Uéno).

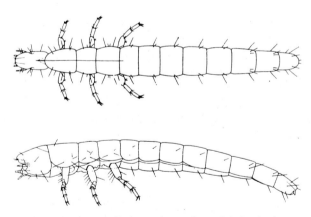

FIG. 51. Larva of *Phreatodytes relictus* (after Uéno).

B. STAPHYLINOIDEA

(a) Catopidae

The Catopidae are a large family of Staphylinoidea containing about a thousand species. A detailed monograph on this family has been written by Jeannel (1936). The Catopidae are made up of saprophages and obscuricoles. In addition many species are myrmecophiles, termitophiles, commensals in the nest of various Hymenoptera, or pholeophiles.

The family is divided into five sub-families, but we shall deal with only one, the Bathysciinae, because it is the unique section which contains true cavernicoles. A few words concerning the ecology of the other Catopidae are all that is required here.

Catops (Catopinae) are frequently found in caves and may be regarded as troglophiles. The significance of the presence of *Choleva* in the caves has been elucidated by observations by Madame Deleurance (1959). The members of this genus spend the summer in the entrances of caves where they enclose themselves in a clay cell and then undergo diapause for some months. The rest of the life cycle takes place outside caves.

The North American genus *Adelops* contains many lucicolous species but there are also cavernicoles which populate a vast area from Kentucky to Guatemala. These are of recent origin. Anophthalmic species (from Tennessee and Alabama) may be observed together with species with eyes (Missouri, Arkansas, Oklahoma) (Jeannel and Henrot, 1949).

(b) Bathysciinae

This sub-family contains about 700 species which are distributed throughout the Mediterranean region with the exception of North Africa. Jeannel (1911, 1924), has produced two important monographs on them. Although many papers concerning this group have appeared since the publication of these monographs, they have not subtracted from the value of the latter, which still form an indispensable guide in the field of biospeological entomology. However, the important works of G. Müller (1901 to 1941) devoted to the Bathysciinae of the Adriatic region must not be overlooked.

Preadaptation to Cavernicolous Life. It may be stated that the Bathysciinae have been preadapted to cavernicolous life because their lucicolous members already show hypogeous characteristics.

(1) All of them are depigmented whether they are lucicolous or obscuricolous. Depigmentation has thus come before their entry into caves.

(2) The great majority of them are completely anophthalmic. This is an invariable rule for the Bathysciinae of Western Europe, whether they are epigeous or cavernicolous. It is true that there are in Eastern Europe a few

lucicolous species which possess eyes but these organs are always more or less reduced. The state of least regression of the eyes is found in *Adelopsella bosnica*, a species found in the great forests of Bosnia. Here the eyes are reduced but probably still functional. They are composed of about ten ommatidia which possess ocular pigment (Jeannel, 1908). A few species of *Bathysciola* have vestigial eyes which are very small but still pigmented.

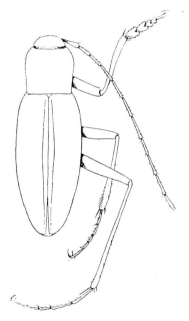

FIG. 52. *Isereus xambeui* (after Jeannel).

This is the case in *B. fausti* (Southern Russia), *B. pusilla* (Caucasia), *B. peyroni* (Lebanon), and *B. persica* (Persia). It is clear that the Bathysciinae have lost their eyes before entering caves.

(3) All the Bathysciinae are apterous whether they are epigeous or hypogeous. Apterism has therefore preceded hypogeous life. However, in the cavernicolous species the elytra are united to each other.

The Subterranean Evolution of the Bathysciinae. It is easy to reconstruct the evolution of the Bathysciinae because its various stages are preserved and thus may be studied.

(1) The initial stage of this evolution is represented by the muscicolous Bathysciinae. The principal example is the Mediterranean genus *Bathysciola*. It has been stated above that this genus already possesses the majority of the characters found in cavernicoles.

It is from this initial stage that the evolution towards subterranean life proceeds. This process has been followed independently, although in a parallel manner, in Western and in Eastern Europe. The reason must be sought in the geographical structure of Europe in the middle Tertiary. Mediterranean Europe was then made up of two emergent land masses, Tyrrhenis to the west and the Aegeid to the east. The first of these was populated by the Tyrrhenian Bathysciinae and the second by the Aegean Bathysciinae. The evolution of the Bathysciinae has proceeded much further in Aegeid than in Tyrrhenis, and has culminated in the production of very strange types.

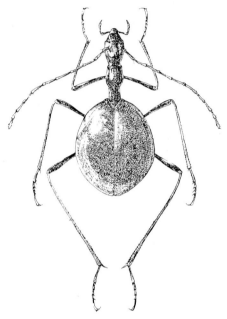

FIG. 53. *Leptodirus hohenwarti* (after Jeannel).

(2) The second stage of evolution is represented by endogeans and little-specialised cavernicoles. They are similar in general form to muscicoles, that is the body is ovoid with short appendages. Jeannel refers to this morphological type by the name "bathyscioid".

A large number of cavernicoles belong to this group. For example, in Western Europe the genera *Speonomus* (Pyrenees), *Diaprysus* (Cévennes), *Cytodromus* and *Royerella* (Alps) belong to this type. In Eastern Europe, bathyscoid types are equally common, for example, *Aphaobius*, *Bathyscio-tes* and *Neobathyscia*, and so on.

(3) In the final stages of evolution the body and the appendages elongate, and at the same time the pronotum becomes thinner and its sides be-come obliterated. This is the "pholeuonoid" type of Jeannel. This type contains specialised cavernicoles such as the genera *Antrocharis* and *Isereus*

(Fig. 52) from Western Europe, and *Pholeuon* and *Apholeuonus* from Eastern Europe.

These specialised types sometimes live in extraordinary habitats, particularly in ice caves. For example, *Isereus xambeui* (Fig. 52) has been taken in the Trou du Glaz in the Dent de Crolles, in the French Alps, where the temperature is between 0 and 1 °C, and *Pholeuon glaciale* inhabits an ice hole (Ghetar de Scarisoara) in Mount Bihar in Transylvania where the summer temperature rarely exceeds 0·8 °C.

(4) Finally there are ultra-evolved forms which are referred to as the "leptodiroid" type (Figs. 53 and 54). Here the legs and antennae are extremely long and the front part of the body is narrow and elongated while the posterior part has a tendency to swell. In the final stage of evolution the abdomen becomes perfectly spherical as is the case in *Leptodirus* (Fig. 53). This phenomenon has been described as false physogastry because there is no real increase in the volume of the abdomen. All that happens is that the elytra fuse together and form a high dome containing a pocket of air over the abdomen.

The leptodiroid type is found only in the Aegean part of the Mediterranean region. It is characteristic of the genera *Leptodirus* (Fig. 53), *Antroherpon*, *Spelaeobates*, *Hadesia*, and so on.

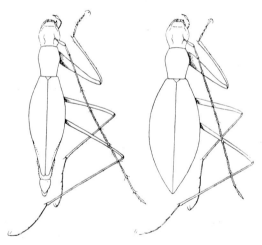

Fig. 54. *Remyella scaphoides*, male (left) and female (right) (after Jeannel).

Leptodirus hohenwarti is a large species which shows physogastry. It is indigenous to the caves of Karst. *Antroherpon*, a form showing a similar stage of ultra-evolution, contains many species which live in caves in Bosnia, Herzegovina, Montenegro and Albania. *Hadesia vašičeki* from Vjeternica pečina in the Popovo Polje lives in immediate contact with water. It is not a truly aquatic form however, as Absolon has maintained, but one which can only survive where there is a shower of minute droplets of water (Remy, 1940).

These evolutionary processes are generally referred to under the name "subterranean evolution", a term which is adopted in this book. In fact they are processes of orthogenetic evolution, which although realised to a great extent in surface forms can only be completed in the subterranean world, which is an essentially conservative environment. The subterranean evolution of animals is certainly not the result of any interference of the hypogeous medium.

(c) Liodidae

Uéno (1957) in his monograph on the cavernicolous dytiscids indicated that Liodidae have been captured in Japanese caves, but they do not seem to have been described.

(d) Leptinidae

Leptinus testaceus has frequently been found in caves. It is in fact a pholeophile whose normal habitat is the nests of rodents, insectivores and also bumble bees *(Bombus lapidarius)*. This insect is blind and depigmented.

(e) Ptiliidae

Ptiliids have frequently been reported from caves. They are trogloxenes. Three species, however, can probably be considered to be cavernicoles, although their ecology is unknown. These species are *Ptenidium caecum* from a cave near Trieste (Joseph, 1882), *P. ponteleccianum* from a cave in Corsica (Strassen, 1955) and *Malkinella cavatica* from the Cango caves in Cape Province (Dybas, 1960). All these species are apterous, depigmented and anophthalmic.

(f) Staphylinidae

In spite of the large number of beetles in this family very few cavernicoles have arisen from them. The staphylinids found in caves belong to three ecological categories.

(1) Endogeans. A large number of staphylinids belong to this category. Some of them are found in caves and of these some are accidental guests while others are permanent inhabitants. The most interesting belong to the genus *Lathrobium* (Paederini) which are microphthalmic or anophthalmic. A number of species are cavernicolous, for example *Lathrobium (Glyptomerus) cavicola* (Carniola), *L. apenninum* (Italy), *L. caecum* (Transylvania)

and *L. lethierryi* (Algeria) (Czwalina, 1888). The larvae of these species are anophthalmic and remarkable for the great length of their appendages.

(2) Guanobia. Many staphylinid species are guanobious and some of them are very common in caves, for example, the European species *Quedius mesomelinus, Atheta subcavicola, A. orcina* and *A. spelaea.*

These species are rarely found outside caves and may therefore be considered to be almost permanent cavernicoles, although they have not undergone any morphological modifications.

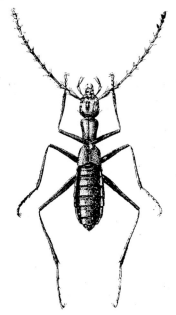

Fig. 55. *Apteranillus rotroui;* dorsal aspect and lateral aspect
(after Scheerpeltz).

However, certain species of *Atheta* (Aleocharinae) are cavernicoles rather than guanobia. For example *A. (Spelaeocolla) absoloni* from the Ponor Turkovici in the Popovo Polje (Herzegovina) is depigmented, microphthalmic and has elongated legs (Ramboussek, 1915; Absolon, 1915).

(3) Troglobia. With the exception of *Typhlomalota glenniei,* an anophthalmic species, collected in a cave, near Simla, in the Himalayan Mountains (Cameron, 1947), the only staphylinids which can be regarded as troglobia belong to the Aleocharinae and are found in North Africa.

Domene camusi†, which has been found in the Grotte de Gorane, at Cape Cantin (Morocco), has reduced eyes, slender antennae and extraordinarily elongated legs (Peyerimhoff, 1949).

† The other species of the genus *Domene* are epigeans or endogeans. One species has been collected in a cave in the province of Jaen (southern Spain) (Coiffait, 1954).

B 8a

Typhlozyras camusi has been found in Torobeit Cave (Rif, Morocco). The antennae are short but, on the other hand, the legs are long and slender and the eyes vestigial (Jeannel, 1959).

The two following species are rather more modified: *Apteranillus rot-rouï*† from the Daya Chiker Cave near Taza (Morocco), which is related to the endogeous members of this genus (Scheerpeltz, 1935), and *Apter-apheanops longiceps* from the Jurjura Caves (Algeria) which is related to *Paraleptusa* (Jeannel, 1907, 1908, 1909). Both species are depigmented, anophthalmic and apterous, and their appendages are extremely elongated (Fig. 55).

(g) Pselaphidae

The pselaphids are a very large group of small beetles (they rarely exceed 3 mm) which are lucifugous and hygrophilic. They are essentially humicoles and endogeans. However, a number of them can be regarded as true troglobia. The European species are the best known and these belong to the genera *Linderia*, *Macrobythus*, *Megalobythus*, *Machaerites* and *Amaurops*, among others.

The cavernicolous pselaphids of North America have been dealt with by Orlando Park (1951, 1953, 1960).

The important work of Jeannel (1952a b, 1953) has brought to light some most interesting data concerning the evolution of the Pselaphidae and the origin of the cavernicolous forms. A brief résumé of this work is given below:

(1) There is often a marked sexual dimorphism. This is frequently shown by the presence of winged males with normal eyes together with apterous females with the eyes reduced or absent.

(2) The morphological sexual dimorphism is linked with a difference in behaviour. The females do not leave the humus in which they live, while the males can fly some distance to search for females in order to fertilise them. A most remarkable fact is that the cavernicolous species of *Gly-phobythus* have retained the same sort of behaviour. Whilst the females always remain underground, the males leave the caves and may be captured by sifting various debris and a male of *G. gracilipes* has even been taken in the Alpes Maritimes by beating some bushes. This type of behaviour is quite unique amongst troglobia.

(3) The males are often poecilandrous, that is to say they are of two distinct forms: those individuals with eyes and wings and those without wings and with reduced eyes. The latter form is generally more common than the former. This is the case, for example, in *Octozethinus leleupi* (Fig. 56). These facts demonstrate that regressive evolution affects the

† Scheerpeltz (1935) had established a new genus for this species, the genus *Antrosemnotes*. Jeannel (1959) recognised that this genus is wrong; it is identical with *Apteranillus* Fairmaire.

females first and only later the males. Therefore, the poecilandry is an indicator of evolutionary instability.

(4) The last stage in this regressive evolution is that in which both males and females are apterous and anophthalmic. This condition is found in the genus *Typhloproboscites*, from the Congo.

(5) Jeannel and Leleup have shown that the regressive evolution which is often somewhat incorrectly referred to as "cavernicolous", began on

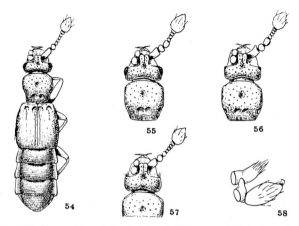

FIG. 56. *Octozethinus leleupi;* 54 and 55, macrophthalmic males; 56 microphthalmic male; 57, female (after Jeannel).

FIG. 57. *Acanthicomus leleupi*, a species with wings and eyes.

the surface within the humus found in the extensive humid and rather cold forests which develop on the sides of high mountains in equatorial Africa. The Itombwee Forest which is of this type, has been thoroughly investigated by Leleup. Microphthalmic and apterous species are particularly frequent in this environment. It cannot be doubted that they have been derived from normal types with wings and eyes; this is proved by the comparison of *Acanthicomus leleupi*, a form with wings and eyes, with *Acanthanops bambuseti* which has neither (Figs. 57 and 58). Thus "sub-

FIG. 58. *Acanthanops bambuseti*, an anophthalmic and apterous species
(after Jeannel and Leleup).

terranean" evolution is still in its early stages in the Congo where there are no known troglobious pselaphids, and has reached its ultimate development in the Mediterranean region where the ancestors of the cavernicolous strains have disappeared.

(h) Histeridae

Other than a few guanobia, only two species of cavernicolous Histeridae are known.

Spelaeacritus anophthalmus has been found in the Grotte de Fersine, in the Taurus Mountains, Asia Minor (Jeannel, 1934). The integument is depigmented, the eyes are absent and the legs rather elongated.

More recently, Patrizi (1955) has described another hypogeous histerid, *Sardulus spelaeus*, which inhabits the Grotte di Toddeitto, near Dorgali in Sardinia. It is apterous and without eyes.

C. CUCUJOIDEA

(a) Dryopidae

R. Jeannel (1950, 1953) has described a remarkable cavernicolous dryopid called *Troglelmis leleupi* which has been found in Kobe Cave near Thysville in the Congo. It is a depigmented beetle, and the eyes which are depigmented are composed of about 10 ommatidia. These insects are also without the layer of air which usually adheres to the ventral surface of species of Elmidae. It has not been seen to respire air on the surface of water as is the rule with epigean Dryopidae. This beetle feeds on bat droppings while the epigeous dryopids are vegetarians.

Zaitzevia uenoi inhabits a cave in the island of Okinoerabu, Japan (Nomura, 1961).

(b) Endomychidae

A single troglobious endomychid, *Cereaxina troglodytes*, is known to science (Jeannel, 1934). It was discovered in the Fersine Cave in the Taurus Mountains. The eyes are represented by only a single depigmented ommatidium.

D. HETEROMERA

(a) Tenebrionidae

The Tenebrionidae are mainly xerophilic insects which have little aptitude for the colonisation of the subterranean environment. However, a few tenebrionids have been collected in caves although none of them are troglobia. They are trogloxenes, or at the most troglophiles. The following species have been found: *Scaurus barbuloti* from Morocco (Reymond, 1956), *Mesostenopa cavatica* from Egypt (Andres, 1925), *Palorus exilis* from the Siju cave, Assam (Blair, 1924) and *Coeloecetes cavernicola* from the Batu Cave, Selangor (Blair, 1929).

(b) Hylophilidae

A few species of Hylophilus have been found in caves in tropical countries. They are listed below:

Hylophilus speluncarum: Kulumuzi Cave, East Africa (Pic, 1914)
H. kempi: Siju Cave, Assam (Blair, 1924)
H. troglodytes: Batu Caves, Selangor (Blair, 1929)
H. sellatus: Batu Caves, Selangor (Blair, 1929).

E. MALACODERMA

(a) Cantharidae (= Telephoridae)

The white, anophthalmic larvae of *Rhagonycha* have frequently been found in caves in Europe and in North America. Their presence underground has frequently been reported (Packard, H. de Bonvouloir, V. Mayet, R. Jeannel). It is known that their metamorphosis takes place in caves but the details of the life-cycle in these Malacoderma are still unknown.

(b) Lampyridae

The glow worm *Lychnocrepis antricola* has been taken in the Batu caves. It has been observed producing its characteristic luminescence some 270 metres from the entrance of a cave. The adults and larvae feed on snails in a similar fashion to that of their epigeous congeners (Blair, 1929).

BIBLIOGRAPHY

CARABOIDEA

Nebriidae

COLAS, G. (1954) Note sur un *Oreonebria* cavernicole. *Notes biospéol.* **IX.**
PEYERIMHOFF, P. DE (1910) Nouveaux Coléoptères du Nord-Africain (Onzième Note: faune cavernicole du Djurdjura). *Bull. Soc. Entomol. France.*
PEYERIMHOFF, P. DE (1911) Nouveaux Coleoptères du Nord-Africain (Quartorzième Note: faune cavernicole du Djurdjura). *Bull. Soc. Entomol. France.*
PEYERIMHOFF, P. DE (1914) Nouveaux Coleoptères du Nord-Africain (Dix-neuvième Note: faune cavernicole du Djurdjura). *Bull. Soc. Entomol. France.*

Scaritidae

BASILEWSKY, P. (1951) Description d'un Scaritide aveugle du Kivu (Col. Carabidae, Scaritidae). *Rev. Zool. Bot. Afric.* **XLIV.**
MILLER, L. (1863) Ein neuer Grottenkäfer aus der Gruppe der Scaritiden. *Wien. Entomol. Monatsschr.* **VII.**
MÜLLER, G. (1939) *Italodytes stammeri*, nuovo genere e nuova specie di Carabidi cavernicoli dell'Italia meridionale. *Boll. Soc. Entomol. Ital.* **LXXI.**

Trechidae

The bibliography of the Trechidae until 1930 is compiled in Jeannel's Monograph. It is not reproduced below.
BARR, TH. C. JR. (1959) New cave beetles (Carabidae; Trechini) from Tennessee and Kentucky. *J. Tennessee Acad. Sci.* **XXXIV.**
BARR, TH. C. (1960) A Synopsis of the cave beetles of the genus *Pseudanophthalmus* of the Mitchell Plain in Southern Indiana (Col., Carabidae). *Amer. Midl. Nat.* **LXIII.**
BOLIVAR Y PIELTAIN, C. (1943) Estudio del primer Trechinae ciego hallada en cavernas de Mexico. *Ciencia* **III.**
BRITTON, E. B. (1958) The New Zealand genus *Duvaliominus* (Col. Carabidae). *Proc. R. Entomol. Soc. Ser. B.* **XXVIII.**
BRITTON, E. B. (1959) Carabidae (Coleoptera) from New Zealand caves. *Proc. R. Entomol. Soc. Ser. B.* **XXVIII.**
BRITTON, E. B. (1960) A new cavernicolous beetle from New Zealand. *Proc. R. Entomol. Soc. Ser. B.* **XXIX.**
BRITTON, E. B. (1961) New genera of beetles (Carabidae) from New Zealand. *Ann. Mag. Nat. Hist.* (13) **IV.**
CERRUTI, M. and HENROT, H. (1956) Nuovo genere e nuova specie di Trechidae troglobio della Sardegna centro-orientale (Col.). *Fragmenta Entomologica.* **II.**
COIFFAIT, H. (1962) Monographie des *Trechinae* cavernicoles des Pyrénées. *Ann. Speleologie* **XVII.**

HABU, A. (1950) On some cave-dwelling Carabidae from Japan. *Mushi.* **XXI.**

JEANNEL, R. (1926, 1927, 1928, 1930) Monographie des Trechinae. I. *L'Abeille*, **XXXII.** II. *Ibid.* **XXXII.** III. *Ibid.* **XXXV.** IV. *Ibid.* **XXXVI.**

JEANNEL, R. (1931) Révision des *Trechinae* de l'Amérique du Nord. *Biospeologica*, **LVI.** *Archiv. Zool. expér. géner.* **LXXI.**

JEANNEL, R. (1953a) Un genre nouveau de Trechini cavernicoles des Monts Cantabriques. *Notes biospéol.* **VIII.**

JEANNEL, R. (1953b) Sur les Trechini cavernicoles du Japon. *Notes biospéol.* **VIII.**

JEANNEL, R. (1962) Les Trechini de l'Extrême-Orient. *Rev. franç. Entomol.* **XXIX.**

JEANNEL, R. and HENROT, H. (1949) Les Coléoptères cavernicoles de la région des Appalaches. *Notes biospéol.* **IV.**

KREKELER, C.H. (1958) Speciation in cave beetles of the genus *Pseudanophthalmus*. *Amer. Midl. Nat.* **LIX.**

KURNAKOV, V.N. (1959) Les *Trechini* de la faune souterraine de l'Abkhazie. *Rev. franç. Entomol.* **XXVI.**

LARSON, S.G. (1939) Entwicklungstypen und Entwicklungszeiten der dänischen Carabiden. *Entomol. Medd.* **XX.**

UÉNO, S. (1954) Studies on the Japanese Trechinae (II) (Col. Harpalidae). *Mem. Coll. Sc. Univ. Kyoto. Ser. B.* **XXI.**

UÉNO, S. (1955) Studies on the Japanese Trechinae (V) (Col. Harpalidae). *Mem. Coll. Sc. Univ. Kyoto. Ser. B.* **XXII.**

UÉNO, S. (1956) New cave-dwelling trechids of *Kurasawatrechus*-group (Coleoptera, Harpalidae). *Mem. Coll. Sc. Univ. Kyoto. Ser. B.* **XXIII.**

UÉNO, S. (1957) Studies on the Japanese Trechinae (VI) (Coleoptera, Harpalidae). *Mem. Coll. Sc. Univ. Kyoto. Ser. B.* **XXIV.**

UÉNO, S. (1958a) A remarkable new cave trechid from Eastern Kyushu of Japan (Coleoptera, Harpalidae). *Mem. Coll. Sc. Univ. Kyoto. Ser. B.* **XXV.**

UÉNO, S. (1958b) The cave beetles from Akiyoshi-dai Karst and its vicinities. I. A new species of the genus *Trechiama. Mem. Coll. Sc. Univ. Kyoto. Ser. B.* **XXV.**

UÉNO, S. (1958c) The cave beetles from Akiyoshi-dai Karst and its vicinities. II. *Uozumitrechus*, a new group of the genus *Rakantrechus. Mem. Coll. Sc. Univ. Kyoto. Ser. B.* **XXV.**

UÉNO, S. (1958d) Two new trechids of *Kurasawatrechus*-group found in the limestone caves of Japan (Coleoptera, Harpalidae). *Japan. J. Zool.* **XII.**

UÉNO, S. (1960a) Occurrence of *Yamautidius* in two limestone caves of Western Shikoku (Coleoptera, Harpalidae). *Trans. Shikoku Entomol. Soc.* **VI.**

UÉNO, S. (1960b) A synopsis of the genus *Kusumia* (Coleoptera, Harpalidae). *Mem. Coll. Sc. Univ. Kyoto. Ser. B.* **XXVII.**

VALENTINE, J.M. (1931) New cavernicole Carabidae of the subfamily *Trechinae* Jeannel. *J. Elisha Mitchell Sci. Soc.* **XLVI.**

VALENTINE, J.M. (1932) A classification of the genus *Pseudanophthalmus* Jeannel (Fam. Carabidae) with description of new species and notes on distribution. *J. Elisha Mitchell Sci. Soc.* **XLVII.**

VALENTINE, J.M. (1937) Anophthalmid beetles (Carabidae) from Tennessee caves. *J. Elisha Mitchell Sci. Soc.* **LIII.**

VALENTINE, J.M. (1945) Speciation and raciation in *Pseudanophthalmus* (cavernicolous Carabidae). *Trans. Connect. Acad. Arts. Sc.* **XXXVI.**

VALENTINE, J.M. (1948) New anophthalmid beetles from the Appalachian region. *Geol. Surv. Alabama. Mus. Pap.* No. 27.

VALENTINE, J.M. (1952) New genera of anophthalmid beetles from Cumberland Caves. *Geol. Surv. Alabama. Mus. Pap.* No. 34.

Harpalidae

BRITTON, E.B. (1962) New genera of beetles (Carabidae) from New Zealand. *Ann. Mag. Nat. Hist.* (13) **IV.**

MOORE, B. P. (1962) Notes on Australian Carabidae (Col.). III. A remarkable cave-frequenting harpaline from Western Victoria. *Entomol. mon. Mag.* **XCVII.**

Pterostichidae

BARR, TH. C. (1960) The cavernicolous beetles of the subgenus *Rhadine*, genus *Agonum* (Col. Carabidae). *Amer. Midl. Nat.* **LXIV.**
BOLIVAR Y PIELTAIN, C. (1944) Descubrimiento de un *Rhadine* afenopsiano en el Estado de Nuevo Leon, Mexico. *Ciencia* **V.**
BRITTON, E. B. (1959) Carabidae (Coleoptera) from New Zealand caves. *Proc. R. ent. Soc. Lond. Ser. B.* **XXVIII.**
DYKE, E. C. VAN (1918) A new genus and species of cave-dwelling Carabidae from the United States. *J. New York Entomol. Soc.* **XXVI.**
FAGNIEZ, C. (1921) Deux nouveaux Coléoptères cavernicoles des Alpes-Maritimes (Carabidae). *Bull. Soc. Entomol. France.*
JEANNEL, R. (1919) *Troglorites Breuili*, nouveau Carabique cavernicole des Pyrénées espagnoles. *Bull. Soc. Entomol. France.*
JEANNEL, R. (1924) Coléoptères nouveaux de Catalogne. *Trab. Mus. Barcelona.* **IV.**
JEANNEL, R. (1931) Insectes Coléoptères et Révision des *Trechinae* de l'Amérique du Nord. *Biospeologica,* **LVI.** *Archiv. Zool. expér. géner.* **LXXI.**
JEANNEL, R. (1950) Sur deux Ptérostichides cavernicoles de Majorque. *Rev. franç. Entomol.* **XVII.**
JEANNEL, R. (1953) Un Ptérostichide cavernicole de Turquie et remarques sur la systématique des *Tapinopterus* Schaum et genres voisins. *Notes biospéol.* **VIII.**
PATRIZI, S. (1955) Nuove genere e nuova specie di Pterostichide troglobio della Sardegna orientale (Col. Caraboidea; Pterostichidae; trib. Molopini). *Fragmenta Entomologica* **II.**

Brachynidae

JEANNEL, R. (1914) Observations sur la Faune des grottes du Kulumuzi. *Biospeologica,* **XXXIII.** *Archiv. Zool. expér. géner.* **LIII.**
REITTER, E. (1904) Ein neuer blinder *Brachynus* aus Deutsch-Ostafrika. *Wien. Entomol. Zeit.* **XXIII.**

Dytiscidae

ABEILLE DE PERRIN, E. (1904) Description d'un Coléoptère hypogé français. *Bull. Soc. Entomol. France.*
BIRSTEIN, Y. A. and BOROUTZKY, E. V. (1950) La vie dans les eaux souterraines. *In La vie des eaux douces de l'URSS.* Tome III; Chapitre 28. Moscou–Leningrad.
COLAS, G. (1946) Une station nouvelle de *Siettitia balsetensis* Ab. de Perrin (Col. Dystiscidae). *L'Entomologiste.* **II.**
GUIGNOT, F. (1925) Description d'un *Siettitia* nouveau du Midi de la France (Col. Dytiscidae). *Bull. Soc. Entomol. France.*
GUIGNOT, F. (1955) Speologia africana. Dytiscidae et Gyrinidae capturés dans les grottes de Guinée. *Bull. Inst. Franç. Afrique Noire. Ser. A.* **XVII.**
GUIGNOT, F. (1958) Un nouveau *Copelatus* (Col. Dytiscidae) des grottes du Moyen-Congo. *Rev. suisse Zool.* **LXV.**
JEANNEL, R. (1906) Remarques sur *Siettitia balsetensis* Abeille et sur la faune aquatique hypogée. *Bull. Soc. Entomol. France.*
PESCHET, R. (1932) Description d'un *Bidessus* nouveau hypogé de l'Afrique occidentale française. *Soc. Entomol. France. Livre du Centenaire.*
REGIMBART, M. (1905) Note sur le *Siettitia balsetensis* Abeille (Col.) *Bull. Soc. Entomol. France.*
UÉNO, S. (1957) Blind aquatic beetles of Japan with some accounts of the fauna of Japanese subterranean waters. *Archiv f. Hydrobiol.* **LIII.**

STAPHYLINOIDEA

Catopidae

DELEURANCE, S. (1959) Sur l'écologie et le cycle évolutif de *Choleva angustata* Fab. et *Fagniezi* Jeann. (Col. Catopidae). *Annal. Spéléol.* **XIV.**

JEANNEL, R. (1936) Monographie des *Catopidae*. *Mém. Mus. Hist. Nat. Paris* N.S. **I.**

JEANNEL, R. and HENROT, H. (1949) Les Coléoptères cavernicoles de la région des Appalaches. *Notes biospéol.* **IV.**

Bathysciinae

The whole bibliography of Bathysciinae is compiled in Jeannel's two monographs. (1911, 1926)

JEANNEL, R. (1908) *Adelopsella*, nouveau genre oculé de la tribu des *Bathysciini*. *Bull. Soc. Entomol. France.*

JEANNEL, R. (1911) Révision des *Bathysciinae*. Morphologie. Distribution géographique. Systématique. *Biospeologica*, **XIX.** *Archiv. Zool. expér. géner.* (5) **VII.**

JEANNEL, R. (1924) Monographie des *Bathysciinae*. *Biospeologica*, **L.** *Archiv. Zool. expér. géner.* **LXIII.**

JEANNEL, R. (1931) *Bathysciinae* nouveaux recueillis par P. Remy dans les grottes du Novi-Pazar. *Bull. Soc. Zool. France* **LVI.**

JEANNEL, R. (1934) *Bathysciinae* recueillis par MM. P. Remy et R. Husson dans le Sandjak de Novi-Pazar et la Macédoine grecque. *Rev. franç. Entomol.* **I.**

JEANNEL, R. (1947) Coléoptères cavernicoles nouveaux de France avec une étude sur la phylogénie des *Speonomus*. *Notes biospéol.* **I.**

JEANNEL, R. (1956) Sur un Bathysciite cavernicole nouveau de la Sardaigne (Col. Catopidae). *Fragmenta Entomol.* **II.**

MÜLLER, J. (1901) Beitag zur Kenntnis der Höhlensilphiden. *Verhandl. k. k. Zool. Bot. Gesell. Wien.* **LI.**

MÜLLER, J. (1903 a) Über neue Höhlenkäfer aus Dalmatien. *Sitzb. K. Akad. Wiss. Wien.-Math. Naturw. Kl.* **CLII.**

MÜLLER, J. (1903 b) Die Koleopterengattung *Apholeuomus* Reitt. *Sitzb. k. Akad. Wiss. Wien.-Math. Naturw. Kl.* **CLII.**

MÜLLER, J. (1904) Zwei neue Höhlensilphiden von der Balkanhalbinsel. *Münchner Koleopt. Zeit.* **II.**

MÜLLER, J. (1908) *Bathyscia khevenhülleri* Mill. und *freyeri* Mill. ihre systematische Stellung und ihre Rassen. *Wiener Entomol. Zeit.* **XXVII.**

MÜLLER, J. (1911) Zwei neue Höhlensilphiden aus den österreischicn Karstländern. *Winer Entomol. Zeit.* **XXX.**

MÜLLER, J. (1937) Nuovo Silfidi cavernicoli delle Balcania. *Atti Mus. Stor. Nat. Trieste.* **XIII.**

MÜLLER, J. (1941) Cinque Silfidi cavernicoli del Carso adriatico e delle Alpi Giulie. *Atti Mus. Stor. Nat. Trieste.* **XIII.**

REMY, P. (1940) Sur le mode de vie des *Hadesia* dans la grotte Vjetrenica (Col. Bathysciinae). *Rev. franç. Entomol.* **VII.**

Leptinidae

JEANNEL, R. (1922) Silphidae Leptinidae (Coléoptères) (1ère Série). *Biospeologica*, **XLV.** *Archiv. Zool. expér. géner.* **LX.**

Ptiliidae

DYBAS, H. (1960) A new genus of blind beetles from a cave in South Africa. *Fieldiana, Zool.* **XXXIX.**

218 BIOSPEOLOGY: THE BIOLOGY OF CAVERNICOLOUS ANIMALS

JOSEPH, G. (1882) *Systematisches Verzeichnis der in den Tropfstein-Grotten von Krain einheimischen Arthropoden.* Berlin.

STRASSEN, R. ZUR (1955) Eine neue Ptiliide aus Korsika (Ins. Col.). *Senck. Biol.* **XXVI.**

Staphylinidae

ABSOLON, K. (1915) Bericht über höhlenbewohnende Staphyliniden der dinarischen und angrenzenden Karstgebiete. *Col. Rundschau* **IV–V.**

CAMERON, M. (1947) A new genus of blind cavernicolous Staphylinidae from India. *Proc. R. ent. Soc. Lond.* (B) **XVI.**

COIFFAIT, H. (1934) Un nouveau *Domene* cavernicole du sud de l'Espagne, indice paléogeographique. *Notes biospéol.* **IX.**

CZWALINA, G. (1888) *Lathrobium (Glyptomerus) cavicola* Müller und *apenninum* Baudi. *Deutsch. Entomol. Zeit.*

JEANNEL, R. (1907) Diagnose d'un Staphylinide cavernicole nouveau de l'Algérie. *Bull. Soc. Entomol. France.*

JEANNEL, R. (1908) A propos d'*Aptaeraphaenops longiceps* Jeannel, Staphylinide cavernicole de l'Algérie. *Bull. Soc. Entomol. France.*

JEANNEL, R. (1909) Coléoptères (2ème Série). *Biospeologica,* **X.** *Archiv. Zool. expér. géner.* (5) **I.**

JEANNEL, R. (1959) Un Staphylinide cavernicole du Maroc. *Bull. Soc. Sc. nat.phys. Maroc.* **XXXIX.**

PEYERIMHOFF, P. DE (1949) Diagnose d'un *Domene* cavernicole du Maroc *(Domene camusi) Rev. franç. Entomol.* **XVI.**

RAMBOUSSEK, FR. (1915) Über eine neue dem Höhlenleben angepaßte *Atheta* (Staphyl.) aus Südherzegowina. *Col. Rundschau.* **IV.**

SCHEERPELTZ, O. (1935) Un genre nouveau et une espèce nouvelle de Staphylinides troglodytes du Maroc (Coléoptères). *Bull. Soc. Sc. Nat. Maroc.* **XV.**

Pselaphidae

JEANNEL, R. (1952) Psélaphides recueillis par M.Leleup au Congo belge. IV. Faune de l'Itombwee et du Rugege. *Annal. Mus. Congo Belge. Sc. Zool.* **XI.**

JEANNEL, R. (1953) Origine et répartition des Psélaphides de l'Afrique intertropicale. *Trans. IX Intern. Congr. Entomol. Amsterdam* **II.**

JEANNEL, R. et LELEUP, N. (1952) L'évolution souterraine dans la région méditerranéenne et sur les montagnes du Kiwu. *Notes biospéol.* **VII.**

PARK, O. (1951) Cavernicolous pselaphid beetles of Alabama and Tennessee, with observations on the family. *Geol. Surv. Alabama. Mus. Pap.* **XXXI.**

PARK, O. (1953) Evolution of American cavernicolous Pselaphidae. *Entomol. Soc. Amer. Proc.*

PARK, O. (1960) Cavernicolous pselaphid beetles of the United States. *Amer. Midl. Nat.* **LIV.**

Histeridae

JEANNEL, R. (1934) Coléoptères de la Grotte de Fersine, en Asie Mineure. *Ann. Soc. Entomol. France* **CIII.**

PATRIZI, S. (1955) *Sardulus spelaeus* n.g.n.sp. (Col. Histeridae). *Fragmenta Entomologica* **II.**

CUCUJOIDEA

Dryopidae

JEANNEL, R. (1950) Un Elmide cavernicole du Congo belge (Col. Dryopidae). *Rev. franç. Entomol.* **XVIII.**

JEANNEL, R. (1953) Note sur le *Troglelmis leleupi* Jeannel (Col. Dryopidae). *Notes biospéol.* **VIII.**

NOMURA, S. (1961) Elmidae found in subterranean waters of Japan. *Akitu. Trans. Kyoto Entomol. Soc.* **X.**

(b) Endomychidae

JEANNEL, R. (1934) Coléoptères de la grotte de Fersine, en Asie Mineure. *Ann. Soc. Entomol.* **CIII.**

HETEROMERA

Tenebrionidae

ANDRES, A. (1925) Notes et description d'un Tentyrinide cavernicole d'Egypte: *Mesostenopa cavatica* n.sp. *Bull. Soc. R. Entomol. Egypte.*

BLAIR, K.G. (1924) Histeridae, Hydrophilidae, Erotylidae, Lathriidae, Tenebrionidae and Hydrophilidae. *In* Coleoptera from the Siju Cave, Garo Hills, Assam. *Rec. Indian Mus.* **XXVI.**

BLAIR, K.G. (1929) Fauna of the Batu Caves, Selangor. XVII. Coleoptera. *J. Feder. Malay States Mus.* **XIV.**

REYMOND, A. (1956) Description d'un nouveau Coléoptère troglophile du Moyen Atlas septentrional: *Scaurus barbuloti* n.sp. (Col. Tenebrionidae). *Bull. Soc. Sci. Nat. Maroc.* **XXXVI.**

Hylophilidae

BLAIR, K.G. (1924) Histeridae, Hydrophilidae, Erotylidae. Lathriidae, Tenebrionidae and Hylophilidae. *In* Coleoptera from the Siju Cave, Garo Hills, Assam. *Rec. Indian Mus.* **XXVI.**

BLAIR, K.G. (1929) Fauna of the Batu Caves, Selangor. XVII. Coleoptera. *J. Feder. Malay States Mus.* **XIV.**

PIC, M. (1914) Insectes Coléoptères. IV. Hylophilidae et Anthicidae. *In* ALLUAUD, CH. and JEANNEL, R., *Voyages en Afrique Orientale, 1911–1912. Résultats scientifiques.*

MALACODERMA

Cantharidae

JEANNEL, R. (1908) Coléoptères (1ère Série). *Biospeologica,* **V.** *Archiv. Zool. expér. géner.* (4) **VIII.**

JEANNEL, R. (1909) Coléoptères (2ème Série). *Biospeologica,* **X.** *Archiv. Zool. expér. géner.* (5) **I.**

Lampyridae

BLAIR, K.G. (1929) Fauna of the Batu Caves, Selangor. XVII. Coleoptera. *J. Feder. Malay States Mus.* **XIV.**

CHAPTER XIV

THE VERTEBRATES

IF THE Agnatha are excluded, for today this group is reduced to a few genera, the five large vertebrate classes all possess representatives in the subterranean world. However, the number of them diminishes as one ascends the vertebrate scale. The most numerous and the most specialised cavernicoles are the fishes. The Amphibia contain a few true cavernicoles but they all belong to only one sub-class, the Urodela. The three classes of the Amniota, the reptiles, birds and mammals, contain a number of species which are found underground but none of them are true cavernicoles.

A. FISH

The fish may be divided into five sub-classes of which two are represented only by fossil forms (acanthodians and placoderms). The elasmobranches which are exclusively marine, and the Choanates (Dipnoi and Crossopterygii which are now very poorly represented in the world, have no cavernicolous members. Consequently all the hypogeous forms belong to the remaining sub-class Actinopterygii or more exactly to the only super-order Teleostei.

(a) Cavernicolous Teleosts

Lists of cavernicolous teleosts have been drawn up on many occasions, more recently by Norman (1926), Hubbs (1938), Thinès (1955), Dearolf (1956) and Bertin (1958). The numbers of species listed by the various authors are very different because some of them have included all more or less depigmented forms which have the eyes more or less regressed. However, many of these cannot be considered to be true cavernicoles. In the present review the following forms are excluded:

(1) Very strange fish have been found in the rapids of the great tropical rivers. They probably live in crevices in the rocks. This is the case, for example, in the two members of the Pygidiidae, *Pygidianops eigenmanni* Myers and *Typhlobelus ternetzi* Myers which have been found in the Sâo Gabriel rapids on the Rio Negro, Brazil. The first is completely anophthalmic and the second possesses vestigial eyes, while both of them are

depigmented (Myers, 1944). *Caecomastacembelus brichardi*, which belongs to the Mastacembelidae, lives in the rapids at the outlet of Stanley Pool in the Congo. This fish is almost totally depigmented and the eyes, which are regressed, are deeply buried in the skin (Poll, 1958).

(2) Those fish which live in burrows may also be excluded from our list. *Typhlogobius californensis* Steindachner, which lives in the holes made by crabs under rocks in San Diego Bay, California, is an example of this group (Eigenmann, 1909).

Certain members of the Clariidae which have very elongated bodies and reduced eyes *(Gymnaballes, Channaballes, Dolichoballes)* are also burrowing forms. One must place in the same category *Typhlosymbranchus boueti* Pellegrin (Symbranchidae), which has been found in rivers near Monrovia (Liberia) and is blackish in colour and anophthalmic (Pellegrin, 1922).

Finally we may place *Phreatobius cisternarum* Goeldi in this ecological category, although its biology is still unknown. This fish was found by Goeldi in 1905, in a reservoir on the island of Marajo which is situated in the mouth of the Amazon (Brazil). The anatomy of this species has been studied in detail by Reichel (1927). There is little pigmentation, the eyes are much reduced and covered by skin, and the optic nerve has degenerrated. The systematic position of this species remains uncertain, although it probably belongs to the Pimelodidae (Eigenmann, 1918; Reichel, 1927).

(3) The genus *Paraphoxinus* which inhabits the Dinaric regions has sometimes been regarded as cavernicolous. However, the species of *Paraphoxinus* are in fact pigmented and possess eyes. They live in surface waters and reproduce there. It is only during periods of drought, when the streams in which they live are low that they take refuge in caves (Karaman, 1923; Spandl, 1926).

Only those fish which spend a completely subterranean life will be dealt with here. They fall into two groups:

(*a*) The true cavernicoles which populate the rivers of freshwater lakes of continental caves.

(*b*) The phreatobia which are found in ground water and which may be collected in wells or springs fed by such waters, or which may be ejected by the artesian wells. These species are usually captured in arid regions of the world: Somalia, Iran, India, Texas, Australia.

Thus defined the list of cavernicolous species is not very long. It includes only 33 species which are distributed in 26 genera. Naturally, this list cannot be regarded as final because other cavernicolous fish will surely be discovered in the future.

(b) The Geographical Distribution of Hypogeous Fish

The hypogeous fish are distributed in all regions of the globe with the exception of Europe and northern Asia. America and central Africa contain the majority of known species.

(c) The Distribution of Hypogeous Fish in the Systematic Classification

(1) The hypogeous fish are represented essentially by two large groups of teleosts belonging to the malacopterygian type, the Cypriniformes and the Siluriformes. These fish possess Weber's Apparatus and they are usually united under the name Ostariophysi; they all inhabit fresh water.

(2) The acanthopterygian teleosts correspond, in modern classification, to the Perciformes. Hypogeous species are found in three families of this group, the Brotulidae, the Eleotridae, and the Gobiidae. The hypogeous Perciformes are much less numerous than the subterranean Ostariophysi.

(3) Finally, two families which are intermediate between the malacopterygian type and the acanthopterygian type, the Amblyopsidae and the Symbranchidae, contain hypogeous forms.

(d) Characidae

The Characidae belong to the Cypriniformes; they inhabit Africa and tropical America. They are generally carnivorous fish which have razor-sharp cutting teeth. The most celebrated are the piranhas *(Serrasalmus)* of the Amazon Basin.

Anoptichthys jordani (Fig. 59) a cavernicolous fish from Mexico belongs to this family. It lives in the Cueva Chica (San Luis Potosi). It has been described by Hubbs and Innes (1936). Two other species from the same region of Mexico have since been described: *A. antrobius* Alvarez 1946 from the Cueva del Pachon and *A. hubbsi* Alvarez 1947, from the Cueva de los Sabinos. *A. jordani* is one of the best-known cavernicolous fish because it can easily maintained in aquaria and even may be bred in captivity. Therefore today this fish is sold commercially. Important research on this fish by Breder and his collaborators, and also by Lüling, has been carried out and will be summarised later.

The cavernicolous forms have evolved from a surface species which may be captured in the Rio Tampaon, this is *Astyanax mexicanus* (Fig. 59,4) which is closely related to *Anoptichthys*. There are in the Cueva Chica intermediate forms, with regard to pigmentation and ocular regression, between the completely depigmented types with reduced eyes and the surface species (Fig. 59) (Breder, 1943). Moreover, *A. jordani* and *Astyanax mexicanus* can hybridise among themselves (Breder, 1943; Sadoglu, 1957).

(e) Cyprinidae

It is convenient to mention first of all four species which are directly related to each other. They are:

Caecobarbus geertsi Boulenger 1921; caves near Thysville, the Congo.
Eilichthys microphthalmus Pellegrin 1929; wells in Somalia (ground water).

Fig. 59. 1, *Anoptichthys jordani* from the Cueva Chica; a completely depigmented specimen. 2, *A. jordani* an individual which is for the most part depigmented. 3, *A. jordani* a partly pigmented specimen. 4, *Astyanax mexicans*, from the Rio Tampaon (after Breder and Gresser).

Phreatichthys andruzii Vinceguerra 1924; a spring in Somalia (ground water).

Iranocypris typhlops Brunn and Kaiser 1943; an underground spring in Iran.

These species are so closely related to each other that the establishment of a particular genus for each appears rather exaggerated (Bruun and Kaiser, 1943). These four cavernicoles are certainly derived from the genus *Barbus* which is particularly well represented in Africa (250 species).

Caecobarbus geertsi is the best-known species (Petit and Besnard, 1937; Heuts, 1952, 1953; Thinès, 1953, 1954; Heuts and Leleup, 1954). One can easily culture this species in aquaria maintained at 25°C; but reproduction of this fish has not yet been obtained in captivity.

Typhlogarra widdowsoni, described by Trewavas (1955) has been extensively studied by Marshall and Thinès (1958). This is also a barb, but is related to the genus *Garra*, which is characterised by the presence of an adhesive organ. It is found in a cave in Iraq near Haditha, in the Euphrates valley.

Puntius microps (Günther, 1868) a barb which has reduced eyes has been collected in a subterranean river in Java. This species is poorly known and further studies are required.

(f) Bagridae

The Bagridae constitute one of the most important families of the Siluroidea. The best-known example is the cat fish *Amiurus nebulosus*.

This family contains two remarkable hypogeous fish: *Trogloglanis pattersoni* Eigenmann, 1919, and *Satan eurystomus* Hubbs and Bailey, 1947. These two fish come from artesian wells which rise in the neighbourhood of San Antonio (Texas). These fish have no eyes and are depigmented. *Trogloglanis* is related to *Amiurus*, and *Satan* to *Pilodictis* (Hubbs and Bailey, 1947).

We must classify in the family Bagridae, *Prietella phreaticola* Carranza, discovered in underground waters in the vicinity of Múzquiz (Estado de Coahuila, Mexico) (Carranza, 1954). This fish is blind and depigmented. It is related to *Satan eurystomus*, of Texas.

(g) Pimelodidae

This family is closely related to the Bagridae and has formerly been confused with them. The "bagres" (genus *Rhamdia*) are widely distributed in Central and Southern America. One of the most common species is *Rhamdia guatemalensis*. This species is lucifugous, and it is not surprising, therefore, that they have been captured in the "cenotes" and caves in Yucatan, Mexico. Some sub-species *(decolor, stygaea)* are more or less decoloured and another sub-species *(stygaea)* also has reduced eyes (Hubbs, 1936, 1938).

Caecorhamdia urichi Norman 1926, differs from the members of the above-mentioned genus in that the eyes are absent. However, this species appears to be closely related to *R. queleni*, a species widespread in the northern areas of South America (Norman, 1926). This species has been collected in the Cueva del Guacharo, Trinidad.

Caecorhamdella brasiliensis Borodin, 1927 from an indeterminate locality in the state of Sâo Paulo, Brazil, is related to two surface-living species *R. foina* and *R. eriarcha*, from which it is distinguished by its anophthalmia (Borodin, 1927).

Pimelodella kronei was described in 1907 by Ribeiro under the name *Typhlobagrus*. However, it is not a "bagre", but a member of the genus *Pimelodella* (Haseman, 1911; Eigenmann, 1917, 1919; Norman, 1926; Borodin, 1927). This fish is depigmented and anophthalmic (its common name is "Ceguinho"). It has been found in two caves in the state of Sâo Paulo, Brazil. An excellent monograph on this species has been written by C. Pavan (1946).

(h) Clariidae

The Clariidae belong to the Siluroidea, and inhabit the fresh waters of Asia and Africa. They seem to have originated in the former continent and spread to the latter at a later date (Menon, 1951 b). These fish show a tendency towards orthogenetic evolution with respect to the elongation of the body, the more specialised types being filiform. This is the case in the genera *Gymnaballes*, *Channaballes* and *Dolichoballes*, which are burrowing forms with reduced eyes.

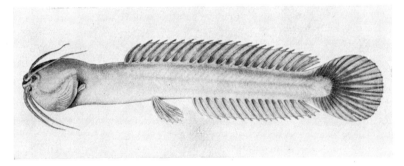

FIG. 60. *Horoglanis krishnai* (after Menon).

Three cavernicolous Clariidae are known: *Clarias* sp. Trewavas, 1936, from caves in south-west Africa. The second species is *Uegitglanis zammaranoi* Gianferrari, 1923; it is anophthalmic and depigmented and has been found near Uegit, Somalia, in wells. Thinès (1958) has studied the biology and ecology of this fish. The third cavernicole, *Horoglanis krishnai*

Menon 1951 (Fig. 60), is also anophthalmic and depigmented and has been taken in wells at Kottayam, Kerala, Southern India (Menon, 1951a).

(i) Symbranchidae

These fish are characterised by the fusion of the branchial openings into one median ventral orifice. At the present time both cavernicolous species are included in the family. *Pluto infernalis*† Hubbs 1938 (Fig. 61) has been captured in a cave in Yucatan, Mexico. It would appear to be related to

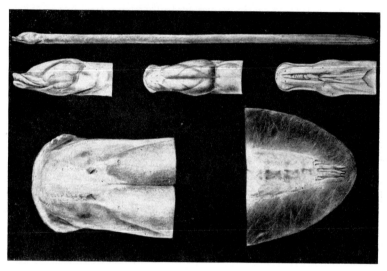

FIG. 61. *Pluto infernalis* (after Hubbs).
The structures of the head are shown enlarged, in three aspects. The further enlarged lower figures of the tip of muzzle and of the tail show respectively the structure of the dermal sense organs and of the last vertebrae and hypurals.

Symbranchus marmovatus, a species widespread in Central and South America. *P. infernalis* is anophthalmic and is only lightly pigmented (Hubbs, 1938).

According to Poey there is a blind symbranchid which inhabits caves in Cuba.

Another symbranchid, which is depigmented and anophthalmic, *Anommatophasma candidum*, has been recently discovered in the underground waters of the North West Cape, Australia (Mees, 1962).

(j) Amblyopsidae

This family belongs to the Cyprinodontiformes. Nearly all of them are cavernicolous. They are native to North America.

† Named *Furmastix infernalis* in the revision of Whitley (1951).

For one hundred and twenty years the amblyopsids have been the subject of many publications (de Kay, Wyman, Tellkampf, Agassiz, Girard, Putnam, Cope, Forbes, Eigenmann, Cox, Hubbs), but two recent publications are of particular interest. These are those of Woods and Inger (1957) and Poulson (1961).

The genus Chologaster. The cavernicolous amblyopsids are derived from the genus *Chologaster* (= *Forbesichthys*). This genus contains two species, one epigeous and the other troglophilic.

C. cornutus, which was described by Agassiz in 1853, is probably ancestral to, or at least closely related to the ancestor of the amblyopsids. It

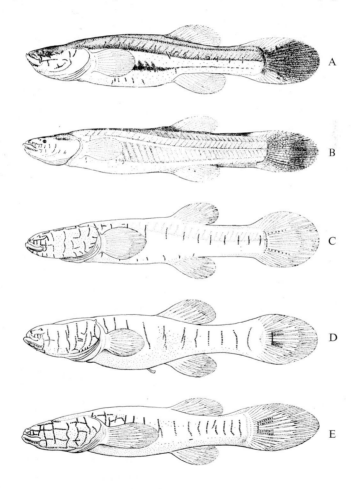

FIG. 62. The representatives of the family *Amblyopsidae*. A, *Chologaster cornutus;* B, *Chologaster agassizi;* C, *Typhlichthys subterraneus;* D, *Amblyopsis spelaeus;* E, *Amblyopsis rosae* (after Woods and Inger).

is well pigmented, the dorsal surface being dark and the sides and ventral surface paler. There is a longitudinal stripe along each side of the body. The fins are more or less pigmented (Fig. 62 A). The eyes are normal. The species is found in marshes in Carolina, Georgia and Virginia (Fig. 63).

C. agassizi Putnam, 1872, is not separable specifically from C. papilli-ferus Forbes 1882 (Woods and Inger, 1957). This fish is pale pink in colour. The lateral pigmented band is absent, but the dorsal surface is slightly darker than the rest of the body (Fig. 62 B). The eyes are slightly reduced.

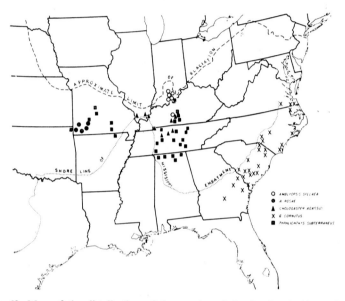

FIG. 63. Map of the distribution of the species of the family *Amblyopsidae* in North America (after Woods and Inger).

C. agassizi may be regarded as a troglophile. It has been taken in caves (e.g. Mammoth Cave) in addition to wells, springs and also marshes. It is distributed over a part of Kentucky, Tennessee, and Illinois (Fig. 63).

The genus Typhlichthys. There is only one species *T. subterraneus* Girard, 1859. The types described under the names *T. wyandotte* Eigenmann, 1897, *T. osborni* Eigenmann, 1905, *T. eigenmanni* Hubbs 1938 are all synonymous with *T. subterraneus* according to Woods and Inger (1957).

Typhlichthys differs from *Chologaster* in that the ocular system has degenerated and the number of sensory papillae increased. The latter are particularly numerous in the region of the head and are disposed in transverse rows on the body; two rows of these papillae are found on the base of the caudal fin (Fig. 62 C).

This species is normally without pigment but if they are kept in the light, pigmentation starts to be acquired within a few months. A lateral band of pigment develops and further pigmentation occurs at the level of the myosepta. Thus the ability of this species to form pigment has not been totally lost.

T. subterraneus is found in subterranean rivers and exceptionally in wells. Its distribution extends throughout Arkansas, Indiana, Kentucky, Missouri, Oklahoma and Tennessee (Fig. 63).

The genus Amblyopsis. This differs from the two preceding genera in that it lacks a cleithrum. There are two species. *A. spelaeus* de Kay 1842 was the first cavernicolous fish to be discovered. It was caught in Mammoth Cave and displayed at the Philadelphia Academy by W.T.Craige in 1842. De Kay described it in his *Natural History of New York*.

This species is pink when alive, due to the colour of the blood. There are a few chromatophores under the skin but these contain very little pigment. The eyes are degenerate and buried under the skin. Sensory papillae are numerous and disposed in several longitudinal rows on the caudal fin (Fig. 62 D).

This species has been found in caves in Indiana and in two others, Mammoth Cave and Cedar Sink, in Kentucky (Fig. 63).

A. rosae Eigenmann 1898 was placed by this author in the new genus *Troglichthys*, but Woods and Inger (1957) have demonstrated that this species is a specialised member of the genus *Amblyopsis*.

A. rosae is indeed more specialised than *A. spelaeus* and differs from it in that the pelvic fins are absent (Fig. 62 E), and the eyes are more degenerate. This species has been taken in caves in Missouri (Fig. 63).

(k) Brotulidae

This family belongs to the Perciformes. A few rare species are found in the shallow marine environment of the continental shelves, but the vast majority of the family are abyssal forms, many of them penetrating to the greatest depths. *Grimaldichthys profundissimus* holds the depth-record for fish, at 6000 metres.

Most Brotulidae, by reason of their habitat, are depigmented; their eyes are at different steps of regression, ending in almost complete anophthalmia. They are marine with the exception of three freshwater cavernicoles.

Two of these have been found in Cuba: *Lucifuga subterraneus* and *Stygicola dentatus* (Fig. 64). They have been described by Poey (1856) in his book *Memorias sobre la Historia Natural de la Isla de Cuba*. They have also been studied by Eigenmann (1909). These fish are related to two genera *(Brosmophycis* and *Ogilbia)* which live along the coast of California. It is quite probable that the cavernicolous forms have been derived from marine littoral species, which inhabit cavities in coral reefs. A number of them would enter subterranean rivers which flow into the sea; they would thus have been able to become progressively adapted to fresh water, and consequently evolve into purely cavernicolous forms. Today cavernicolous brotulids are found in freshwater underground streams at an altitude of 1–30 metres above sea level.

FIG. 64. *Stygicola dentatus* (after Eigenmann).

The third cavernicolous species of the Brotulidae is *Typhlias pearsi†* Hubbs 1938, which is related to the other two genera, but is more specialised. It is completely depigmented and anophthalmic. It has been captured in caves in Yucatan, Mexico (Hubbs, 1938).

(l) Eleotridae

The Eleotridae have formerly been confused with the Gobiidae to which they are closely related. There are three cavernicolous species. They are anophthalmic, more or less depigmented and are covered with peculiarly thin scales which are the texture of papyrus. The first cavernicolous species is *Typhleotris madagascariensis* Petit 1933, which has been found in subterranean streams in the calcareous regions of south west Madagascar (Petit, 1933; Angel, 1949). The second is *T. pauliani* Arnoult 1959, which was captured in pools in Andalambezo Cave in the west of Madagascar (Arnoult, 1959) and the third, *Milyeringa veritas* Whitley 1954, lives in subterranean streams of North West Cape, Australia.

(m) Gobiidae

The discovery of cavernicolous Gobiidae is a recent event due to the work of the Japanese zoologist Tomiyama (1936). These fish come from Japan and there are two species (Regan, 1940), *Luciogobius pallidus* Regan which was ejected from an artesian well, and *L. albus* Regan, which was discovered in a cave. The eyes are reduced in both species; in the first the colouration is pale while the second is quite depigmented. These cavernicoles are related to the marine species *Luciogobius guttatus* which is widespread on the coasts of Japan and Korea.

† Named *Typhliasina pearsi* in the revision of Whitley (1951).

(n) The Evolution of Cavernicolous Fish

(1) The Origin of Cavernicolous and Phreatobious Fish. It is usual to find that hypogeous fish are related to surface forms living in the same region. It must be concluded that their origins are relatively recent. All the truly hypogeous species are freshwater forms, and the large majority of these *(Ostariophysi, Symbranchidae, Amblyopsidae* and *Eleotridae)* have evolved from epigeous freshwater fish.

It is, however, a remarkable fact that the cavernicolous fish belonging to Brotulidae and the Gobiidae have a marine origin. In the Brotulidae the only freshwater species known are also the three cavernicolous species.

(2) Regressive Evolution in Cavernicolous Fish. Two characteristics of cavernicolous fish are most striking:

(i) Depigmentation of the body.
(ii) The progressive reduction of the eyes towards a state of almost complete disappearance.

We will consider these regressive characteristics later, but it should be pointed out that they are only two amongst a large number of other degenerative manifestations. In addition to these the following may be mentioned:

(iii) The frequent disappearance of scales *(Eilichthys, Phreatichthys, Pluto)*; where they persist they are frequently thin, soft and little apparent *(Caecobarbus, Typhleotris, Milyeringa).*
(iv) The loss of teeth has been reported in the genus *Trogloglanis.*
(v) The swim bladder is reduced in *Trogloglanis* and *Satan.*

(3) Regressive Evolution in Abyssal and Parasitic Fish. Not only cavernicolous fish live in darkness. This condition is shared with the abyssal and the parasitic fish. It may appear artificial to relate these three types of fish in which the ways of life are so profoundly different. However, this comparison is less arbitrary when one considers that it is only by adopting such peculiar modes of life, that these groups have been able to maintain themselves, in spite of their regressive and degenerate evolution.

Abyssal fish frequently show regressive characters such as incomplete ossification, retarded development, often arrested prior to the final stage, etc. They share with cavernicolous fish such obviously regressive characters as reduced eyes and depigmented skins.

However, many abyssal fish are notably different from cavernicoles. They are often dark in colour, brown or black, and they may have enormous, sometimes telescopic eyes. Such manifestations, which are unknown in cavernicoles, are due to the presence of numerous bioluminescent species in abysses whilst bioluminescence is exceptional in the subterranean world (p. 304).

The affinities which exist between cavernicolous and abyssal fish are clearly demonstrated in the Brotulidae. This family, as noted above, contains a majority of bathophilous species, although rare littoral forms belonging to it have been described. All the bathophilous forms are depigmented and the eyes are generally reduced or entirely absent *(Tauredophidium)*. However, three remarkable cavernicolous species belong to this family. That they are all relict species of an ancient fauna, preserved in a conservative media, cannot be doubted.

The truly parasitic fish belong to the Pygidiidae (Siluroidea). Some of them suck blood *(Vandellia)*, but others *(Stegophilus, Branchioica)* live in the branchial cavities of large siluroids *(Platysoma)* and feed on the gills. Other Pygidiidae are light avoiding and burrowing forms *(Pygidianops, Phreatobius)*.

Workers who have studied these fish (Eigenmann, 1918; Myers, 1944) have concluded that both cavernicoles and parasites have had a common origin and have proceeded from that point along different lines of regressive evolution.

B. AMPHIBIA

All cavernicolous amphibians belong to the Urodela; there are no hypogeous Anura, and the Apoda are burrowing endogeans, but not cavernicolous animals.

With a number of exceptions *(Proteus, Hydromantes)*, all the cavernicolous Urodela are North American and belong to the family Plethodontidae. The reader will find an excellent guide to these creatures in S. Bishop's *Handbook of Salamanders*, which is, unhappily, already dated.

The cavernicolous urodeles may conveniently be divided into three ecological groups:

(*a*) lucifugous salamanders found in the entrances of caves,

(*b*) species intermediate to (*a*) and (*c*),

(*c*) true cavernicoles.

(a) Lucifugous Salamanders

In North America these belong to the genera *Eurycea*, *Gyrinophilus* and *Plethodon* (Cooper, 1962). They are not true cavernicoles. Their eyes are normal and their pigmentation often vivid. They are twilight species. The following species may be listed:

Eurycea longicauda (Green) — central and eastern United States
E. lucifuga (Rafinesque) — central United States (Hutchinson, 1958)
E. troglodytes (Baker) — Texas (Baker, 1957)
Gyrinophilus danielsi (Blatchley) and *G. porphyriticus* (Green) — northeast United States.

(b) Intermediate Types

Members of the genus *Typhlotriton* are examples of types intermediate between the lucifugous salamanders and cavernicolous ones.

Typhlotriton spelaeus Stejneger 1892 is a specialised plethodontid from the caves of Missouri, Oklahoma, Kansas and Arkansas (Ozark Plateau). Although this form is pale in appearance, the skin still contains a few chromatophores. This salamander undergoes complete metamorphosis. The larvae are strongly pigmented and possess well developed eyes. They are found at the entrances of caves. At the time of the metamorphosis, they penetrate deep into the caves. Here the eyes start to regress. The lids fuse with each other and the rods and cones in the retina degenerate.

If, however, the larvae are maintained in light the body pigmentation persists and fusion of the eyelids does not occur. But, degeneration of the retina still occurs, showing that this regression is not induced by the absence of light (Noble and Pope, 1928).

Another species found in the Ozark Plateau (Missouri, Arkansas, Oklahoma) was first thought to be the larval form of *Typhlotriton spelaeus*. It is, in fact, a neotenous species which has since been given the name *T. nereus* (Bishop, 1944). The pigmentation of this salamander is in a state of instability. The specimens living outside are pigmented; those taken from caves exhibit a pale colour. The eyes of the larva are normal; those of the adult are more or less reduced.

(c) Troglobious Urodeles

As there are more urodeles in North America than in Europe, it would be expected that troglobious urodeles would be more numerous there. Thus while only a single cavernicolous urodele has been found in Europe four are known in North America. The latter belong to the Plethodontidae whilst the European form belongs to the Proteidae.

All the troglobious urodeles are neotenous. They do not undergo metamorphosis and thus retain their larval characters throughout life.

(1) Gyrinophilus palleucus. This species has been collected in several caves in the Cumberland Plateau in Tennessee (MacCrady, 1954). This urodele is exclusively cavernicolous and probably neotenous. It is mainly pale in colour, but some grey pigment can be seen on the dorsal and lateral regions of the body. This pigmentation does not undergo modification when the animal is placed in light and would therefore appear to be stable. The eyes contain pigment but are reduced in size.

(2) Eurycea latitans and E. troglodytes. These two species have been discovered in caves in the Edwards Plateau in central Texas. They are cavernicolous and neotenous forms. The eyes of *Eurycea latitans* are slightly reduced but not covered by the skin (Smith and Potter, 1950). *E. troglodytes*

has reduced eyes which are completely or partially covered by skin (Baker, 1957).

(3) Typhlomolge rathbuni. *Typhlomolge rathbuni* Stejneger 1896 is found in subterranean waters of the San Antonio region of Texas. This species can be collected either in caves or in ground waters. It is sometimes thrown up by the water gushing from artesian wells.

T. rathbuni has formerly, but erroneously, been classified near to *Proteus.* However, it belongs to the Plethodontidae. It is a neotenous form, with external gills, which resembles the larvae of the Plethodontidae (Fig. 65).

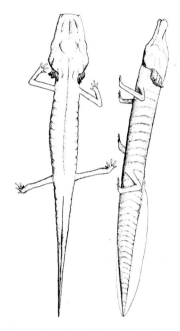

FIG. 65. *Typhlomolge rathbuni* (after Eigenmann).

This urodele shows several degenerative characteristics. The legs are very long and slender. Body pigmentation is absent. The eyes are reduced and have undergone more regression than those of *Proteus.* The eye muscles, the crystalline lens and the vitreous body are absent. The retina has become profoundly disorganized (Eigenmann, 1900). However, contrary to that stated by Emerson (1905) and Uhlenhuth (1923), the thyroid gland is present (Gorbman, 1957).

(4) Haideotriton. *Haideotriton wallacei* is known by a single female specimen which contained numerous eggs (Carr, 1939). This individual was thrown up by the pump of a well drilled near Albany, Georgia. It has external gills. The skull is for the most part cartilaginous. The animal is white

and the eyes appear to be extremely regressed, but their detailed structure is unknown.

Other specimens belonging to the genus *Haideotriton*, but probably to a different species, have been captured in a cave in Florida (Pylka and Warren, 1958). These forms differ from *H. wallacei* in that melanophores and xanthophores are present in the dorsal and lateral regions of the body. The eyes, although pigmented, are probably not functional.

(5) Proteus anguinus. *Proteus anguinus* Laurenti 1768 was the first cavernicole to be recognised. It was reported by Baron Johann Weichard Valvasor in 1689 in his book *Die Ehre des Herzogthums Crain*. Since then it has been featured in innumerable papers and reviews, the principal ones of which are given in the recent publication by Vandel and Bouillon (1959).

Proteus is most closely related to the genus *Necturus* and together they constitute the family Proteidae. *Necturus* is indigenous to North America, where it is represented by two species, *N. punctatus* and *N. maculosus.*

Proteus is the only cavernicolous urodele found in Europe. Its presence in the Old World shows that there must at sometime have been a connexion between Europe and North America, a conclusion also supported by the presence in Europe of members of the genus *Hydromantes*.

Proteus anguinus is found in the Karstic regions which border the Adriatic Sea from Venetia Juliana to Herzegovina. The qualification of "relict" also applies to *Proteus*. Its localisation in the Dinaric region is associated with the great dimensions of the subterranean cavities of Carniola and the abundance of food, both endogenous *(Troglocaris, Niphargus, Asellus, Titanethes)* and exogenous, the latter being carried into the caves when the subterranean rivers flood. The same conditions are to be found in the great "polje" of Herzegovina.

Proteus is a form with external gills. These gills persist throughout life, and appear red due to the blood which flows through them. It is related to fossil forms, for example *Hylaeobatrachus croyi* Dollo, of the Cretaceous (Wealdian) of Belgium, and *Palaeoproteus klatti* Herre of the middle Eocene in Europe (Herre, 1935). The first possessed three pairs of bony branchial arches, as does *Proteus*, while the latter possessed four pairs. It can be stated with some certainty that representatives of these two genera had gills and thus led an aquatic life. The neotenous state in the Proteidae thus goes back at least to the Cretaceous.

The adult *Proteus* is white or pink in colour although chromatophores persist in the skin and can multiply if the animal is kept in the light. The eyes are regressed in the adult *Proteus* (Kammerer, 1912; Hawes, 1946; Gostojeva, 1949), but it is better developed in the larva (Durand, 1962).

Fertilisation is brought about by a spermatophore which has recently been described by Briegleb (1961, 1962). *Proteus* is normally oviparous (Zeller, 1888, 1889; Vandel and Bouillon, 1959). The presence of numerous chromatophores in the larva gives it a dark appearance (Pl. IV). This pigmentation persists for about a year. The eyes are apparent and above the surface of the skin (Pl. IV). The ocular apparatus is still visible as a dark spot in young *Proteus* of three years of age.

C. REPTILES

Apodous, vermiform lizards are known which spend their life underground, and burrow in a similar manner to lumbricids, for example Anniellidae, Amphisbaenidae, Anelytropidae, Dibamidae. Similarly some ophidians belonging to the Typhlopidae have a habitat and behaviour

PLATE IV. The larva of *Proteus anguinus*, at birth.

resembling those of the vermiform lizards and, among the Amphibia, the Apoda. None of these burrowing forms can be considered to be cavernicoles.

One ophidian can perhaps be identified as a cavernicole or at least as troglophile. This is *Coluber (= Elaphe) taeniura* which is indigenous to tropical Asia. Its varieties *ridleyi* and *grabowskyi* are found in caves in Burma and Malaya which they enter in order to find the bats on which they feed. Representatives of these troglophilous varieties are rather pale due to the disappearance of brown and red pigments; however, black pigment persists (Annandale, Brown and Gravely, 1914; Kloss, 1929).

It is well known that geckos are essentially nocturnal. Two species have been captured in the caves of the Ryukyu Archipelago: *Gymnodactylus yamashinae* (Okada, 1936) and *Eublapheris splendens* (Nakamura and Uéno, 1959). More research is necessary before these species can be considered to be cavernicoles.

D. HOMOIOTHERMIC VERTEBRATES

Not one true cavernicole is known among the homoiothermic vertebrates, that is to say the birds and mammals. A few species belonging to these classes temporarily inhabit caves, either during the day, or to pass the winter there, or for the purposes of nidation. These cannot be considered as troglophiles because their higher organisation confers upon them a more complex behaviour which does not involve simple attraction or repulsion with respect to physical factors.

Nearly all these vertebrates possess the remarkable ability of finding their way about in darkness. This is as a result of a very complicated sensory system which has been given the name "echolocation" to which a later chapter has been devoted (Chap. XXVII). It is for this reason that we propose to designate the birds and mammals which regularly frequent caves under the name "echolocators".

E. BIRDS

Three groups of birds are known which contain species which frequent subterranean cavities and nest there.

(a) Salanganes

The salanganes *(Collocalia)* belong to the Apodidae and are represented by several species in Asia and tropical Indonesia. They are particularly abundant in Ceylon, Malaya, Cambodia and Borneo. These birds fly in the open air to obtain their food and cannot therefore be considered as cavernicoles. They enter caves only at nesting time (Medway, 1962). The salanganes construct the famous nests from which "birds' nest soup" is made. These are often a great distance from the entrance of the caves, sometimes even several hours' walking distance (Annandale, Brown, and Gravely, 1914; Tichelman, 1925; Stresemann, 1926; Novick, 1959; Medway, 1959). It is therefore necessary for these birds to have a method of finding their direction in the dark. In Chapter XXVII of this book we shall explain how their flight is guided by a system of echolocation.

(b) Petrochelidon

Petrochelidon belongs to the family Hirundinidae and is found in America, South Africa, India and Australia. A tropical species, *Petrochelidon fulva*, just penetrates into the United States, to New Mexico and Texas. In the most northern region of its habitat this species nests in caves. For this reason the Americans call it the "cave swallow". This behaviour has become established, it would seem, due to competition between this species and a swallow belonging to the same genus, *P. pyrrhonata*, which nests on cliffs, and is thus called the "cliff swallow". Indeed, *P. fulva* nests outside in the more southern lands, for instance in Mexico.

(c) The Guacharo

The Guacharo (*Steatornis caripensis* Humboldt) is the only representative of the Steatornidae which is related to the Caprimulgidae. This bird has been known to the Indians at all times; it is sought by them for the fat, which is especially abundant in the chick and from which the name *Steatornis* is derived. The bird was first reported by the celebrated naturalist Alexandre von Humboldt in 1799, during his travels in America. He provided the scientific name and a full description. This bird is indigenous to Venezuela, Trinidad, Guiana, Columbia, Ecuador, and Northern Peru. It is nocturnal and frugivorous, going out at night in search of food. During the day it shelters in large caves where it also builds its nests. These nests are often a long distance from the entrance, for example 600–800 metres in the famous Cueva del Guacharo, at Caripe in Venezuela. In Chapter XXVII we shall describe how this bird finds its way in darkness by means of echolocation.

F. MAMMALS

(a) Neotoma

The rats of North American caves *(Neotoma magister* and *N. pennsylvanica)* appear to be adapted to the subterranean environment, or at least are not often found at the surface. Contrary to that which has sometimes been stated, their eyes are normal (Eigenmann, 1909). They are not blind although their sight is mediocre. However, this condition is frequently found in rodents. The details of their behaviour as yet are wanting.

(b) Bats

The Chiroptera or bats are twilight and nocturnal animals but cannot be considered to be cavernicoles. Most bats never enter caves. A few species are regularly found in caves, but these places form only a temporary or seasonal refuge. We shall not deal in any detail with a subject which is not a true part of biospeology. Moreover, research on Chiroptera has given rise to a vast bibliography which is impossible to review in a few lines. We refer the reader to the more recent works listed in the bibliography.

The Megachiroptera hang in trees during the day. However, representatives of the genera *Rousettus* and *Cynonycteris* shelter during the day at the entrances of caves or in cavities of various depths.

The behaviour of members of the Microchiroptera varies according to their ability to resist the cold.

Many species of the genus *Vesperus* are found in Europe *(V. serotinus, V. noctula, V. pipistrellus)* but they are only exceptionally found in caves. They shelter in bell towers, barns or hollow trees during the winter.

Vespertilio and *Plecotus* hibernate in caves but are not found there during the summer.

Finally, *Rhinolophus* and *Miniopterus* are much more sensitive to low temperatures, and shelter during both the summer and winter in caves. They are regular inhabitants of subterranean cavities.

The bats lead a gregarious life. They live in groups which are sometimes exceedingly large. The areas they occupy in caves are covered with layers of guano which reach great depths in tropical caves. The role of this guano in the nutrition of cavernicoles is to be considered in a later chapter.

BIBLIOGRAPHY

FISH

ANGEL, F. (1949) Contribution à l'étude de *Typhleotris madagascariensis*, poisson aveugle cavernicole du Sud-ouest de Madagascar. *Bull. Mus. Hist. Nat. Paris* (2) **XXI**.

ARNOULT, J. (1959) Une nouvelle espèce de Poisson aveugle de Madagascar: *Typhleotris pauliani* n.sp. *Mem. Inst. Sci. Madagascar. Ser. A.* **XIII**.

BERTIN, L. (1958) Poissons cavernicoles. *In* GRASSÉ, P.P., *Traité de Zoologie* **XIII**.

BORODIN, N.A. (1927) A new blind catfish from Brazil. *Americ. Mus. Nov.* No. 263.

BOULENGER, G.A. (1921) Description d'un poisson aveugle découvert par M.G.Geerts dans la grotte de Thysville (Bas-Congo). *Rev. Zool. Afric.* **IX**.

BREDER, C.M. JR. (1943) Apparent changes in phenotypic ratios of the characins in the type locality of *Anoptichthys jordani* Hubbs and Innes. *Copeia*, No. 1.

BRUUN, A.FR. and KAISER, E.W. (1943) Iranocypris typhlops n.g., n.sp., the first true cave fish from Asia. *Danish Sc. Investig. in Iran.* Part IV. Copenhagen.

CARRANZA, J. (1954) Descripcion del primer bagre anophthalmo y depigmentado en aguas mexicanes. *Ciencia* **XIV**.

DEAROLF, K. (1956) Survey of North American cave vertebrates. *Proc. Pennsylvania Acad. Sci.* **XXX**.

EIGENMANN, C.H. (1898) *Amblyopsidae* and the eyes of blind fishes. *Proc. Ind. Acad. Sci.*

EIGENMANN, C.H. (1905) Divergence and convergence in fishes. *Biol. Bull.* **VIII**.

EIGENMANN, C.H. (1909) Cave vertebrates of America. A Study of degenerative evolution. *Carnegie Inst. Washington Public.* No. 104.

EIGENMANN, C.H. (1917) *Pimelodella* and *Typhlobagrus*. *Mem. Carnegie Mus.* **VII**.

EIGENMANN, C.H. (1918) The Pygidiidae, a family of South American catfishes. *Mem. Carnegie Mus.* **VII**.

EIGENMANN, C.H. (1919) *Trogloglanis pattersoni*, a new blind fish from San Antonio, Texas. *Proc. amer. phil. Soc.* **LVIII**.

FORBES, S.A. (1882) The blind cave fishes and their allies. *Amer. Nat.* **XVI**.

GIANFERRARI, L. (1923) *Uegitglanis zammaronoi*, un nuevo Siluride cieco africano. *Atti Soc. Ital. Sc. Nat. Milano* **LXII.**

GÜNTHER, A. (1868) Catalogue of the Fishes in the British Museum. **VII.**

HASEMAN, J.D. (1911) Description of some new species of fishes and miscellaneous notes on others obtained during the expedition of the Carnegie Museum to Central South America. *Ann. Carnegie Mus.* **VII.**

HEUTS, M.J. (1952) Ecology, variation and adaptation of the blind cave fish, *Caecobarbus geertsii* Boulenger. *Ann. Soc. Zool. Belgique* **LXXXII.**

HEUTS, J.M. (1953) Regressive evolution in cave animals. *Evolution* **VII.**

HEUTS, M.J. and LELEUP, N. (1954) La géographie et l'écologie des grottes du Bas-Congo. Les habitats de *Caecobarbus geertsi* Blgr. *Ann. Mus. R. Congo Belge.* **XXXV.**

HUBBS, C.L. (1936) Fishes from the Yucatan Peninsula. *Carnegie Inst. Washington Public.* No. 457.

HUBBS, C.L. (1938) Fishes from the caves of Yucatan. *Carnegie Inst. Washington Public.* No. 491.

HUBBS, C.L. and BAILEY, R.M. (1947) Blind catfishes from artesian waters of Texas. *Occas. Papers Mus. Zool. Univ. Michigan* No. 499.

HUBBS, C.L. and INNES, W.T. (1936) The first known blind fish from Mexico. *Occas. Papers Mus. Zool. Univ. Michigan* No. 342.

KARAMAN, ST. (1923) Über die Herkunft der Süßwasserfische unseres Karstes. *Glasnik zem Mus. Bosn. Herz.* **XXXV.**

MARSHALL, N.B. and THINÈS, G.L. (1958) Studies of the brain, sense organs and light sensibility of a blind cave fish *(Typhlogarra widdowsoni)* from Iraq. *Proc. Zool. Soc. Lond.* **CXXXI.**

MEES, G.F. (1962) The subterranean fauna of Yardie Creek station, North West Cape, Western Australia. *J. R. Soc. West. Austral.* **XLV.**

MENON, A.G.K. (1951a) On a remarkable blind siluroid fish of the family Clariidae from Kerala (India). *Rec. Indian Mus.* **XLVIII.**

MENON, A.G.K. (1951b) Distribution of clariid Fishes, and its significance in zoogeographical studies. *Proc. Nat. Inst. Sci. India.* **XVII.**

MYERS, G.S. (1944) Two extraordinary new blind nematognath fishes from the Rio Negro representing a new subfamily, with a rearrangement of the genera of the family and illustrations of some previously described genera and species from Venezuela and Brazil. *Proc. California Acad. Sci.* (4) **XXIII.** No. 40.

NORMAN, J.R. (1926) A new blind catfish from Trinidad, with a list of the blind cave-fish. *Ann. Mag. Nat. Hist.* (9) **XVIII.**

PAVAN, C. (1946) Observations and experiments on the cave fish, *Pimelodella kronei* and its relatives. *Amer. Nat.* **LXXX.**

PELLEGRIN, J. (1922) Sur un nouveau Poisson aveugle des eaux douces de l'Afrique occidentale. *Compt. rend. Acad. Sc. Paris.* **CLXXIV.**

PELLEGRIN, J. (1929) L'*Eilichthys microphthalmus* Pellegrin, Poisson Cavernicole de la Somalie italienne. *Bull. Mus. Hist. Nat. Paris* (2) **I.**

PETIT, G. (1933) Un Poisson cavernicole aveugle des eaux douces de Madagascar: *Typhleotris madagascariensis* gen. et sp. nov. *Compt. rend. Acad. Sci. Paris* **CXCVII.**

PETIT, G. (1938) Sur *Typhleotris madagascariensis* G. Petit. *Bull. Mus. Hist. Nat. Paris* (2) **X.**

PETIT, G. and BESNARD, W. (1937) Sur le comportement en aquarium du *Caecobarbus geertsii* Blgr. *Bull. Mus. Hist. Nat. Paris* (2) **IX.**

POEY, F. (1856) Peces ciegos de la isla de Cuba comparados con algunas especies de distinto genero. *Memorias sobre la Historia Natural de la Isla de Cuba.* Habana. **II.**

POLL, M. (1958) Description d'un poisson aveugle du Congo belge appartenant à la famille des *Mastacembelidae. Rev. Zool. Bot. Afric.* **LVII.**

POULSON, T.L. (1961) Cave adaptation in amblyopsid fishes. *Thesis of the University of Michigan.*

PUTNAM, F.W. (1872) The blind fishes of the Mammoth Cave and their allies. *Amer. Nat.* **VI.**

REGAN, T. (1940) The fishes of the gobiid genus *Luciogobius* Gill. *Ann. Mag. Nat. Hist.* (11) **V.**

REICHEL, M. (1927) Étude anatomique du *Phreatobius cisternarum* Goeldi, Silure aveugle du Brésil. *Rev. suisse Zool.* **XXXIV.**

SADOGLU, P. (1957) Mendelian inheritance in the hybrids between the Mexican blind cave fishes and their overground ancestors. *Verh. deutsch. zool. Gesell.* Graz.

SPANDL, H. (1926) *Die Tierwelt der unterirdischen Gewässer.* Wien.

THINÈS, G. (1953) Recherches expérimentales sur la photosensibilité du poisson aveugle, *Caecobarbus geertsii* Blgr. *Ann. Soc. R. Zool. Belgique* **LXXXIV.**

THINÈS, G. (1954) Étude comparative de la photosensibilité des poissons aveugles, *Caecobarbus geertsii* Blgr. et *Anoptichthys jordani* Hubbs et Innes. *Ann. Soc. R. Zool. Belgique.* **LXXXV.**

THINÈS, G. (1955) Les Poissons aveugles (I). Origine, Taxonomie, Répartition géographique, Comportement. *Ann. Soc. R. Zool. Belgique* **LXXXVI.**

THINÈS, G. (1958) Observations sur les habitats de *l'Uegitglanis zammaranoi* Gianferrari 1923, Clariidae de la Somalie italienne. *Rev. Zool. Bot. Afric.* **LVII.**

TOMIYAMA, I. (1936) *Gobiidae* of Japan. *Jap. J. Zool.* **VII.**

TREWAVAS, E. (1936) Dr. Karl Jordan's expedition to South-West Africa and Angola: the freshwater fishes. *Novit. Zool.* **XL.**

TREWAVAS, E. (1955) A blind fish from Irak related to *Garra. Ann. Mag. Nat. Hist.* (12) **VIII.**

VINCIGUERRA, D. (1924) Descrizione di un ciprinide cieco proveniente della Somalia italiana. *Ann. Mus. Civ. Stor. Nat. Genova* **LI.**

WHITLEY, G.P. (1954) Some freshwater gudgeons, mainly from tropical Australia. *Austral. Mus. Magazine* **XI.**

WOODS, L.P. and INGER, R.F. (1957) The cave, spring and swamp fishes of the family Amblyopsidae of Central and Eastern United States. *Amer. Midl. Nat.* **LVIII.**

AMPHIBIA

Lucifugous Salamanders

BAILEY, V. (1933) Cave life of Kentucky, mainly in the Mammoth Cave Region. *Amer. Midl. Nat.* **XIV.**

BAKER, J.K. (1957) *Eurycea troglodytes:* a new blind cave salamander from Texas. *Texas J. Sci.* **IX.**

BANTA, A.M. and MACATEE, W.L. (1906) The life history of the cave salamander, *Spelerpes maculicaudus* (Cope). *Proc. U.S. Nat. Mus.* **XXX,** No. 1443.

BISHOP, SH.C. (1943) *Handbook of Salamanders.* New York.

COOPER, J.E. (1962) Cave records for the salamander *Plethodon r. richmondi* Pope, with notes on additional cave-associated species. *Herpetologia.* **XVII.**

DUNN, E.R. (1926) The salamanders of the family *Plethodontidae.* Norhampton.

EIGENMANN, C.H. (1901) Description of a new cave salamander, *Spelerpes stejnegeri,* from the caves of southwestern Missouri. *Trans. Amer. Microsc. Soc.* **XXII.**

HUTCHINSON, V.H. (1958) The distribution and ecology of the cave salamander, *Eurycea lucifuga. Ecol. Monogr.* **XXVIII.**

Intermediate Types

BISHOP, SH.C. (1944) A new neotenic plethodon salamander, with notes on related species. *Copeia.*

NOBLE, G.K. and POPE, S.H. (1928) The effects of light on the eyes, pigmentation and behaviour of the cave salamander, *Typhlotriton. Anat. Rec.* **XLI.**

STEJNEGER, L. (1892) Preliminary description of a new genus and species of blind cave salamander from North America. *Proc. U.S. Nat. Mus.* **XV.**

Gyrinophilus palleucus

MacCRADY, ED. (1954) A new species of *Gyrinophilus (Plethodontidae)* from Tennessee caves. *Copeia.*

Eurycea

BAKER, J.K. (1957) *Eurycea troglodytes:* a new blind cave salamander from Texas. *Texas J. Sci.* **IX.**

SMITH, H.M. and POTTER, F.E. (1950) Another neotenic salamander from the Edwards Plateau. *Proc. Biol. Soc. Washington* **LXIII.**

Typhlomolge

EIGENMANN, C.H. (1900) The eyes of the blind vertebrates of North America. II. The eyes of *Typhlomolge rathbuni* Stejneger. *Trans. Amer. Microsc. Soc.* **XXI.**

EMERSON, E.T. (1905) General anatomy of *Typhlomolge rathbuni. Proc. Soc. Nat. Hist. Boston* **XXXII.**

GORBMAN, A. (1957) The thyroid gland of *Typhlomolge rathbuni. Copeia.*

STEJNEGER, L. (1896) Description of a new genus and species of blind tailed Batrachians from subterranean waters of Texas. *Proc. U.S. Nat. Mus.* **XVIII.**

UHLENHUTH, E. (1923) The endocrine system of *Typhlomolge rathbuni. Biol. Bull.* **XLV.**

Haideotriton

CARR, A.F. (1939) *Haideotriton wallacei*, a new subterranean salamander from Georgia. *Occas. Papers. Boston Soc. Nat. Hist.* **VIII.**

PYLKA, J.M. and WARREN, R.D. (1958) A population of *Haideotriton* in Florida. *Copeia.*

Proteus

BRIEGLEB, W. (1961) Die Spermatophore des Grottenolmes. *Zool. Anz.* **CLXVI.**

BRIEGLEB, W. (1962) Zur Biologie und Ökologie des Grottenolms (*Proteus anguinus* Laur. 1768). *Zeit. Mozphol. Ökol. Tiere.* **LI.**

DURAND, J.P. (1963) Structure de l'oeil et de ses annexes chez la larve de *Proteus anguinus. Compt. Rend. Acad. Sc. Paris.* **CCLV.**

GOSTOJEVA, M.N. (1949) New data on the problem of reduction of the eye in *Proteus. Compt. Rend. Acad. Sc. Moscou. N.S.* **LXVII.**

HAWES, R.S. (1946) On the eyes and reactions to light of *Proteus anguinus. Quart. J. Microsc. Sci. N.S.* **LXXXVI.**

HERRE, W. (1935) Die Schwanzlurche der mitteleocänen (oberluttischen) Braunkohle des Geiseltales und die Phylogenie der Urodelen unter Einschluß der fossilen Formen. *Zoologica.* **XXXIII.**

KAMMERER, P. (1912) Experimente über Fortpflanzung, Farbe, Augen und Körperreduktion bei *Proteus anguinus* Laur. *Archiv. f. Entwicklungsm.* **XXXIII.**

VANDEL, A., and BOUILLON, M. (1959) Le Protée et son intérêt biologique. *Ann. Spéléo-logie* **XIV.**

ZELLER, E. (1888) Über die Larve des *Proteus anguinus. Zool. Anz.* **XI.**

ZELLER, E. (1889) Über die Fortpflanzung des *Proteus anguineus* und seine Larve. *Jahres-hefte d. Ver. f. Vaterl. Naturkunde i. Württemberg.*

REPTILES

ANNANDALE, N., BROWN, J. C. and GRAVELY, F. H. (1914) The limestone caves of Burma and the Malay Peninsula. *J. Proc. Asiatic Soc. Bengal N.S.* **IX.**

KLOSS, C. B. (1929) Mammalia. *In* Fauna of the Batu Caves, Selangor. *J. Feder. Malay States Mus.* **XIV.**

NAKAMURA, K. and UÉNO, S. I. (1959) The Geckos found in the limestone caves of the Ryu Kyu Islands. *Mem. Coll. Sc. Univ. Kyoto Ser. B.* **XXVI.**

OKADA, Y. (1936) A new cave-gecko, *Gymnodactylus yamashinae*, from Kumejima, Okinawa group. *Proc. Imper. Acad. Japan* **XII.**

BIRDS

Salanganes

ANNANDALE, N., BROWN, J. C. and GRAVELY, F. H. (1914) The limestone caves of Burma and the Malay Peninsula. *J. Proc. Asiatic Soc. Bengal N.S.* **IX.**

MEDWAY, LORD (1959) Echolocation among *Collocalia. Nature, Lond.* **CLXXXIV.**

MEDWAY, LORD (1962) The swiftlets (Collocallia) of Niah Cave, Sarawak, 11. Ecology and the regulation of breeding. *Ibis.* **CIV.**

NOVICK, A. (1959) Acoustic orientation in the Cave Swiftlet. *Biol. Bull.* **CXVII.**

STRESEMANN, E. (1926) Zur Kenntnis der Salanganen S. O. Borneos. *Ornithol. Monatsber.* **XXXIV.**

TICHELMAN, G. L. (1925) Die Vogelnestgrotte Tamaloeang (S. O. Borneo). *Ornithol. Monatsber.* **XXXIII.**

Petrochelidon

BAKER, J. K. (1960) The Cave Swallow. *Nat. Speleol. Soc. News.* **XVIII.**

VAN TYNE, J. (1938) The Yucatan form of the West Indian Cliff Swallow *(Petrochelidon fulva). Occas. Papers Mus. Zool. Univ. Michigan* No. 385.

WHITAKER, L. (1959) Cave Swallow nesting in building near Cuatro Cienegas, Coahuila, Mexico. *The Condor* **LXI.**

The Guacharo

BELLARD PIETRI, E. DE (1960) Un Oiseau troglodyte: le Guacharo. *Bull. Comité Nat. Spéléol.* **X.**

HUMBOLDT, A. de (1833) Mémoire sur le Guacharo de la caverne de Caripe. Nouveau genre d'Oiseaux nocturnes de la famille des Passereaux. In *Recueil d'Observations de Zoologie et d'Anatomie comparée faites dans l'Océan Atlantique, dans l'intérieur du nouveau Continent et dans la mer du Sud pendant les années 1799, 1800, 1801, 1802 et 1803*, by A. de Humboldt and A. Bonpland. Paris, F. Schoell et D. Dufour. Vol. 2.

SNOW, D. W. (1961) The natural history of the Oil Bird, *Steatornis caripensis*, in Trinidad, W. I. – I. General behaviour and breeding habits. *Zoologica.* **XLVI.**

This publication includes the complete bibliography on the Guacharo.

MAMMALS

Neotoma

EIGENMANN, C. H. (1909) Cave vertebrates of America. *Carnegie Inst. Washington Public.* No. 104.
BAILEY, V. (1933) Cave life of Kentucky, mainly in the Mammoth Cave region. *Amer. Midl. Nat.* **XIV.**

Bats

EISENTRAUT, M. (1957) *Aus dem Leben der Fledermäuse und Flughunde.* Jena, Fischer.
GRASSÉ, P. P. (1955) Ordre des Chiroptères (Chiroptera Blumenbach 1774). In GRASSÉ, P. P. *Traité de Zoologie.* Tome **XVII,** fasicule **II.** Paris.
TOSCHI, A. and LANZA, B. (1959) Mammalia. Generalità. Insectivora; Chiroptera. In *Fauna d'Italia.* **IV.** Bologna.

CHAPTER XV

PHORETIC AND PARASITIC FORMS

A. INTRODUCTION

Both cavernicoles and epigeans are hosts to external and internal parasites. However, the characteristics of parasitic life are identical in the two
groups and there is therefore no point in dealing with hypogeous parasitism
as a special case. Therefore, the subject matter of this present chapter
does not belong to biospeology but rather to parasitology. It seemed,
however, useful to give the reader an idea, however summarily, of the
parasites which attack cavernicoles.

Phoresy or epizoism, that is to say the fixation of a plant or animal
species on to the surface of another organism without the first living at the
expense of the second, provides some of the more interesting cases. Many
cavernicoles are hosts to epizoic forms which exist only on these species.
However, this is more a case of parasitic or symbiotic specificity than
cavernicolous specialisation.

B. PARASITIC FUNGI

The fungi found as parasites on cavernicoles may be divided into the
five following groups:

  (a) Saprolegniales
  (b) Entomophthorales
  (c) Hyphomycetes (Fungi imperfecti)
  (d) Fungi of uncertain affinities. *Histoplasmosis*
  (e) Laboulbeniaceae (belonging to the order of Ascomycetes).

(a) Saprolegniales

Lagarde (1922) reported the presence of *Saprolegna thureti* on an unknown cavernicolous beetle from Spain.

(b) Entomophthorales

An unidentified species of this group was observed by Theodoridès
(1954) on the bathysciine beetle *Speonomus delarouzei.*

(c) Hyphomycetes (Fungi imperfecti)

Lagarde (1913, 1917, 1922) has reported the presence of *Isaria guignardi* on various cavernicoles *(Aphaenops, Speonomus, Diaprysus)*; *Isaria densa*, *Beauveria globulifera*, and *Verticillium* sp. on various Diptera; *Isaria* sp. on Lepidoptera and Trichoptera; and *Isaria arachnophila* on a spider.

In North America, Call (1897a, b) showed the presence of *Isaria (Sporotrichum) densa* on a dead specimen of the orthopteran *Hadenoecus subterraneus* from the Mammoth Cave.

(d) Fungi of Uncertain Affinities. Histoplasmosis

Histoplasmosis (Darling's disease), was discovered in 1906 by S. T. Darling in Panama. It is due to a parasite to which he gave the name *Histoplasma capsulatun* by reason of the thick capsule which surrounds it. Darling thought that this parasite was a protozoan, but in 1912, H. da Roch-Lima recognised it as a fungus. Histoplasmosis is widespread in warm regions of the world. It appears as a pulmonary infection resembling tuberculosis. It is usually benign, deaths from it being very rare.

Only the relationship which exists between this infection and the subterranean medium will be considered here. It appears that certain caves are reservoirs of *Histoplasma* and most visitors become infected. It is possible that the death of Lord Carnavon in 1923 following his exploration of the tomb of Tutankhamen in upper Egypt was a result of cave sickness.

All the caves containing *Histoplasma* are situated in the warm regions of America (south east of the United States, Venezuela, Peru) and Africa (Rhodesia, Tanganyika, Transvaal). Histoplasmosis is unknown in temperate countries. The fungi can be found in either the mycelial or conidial stage in warm, humid soil, and also in cave deposits, particularly in those which are rich in guano. It would be possible that bats serve as a reservoir of histoplasmosis (Shacklette, Diercks and Gale, 1962).

(e) Laboulbeniaceae

The Laboulbeniaceae or Laboulbeniales are parasitic Ascomycetes of microscopic size (less than 1 mm). These fungi are attached to the integuments of insects or more rarely of mites. The great majority of species live on Coleoptera; the carabids and staphylinids are the groups which support the greatest number of them.

Cavernicolous as well as epigeous Coleoptera are infected with these fungi. The first cavernicolous species was described in 1896 by Istvanffi. This is *Laboulbenia gigantea* which was found fixed to *Antisphodrus cavicola*.

The cavernicolous Laboulbeniaceae belong to the genera *Laboulbenia* and *Rhachomyces*. The first genus contains numerous species of which the majority are parasites on epigeous genera. However, *Laboulbenia sub-terranea* is a frequent parasite of cavernicolous Coleoptera belonging to the genera *Anophthalmus* (Venetia Juliana, Istria, Slovenia, Croatia) and *Trechus*. This species has also been found on the American genera *Pseudanophthalmus* and *Neaphaenops*. *Laboulbenia gigantea*, a species related to *L. flagellata*, but according to Picard quite distinct, is parasitic on troglophilic carabids belonging to the genus *Antisphodrus*.

Species of the genus *Rhachomyces* are of the most part restricted to cavernicolous Coleoptera *(Trechus, Speotrechus, Trichaphaenops, Duvalius, Geotrechus, Aphaenops)*. Lepesme (1942) has produced a revision of the European and North African species. A single species, *R. speluncalis*, is known in North America which attaches itself to *Pseudanophthalmus pusio*. Lepesme considers that the penetration of the *Rhachomyces* into the subterranean environment is more recent than that of *Laboulbenia*.

C. GREGARINA

A list of gregarines which are parasitic on cavernicoles is given below but this should not be considered to be complete.

Host	Site of Infection	Species	Family	Author
Oligochaeta				
Phreatothrix pragensis	Seminal vesicles	?	?	Vejdovsky, 1882
Pachydrilus pagenstecheri		*Gonospora pachydrili*	Urosporidae	Vejdovsky, 1882
Pachydrilus subterraneus		*Gonospora sp.*	Urosporidae	Moniez, 1889
Eisenia sp.		*Clepsinidra sp.*	Gregarinidae	Wetzel, 1929
Crustacea				
Canthocamptus minutus	Digestive tract	*Monocystis lacrima*	Monocystidae	Vejdovsky, 1882
Cyclops macrurus	Digestive tract	*Monocystis* aff. *lacrima*	Monocystidae	Chappuis, 1920
Coleoptera				
Ceutosphodrus oblongus	Intestine	*Actinocephalus permagnus*	Actinocephalidae	Theodoridès, 1954

D. CNIDOSPORIDIA
Myxosporidia

H. Joseph (1905, 1906) described a myxosporidian which is parasitic in the kidneys of *Proteus* and gave it the name *Chloromyxum protei*.

Microsporidia

Two microsporidians have been described by Poisson (1924), which were found by the present author in the adipose tissue of *Niphargus stygius*, found near Paris. These two protozoans are *Theolania vandeli* and *Bacillidium niphargi* (formerly described under the name *Mrazekia niphargi*).

Menozzi (1939) found in larva of *Parabathyscia doderi* (Col.), a parasite which he associated with the Gregarina, but which according to Theodoridès belongs to the Microsporidia.

E. CILIATES

The ciliates may be divided into internal parasites, and epizoic or phoretic types which do not live at the expense of their hosts.

(a) Parasitic Ciliates

Two parasitic ciliates belonging to the order Astomata and the genus *Anophlophrya* have been found in cavernicoles. They are *Anoplophrya branchiarum*, which has been found in the circulatory system of *Niphargus* in the north of France (Moniez, 1889), and an unidentified species of the Anoplophryidae found to be parasitic in the digestive tract of the oligochaetes *Stylodrilus parvus* and *Bythonomus lemani*, which have been collected in the Grotte de Corveissiat, in the department of Ain (Juget, 1959).

(b) Epizooic Ciliates

With the exception of an unidentified hypotrich found on the cavernicolous dytiscid *Phreatodytes relictus* by Uéno (1957), all other epizooic ciliates fixed to hypogeous forms belong to the order Peritricha and the sub-order Dexiotricha.

Trichodina sp. found on the carapace of ostracods of the genus *Candona* (Moniez, 1889).

Lagenophrys monolistrae, attached to the pleopods of various species of spheromid isopods of the genus *Monolistra* (Racovitza, 1910; Stammer, 1935a, b; Remy, 1948).

Folliculina sp. which lives in the tubes of the polychaete annelid, *Marifugia cavatica*, from Slovenia and Herzegovina (unpublished observation).

Vorticella microstoma, V. campanula and *V. monilata* fixed on copepods of the genus *Canthocamptus* (Wetzel, 1926).

Vorticella sp. attached to *Niphargus* (Moniez, 1889).

Epistylis digitalis, attached to copepods of the genus *Canthocamptus* (Wetzel, 1929).

Ballodora marceli, on the pleopods of the terrestrial isopod, *Titanethes albus* (Matthes, 1955).

(c) Suctoria

Spelaeophrya troglocaridis, attached to the antennae of the shrimp, *Troglocaris schmidti* (Stammer, 1935a, b; Matjašič, 1956).

Podophrya cyclopum, on the copepods of the genus *Cyclops* (Moniez, 1889).

Podophrya niphargi, attached to *Niphargus* (Lachmann, 1859; Pratz, 1866; Moniez, 1889; Strouhal, 1939).

Tokophrya cyclopum on the copepod *Moraria varica* Chappuis, 1920.)

Tokophrya bathynellae, on syncarids of the genus *Bathynella* (Chappuis, 1947).

Tokophrya sp. on *Parabathynella caparti* (Fryer, 1957).

Tokophrya phreaticum on *Bathynella inlandica* (Uéno, 1962).

Tokophrya microcerberi on the anthurid, *Microcerberus remyi* (Delamare-Deboutteville and Chappuis, 1956).

Tokophrya stammeri on amphipods of the genus *Niphargus* (Strouhal, 1939).

Acineta puteana on *Niphargus* Moniez, 1889).

Dendrocometes paradoxus on *Niphargus* (Lachmann, 1859).

An unidentified suctorian on the parasellid isopod *Microcharon profundalis* (Karaman, 1940).

An unidentified suctorian on the isopod *Stenasellus hungaricus* (Remy, 1948).

An unidentified suctorian on the amphipod *Bogidiella albertimagni* (Hertzog, 1936).

Unidentified suctorians on amphipods of the genera *Niphargus* and *Crangonyx* (Husmann, 1956).

F. TEMNOCEPHALA

In 1906, Al Mrazek discovered the first European species of the Temnocephala in the region near Lake Scutari. It was attached to a freshwater shrimp, *Atyaephyra desmaresti*. He named it *Scutariella didactyla*. In 1912, Annandale established the family Scutariellidae in which he included in addition to the genus *Scutariella*, an apparently closely related one, *Cari-*

dinicola. The latter type lives on various caridines of the Indian region. Representatives of this family are characterised by the presence of only two tentacles. They are closely related to the Rhabdocoela, and Monticelli has even placed them, but wrongly, in this order.

H.J.Stammer (1933, 1935) found on the troglobious shrimp *Troglocaris schmidti,* a temnocephalan which he believed identical to the species which lives on *Atyaephyra desmaresti,* that is *Scutariella didactyla.*

Matjašič (1957, 1958, 1959) showed that this conclusion was incorrect. The slovenian zoologist established the presence of an, until then, unsuspected polymorphism in temnocephalan parasites of subterranean shrimps. This polymorphism prompted him to establish a number of new genera and species.

Most Temnocephala living in underground waters are found attached to *Troglocaris schmidti (Troglocaridicola cestodaria, T. istriana, T. maxima, T. cervaria* and *T. capreolaria; Bubalocerus pretneri; Subtelsonia perianalis).* On the other hand, *Scutariella stammeri* is confined to *Troglocaris herzegovinensis* and *T. anophthalmus;* and *Troglocaridicola speleocaridis* to *Speleocaris pretneri* and *Troglocaris herzegovinensis.* Finally, Matjašič discovered a temnocephalan which lives on one of the largest subterranean amphipods, *Niphargus (Stygodytes)balcanicus;* this species is *Stygodyticola hadzii.*

While the majority of Temnocephala are epizooic forms which feed on small organisms suspended in the water, Mrazek has stated that *Scutariella didactyla* feeds at the expense of its host. Matjašič confirmed that these Temnocephala are true parasites which suck the body fluids of the shrimps to which they are attached.

Most representatives of the family Scutariellidae live in the branchial cavity of Atyidae. However, *Bubalocerus pretneri* is found fixed at the base of the antennae and the antennules as well as on the first pereiopods of the host. *Subtelsonia perianalis* attaches itself to the anal regions of the shrimp. Finally, *Stygodyticola hadzii* is fixed on the pleopods of *Niphargus.*

G. TREMATODES

Parasites of Molluscs

Cercaria and redia of an unidentified trematode have been observed in the cavernicolous gasteropod *Vitrella quenstedti* (Seibold, 1904).

Parasites of Crustacea

The metacercarial stage of a trematode has been found in *Troglocaris* from the Vjetrenica, in Herzegovina (Matjašič, 1956). It is probable that the adult trematode is parasitic on fish of the genus *Paraphoxinus.*

Parasites of Fish

A specimen of *Derogenes tropicus* has been found on the gills of *Rhamdia guatemalensis* from the cenotes of Yucatan (Manter, 1936). *Acanthostomum minimum* and *Clinostomum intermedialis* has been collected on fish of the same species from a cave in the Yucatan (Stunkard, 1938).

Parasites of Proteus

An adult trematode has been found in the intestine of *Proteus* (Mat-jašič, 1956).

H. CESTODA

Adult Forms

A specimen of *Bothriocephalus sp.* (related to *B. scorpii*) has been found in the intestine of the fish *Rhamdia guatemalensis* from the cenotes of Yucatan (Pearse, 1936).

Larval Forms

Cysticercoids of *Hymenolepsis microstoma* have been found in the tro-glophilic orthopteran *Dolichopoda linderi*, which lives in caves of the Pyré-nées-Orientales (Dollfus, 1950; Delamare-Deboutteville, 1952, 1960). The adults are found in various species of rodents.

Vejdovsky (1882) reported, in his well-known book devoted to the fauna of wells of the town of Prague, a larva of *Distome* but the figure he gives would rather appear to resemble a coracid of *Bothriocephalus.*

I. ROTIFERA

Juget (1959) observed a rotifer, *Drilophaga* sp. attached to the integument of the oligochaete *Bythonomus lemani.*

J. NEMATOMORPHA

(a) Phoretic Nematoda

The nematode *Cheilobius quadrilabiatus* is an anguillulid related to *Rhabditis*. It is found in decomposing organic material, in excrement, or in guano. The larvae which insure the dispersal of the species (Dauerlarven of German authors) are phoretic and attach themselves to a variety of arthro-pods (terrestrial isopods, Acarina, Collembola, campodeids, Diptera,

Coleoptera, and so on). They are found particularly frequently on coprophagous or guanobious arthropods. The larvae arrange themselves in a
spiral and are fixed to the host with the aid of a short anterior stalk. These
larvae are not cavernicoles but their ways are such that they are most often
found on guanobious arthropods which live in caves, and in particular on
Collembola (Delamare-Deboutteville and Theodoridès, 1951; Delamare-
Deboutteville, 1952).

(b) Parasitic Nematoda

Nematodes which are Parasitic on Invertebrates. It is probable that many
cavernicolous invertebrates act as hosts for parasitic nematodes. They have
not, however, held the attention of zoologists. Matjašič (1956) reported
the presence of nematodes in the body cavity of subterranean shrimps
(Troglocaris), and Remy (1946) found encysted nematodes in the commensal ostracod, *Sphaeromicola hamigera*.

Nematodes which are Parasitic on Fish. Two ascarids, *Contracaeum sp.* and
Dujardinia cenotae have been found in the troglophilic fish *Rhamdia
guatemalensis* (Pearse, 1936). A spiruroid, *Rhabdochona kidderi* was found
in the intestine of two fish *(Rhamdia guatemalensis* and *Typhlias pearsi)*
which live in the cenotes and caves of the Yucatan (Pearse, 1936; Chitwood, 1938).
Nematodes which are Parasitic on Proteus. Schreiber (1933) reported the
presence of encysted nematodes in the pancreas of *Proteus*.

(c) Mermithidae and Gordiidae

Mermithidae. Mermithidae are parasitic during their larval stage, but
they lead a free life when they become adult. Some are terricoles, others
are aquatic. The majority of aquatic forms inhabit ground waters (Husmann, 1956, Coman, 1961). The hosts on which the larval forms feed in
the subterranean world, are poorly known. *Mermis* has been obtained
from *Niphargus* (Chappuis, 1920) and from the dipteran *Blepharoptera
serrata* (Lengensdorf 1929).
Gordiidae. The Gordiidae possess a parasitic larval phase and a free-living
adult phase in the same way as the Mermithidae. Gordiidae belonging to
the genera *Gordius, Gordionus, Chordodes, Parachordodes* have been reported in subterranean waters of Europe and Africa (Sanfilippo, 1950;
Pavan, 1951; Sciacchitano, 1952, 1955).
The larvae of *Parachordodes alpestris* have been found in the troglophilic orthopteran *Dolichopoda linderi* (Delamare-Deboutteville, 1952,
1960). The larvae of *Chordeudes capensis* from Africa are parasites of
cockroaches (Sciacchitano, 1955).

K. OLIGOCHAETA AND HIRUDINEA

A branchiobdellid, *Cambarincola macrodonta*, lives on a cavernicolous crayfish, *Cambarus blandingi cuevachicae* from the Cueva Chica in Mexico (Hobbs, 1941). It is probable that branchiobdellids are to be found on cavernicolous crayfish in the United States. Representatives of this group *(Bdellodrilus illuminatus, Cambarincola meyeri)* have been reported on *Cambarus bartoni*, a species which possesses many troglophilic races (Goodnight, 1940, 1941, 1942).

A leech, *Cystobranchus* sp. has been found on the gills of fish *Rhamdia guatemalensis* captured in the cenotes and caves of Yucatan (Moore, 1936; Pearse, 1936).

L. COPEPODA

A lernaeopodid, *Cauloxenus stygius*, has been found attached to the lips of specimens of the blind fish *Amblyopsis spelaeus*, from the Wyandotte Cave (Indiana), or in the immediate neighbourhood (Cope, 1871, 1872; Packard, 1886). The genus *Cauloxenus* is closely related to, or even identical with *Achtheres*.

Branchiura have been discovered attached to specimens of the troglophilic fish *Rhamdia guatemalensis* from the cenotes of Yucatan; *Argulus chromidis* was found in the intestine, and *A. rhamdiae* in the branchial cavity (Pearse, 1936; Wilson, 1936).

M. OSTRACODA

Two genera of Ostracoda, both belonging to the Cytheridae, are commensals on other Crustacea. They are the genera *Sphaeromicola* and *Entocythere* which form the sub-family Entocytherinae.

(a) Sphaeromicola

Commensal ostracods of cavernicolous isopods were first observed and reported by Racovitza in 1910 but they were not described until six years later by Paris (1916) who established the genus *Sphaeromicola*.

Seven species of *Sphaeromicola* are known today. All are commensal on Crustacea but their hosts are very varied. The most primitive species of the genus is *S. dudichi*, a marine form from the Mediterranean which is commensal on the amphipod *Chelura terebrans*.

All the other species of *Sphaeromicola* are cavernicolous freshwater forms, commensal on isopods. The hosts of three species are cirolanids.

S. cirolanae has been found attached to *Cirolana pelaezi* from the Cueva de los Sabinos in Mexico. *S. cebennica* is commensal on *Sphaeromides raymondi* and *S. sphaeromidicola* attaches itself to *Sphaeromides virei*.

The last three species are found on spheromids. *S. topsenti* is a parasite of *Caecosphaeroma burgundum*, *S. hamigera* belongs to *Caecosphaeroma virei* and *S. stammeri* has been collected on several species of *Monolistra*.

The marine species has pigment and eyes, but the cavernicolous species are depigmented and transparent. Remnants of the eyes can be found in *S. sphaeromidicola* and *S. cirolanae*. The other species are anophthalmic.

(b) Entocythere

The genus *Entocythere*, established by Marshall in 1903, comprises species which live in the branchial cavity of crayfish belonging to the genera *Cambarus* and *Pacifastacus*. Many species of this genus are known and are widely dispersed throughout the United States, Mexico and Cuba.

The majority of species are epigeous. However, in the last years, some species of Entocythere have been described, which live upon cavernicolous crayfish (*Cambarus, Orconectes*) from the United States and Mexico (Klie, 1931; Rioja, 1942; Walton and Hobbs, 1959; Hart and Hobbs,1961).

N. DIPTERA

A tachinid *Actia (Thyrptocera)* sp. is possibly a parasite of cave crickets *(Arachnomimus microphthalmus)* from the Batu caves in Malaya (Edwards, 1929).

A phorid, *Megaselia melanocephala*, is a parasite of the cocoons of the troglophilic araneid, *Meta menardi* (Decou-Burghele, 1961).

O. ACARINA

The deutonymphs of gamasid and sarcoptiform mites are carried by various arthropods. They are particularly frequent on coprophages, and underground, on guanobia. Troglophilic sphodrids *(Ceutosphodrus, Laemostenus)* are often covered with these nymphs. They are more rarely present on Trechinae. It is probable that most of the adults of these acarina live on guano (Jeannel, 1926; Theodoridès, 1954), and the nymph of *Anoetus digitiferus* to the terrestrial isopod, *Mesoniscus alpicola* (Vitzthum, 1925).

Sarcoptiform mites of the genus *Anoetus* have been found attached to Diptera *(Borboridae, Helomyzidae)* (Leruth, 1939).

Some Pterygosomidae have been captured around the eyes of the troglophilic gecko *Eublepharis splendens* (Nakamura and Uéno, 1959).

P. PARASITES OF BATS

It has already been stated that bats cannot be considered to be true caverni-coles. Also, the study of their parasites which are very numerous and has been the subject of much work, resulting in an abundant bibliography. A detailed consideration of this topic cannot be given here. However, the different types of parasites found on bats will be rapidly enumerated.

(a) Laboulbeniaceae

Anthrorhynchus nycteribiae has been found on the nycteribid parasites of bats.

(b) Trypanosomes

The presence of *Trypanosoma* has frequently been reported from bats. An older bibliography has been given by Laveran and Mesnil (1912). The trypanosomes found in Europe and North Africa most often belong to *Schizotrypanum vespertilionis* (=*pipistrelli*). These parasites multiply in the organs rather than in the blood. They are transmitted either by the bug *Cimex pipistrelli* or by Acarina. *Trypanosoma leleupi* is a parasite of the bat *Hipposideros caffer-centralis* from Katanga. The vector of this haemo-flagellate is, according to Leleup, the bug, *Afrocimex leleupi*.

(c) Haemosporidia

Bats from Africa contain haemosporidia which are divided between the four genera *Plasmodium, Polychromophius, Hepatocystis* and *Nycteria* (Garnham, 1953; Lips, 1960). The genus *Plasmodium* contains a single parasite of the Chiroptera. This is *P. rousetti* which is found in *Rousettus leachi* (Van Riel and Hiernaux, 1951). The vector is probably *Anopheles faini*. The genus *Hepatocystis* contains several species parasitic on Chiro-ptera. It seems probable that they are all transmitted by *Anopheles*, but this has not been shown experimentally.

(d) Platyhelminthes

Trematodes are frequently found in the intestines of bats. For the most part they belong to the Plagiorchiidae (and to the genus *Plagiorchis*), and to the Lecithodendriidae (and to the genera *Lecithodendrium* and *Prostho-dendrium*). Many publications have been devoted to the study of these endoparasites. A very complete bibliography has been given by G. Dubois (1955, 1956, 1960).

The cestodes are much less frequent in the Chiroptera. There are also less varieties. However, a number of species of the genus *Hymenolepis* have been reported in the intestine of Microchiroptera. A cysticercoid described by Van Beneden, coming from the intestine of varied Microchiroptera is probably the larval form of *Hymenolepis grisea*. If this is correct, the development of this tapeworm would be direct, that is, not requiring an intermediate host.

(e) Nematodes

Zoologists have frequently established the presence of nematodes in the Chiroptera (B.G.Chitwood, J. Lins de Almeida, M.A.Ortlepp, J.H.Sandground, B.Schwartz, T.S.Skarbilovitch).

(f) Acarina

There are many Acarina which parasitise bats. Some of them are hyperparasites such as *Monunguis streblida* which is found attached to Streblidae (Diptera) (Wharton, 1938). The parasitic Acarina of bats belong to very diverse sub-orders and families. Reference to the works of Vitzthum (1940 to 1942), Baker and Wharton (1952) and Hughes (1959) help one to clarify the complex classification of the Acarina.

Sub-order Parasitiformes: Mesostigmata, Gamasides. The family Spinturnicidae contains numerous species which live on bats. Many of them have been known for a long time: *Spinturnix vespertinilionis* (Linné), *S. plecotinus* (C.L.Koch), *S. murinus* Walkenaer, *S. psi* Kolenati. A revision of this family, with a complete bibliography, has been given by A.Rudnick (1960). The Acarina are found exclusively on the alary and caudal membranes and not on the body itself. Representatives of the genera *Ancystropus* and *Meristaspis* are fixed to the eyelids of Chiroptera. These parasites are found both on bats which enter caves, and on those which do not. They are thus not cavernicoles. For this reason, Cooreman (1954) gave them the name trogloxenous parasites.

The family Spelaeorhynchidae consists entirely of parasites of bats. Its representatives are confined to Central and South America.

The family Laelaptidae contains several parasitic species on bats including *Neolaelaps magnistigmatus*, a parasite of *Pteropus*, and several species of the genus *Ornithonyssus* (most representatives of this genus are parasites of birds).

Metastigmata, Family Ixodidae. An ixodid parasitic on bats has been found frequently in Europe, Asia and Africa; it has even been found in Australia. This species is *Ixodes vespertilionis* and attaches itself to numerous species of Microchiroptera. It is found particularly frequently in Europe on rhinolophids, bats which are found mostly underground. This ixodid could be considered to be a true troglobian. The larvae and nymphs live as parasites of bats. The female fixes itself to bats to obtain a blood meal. The adult males, however, seem to have no need of food. They are found on the walls of the caves where they are sometimes very common (Neumann, 1916).

As all other species of *Ixodes*, *I. vespertilionis* is without eyes. It is characterised by the extreme length and slenderness of its legs. Another ixodid, *Ixodes simplex*, is also parasitic on bats. It is also widely distributed geographically.

Family Argasidae. Some species belonging to the genera *Ornithodorus*, *Amblyomma* and *Argas* are parasites of bats. The life cycle of *Ornithodorus kelleyi* is known (Sonenshine and Anastos, 1960).

Sub-order Trombidiformes: Family Spelaeognathidae. Some representatives of this family live in the nostrils of Chiroptera. *Spelaeognathopsis bastini* is a parasite of Microchiroptera in Europe and the United States and *S. chiropteri* lives on African Megachiroptera. These two are the best-known species (Fain, 1955, 1958; Hyland and Ford, 1961).

Family Myobiidae. Myobia pipistrelli is parasitic on *Pipistrellus pipistrellus* (Radford, 1939), a bat which is found only exceptionally in caves.

Family Trombiidae. The larvae of certain species of *Trombicula* attack bats. Such examples are *Trombicula penetrans* and *T. russica*. The adults live in guano (Vitzthum, 1932).

Sub-order Sarcoptiformes: Family Sarcoptidae. The Belgium parasitologists Rodhain and Gedoelst, and more recently Fain (1959–1961) have described many species of the Sarcoptidae which are parasitic on bats. For example, members of the genus *Nycteridocoptes* are widespread on bats of the Old World. *N. poppei* causes mange on *Vespertilio murinus* (Fain, 1960) and *N. pteropodi* has a similar effect on the African species, *Eidelon helvum* (Rodhain, 1921, 1923). *Teinocoptes epomophori* is a parasite of *Micropteropus pusillus* and *Epomorphus walhbergi* (Rodhain, 1923). *Chirnyssoides* is found on bats of neo-tropical regions.

(g) Dermaptera

Arixenia esau and *A. jacobsoni* are Dermaptera which lack wings and have reduced eyes. They are parasitic on bats of the Indo-Malayan region. The early studies of Jordan (1909) and Burr (1912) have been extended by the more recent ones of Cloudsley-Thompson (1957, 1959), Lord Medway (1958) and Giles (1961).

(h) Hemiptera

The Cimicidae contain a number of species which are parasitic on bats: *Cimex pipistrelli* and *C. stadleri* in Europe, and *Afrocimex leleupi* in Africa.

The Polyctenidae *(Polyctenes, Eoctenes, Adroctenes, Hesperoctenes)* are very small bugs, parasitic on bats, where they live buried in the fur of their hosts. The hemelytra are reduced and the eyes absent. These parasites are superficially similar to Mallophaga.

(i) Coleoptera

Platypus rugulosus (Platypodidae) has been collected from the bat *Natalus mexicanus* from a cave in Yucatan (Mexico).

(j) Siphonaptera

Numerous fleas are to be found on bats and these have been revised by Hopkins and Rothschild (1956). For the most part they belong to the Ischnopsyllidae. Host specificity appears to be weak. However, the species of the genus *Rhinolophopsylla* are exclusively parasites of the Rhinolophidae. The larvae of *Ischnopsyllus intermedius* lives in guano (Hurka, 1956). This is probably the case with all larvae of other species found on bats.

(k) Pupipara

The Pupipara consist of a heterogenous group of parasitic, hematophagous forms all of which have the same method of reproduction, that is the larvae develop completely inside the maternal organism. This group consists of three families, the Hippoboscidae, Streblidae and Nycteribiidae. The species of the last two families are all parasitic on bats. They have probably been derived from guanobious Diptera. The Streblidae would appear to be related to the Borboridae and the Nycteribiidae to the Helomyzidae.

The Nycteribiidae seem to have arisen in the Indo-Malayan area. From this place of origin they have spread throughout a large part of the world. The Streblidae are found in Africa, in the Indo-Malayan region and the New World. One species *Nycteribosca kollari (= africana)*, has penetrated into the Mediterranean region.

Host specificity seems to be weak among the Pupipara parasitic on Chiroptera. According to Jeannel this specificity is somewhat strengthened in caves.

Streblidae usually retain their wings, although often in a more or less advanced state of reduction. They are, however, rarely completely absent *(Paradyschiria)*. After copulation, the females of *Ascodipteron* lose their legs and wings and transform themselves into a sac of eggs.

All Nycteribiidae are apterous. Their legs are very long and inserted almost dorsally which gives these insects a strange appearance.

In both families the eyes are often reduced or absent. The successive stages of regression of the eyes cannot be followed in the Streblidae. The primitive types *(Trichobius, Strebla)* still possess eyes composed of several ommatidia. In *Nycteribosca* the eyes are reduced to single ocelli. These are still large in *N. alluaudi* from East Africa, but much smaller in *N. kollari* (Sadoglu, 1956). Finally, species of the genus *Raymondia* are totally anophthalmic. This regression of the eye is related to parasitism and not to life in darkness.

It is remarkable that in spite of regression of the eye *(Nycteribosca)* or its complete absence *(Raymondia)*, the wings of these two genera are of normal size and functional. The usual correlation of ocular regression with alary regression does not, therefore, apply in this case.

BIBLIOGRAPHY

Introduction

PEARSE, A. S. (1936) Parasites from Yucatan. *Carnegie Inst. Washington Public.* No .457.
SPREHN, C. (1935) Parasiten der Wirbeltiere des Glatzer Schneeberges. 1. Helminthen aus Kleinsäugern und Amphibien vom Glatzer Schneeberg. *Beiträge zur Biologie des Glatzer Schneeberges.* **I.**
THEODORIDÈS, J. (1954) Parasites et phorétiques des Coléoptères des grottes de l'Ariège et autres cas connus de parasitisme chez des Coléoptères cavernicoles. *Notes biospéol.* **IX.**

Fungi (general)

CALL, E. (1897a) Note on the flora of Mammoth Cave. *J. Cincinnati Soc. Nat. Hist.* **XIX.**
CALL, E. (1897b) Some notes on the flora and fauna of Mammoth Cave, Ky. *Amer. Nat.* **XXXI.**
LAGARDE, J. (1913) Champignons (1ère Série). *Biospeologica,* **XXXII,** *Archiv. Zool. expér. géner.* **LIII.**
LAGARDE, J. (1917) Champignons (2ème Série). *Biospeologica,* **XXXVIII,** *Archiv. Zool. expér. géner.* **LVI.**
LAGARDE, J. (1922) Champignons (3ème Série). *Biospeologica,* **XLVI,** *Archiv. Zool. expér. géner.* **LX.**
THEODORIDÈS, J. (1954) Parasites et Phorétiques des Coléoptères des grottes de l'Ariège et autres cas connus de parasitisme chez des Coléoptères cavernicoles. *Notes biospéol.* **IX.**

Histoplasmosis

ASPIN, J. and BELLARD PIETRI, E. DE (1959) Cave sickness. Benign pulmonary histoplasmosis. *Trans. Cave Research Group, Great Britain.* **V,** No. 2.
BELLARD PIETRI, E. DE, BRICENO IRAGORRY, L., PARDO, I., REQUENA, A. and POLLAK, L. (1956) La Histoplasmosis en Venezuela. Observaciones. *Gaceta Medica.* No. 3, 4, 5.
BELLARD PIETRI, E. DE (1957) La Histoplasmosis y las Cuevas de Venezuela. *Bol. Soc. Venezol. Cienc. Natur.* **XVIII.**
SHACKLETTE, M. H., DIERKS, F. H. and GALE, N. B. (1962) *Histoplasma capsulatum* from Bat tissues. *Science.*
STENN, FR. (1960) Cave Disease or Speleonosis. *A.M.A. Arch. Intern. Medicine* **CV.**

Laboulbeniacea

CALL, R. E. (1897) Note on the flora of Mammoth Cave, Kentucky. *J. Cincinnati Soc. Natur. Hist.* **XIX.**
CÉPÈDE, C. and PICARD, FR. (1908) Contribution à la biologie et à la systématique des Laboulbéniacées de la flore française. *Bull. Sc. France Belgique* **XLII.**

HOVEY, H. C. and CALL, R. E. (1897) *Mammoth Cave of Kentucky*. An illustrated manual. Louisville.

ISTVANFFI, G. DE (1896) Eine auf höhlenbewohnenden Käfern vorkommende neue Laboulbeniaceae. *Ann. Naturhist. Mus. Budapest.*

LEPESME, P. (1941) Catalogue des Laboulbeniales de la collection François Picard. *Bull. Mus. Hist. Nat. Paris*, **XIII.**

LEPESME, P. (1942) Révision des *Rhachomyces* paléarctiques. *Bull. Soc. Mycol. France* **LVIII.**

LEPESME, P. (1944) Note sur quelques Laboulbeniacées de France. *Bull. Soc. Entomol. France.*

LEPESME, P. and TEMPÈRE, G. (1947) Description d'un *Rhachomyces* nouveau des Basses-Pyrénées (Laboulbeniacées). *Bull. Soc. Mycol. France* **LXIII.**

MAIRE, R. (1912) Contribution à l'étude des Laboulbeniacées de l'Afrique du Nord. *Bull. Soc. Hist. Nat. Afrique du Nord* **IV.**

PICARD, FR. (1913) Contribution à l'étude des Laboulbeniacées d'Europe et du Nord de l'Afrique. *Bull. Soc. Mycol. France* **XXIX.**

THAXTER, R. (1895) Contribution towards a monograph of the Laboulbeniaceae. Part I. *Mem. Amer. Acad. Arts. Sc.* **XII,** No. 3.

THAXTER, R. (1908) Contribution towards a monograph of the Laboulbeniaceae. Part II. *Mem. Amer. Acad. Arts. Sc.* **XIII,** No. 6.

THAXTER, R. (1931) Contribution towards a monograph of the Laboulbeniaceae. Part V. *Mem. Amer. Acad. Arts. Sc.* **XVI.**

Gregarina

CHAPPUIS, P. A. (1920) Die Fauna der unterirdischen Gewässer der Umgebung von Basel. *Archiv. f. Hydrobiol.* **XIV.**

MONIEZ, R. (1889) Faune des eaux souterraines du département du Nord, et en particulier de la ville de Lille. *Rev. biol. Nord. France* **I.**

THEODORIDÈS, J. (1954) Parasites et phorétiques des Coléoptères des grottes de l'Ariège et autres cas connus de parasitisme chez des Coléoptères cavernicoles. *Notes biospéol.* **IX.**

VEJDOVSKY, FR. (1882) *Thierische Organismen der Brunnenwässer von Prag.* Prag.

WETZEL, A. (1929) Die Protozoen der Schneeberger Erzbergwerke. *Zool. Jahrb. Abt. System.* **LVI.**

Cnidosporidia

JOSEPH, H. (1905) *Chloromyxum protei* n. sp. *Zool. Anz.* **XXIX.**

JOSEPH, H. (1906) *Chloromyxum protei* n.sp., ein in der Niere des Grottenolmes parasitierendes *Myxosporidium. Archiv. f. Protistenkd.* **VIII.**

MENOZZI, C. (1939) La fauna della grotte Suja sul Monte Fascie (Genova) ed osservazioni biologiche sulla *Parabathyscia doderoi* Fairm. (Col. Catopidae) con descrizione della larva e delle caracterische morfologiche del suo intestino e di quello dell'adulto. *Mem. Soc. Entomol. Ital.* **XVIII.**

POISSON, R. (1924) Sur quelques Microsporidies parasites d'Arthropodes. *Compt. rend. Acad. Sci. Paris* **CLXXVIII.**

Ciliates

CHAPPUIS, P. A. (1920) Die Fauna der unterirdischen Gewässer der Umgebung von Basel. *Archiv. f. Hydrobiol.* **XIV.**

CHAPPUIS, P. A. (1947) *Tokophrya bathynellae*, une nouvelle Acinète troglobie. *Bull. Acad. Roumaine Sect. Sc.* **XXIX.**

DELAMARE-DEBOUTTEVILLE, CL. and CHAPPUIS, P. A. (1956) Présence d'un Acinétien sur un Microcerberinae des eaux souterraines du Saraha occidental. *Notes biospéol.* **XI.**

FRYER, G. (1957) A new species of *Parabathynella* (Crust. Sync.) from the psammon of Lake Bangweulu, Central Africa. *Ann. Mag. Nat. Hist.* (10) **X.**

HERTZOG, G. (1936) Crustacés de biotopes hypogés de la vallée du Rhin d'Alsace. *Bull. Soc. Zool. France.* **LXI.**

HUSMANN, S. (1956) Untersuchungen über die Grundwasserfauna zwischen Harz und Weser. *Archiv. f. Hydrobiol.* **LII.**

JUGET, J. (1959) Recherches sur la faune aquatique de deux grottes du Jura méridional français: la grotte de la Balme (Isère) et la grotte de Corveissiat (Ain). *Ann. Spéléol.* **XIV.**

KARAMAN, ST. (1940) Die unterirdischen Isopoden Südserbiens. *Glasnik. Bull. Soc. Sc. Skoplje* **XXII.**

LACHMANN, J.C. (1859) Über einige Parasiten des Brunnenflohkrebses. *Sitzber. Niederrhein. Gesell.* Bonn.

MATJAŠIČ, J. (1956) Observations on the cave suctorian, *Spelaeophrya troglocaridis* Stammer. *Biol. Vestnik.* Ljubljana. **V.**

MATTHES, D. (1955) Neues zur Kiemenfauna der Landisopoden. *Verhandl. Deutsch. Zool. Gesell.* Erlangen.

MONIEZ, R. (1889) Faune des eaux souterraines du département du Nord et, en particulier, de la ville de Lille. *Rev. biol. Nord. France.* **I.**

PRATZ, ED. (1866) Über einige im Grundwasser lebende Thiere. Beiträge zur Kenntnis der unterirdischen Crustaceen. *Dissert. St. Petersbourg.*

RACOVITZA, E.G. (1910) Sphéromiens (Première Série) et Révision des *Monolistrini* (Isopodes Sphéromiens). *Biospeologica,* **XIII.** *Archiv. Zool. expér. géner.* (5) **IV.**

REMY, P. (1948) Sur quelques Crustacés cavernicoles d'Europe. *Notes biospéol.* **III.**

STAMMER, H.J. (1935a) Zwei neue troglobionte Protozoen: *Spelaeophrya troglocaridis* n. gen. von den Antennen der Höhlengarnele *Troglocaris schmidti* Dorm. und *Lagenophrys monolistrae* n. sp. von den Kiemen der Höhlenassel *Monolistra. Archiv. Protistenk.* **LXXXIV.**

STAMMER, H.J. (1935b) Untersuchungen über die Tierwelt der Karsthöhlengewässer. *Verhandl. Intern. Ver. Theor. Angew. Limnologie.* **VII.**

STROUHAL, H. (1939) Die in den Höhlen vom Warmbad Villach, Kärnten, festgestellten Tiere. *Folia Zool. Hydrobiol.* **IX.**

UÉNO, M. (1962) A new suctorian parasite to a Japanese Bathynellid. *Hydrobiologia* **XX.**

UÉNO, S.I. (1957) Blind aquatic beetles of Japan, with some accounts of the fauna of Japanese subterranean waters. *Archiv. f. Hydrobiol.* **LIII.**

WETZEL, A. (1929) Die Protozoen der Schneeberger Erzbergwerke. *Zool. Jahrb. Abt. System.* **LVI.**

Temnocephala

MATJAŠIČ, J. (1957) Biologie und Zoogeographie der europäischen Temnocephaliden. *Verh. detsch. Zool. Gesell.* Graz.

MATJAŠIČ, J. (1958) Vorläufige Mitteilungen über europäische Temnocephalen. *Biol. Vestnik, Ljubljana* **VI.**

MATJAŠIČ, J. (1959) Morphologie, Biologie et Zoogéographie des Temnocéphales européens, et leur situation systématique. *Razprave. Dissertationes.* Ljubljana. **V.**

MRAZEK, AL. (1906) Ein europäischer Vertreter der Gruppe Temnocephaloidea. *Sitzb. k. böhm. Gesell. Wiss.* **XI.**

STAMMER, H.J. (1933) Einige seltene oder neue Höhlentiere. *Zool. Anz.* **VI.**

STAMMER, H.J. (1935) Untersuchungen über die Tierwelt der Karsthöhlengewässer. *Verhandl. Intern. Ver. Theor. angew. Limnologie.* **VII.**

Trematoda

MANTER, H.W. (1936) Some trematodes of cenotes fishes from Yucatan. *In* PEARSE, A.S., CREASER, E.P. and HALL, F.G. The cenotes of Yucatan, a zoological and hydrographic Survey. *Carnegie Inst. Washington Public.* No. 457.

MATJAŠIČ, J. (1956) A trematode from *Troglocaris*. *Biol. Vestnik. Ljubljana* V.
SEIBOLD, W. (1904) Anatomie von *Vitrella Quenstedtii* (Wiedersheim) Clessin. *Jahreshefte Ver. f. Vaterl. Naturk. Württemberg* LX.
STUNKARD, H.W. (1938) Parasitic flatworms from Yucatan. *In* PEARSE, A.S. Fauna of the caves of Yucatan. *Carnegie Inst. Washington Public.* No. 491.

Cestoda

DELAMARE-DEBOUTTEVILLE, CL. (1952) Données nouvelles sur la biologie des animaux cavernicoles. *Notes biospéol.* VII.
DELAMARE-DEBOUTTEVILLE, CL. (1960) Remarques sur la faune de la Grotte de Can Brixot, commune de Prats de Mollo (Pyrénées-Orientales). *Ann. Spéléol.* XV.
DOLLFUS, R.PH. (1950) Cysticercoides d'*Hymenolepis* chez un Orthoptère cavernicole. *Vie et Milieu* I.
PEARSE, A.S. (1936) Parasites from Yucatan. *In* PEARSE, A.S., CREASER, E.P., and HALL, F.G. The cenotes of Yucatan, a zoological and hydrographic survey. *Carnegie Inst. Washington Public.* No. 457.
VEJDOVSKY, FR. (1882) *Thierische Organismen der Brunnenwässer von Prag.* Prag.

Rotifera

JUGET, J. (1959) Recherches sur la faune aquatique de deux grottes du Jura méridional français: la grotte de la Balme (Isère) et la grotte de Corveissiat (Ain). *Ann. Spéléol.* XIV.

Phoretic Nematoda

DELAMARE-DEBOUTTEVILLE, CL. and THEODORIDÈS, J., (1951) Sur la constance de l'association entre Nématodes phorétiques et Collemboles. *Vie et Milieu* II.
DELAMARE-DEBOUTTEVILLE, CL. (1952) Données nouvelles sur la biologie des animaux cavernicoles. *Notes biospéol.* VII.

Parasitic Nematoda

CHITWOOD, B.G. (1938) Some nematodes from the caves of Yucatan. *In* PEARSE, A.S. Fauna of the caves of Yucatan. *Carnegie Inst. Washington Public.* No. 491.
MATJAŠIČ, J. (1956) A trematode from *Troglocaris. Bioloski Vestnik. Ljubljana* V.
PEARSE, A.S. (1936) Parasites of Yucatan. *In* PEARSE, A.S., CREASER, E.P., and HALL, F.G. The cenotes of Yucatan, a zoological and hydrographic survey. *Carnegie Inst. Washington Public.* No. 457.
REMY, P. (1946) *Sphaeromicola stammeri* Klie var. *hamigera* n. var., Ostracode commensal du Sphéromide obscuricole *Caecosphaeroma* (C.) *virei* Dollfus. *Coll. Mus. Zool. Nancy* I.
SCHREIBER, G. (1933) Sui Nematodi parassiti nel pancreas del Proteo. *Atti I Congresso Speleol. Naz. Trieste.*

Mermithidae

CHAPPUIS, P.A. (1920) Die Fauna der unterirdischen Gewässer der Umgebung von Basel. *Archiv. f. Hydrobiol.* IV.
COMAN, D. (1961) *Mermithidae.* In *Fauna Republicii Populare Romîne.* Nematoda. Vol. II; Fas. 3. Bucuresti.
HUSMANN, S. (1956) Untersuchungen über die Grundwasserfauna zwischen Harz und Weser. *Archiv. f. Hydrobiol.* LII.
LENGENSDORF, FR. (1929) Biologisch interessante Funde aus westfälischen Höhlen. *Mitteil. Höhl. Karstf.*

Gordiidae

DELAMARE-DEBOUTTEVILLE, CL. (1952) Données nouvelles sur la biologie des animaux cavernicoles. *Notes biospéol.* **VII.**

DELAMARE-DEBOUTTEVILLE, CL. (1960) Remarques sur la faune de la grotte de Can Brixot, commune de Prats de Mollo (Pyrénées-Orientales). *Ann. Spéléol.* **XV.**

PAVAN, M. (1951) La fauna della "Arma Pollera" No. 24 Li, presso Finale Ligure. *Rassegna Speleol. Ital.* **III.**

SANFILIPPO, N. (1950) Le grotte della provincia di Genova e la loro fauna. *Club alpino Ital. Mem. Comitato Sc. Centrale* **II.**

SCIACCHITANO, I. (1952) Irudinei e Gordii cavernicoli in Italia. *Archiv. Zool. Ital.* **XXXVII.**

SCIACCHITANO, I. (1955) Su due Gordii cavernicoli del Congo Belga. *Rev. Zool. Bot. Africa* **LI.**

Oligochaeta and Hirudinea

GOODNIGHT, CL. J. (1940) The Branchiobdellidae (Oligochaeta) of North American crayfishes. *Illinois Biol. Monographs* **XVII.**

GOODNIGHT, CL. J. (1941) A new species of branchiobdellid from Kentucky. *J. Parasitology, Urbana;* **XXVII,** Suppl. 6.

GOODNIGHT, CL. J. (1942) A new species of branchiobdellid from Kentucky. *Trans. Amer. microsc. Soc. Menasha* **LXI.**

HOBBS, H. H. (1941) A new crayfish from San Luis Potosi, Mexico (Decapoda, Astacidae). *Zoologica* **XXVI.**

MOORE, J. P. (1936) Hirudinea from Yucatan. *In* PEARSE, A. S., CREASER, A. P. and HALL, F. G. The cenotes of Yucatan, a zoological and hydrographic survey. *Carnegie Inst. Washington Public.* No. 457.

PEARSE, A. S. (1936) Parasites from Yucatan. *In* PEARSE, A. S., CREASER, E. P. and HALL, F. G. The cenotes of Yucatan, a zoological and hydrographic survey. *Carnegie Inst. Washington Public.* No. 457

Copepoda

COPE, E. D. (1871) Life in the Wyandotte Cave. *Ann. Mag. Nat. Hist.* (4) **VII.**

COPE, E. D. (1872) On the Wyandotte Cave and its fauna. *Amer. Nat.* **VI.**

PACKARD, A. S. (1886) The cave fauna of North America, with remarks on the anatomy and origin of the blind species. *Mem. Nat. Acad. Sci. Washington.* **IV.**

PEARSE, A. S. (1936) Parasites from Yucatan. *In* PEARSE, A. S. The Cenotes of Yucatan, a zoological and hydrographic survey. *Carnegie Inst. Washington Public.* No. 457.

WILSON, C. B. (1936) Copepods from the cenotes and caves of the Yucatan Peninsula, with notes on cladocerans. *In* PEARSE, A. S. The cenotes of Yucatan, a zoological and hydrographic survey. *Carnegie Inst. Washington Public.* No. 457.

Sphaeromicola

HUBAULT, E. (1937) *Monolistra hercegovinensis* Absolon, Sphéromien cavernicole d'Herzégovine et *Sphaeromicola stammeri* Klie, son commensal. *Archiv. Zool. expér. géner.* **LXXVIII.**

HUBAULT, E. (1938) *Sphaeromicola sphaeromidicola* nov. sp., commensal de *Sphaeromides virei* Valle, en Istrie et considérations sur l'origine de diverses espèces cavernicoles périméditerranéennes. *Archiv. Zool. expér. géner.* **LXXX.**

KLIE, W. (1930) Über eine neue Art der Ostracodengattung *Sphaeromicola. Zool. Anz.* **LXXXVIII.**

KLIE, W. (1938) *Sphaeromicola dudichi* n. sp. (Ostr.). Ein Kommensale des Bohramphipoden *Chelura terebrans. Zool. Anz.* **CXXI.**

PARIS, P. (1916) *Sphaeromicola topsenti* n.g.n.sp., Ostracode commensal d'Isopodes troglobies du genre *Caecosphaeroma. Compt. rend. Acad. Sci. Paris* **CLXIII.**

PARIS, P. (1920) Ostracodes (Première Série). *Biospeologica*, XLI. *Archiv. Zool. expér. géner.* LVIII.

REMY, P. (1943) Nouvelle station du Sphéromien troglobie, *Caecosphaeroma (Vireia) burgundum* Dollfus, var. *rupis fucaldi* Hubault et de l'Ostracode commensal, *Sphaeromicola topsenti* Paris. *Bull. Soc. Zool. France* LXVIII.

REMY, P. (1946) *Sphaeromicola stammeri* Klie, var. *hamigera* n. var., Ostracode commensal du Sphéromide obscuricole *Caecosphaeroma (Caecosphaeroma) virei* Dollfus. *Coll. Mus. Zool. Nancy* I.

REMY, P. (1948a) Sur quelques Crustacés cavernicoles d'Europe. *Notes biospéol.* III.

REMY, P. (1948b) Sur la distribution géographique des *Sphaeromidae* Monolistrini et des *Sphaeromicola*, leurs Ostracodes commensaux. *Bull. Mus. Hist. Nat. Marseille* VIII.

REMY, P. (1948c) Description de *Sphaeromicola cebennica* n. sp., Ostracode Cythéride commensal de l'Isopode Cirolanide cavernicole, *Sphaeromides raymondi* Dollfus. *Bull. Soc. linn. Lyon.* XVII.

REMY, P. (1951) Stations de Crustacés obscuricoles. *Biospeologica*, LXXII, Appendice. *Archiv. Zool. expér. géner.* LXXXVIII.

RIOJA, E. (1951) Estudios carcinologicos. XXV. El Hallazgo del genere *Sphaeromicola* en America (Ostracodos, Citeridos) y descripcion de una nueva especie. *Anales Inst. Biol. Mexico* XXII.

STAMMER, H.J. (1935) Untersuchungen über die Tierwelt der Karsthöhlengewässer. *Verhandl. Intern. Ver. Theor. angew. Limnologie* VII.

Entocythere

HART, C.W. JR. and HOBBS, H.H. JR. (1961) Eight new troglobitic ostracods of the genus *Entocythere* (Crustacea, Ostracoda) from the eastern United States. *Proc. Acad. Nat. Sc. Philadelphia* CXIII.

KLIE, W. (1931) Crustacés Ostracodes. Campagne spéologique de C.Bolivar et R.Jeannel dans l'Amérique du Nord (1928). *Biospeologica*, LVI, No. 3. *Archiv. Zool. expér. géner.* LXXI.

RIOJA, E. (1942) Descripcion de una especia y una subespecie nuevas del genero *Entocythere* Marshall procedentes de la Cueva Chica (San Luis Potosi, Mexico). *Ciencia* III.

WALTON, M. and HOBBS, H.H. IR. (1959) Two new eyeless ostracods of the genus *Enthocythere* from Florida. *Quart. J. Florida Ac. Sc.* XXII.

Diptera

DECOU-BURGHELE, A. (1961) Sur la biologie de *Megaselia melanocephala* von Roser, Phoride parasite des cocons de *Meta menardi* Latr. *Annal. Lab. souterrain Han-sur Lesse.* II.

EDWARDS, F.W. (1929) Fauna of the Batu Caves, Selangor. XV, Diptera. *J. Fed. Malay States Mus.* XIV.

Acarina

JEANNEL, R. (1926) *Faune cavernicole de la France*. Paris.

LERUTH, R. (1939) La Biologie du Domaine souterrain et la Faune cavernicole de la Belgique. *Mém. Mus. R. Hist. Nat. Belgique* LXXXVII.

NAKAMURA, K. and UÉNO, S.I. (1959) The geckos found in the limestone caves of the Ryu-Kyu Islands. *Mem. Coll. Sc., Univ. Kyoto Ser. B.* XXVI.

THEODORIDÈS, J. (1954) Parasites et phorétiques des Coléoptères des grottes de l'Ariège et autres cas connus de parasitisme chez les Coléoptères cavernicoles. *Vie et Milieu* IX.

VITZTHUM, H. (1925) Die unterirdische Acarofauna. *Jen. Zeit. Naturw.* LXII.

Parasites of Bats (general)

KOLENATI, F. A. (1856–1857) *Die Parasiten der Chiropteren.* Brünn and Dresden.
STILES, C. W. and NOLAN, M. O. (1931) Key catalogue of parasites reported for Chiroptera (bats) with their possible public health importance. *Bull. Nat. Inst. Health, Washington* **CLV**.

Ireland

CARLISLE, R. W. and SKILLEN, S. (1960) Parasites of Irish Bats. *Irish Natural. J.* **XIII**.

Belgium

BENEDEN, P. J. VAN (1873) Les parasites des Chauves Souris de Belgique. *Mem. Acad. R. Sc. Belgique* **XL**.
LERUTH, R. (1939) La Biologie du Domaine souterrain et la Faune cavernicole de la Belgique. *Mém. Mus. R. Hist. Nat. Belgique* **LXXXVII** (pp. 81–84: Parasites de Chauves-Souris).

France

JEANNEL, R. (1926) Faune cavernicole de la France. *Encyclopédie Entomologique*, **VII**. Paris (pp. 57–61: Parasites des Chauves Souris).
NEVEU-LEMAIRE, M., JOYEUX, C., LARROUSSE, I., ISOBE, M. and LAVIER, G. (1924) Parasites de Chauves-Souris de la Côte d'Or. *Compt. rend. Congr. Soc. Sav.*

Central Africa

LELEUP, N. (1956) La Faune Cavernicole du Congo belge, et considérations sur les Coléoptères reliques d'Afrique intertropicale. *Ann. Mus. R. Congo belge. Sc. Zool.* **XLVI**.

Trypanosomes

CHATTON, E. and COURRIER, R. (1921) Sur un Trypanosome de la Chauve-Souris *Vesperugo pipistrellus*, à formes crithidiennes intratissulaires et cystigènes. *Compt. rend. Acad. Sci. Paris* **CLXXII**.
LAVERAN, A. and MESNIL, F. (1912) *Trypanosomes et Trypanosomiases.* Paris.
RODHAIN, J. (1942) Au sujet du développement intracellulaire du *Trypanosoma pipistrelli* (Chatton et Courier) chez *Ornithodorus moubata. Acta biol. belg.* **II**.
RODHAIN, J. (1951) *Trypanosoma releupi* n. sp. parasite de *Hipposideros caffer* au Katanga. *Ann. Parasitologie* **XXVI**.

Haemosporidia

GARNHAM, P. C. C. (1953) Types of Bat Malaria. *Riv. d. Malarologia* **XXXII**.
LIPS, M. (1960) Anophèles du Congo. 3. Faune des Grottes et des Anfractuosités. *Riv. d. Parasitologia* **XXI**.
VAN RIEL, D. and HIERNAUX, C. R. (1951) Description of *Plasmodium* found in a bat, *Rousettus leachi. Parasitology* **XLI**.

Platyhelminthes

DUBOIS, G. (1955) Les Trématodes de Chiroptères de la collection Villy Aellen. Étude suivie d'une révision du sous-genre *Prosthodendrium* Dollfus 1937 (*Lecithodendriinae* Lühe). *Rev. suisse Zool.* **LXII**.
DUBOIS, G. (1956) Contribution a l'étude des Trématodes des Chiroptères. *Rev. suisse Zool.* **LXIII**.
DUBOIS, G. (1960) Contribution a l'étude des Trématodes des Chiroptères. Révision du sous-genre *Prosthodendrium* Dollfus 1931 et des genres *Lecithodendrium* Loos 1896 et *Pycnoporus* Loos 1899. *Rev. suisse Zool.* **LXVII**.

Acarina

ARTHUR, D.R. (1956) The *Ixodes* ticks of Chiroptera (Ixodoidea, Ixodidae). *J. Parasitol.* **XLII.**

BAKER, ED.W. and WHARTON, G.W. (1952) *An tntroduction to Acarology.* New York.

COOREMAN, J. (1954) Acariens recueillis dans les grottes de l'Ariège (Ph.Cauchois et J.J.Theodoridès, avril 1951). *Notes biospéol.* **IX.**

FAIN, A. (1955) Sur le parasitisme des fosses nasales chez les Mammifères et les Oiseaux par les Acariens de la famille des *Spelaeognathidae* (Acarina). Description d'une espèce nouvelle chez la Chauve-Souris. *Ann. Soc. belge Medec. trop.* **XXXV.**

FAIN, A. (1958) Un nouveau Spelaeognathe (Acarina-Ereynetidae) parasitant les fosses nasales du murin (*Myotis myotis* (Borkh) en Belgique: *Spelaeognathopsis bastini* n. sp. *Bull. Ann. Soc. R. Entomol. Belgique* **XCIV.**

FAIN, A. (1960) Les Acariens psoriques parasites des Chauves-souris. XIV. Influence du sommeil hibernal sur l'évolution de la gale sarcoptique. *Bull. Ann. Soc. R. Entomol. Belgique* **XCVI.**

HUGHES, T.S. (1959) *Mites, or the Acari.* London.

HYLAND, K.E. and FORD, H.G. (1961) The occurrence of the nasal mite, *Spelaeognathopsis bastini* Fain *(Spelaeognathidae)* from the big brown bat, *Eptesicus fuscus* (Beauvais). *Entomol. News* **LXXII.**

NEUMANN, G. (1916) Ixodidei (Acariens) (1ère série). *Biospeologica,* **XXXVII.** *Archiv. Zool. expér. géner.* (5) **LV.**

RADFORD, C.D. (1939) Notes on some new species of parasitic mites. *Parasitology* **XXX.**

RODHAIN, J. (1921) Un Sarcoptide nouveau, parasite de la Roussette africaine (*Eidelon helvum* Kerr.). *Compt. rend. Soc. Biol.* **LXXXIV.**

RODHAIN, J. (1923) Deux Sarcoptides psoriques parasites de Roussettes Africaines au Congo. *Rev. Zool. Afric.* **XI.**

RUDNICK, A. (1960) A revision of the mites of the family *Spinturnicidae* (Acarina). *Univ. California Public. Entomol.* **XVII.**

SONENSHINE, D.E. and ANASTOS, G. (1960) Observations on the life history of the bat tick, *Ornithodorus kelleyi* (Acarina, Argasidae). *J. Parasitol.* **XLVI.**

VITZTHUM, H. (1932) Acarinen aus dem Karst (excl. Oribatei). *Zool. Jahrb. Abt. System.* **LXIII.**

VITZTHUM, H.G. (1940–42) Acarina. *Bronn's Klassen und Ordnungen des Tierreiches.* 5.Sect., 4.Buch.

WHARTON, G.W. (1938) Acarina of Yucatan caves. *In* PEARSE, A.S. Fauna of the caves of Yucatan. *Carnegie Inst. Washington Public.* No. 491.

Dermaptera

BURR, M. (1912) A new species of *Arixenia. Ent. monthl. Mag.* (2) **XXIII.**

CLOUDSLEY-THOMPSON, J.L. (1957) On the habitat and growth stages of *Arixenia esau* Jordan and *A. jacobsoni* Burr (Dermaptera; Arixenoidea) with description of the unknown adults of the former. *Proc. R. ent. Soc. Lond. Ser. A.* **XXXII.**

CLOUDSLEY-THOMPSON, J.L. (1959) The Growth Stages of *Arixenia* (Dermaptera). *Proc. R. ent. Soc. Lond. Ser. A.* **XXXIV.**

GILES, E.T. (1961) Further studies on the growth stages of *Arixenia esau* Jordan and *Arixenia jacobsoni* Burr (Dermaptera, Arexiniidae), with a note on the first instar of *Hemimerus talpoides* Walker (Dermaptera, Hemimeridae). *Proc. R. ent. Soc. Lond. Ser. A.* **XXXVI.**

JORDAN, K. (1909) Description of a new kind of apterous earwig apparently parasitic on a bat. *Novit. Zool.* **XVI.**

MEDWAY, LORD (1958) On the habit of *Arixenia esau* Jordan (Dermaptera). *Proc. R. ent. Soc. Lond. Ser. A.* **XXXIII.**

Hemiptera

Cooreman, J. (1951) Sur la présence au Congo belge du genre *Eoctenes* Kirkaldy (Polyctenidae, Cimicoidea). *Rev. Zool. Bot. Afric.* **XLIV.**

Cooreman, J. (1955) Un Polycténide nouveau pour la faune du Congo belge (Hemiptera, Cimicoidea). *Rev. Zool. Bot. Afric.* **LI.**

Jordan, K. (1913) Contribution to our knowledge of the morphology and systematics of the Polyctenidae, a family of Rhynchota parasitic on Bats. *Novit. Zool.* **XVIII.**

Schoudeten, H. (1951) Un genre nouveau de Cimicide du Katanga. *Rev. Zool. Bot. Afric.* **XLIV.**

Coleoptera

Pearse, A. S. (1938) Insects from Yucatan Caves, *in* Pearse, A. S., Fauna of the caves of Yucatan. *Carnegie Inst. Washington Public.* No. 491.

Siphonaptera

Aellen, W. (1960) Notes sur les Puces des Chauves-Souris, principalement de la Suisse (Siphonaptera: Ischnopsyllidae). *Bull. Soc. neuchâtel. Sc. nat.* **LXXXIII.**

Hopkins, G. H. E. and Rothschild, M. (1956) An illustrated catalogue of the Rothschild Collection of fleas (Siphanoptera) in the British Museum. London.

Hurka, K. (1956) Die Larve des Fledermausflohes, *Ischnopsyllus intermedius* (Rotsch.) (Aphaniptera, Ischnopsyllidae). *Acta Soc. Zool. Bohemoslov.* **XX.**

Weidner, H. (1937) Beiträge zur Kenntnis der Biologie des Fledermausflohes, *Ischnopsyllus hexactenus. Kol. Zeit. f. Parasitenk.* **IX.**

Pupipara

Falcoz, L. (1923) Pupipara (Diptères) (1ère Série). *Biospeologica,* **XLIX.** *Archiv. Zool. expér. géner.* **LXI.**

Jobling, B. (1954) Streblidae from the Belgian Congo, with a description of a new genus and three new species (Diptera). *Rev. Zool. Bot. Afric.* **L.**

Jobling, B. (1955) New species of *Raymondia* from the Belgian Congo (Diptera, Streblidae). *Rev. Zool. Bot. Afric.* **LI.**

Rodhain, J. and Bequaert, J. (1916) Observations sur la biologie de *Cyclopodia Greeffi* Karsch, Nyctéribiide parasite d'une Chauve-Souris congolaise. *Bull. Soc. Zool. France* **XL.**

Sadoglu, P. (1956) Ocular reduction of the Bat-fly, *Nycteribosca africana* (Dipt. Streblidae) and its comparison with those seen in some nycteribiids. *Istanbul Univ. Fen. Fak. Mecmuasi.* B-**XXI.**

Speiser, P. (1900a) Über die Strebliden, Fledermausparasiten aus der Gruppe der pupiparen Dipteren. *Archiv. f. Naturg.* **LXVI.**

Speiser, P. (1900b) Über die Art der Fortpflanzung bei den Strebliden, nebst synonymischen Bemerkungen. *Zool. Anz.* **XXIII.**

Speiser, P. (1901) Über die Nycteribien, Fledermausparasiten aus der Gruppe der pupiparen Dipteren. *Archiv. f. Naturg.* **LXVII.**

Theodor, O. (1954 a) Nycteribiidae. *In* E. Linder, *Fliegen. Palaearktischen Region.* **CLXXIV,** 66a.

Theodor, O. (1954b) Streblidae. *In* E. Linder, *Fliegen. Palaearktischen Region.* **CLXXIV,** 66b.

PART 3

Geographical Distribution and Ecology of Cavernicoles

CHAPTER XVI

GEOGRAPHICAL DISTRIBUTION OF CAVERNICOLES

IN THE *Liste générale des Articulés cavernicoles de l'Europe*, published in 1875, Bedel and Simon brought for the first time to the attention of naturalists the unequal distribution of hypogeous animals in the different regions of the world. In fact, they considered the areas inhabited by cavernicoles to be very narrow, since they believed that it was limited to the Alps and Pyrenees in Europe, and to mountain ranges in the United States which were of similar latitude to the two great European mountain systems. It is known today that these statements are incorrect. One can no longer speak of a world distribution of cavernicoles. Such a consideration is without significance. In order to make our study more precise it is necessary to divide the cavernicoles into three categories:

A. Aquatic cavernicoles.
B. Terrestrial troglophiles.
C. Terrestrial troglobia.

A. AQUATIC CAVERNICOLES

(a) The Distribution of Aquatic Cavernicoles throughout the World

Aquatic cavernicoles have been found in almost all regions of the world, but they are absent in the northern areas of Europe, Asia and America.

The absence of cavernicoles in these northern regions has generally been attributed to past glacial conditions. It was probably these phenomena which brought about the destruction of fauna which inhabited the circumboreal regions during the Tertiary. Leruth (1939) has noted the striking relationship between the northern limits of distribution of cavernicolous aquatic Crustacea (Copepoda, Ostracoda, *Asellus, Niphargus, Crangonyx*) and the limit of maximum extension of the quaternary glaciers. Ruffo (1956) drew a map of the distribution of the species of *Niphargus* (Fig. 66) which clearly shows this correlation.

The geographical distribution of aquatic cavernicoles is always wider than that of terrestrial troglobia. The restricted distribution which is very frequent in subterranean insects is not found in aquatic forms. The reason

271

is that far greater possibilities of dispersal are offered by the aquatic medium, which comprises complex systems of subterranean streams and ground waters with imprecise limits (Boettger, 1958).

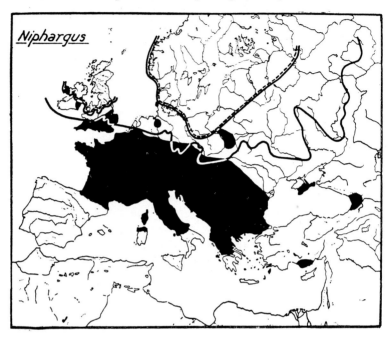

FIG. 66. Map of the distribution of the genus *Niphargus*. The dotted line marks the limit of the extension of the Mindel Glaciers. The continuous line represents the limit of the distribution of the Würm glaciers (after Ruffo).

(b) The Double Origin of the Aquatic Cavernicoles

It is almost certain that aquatic cavernicoles, like all other hypogeous forms, represent the descendants of epigeous animals. It is logical to consider the aquatic cavernicoles as having originated from those river or lake dwelling animals which penetrated the subterranean environment.

Such a descent is certain where the cavernicoles belong to groups entirely composed of freshwater animals. This is the case with the paludicole Tricladida, Crustacea (Syncarida, Phreatoicidea, Atyidae, Astacidae), and fish (Cyprinidae, Clariidae, Cyprinodontidae).

However, strangely enough, the majority of aquatic cavernicoles have not been derived from freshwater epigeous forms, but from marine forms. T. Fuchs (1894) was probably one of the first to bring to the notice of biospeologists the fact that numerous cavernicoles belong to groups which are entirely, or in the majority, marine. Numerous examples of this relationship have been described since then. A few examples will be mentioned here but for more details the reader should refer to the systematic section of this book:

Foraminifera
Nematodes: *Desmoscolecidae*
Annelids: *Troglochaetus, Marifugia*
Molluscs: *Lartetia*
Porohalacarids
Copepods: *Nitocra, Nitrocrella*
Thermosbenacea: *Monodella*
Mysids
Cirolanids
Sphaeromids
Parasellids
Microcerberids
Amphipods: *Ingolfiella, Hadzia, Niphargus,*
    Pseudoniphargus, Salentinella, Metacrangonyx, etc.
Palaemonids
Galatheids: *Munidopsis*
Fish: Brotulidae.

The close systematic relationship which links the cavernicoles in the above list and the connected marine forms constitutes one of the arguments in favour of a marine origin for these subterranean aquatic species. The second argument is based on ecological and physiological considerations. A few cavernicolous species are found normally or occa-

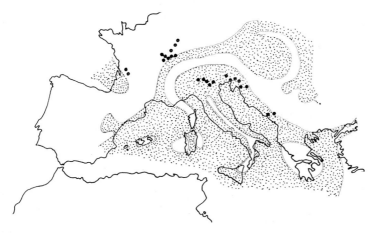

Fig. 67. Map of the distribution of the cavernicolous sphaeromids (the genera *Caecosphaeroma* and *Monolistra*) which has been superimposed on a map of the probable configuration of the areas of the land and sea during the Miocene. The paleogeographic map has been taken from R. Furon.

sionally in brackish waters. Also, many Crustacea living in fresh water, for example *Niphargus* and *Caecosphaeroma*, show a remarkable tolerance to salt water. This condition would be inexplicable had they been derived from types which had been for a long time adapted to fresh water. This idea will be extended in the physiological section of this book (p. 345).

A third, biogeographical argument can be added to the first two. Subterranean evolution of certain aquatic types appears to have taken place, on the spot. That is to say, their geographical distribution today closely approaches the localisation of ancient seas which made up their primitive environment. A map of the distribution of the cavernicolous sphaeromids (Fig. 67) can easily be superimposed on the paleogeographical reconstruction of the Miocene epoch made by geologists. The ancestors of *Caecosphaeroma* and *Monolistra* must have populated the two branches of the ancient Mediterranean which was cut in two by the formation of the Alpine range.

Finally, the circumstances which allowed candidates for cavernicolous life to pass from the sea to the subterranean environment must be considered. It is almost certain that several methods were involved, and each must constitute a distinct study. Nevertheless, the interstitial medium which offers a means of passage from salt water to fresh water, from light to darkness, from an environment undergoing seasonal variations in temperature to one of constant temperature, makes an ideal biotope for facilitating the transition between the marine-littoral environment and that of ground water. The inhabitation of the pools in subterranean cavities is the last stage of migration into the hypogeous environment. The many arguments put forward by Ruffo (1956) and Balazuc (1957) concerning the amphipods, and by Birstein (1961) on the parasellids, give credibility to the above hypothesis.

B. TERRESTRIAL TROGLOPHILES

In a study of terrestrial cavernicoles it is necessary to treat the troglophiles and troglobia separately.

Troglophiles are to be found in most regions of the world. They are absent only in the circumboreal regions. They are found equally in temperate and tropical zones, although these different zones do not contain the same troglophilic species. The cavernicolous blattoids and the majority of underground rhaphidophorids and gryllids are tropical forms.

However, while the subterranean cavities of the temperate zone contain both troglophiles and troglobia, only the troglophiles make up the cavernicolous fauna of tropical regions. The many animal species reported in the caves of Central Africa and tropical Asia are troglophiles (Jeannel, 1959). This is to say that subterranean evolution is only beginning in tropical regions whilst well advanced in temperate ones. In the next paragraph an explanation of the diversity of subterranean populations of these regions will be given.

C. TERRESTRIAL TROGLOBIA

For many years biospeologists have recognised the marked difference between the distribution of terrestrial troglobia and that of aquatic cavernicoles. In particular, the northern limits of distribution of terrestrial subterranean forms are always situated at more southerly latitudes than those of aquatic forms. Thus countries such as Great Britain, Belgium and Germany possess very few terrestrial troglobia† while they do possess a rather varied subterranean aquatic fauna (Leruth, 1939; Arndt, 1940).

Terrestrial troglobia are limited to temperate regions of the northern hemisphere: the Mediterranean region in Europe, the Japanese Archipelago in Asia, the east of the United States and Mexico (Jeannel, 1959).

Troglobia become rare towards the south of the Mediterranean region and are replaced by troglophiles of recent origin. Thus in North Africa the caves are mainly populated by microphthalmic troglophiles and anophthalmic troglobia are found rarely, for example staphylinids of the genus *Apteranillus*.

In the southern hemisphere the caves of South Africa, Australia, Tasmania and New Zealand contain troglobia. These are but little specialised, for example *Duvaliomimus* from New Zealand and the troglophile *Idacarabus* from Tasmania). These species still possess eyes, save the genus *Neanops*.

The distribution of terrestrial troglobia raises two questions which must be considered:

(1) Why are terrestrial troglobia absent from the north of America and Europe, that is to say in regions where aquatic cavernicoles are still abundant?

(2) Why are terrestrial troglobia almost completely absent from tropical regions?

In answering the first question the considerable development of glaciers in the northern hemisphere during the Quaternary may be regarded as a causal factor. We have already said that this interpretation satisfactorily accounts for the limits of distribution of aquatic cavernicoles. However, it does not provide an adequate explanation for the distribution of terrestrial troglobia.

The ice cap which surrounded the pole completely destroyed all higher animal populations. It was only when the covering of ice disappeared that the boreal regions could be repopulated. A similar situation existed

---

† It appears that Belgium possesses only one troglobious coleopteran, a psleaphid *Collartia belgica*, discovered by Collart and Leleup in the Engiboul cave.

in the central part of the Alpine chain. This is why the caves of the Central Alps are today quite azoic or populated with very rare cavernicoles *(Isereus, Arctaphaenops)* which maintain themselves with difficulty in this environment. They are at the limit where life can only just be maintained.†

It must be mentioned that the interglacial periods, characterised by a marked drying of the climate, contributed as much as the glacial phases to the destruction of the preglacial fauna (Peyerimhoff, 1906; Laneyrie, 1952).††

The destruction of fauna by glaciers of the Quaternary provides only a partial explanation of the distribution of terrestrial troglobia. Neither Belgium nor the west of France has been covered by glaciers although these regions are completely deprived of terrestrial troglobia. On the other hand, the borders of the Alps and the Pyrenees and more generally, all the mountain chains rising to the north of the Mediterranean, are regions rich in cavernicoles.

It must be concluded that these mountainous regions which were covered, during the Tertiary, with large humid forests composed of evergreens, might harbour the ancestors of the present cavernicoles. Then, there were orophilous creatures. During this period these animals must have inhabited the layers of humus which covered the forest floor. An analogous humicolous fauna which can be named "precavernicolous" is found today in the mountainous forest of Central Africa. One must consider the reasons why such a form has remained in the humus in Africa, while it has become cavernicolous in Europe. This is very directly related to the second question stated above. The European cavernicoles and the African humicoles are the relics of ancient faunae. The numerous regressive peculiarities of these relics point out their ancient origin. The decrease of their adaptive capacities compel these creatures to live in the places where the climatic variations are very attenuated. These conditions are present in caves of the Mediterranean region, and in the humus of the equatorial forests, between an altitude of 2000 and 3000 m. Both the biotopes are very alike in respect of the constancy of the temperature and the humidity (Leleup, 1956).

The study of this problem will be reserved until later. It will be considered in the last part of the book, where, with all the facts concerning cavernicoles at our disposal, an attempt to understand their evolution and origin will be made.

† An exceptional case is that of the spiders of the genus *Porhomma* which are distributed in northerly areas. During the Quaternary these animals must have lived on the edge of the glacial region. With the post-glacial rewarming they buried themselves in crevices and have thus become cavernicolous (Fage, 1931).

†† In South Africa, the subterranean migration probably results from the progressive drying of this country.

D. CAVERNICOLES AND PALAEOGEOGRAPHY

We have mentioned several times in the systematic part of this book the relict character of the troglobia. The true cavernicoles represent today an ancient fauna which has disappeared from the surface of the Earth. Thus the distribution of the cavernicoles furnishes material which can contribute to the progress of palaeogeography, the science which attempts to reconstitute the physiognomy of the Earth during the past.

The methods and results of palaeogeography can be mentioned only briefly but the reader can refer to the list of publications at the end of this chapter. The contribution made by biospeology to palaeogeography will be discussed here.

(a) The North Atlantic Continent

There are innumerable examples of families and genera which are common to Europe and America. They leave no doubt as to the presence of land-links between Europe and North America during the Tertiary. This free exchange of fauna has continued by a northern route until a relatively recent period.

A few examples of this state of affairs are given here which have been already mentioned in the systematics section. The list comprises cavernicoles from Europe which have their counterparts in North America.

	Europe	North America
Opilionids	*Scotolemon*	*Phalangodes*
Diplopods	Blaniulidae	Paraiulidae
Orthoptera	*Dolichopoda*	*Hadenoecus*
	Troglophilus	*Ceutophilus*
Coleoptera	*Trechoblemus*	*Pseudanophthalmus*
(Trechidae)	*Duvaliopsis*	*Neaphaenops*
Urodela	*Proteus*	*Necturus*

(b) Connexions between North America and Japan

The affinities between the cavernicoles of Japan and the United States have been reported on many occasions. A few examples follow:

	North America	Japan
Copepods	*Attheyella idahoensis*	*A. coiffaiti*
Isopods	*Asellus henroti, A. simonini,*	*A. miurai*
	A. vandeli	
Grylloblattids	*Grylloblatta*	*Galloisiana*
Coleoptera (Trechini)	*Pseudanophthalmus*	*Kurasawatrechus*

(c) The Continent of Gondwanaland

Examples of Gondwanian distribution are abundant in biospeology, zoogeography and phytogeography. For example, the case of those cavernicolous or phreaticolous Crustacea, the Phreatoicidae, may be cited. These Isopoda, which resemble Amphipoda, are distributed in three regions which are widely separated:

(1) Australia, Tasmania and New Zealand (most species)
(2) India *(Nichollsia)*
(3) South Africa *(Phreatoicus capensis)*.

(d) India and Tropical Africa

Reduviidae (Hemiptera) of the sub-family Emesiinae *(Myiophanes, Bagauda)* originated in tropical Asia where they are widely distributed. They are found in tropical Africa but only in caves (Jeannel, 1919; Kemp, 1924; Lhoste, 1939).

BIBLIOGRAPHY

Aquatic Cavernicoles

BALAZUC, J. (1957) Notes sur les Amphipodes souterrains. *Notes biospéol.* **XII.**
BIRSTEIN, J. A. (1961) *Microthambema tenuis* n. gen., n. sp. (Isopoda, Asellota) and relations of some asellote isopods. *Crustaceana.* **II.**
BOETTGER, C. R, (1958) Höhlenfauna und Glazialzeiten. *Zool. Anz. Suppl.* Bd. **XXI.**
FUCHS, TH. (1894) Über Tiefseetiere in Höhlen. *Ann. K. K. Naturh. Hofmuseum. Wien.* **IX.** *Notizer.*
LERUTH, R. (1939) La Biologie du Domaine souterrain et la Faune cavernicole de la Belgique. *Mém. Mus. R. Hist. Nat. Belgique* No. 87.
RUFFO, S. (1956) Lo stato attuale delle conoscenze sulla distribuzione geografica degli Anfipodi delle acque sotteranee europee e dei paesi mediterranei. *Premier Congr. Intern. Spéléologie Paris* **III.**

Terrestrial Troglobia

ARNDT, W. (1940) Die Anzahl der bisher in Deutschland (Altreich) in Höhlen und im Grundwasser lebend angetroffenen Tierarten. *Mitteil. Höhl. Karstf.*
FAGE, L. (1931) Remarques sur la distribution géographique actuelle des Araignées du genre *Porhomma. Compt. rend. Soc. Biogeogr.* No. 68.
JEANNEL, R. (1959) Situation géographique et peuplement des cavernes. *Annales. Spéléologie* **XIV.**
LANEYRIE, R. (1952) Nouvelles Notes sur les *Duvalius* de Provence (Col. Trechidae). *Notes biospéol.* **VII.**
LELEUP, N. (1956) La Faune cavernicole du Congo belge et considérations sur les Coléoptères reliques d'Afrique intertropicale. *Ann. Mus. R. Congo belge.* **XLVI**
LERUTH, R. (1939) La Biologie du Domaine souterrain et la Faune cavernicole de la Belgique. *Mém. Mus. R. Hist. Nat. Belgique* No. 87.
PEYERIMHOFF, P. DE (1906) Recherches sur la faune cavernicole des Basses-Alpes. Considérations sur les origines de la faune souterraine. *Ann. Soc. Entomol. France* **LXXV.**

Palaeogeography

ARLDT, TH. (1907) *Die Entwicklung der Kontinente und ihrer Lebewelt. Ein Beitrag zur vergleichenden Erdgeschichte.* Leipzig.

ARLDT, TH. (1917) *Handbuch der Paläogeographie.* Leipzig.

DACQUÉ, ED. (1915) *Grundlagen und Methoden der Palaeogeographie.* Jena.

FURON, R. (1941–1959) *La Paléogéographie.* Paris. 1 Edit., 1941; 2 Edit., 1959.

JEANNEL, R. (1919) *Insectes Hémiptères. III.* Henicocephalidae *et* Reduviidae. *Voyage de Ch. Alluaud et R. Jeannel en Afrique Orientale* (1911–1912). *Résultats scientifiques.* Paris.

JEANNEL, R. (1942) La Genèse des Faunes terrestres. *Éléments de Biogéographie.* Paris.

JEANNEL, R. (1943) *Les Fossiles vivants des cavernes.* Paris. (Chapitre XVI: Paléogeographie).

JOLEAUD, L. (1941) *Atlas de Paléogéographie.* Paris.

KEMP, S. (1924) Rhynchota of the Siju Cave, Garo Hills, Assam. *Rec. Indian Mus. Calcutta* **XXXVI**.

LHOSTE, J. (1939) Études biospéologiques. XI. Espèces africaines du genre *Bagauda* Bergroth (Hemiptera, Reduviidae). *Bull. Mus. R. Hist. Nat. Belgique* **XV**.

VANDEL, A. (1958, 1962) La Répartition des cavernicoles et la paléogéographie. *Compt. rend. IIème Congr. Intern. Speleol. Bari.*

CHAPTER XVII

THE DISTRIBUTION OF CAVERNICOLES IN THE SUBTERRANEAN WORLD

A. SUBTERRANEAN ECOLOGY

Ecology is that section of biology which is concerned with the relationships between living organisms and their environment. In a book on biospeology the relationships established between the cavernicoles and the subterranean medium must be included. This study has received the name "subterranean ecology".

This science is still in its infancy. The information available is fragmentary and often superficial and indeed, too often incorrect. This information is restricted to a few caves in the temperate regions of Europe and North America. Except for the caves in the region of Thysville, in the Congo, which were studied by Heuts and Leleup (1954), there are no data available concerning caves of the tropical regions.

The collected data on this subject are so rudimentary that the problems of ecological biospeology, although relatively simple, still require to be answered. For example, biospeologists who frequently visit the same caves know that the cavernicolous fauna is not uniformly distributed. It is localised in certain regions such that the troglobia are always found in certain areas. The reason for this localisation remains unknown. It could be accounted for by an abundance of food, or by the high humidity. It is certain that these places provide favourable conditions for reproduction. However, this interpretation is based on impressions rather than on facts.

This chapter will be devoted to an examination of cavernicoles in the subterranean world. In the first part the different modes of life of hypogeous forms will be considered and in the second part the various subterranean biotopes will be listed. This will illustrate the underground synusia and biocoenoses. The following chapter will deal with the physical, chemical, and meteorological factors which constitute the subterranean climate, and the action of these factors on cavernicoles.

B. THE DIFFERENT HABITATS OF HYPOGEOUS ORGANISMS

The development of ecology, of its methods, and its nomenclature, was the work of botanists. It was not until later that zoologists attempted to find in the animal world, similar conditions to those in the plant

280

kingdom. The reason for this is easy to understand. Plants are immobile while animals move about, not only to find food and a mate but also to complete their developmental cycles. The young or larvae often develop in a different medium from the one in which the adults live. Animals are never strictly localised as are plants, and the ecology of animals can thus never be so exact as plant ecology.

(a) Epigeous and Cavernicolous Life

One might believe that cavernicoles are incapable of living outside the underground world. In fact, this is not the case as is shown by the following examples:

(1) Many species of *Niphargus* have a tendency to live an epigeous life (Karaman, 1950). *N. aquilex* is often found in forest pools (Godon, 1912; Müller, 1914; Schellenberg, 1935). This is also the case with *N. tatrensis* (Schäferna, 1933). Cavernicolous species of *Asellus* often penetrate into surface waters (Leonard and Ponder, 1949; Dexter, 1954; Minckley, 1961).

(2) The interstitial media, and in particular the hypotelminorheic biotope provides a passage between the surface habitat and the subterranean environment.

(3) Many terrestrial isopods are found both in caves and on the surface under rocks or in clay crevices (an endogeous habitat). *Nesiotoniscus corsicus* was discovered in Corsica in the Pietralbello Cave, near Ponte-Leccia. It has also been found under buried stones on the banks of the River Golo (Vandel, 1953). *Oritoniscus vandeli* and *Phymatoniscus propinquus*, two species from the southern edge of the Massif Central, are found as frequently under buried stones as they are in subterranean cavities (Vandel, 1960). *Mesonicus alpicola* which populates a large area of the Alps from Lake Maggiore region to Transylvania lives in caves, but is also to be found outside under stones, where it leads an endogeous mode of life. However, in the western part of its distribution (Ticino, Italian Alps) it is more often cavernicolous than endogeous.

(4) More frequently, certain species which are cavernicolous in some regions lead an endogeous life in other parts of their distribution. *Trichoniscoides modestus* which is a cavernicole in Aquitano-Languedocian regions of France, becomes an endogean in the Massif des Albères, the most easterly chain of the Pyrenees (Vandel, 1952). *Trichoniscus fragilis* and *T. halophilus* which are halophilic forms in France, are cavernicoles in North Africa (Vandel, 1955). The opilionid, *Ischyropsalis luteipes*, which is widespread in France, is epigeous in the Massif Central and cavernicolous in the Causses and the Pyrenees (Dresco, 1952).

(5) Finally, many species which are found in mountainous regions under rocks, are cavernicolous in regions of lower altitude.

The terrestrial isopod, *Oritoniscus fouresi*, is found under stones in the western Pyrenees between 1400 and 1900 metres altitude. But the type of the species comes from the abyss of La Palle, near Rieulhès, which opens at 350 metres altitude (Vandel, 1960).

A campodeid, *Plusiocampa caprai*, is cavernicolous around Innsbrück in the Tyrol, but it leads an endogeous life in the Brenner Alps at 2300 metres altitude (Condé, 1955). The grylloblattid, *Galloisiana*, is found on the surface in the mountains of north Japan; it is exclusively cavernicolous in the south of the Japanese Archipelago.

A carabid, *Oreonebria ratzeri*, is a nivicolous species, from the higher regions of the Jura around 1500 metres altitude. This species has been found in a swallow-hole in the Jura Plateau, near Vercel at 500 metres altitude (Colas, 1954). Another carabid, *Duvalius scarisoarae*, lives under stones on Mount Bihar in Transylvania at 1800 metres altitude, it is endogeous at 1000 metres, and is found only in caves at 500 metres (Jeannel and Leleup, 1952).

In conclusion, it would be incorrect to consider the cavernicolous fauna in isolation from the rest of the world. Cavernicoles are linked with the surface in many ways and the division of epigeous fauna and hypogeous fauna is often arbitrary.

(b) Aerial and Aquatic Life

1. *Terrestrial Forms leading an Amphibious Life*. Terrestrial animals which populate caves live in a saturated atmosphere. It is therefore not surprising that some of them penetrate into water and are capable of existing there for some time without harm.

The beetle, *Hadesia vasiceki* (Bathysciinae) from the Vjetrenica pečina in Herzegovina, is not an aquatic form as has been stated by Absolon. Remy (1940) has seen this animal living under the spray caused by a stream of water cascading on to a stalagmite.

A cavernicolous trechinid, *Ryugadous ishikawai*, found in the largest and most complex cave in the Japanese Archipelago, the Ryûga-dô Cave, is frequently found in water in gours. This is also the case for the related form *Kusumia takahasii* (Uéno, 1955).

Several observations prove that certain Collembola, in particular representatives of the genus *Arrhopalites*, are sometimes found at the bottom of subterranean pools (Franciscolo, 1951; Delamare-Deboutteville, 1952). The same is true for *Machilis* (Pavan, 1940).

Other cavernicoles can be considered to be amphibious forms, as they live indifferently either above or below the water. This is seen particularly clearly in several species of terrestrial isopods. This is not surprising, as in the lower oniscoids the respiratory organs have retained the characteristic structure and physiology of aquatic forms (Remy, 1925).

A terrestrial isopod, *Titanethes albus*, can be considered as a true amphibian, a condition which has already been recognised by Verhoeff (1908), Jeannel and Racovitza (1918) and Chappuis (1927). From observations of the author in the Jama Pod Gradom (resurgent of the Piuka), in the cave of Predjama, and from their behaviour in aquaria at the Postojna Experimental Station, and in the cave laboratory at Moulis, it has been shown that this isopod can live indifferently in water or in humid clay.

Another terrestrial isopod, *Mesoniscus alpicolus graniger*, shows analogous behaviour. This has been observed by Dudich (1932) in the Baradla. A cavernicolous trichoniscid from the Pyrenees, *Scotoniscus macromelos*, can survive submersion for many days (Derouet, 1956). This fact accounts for the abundance of these animals in the Goueil di Her, a cave which is periodically flooded.

Other terrestrial cavernicoles show more or less amphibious habits. These include certain diplopods, and Acarina of the genus *Schwiebea* (see p. 99).

The next stage in the evolution of terrestrial forms which have become secondarily amphibious is to assume an exclusively aquatic habit. This is seen in the terrestrial isopod* *Typhlotricholigidioides aquaticus*, a Mexican trichoniscid which is exclusively aquatic (Rioja, 1953).

Phreaticolous Acarina of the family Stygothrombiidae are also entirely aquatic (p. 100).

2. Aquatic Forms leading an Amphibious Life. Inversely, aquatic cavernicoles can exist above water for a long time in the atmosphere of caves which is always of high humidity.

Niphargus are capable of leaving the gours and crawling across the damp clay floor to neighbouring pools. Derouet-Dresco recognised that aquatic Crustacea and in particular amphipods, can exist out of water for several weeks. After a temporary increase in respiratory rate they adapt, and return to normal rates of respiration.

Analogous behaviour is exhibited by cavernicolous asellids (e. g. *Stenasellus*, *Asellus*). A cavernicolous sphaeromid, *Monolistra berica*, frequently leaves the water, whether in caves or in experimental aquaria (Viré and Alzona, 1902).

Specimens of *Monolistra caeca* cultured in the cave laboratory at Moulis are often to be seen resting on rocks out of the water.

Certain harpacticid copepods can lead a more or less amphibious existence.† This property is not general among the cavernicolous harpacticids however, but is seen in most hypogeous species derived from muscicolous epigeans. Many species of the genus *Moraria*, for example *M. subterranea* (Carl, 1904) and *M. varica* (Chappuis, 1920), can live as well outside water as they can in aquatic media. This is also the case with species of the genus

---

* This also occurs in non-cavernicoles. For example, three species of the genus *Haloniscus* are permanent inhabitants of Australian lakes.

† Likewise, this property is seen in one species of cyclopid: *Paracyclops fimbriatus*.

Bryocamptus, in particular *B. pygmaeus* and *B. zschokkei* which are found abundantly in certain caves, on dead and humid wood-debris (Chappuis, 1933).

3. Conclusion. Thus, many either aquatic or terrestrial cavernicoles lead an amphibious existence. It was R.Schneider (1886) who brought this peculiar behaviour of cavernicoles to the attention of biospeologists.

Thus aquatic and terrestrial forms can be found living side by side, underground. Bythinellids of the genus *Pseudamnicola* and the pulmonate gasteropod, *Vitrea spelaea* live together in the tubes of the annelid *Marifugia cavatica* which cover the walls of the Crnulja Cave (Popovo Polje, Herzegovina) (Bole, 1961).

In some cases it becomes impossible to classify cavernicoles as terrestrial or as aquatic. For example, nematodes (Hnatenwytsch, 1929) and certain enchytraeids such as *Enchytraeus buchholzi* live either in water or in humus. This conclusion can be extended to include all the oligochaetes, as these animals can live for many months in water (Combault, 1909). The reason that they suffocate in too humid soil is that under these conditions there is no way of renewing the air.

(c) Subterranean Waters and Ground Water

The interstitial medium was mentioned in the first chapter of this work. It is formed by the ground water. This medium contains a distinct population whose composition differs from that of the gours, lakes and rivers of subterranean cavities. Many examples of the different faunae to be found in these two populations have been given in the systematic section of this book. There is, however, constant interchange between the interstitial medium and the water of caves. Numerous species are found in both media, although they generally show a preference for one or the other.

(d) Subterranean Waters and the Bottoms of Lakes

Some species are known which populate both subterranean waters and the bottoms of lakes. This is not surprising as these two biotopes possess numerous physical characteristics in common (darkness, constant temperature, a silty floor). Examples of species possessing this double habitat are the oligochaetes, *Bythonomus lemani* and *Dorylidrilus michaelseni.*

(e) Niphargus foreli

Niphargus foreli may be given in example of ubiquitous organism, which can colonise a large number of very different habitats: gours in caves, springs (and probably ground water), the bottoms of subalpine lakes and the littoral zones of high mountain lakes (Schellenberg, 1934).

(f) Conclusion

Although the habitat of cavernicoles is more limited than that of surface forms they show a certain ecological flexibility. It is incorrect to attribute to cavernicoles a strictly defined biotope. Numerous aquatic forms especially, are capable of maintaining themselves in very different media.

C. SUBTERRANEAN BIOTOPES
CAVERNICOLOUS SYNUSIA AND BIOCOENOSES

The definition of the three terms classically used in ecology are as follows:

The *biotope* is the fundamental ecological unit. The extent of a region which is given the name biotope can vary, but it has uniform physical, chemical and climatic characteristics. The term "ecological niche" is often used in an analogous sense.

Each biotope contains an animal population which is called the *synusium*.

The various parts of a synusium interact with one another which generally leads to more or less stable equilibrium. These associations have been given the name *biocoenoses*.

The biospeologist must therefore distinguish the different biotopes of the subterranean world, and recognise the synusia belonging to each of them.

The classification adopted by Jeannel (1926) in his *Faune cavernicole de France* is based on unequalled experience of the subterranean habitat and its fauna, and is, in the author's view, the best which has been proposed. This classification has been revised and extended by Leruth (1939) and Coiffait (1956).

Six principal biotopes and the corresponding six synusia can be distinguished in the subterranean world:

(*a*) Entrances to caves and the parietal association.
(*b*) The endogeous medium and the endogeous fauna.
(*c*) Guano and guanobia.
(*d*) The walls with stalagmites, crevices, and clay strata shelter the terrestrial troglobious fauna.
(*e*) The liquid medium and the aquatic cavernicoles.
(*f*) The interstitial medium and the phreatobia.

(a) Entrances to Caves and the Parietal Association

The entrances to caves are often quite small apertures, but in many cases the subterranean cavities open through wide "porches", often larger than the interior galleries. This is the result of erosion by atmospheric conditions which have a much greater effect at the entrances of caves than in their depths.

This area shows characters intermediate between the outside world and the interior of the cave. The amount of light decreases progressively from the entrance. This is the twilight zone. This is also the only region where green plants can grow; in particular those which are shade-loving are well represented. The fluctuations of temperature, although noticeable, are less so than at the exterior. The humidity level is also higher than at the exterior, but less than in the interior of the cave.

The entrances of caves are inhabited by a diverse range of animals. They are primarily muscicoles, humicoles and detriticoles which live in the dead leaves, mosses and detritus which accumulate there, often in large amounts.

Those endogeans which are found under buried rocks are abundant at the entrances to caves. These animals will be considered in the following paragraphs.

Finally, on the walls of the cave entrance are found a diversity of arthropods, which for a long time were ignored as they are not true cavernicoles. However, these animals have not, as was thought, arrived there accidentally. The same species and genera are regularly found at cave entrances. This is why Jeannel has classified them as *regular trogloxenes*. The total of these animals constitutes the *parietal association*. They have been the subject of some interesting research carried out by Jeannel (1926) and Leruth (1939).

Because of its constancy, the parietal association merits the attention of biologists. It raises many problems which are different from those which are concerned with cavernicoles, but which are no less interesting. A solution has not yet been found to many of these problems (Tercafs, 1960).

We shall successively examine (1) the composition of the parietal association and (2) the factors which bring about these associations.

The composition of the parietal association. The parietal association is composed exclusively of arthropods, or more exactly of insects and arachnids.

Diptera are the most numerous members of this association (Leruth, 1939; Tollet, 1959). In Europe, there are about thirty species which are regularly found at the entrances of caves. These species belong to the six families: Mycetophilidae, Culicidae, Limnobiidae, Trichoceridae, Helomyzidae and Borboridae.

Lepidoptera of the parietal association are less numerous with regard to the number of species and individuals. The five species which are most commonly found in Europe are:

Geometridae: Triphosa dubitata and *T. sabaudiata*
Noctuidae: Scoliopteryx libatrix
  (This species is also found at the entrances of caves in North America; Dearolf, 1937).

Microlepidoptera; Orneodidae: Orneodes hexadactyla
                    Acrolepiidae: Acrolepia granitella (Demaison, 1907,
                    1910, 1911; Codina, 1911).

Trichoptera are regularly represented at the entrances of caves by species
belonging to the genera *Micropterna* and *Stenophylax.*

Hymenoptera form part of the parietal association and in particular the
large ichneumonids belonging to the genus *Amblyteles.*

The Orthoptera which populate the caves of Europe, that is species of
Dolichopoda and *Troglophilus*, never penetrate to the depths of caves. They
remain not far from the entrances where they are to be found on the walls
or roofs, or hidden in crevices.

Among the Coleoptera, *Choleva* is frequently found at the entrances.

No representatives of the Hemiptera are found in the parietal association
of European caves. They are frequent, however, in tropical caves (Central
and East Africa, Madagascar, Assam, Malaya). They are carnivorous
forms with filiform antennae and very long legs. They sometimes perform
"bobbing" movements which resemble those carried out by some tipulids.
The eyes may be reduced. They belong to the family Reduviidae and the
sub-families Emesinae *(Myiophanes, Bagauda, Berlandina, Gardenoides)*,
Acanthaspidinae *(Macrospongus)* and Physoderinae *(Paulianocauris)* (Dover, 1929; Jeannel, 1912, 1919; Kemp, 1924; Lhoste, 1939; MacAtee and
Malloch, 1926; Paiva, 1919; Villiers, 1949, 1953a, b, 1962).

Finally, many arachnids are regularly found at caves entrances. In
Europe, these are *Meta menardi* and *M. merianae, Nesticus cellulanus,
Tegenaria silvestris*, and *Leptyphantes leprosa.*

Factors responsible for the existence of the parietal association. The parietal
association consists of a number of very diverse elements, and the factors
which attract regular trogloxenes to the entrances of caves are certainly
different. Information is available for a few of them.

Carnivores: Spiders and reduviids are lucifugous, carnivorous forms.
They find suitable conditions at the entrances to caves which include an
abundance of prey in the form of the Diptera and Lepidoptera of the
parietal association.

Orthoptera: The cavernicolous Orthoptera of Europe, *Dolichopoda* and
Troglophilus, and also the American species, *Hadenoecus subterraneus* (Nicholas, 1962), are troglophiles which spend most of their lives in caves, and
indeed reproduce there. They regularly leave the caves, however, to search
for food. They usually emerge during the night but *Troglophilus* is also
found during the day in regions covered by large forests. It can be concluded that temperature and humidity influence the movements of these
cavernicolous Orthoptera but light does not (Chopard, 1917, 1959).

Culex pipiens: Mosquitoes belonging to the species *Culex pipiens* are
sometimes found in large numbers at the entrances of caves. This pheno-

menon is observed only in winter†, and then only females are found (Leruth, 1939; Tollet, 1959).

The reason for these assemblies has been discovered by Roubaud (1933) and stated in his extensive work devoted to the biology of these mosquitoes. The culicids which invade the caves belong to the sub-species *Culex pipiens pipiens:* they correspond to the primitive form which Roubaud has called the "heterodyname" or "anautogene" type. In this race hibernation is spontaneous. It is not simply a torpor induced by the decrease of temperature. Hibernation is obligatory and corresponds to a phase of purifying which reactivates the species and allows it to reproduce anew. The fat content of the female increases during hibernation although the insect does not feed; this is because the reserves of the larvae are used and the abdominal muscles digested. These transformations occur only when the mosquito is in a humid medium with a slightly low temperature. These are the conditions found at the entrances of caves.

Choleva: The life cycle of the beetles of the genus *Choleva* is in some ways the inverse of that in *Culex pipiens*. Jeannel noticed that earth chambers have often been found on the walls of cave entrances. During the summer these chambers contain adult *Choleva*; they are empty during the winter. Jeannel interpreted these chambers as pupation chambers.

Deleurance (1959) proved by studies in the field and by breeding in the laboratory that this interpretation is incorrect. The first fact which contradicts Jeannel's hypothesis is that these earth chambers contain only imagines, never larvae, pupae, or exuviae.

The life cycle of *Choleva (angustata* and *fagniezi)* takes place in the external medium. This includes pupation which takes place in a chamber of earth. The newly emerged insects fly towards the entrances of caves and construct earth chambers very similar to the pupation chambers but which serve a different function. Each chamber contains an adult undergoing summer diapause. This diapause is necessary for the completion of the normal reproduction cycle (as is hibernation for *Culex pipiens pipiens*). In autumn the *Choleva* adults leave their chambers and copulate. Egg-laying and larval development takes place during the winter.

(b) Guano and Guanobia

Guanobia are not true cavernicoles. The majority of them are pigmented, winged and possess eyes. Jeannel (1943) states: "immersed in the huge mass of food material consisting of guano, the guanobia are biologically unaware that they live in caves". Most guanobia seem to be derived from pholeo-

---

† Aggregations of *Anopheles punctipennis* are observed during winter at the entrances of caves in America (Banta, 1907). This is a phenomenon of hibernation but the factors involved are unknown.

philes (Leleup, 1956). By reason of their coprophagous diet pholeophiles would appear to be forms preadapted to a guanobious existence.

Although guanobia cannot be categorised as cavernicoles, they comprise, as do the members of the parietal association, a very characteristic biocoenose of the subterranean medium.

Guano caves are found in temperate as well as in tropical regions. Most frequently the guano is derived from bats. In Europe it comes principally from the vespertilionids. Horse-shoe bats which live singly do not give rise to large areas of guano. Other animals can form guano: for example *Hyrax (Procavia)* from the caves in Kenya.

Guano is an important food source and may support a swarming population consisting of millions of individuals. Although guanobia are numerous in individuals, the number of guanobious species is relatively low.

The guanobious association consists essentially of Collembola belonging to the families *Hypogastruridae* and *Onychiuridae*, and of numerous Diptera. In African and Asian tropical regions, blattids, caterpillars of tineids, and molluscs of the genus *Opeas*, can be added to this list.

Finally, the guanobia provide food for numerous predators: chilopods of the genus *Lithobius*, carabids of the genus *Laemosthenus*, staphylinids (and in particular in Europe, *Atheta subcavicola* and *Quedius mesomelinus*) and also reduviids (e.g. *Reduvius gua*, from the Batu Caves, Malaya).

(c) The Endogeous Medium and the Endogeous Fauna

The endogeous medium has already been considered in the first chapter of this book. It is sufficient to add that the entrances to caves and in particular those which are filled with a mixture of clay and fallen rocks are particularly rich in endogeans. Endogeans can populate the interiors of caves, however. This leads to a distinction between endogeans of the entrances of caves and those of the dark zone. The latter fauna may be collected under the stalagmitic floor which covers the clay layer.

Examples of endogeans are representatives of the genus *Geotrechus*. They are divided into species which live in forests under buried rocks *(G. orpheus, G. consorranus)*, at the entrances of caves (most species), and in subterranean cavities under the stalagmitic floor *(G. orcinus, G. vulcanus)*.

(d) Stalagmitic Walls, Crevices, and Clay Strata Supporting the Troglobious Fauna

The different associations which have been discussed in the preceding paragraphs are made up of species which cannot be considered to be true cavernicoles.

It is necessary to consider at this point the habitat of the true terrestrial cavernicoles or troglobia. This fauna consists essentially of isopods (Tri-

choniscidae), spiders (Leptonetidae), Diplopoda, Collembola (Entomobryidae), Diplura (Campodeidae) and beetles (Trechinae and Bathysciinae).

The true cavernicoles are generally found in the regions of caves where
it is dark, the temperature is constant and the atmosphere is still and
saturated with water vapour†.

These regions usually contain calcareous formations (stalactites and
stalagmites); the walls of caves are often covered by a stalagmitic facing.
The cave floor consists of a thick layer of clay which may be covered in
places by a stalagmitic floor.

The authors of biospeological monographs have generally divided the
"dark zone" into several biotopes: the stalagmitic walls, the clay layers,
and the crevices. In fact these divisions are not justified either speologically
or biologically. Crystallizations and clay deposits are nearly always associated as both result from the decomposition of calcareous rocks. The same
cavernicolous species are found on the stalagmitic walls and on the clay.

With respect to the crevices it is known that all rocks and in particular
calcareous ones are traversed by *diaclases* (Daubrée), due to fractures resulting initially from orogenic movements but often widened by the action
of water.

Biospeologists (Garman, 1892; Racovitza, 1907; Jeannel, 1926; Leruth,
1939) have frequently spoken of the "domaine of crevices" which they
consider as one of the most characteristic biotopes of the subterranean
medium. However, extensive consideration of this "domaine" appears
worthless as it leads to the conclusion that, being inaccessible to man, it
remains unknown.

The idea of an immense network of crevices populated by troglobia finds
no place in reality. It seems just possible that cavernicoles can spread into
calcareous masses, but such migrations would demand a long time of
which it would be almost impossible to estimate the duration. In any
case, deep crevices are not normally inhabited by cavernicoles, because they
do not offer any food.

It is necessary to replace these negative conclusions by positive information. The first fact, which has been recognised by biospeologists for many
years, is based on the strict localisation of troglobia. Biospeologists know
that it is extremely rare to find cavernicoles distributed evenly in the larger
caves. Three factors would seem to be responsible for the localisation of
cavernicoles: high humidity (in particular the presence of a dripping
wall); abundant food; an area favourable for reproduction and development of the species.

The author, in collaboration with Michael Bouillon has studied the
behaviour of Trechinae of the genus *Aphaenops* which are certainly the
most highly specialised cavernicoles known in the Pyrenees.

† Absolute rules cannot be established in biospeology. Troglobia can occasionally be
collected at the entrances of caves but this is exceptional.

In the Moulis Cave which measures 900 metres in length *Aphaenops pluto* is found in two extremely limited areas: on the "cascade pétrifiée" which rises above the "grand bassin des Protées" and around the waterfall which has been used to supply the needs of the cave laboratory. In these two regions a thin layer of water covers the stalagmitic walls and the constant flow of water brings in organic material from the exterior which nourishes the Collembola and nematoceran Diptera on which *Aphaenops* preys. Thus the conditions of humidity and food are combined and offer favourable conditions for *Aphaenops*.

The third factor must be examined, that is the biotope necessary for the reproduction of *Aphaenops*. For this study we selected a cave which opens seven kilometres from Moulis, the Grotte de Sainte-Catherine, at Balaguères. In this large cave which consists of several chambers, one above the other, *Aphaenops cerberus bruneti* is almost exclusively localised in the small chamber which precedes the last shaft. One of the sides of this chamber is formed by the stalagmitic wall which is covered by a thin layer of water which flows from an opening in the roof. The stalagmitic covering is not directly opposed to the Jurassic dolomite, the rock from which the cave has been hollowed. It is separated by a space a few millimetres in width which contains the products of dissolution of the rock: clotted red and black clay, and black magnesium deposits.

When one enters this chamber three or four *Aphaenops* can usually be seen running on the wall in search of food. If the animals are frightened they rapidly disappear, but if they are caught others soon appear. Closer examination soon explains this phenomenon. The calcareous wall is sometimes formed by stalagmitic columns arranged parallel to each other (Plate V). Between these columns small holes have been hollowed and it is into these that *Aphaenops* disappears. If the stalagmitic covering is broken it can be seen that the holes open into narrow fissures filled with the products of dissolution which have just been described. These are the galleries to which the *Aphaenops* return after searching for food.

In the particulate layer of decomposed rock we have found capsules made of similar material. We consider these to be metamorphosis chambers. Most of them were empty but one contained a specimen of *Hydraphaenops ehlersi* which had just undergone metamorphosis. It appears that the space between the rock and the stalagmitic wall constitutes the biotope in which *Aphaenops* lives when not in search of food, and in which they reproduce.

These observations show that:

(1) The crevices inhabited by these troglobia are easily accessible to man as they are situated only a few centimetres from the surface;

(2) That the rock–clay–concretion complex is a unit of which the different elements cannot be dissociated;

PLATE V. The small terminal chamber of the Grotte de Sainte-Catherine at Balaguères (Ariège). The stalagmitic wall is the habitat of *Apheanops cerberus bruneti* (Photo by Bouillon).

(3) That this complex is the biotope characteristic of troglobia.

(4) The clay has very peculiar properties. It absorbs organic matter; it neutralizes vitamins; it stops the development of mushroom mycelia.

(e) Liquid Medium and Aquatic Cavernicoles

The liquid medium of subterranean regions can be subdivided, as can the surface waters, into several distinct biotopes. The terms established to describe the biocoenoses of surface waters will be applied to subterranean limnology.

1. *The neuston:* The name neuston is given to the collection of organisms which float on the surface of the water. Underground, the neuston is composed essentially of Collembola belonging to the family Sminthuridae (*Arrhopalites* and related genera).

2. *The plankton:* The American biospeologists (Tellkampf, 1845; Packard, 1886; Kofoid, 1900; Scott, 1909) have investigated the plankton of the large subterranean rivers in the central United States. This plankton is rich but composed exclusively of epigeous organisms from surface lakes and rivers which flow underground. They contain very diverse types, corresponding to various groups of freshwater planktonic organisms. Most abundant are the cyclopids.

The fauna brought in from the surface can exist underground for a certain period of time, but disappears sooner or later. Chlorophyll-containing organisms cannot exist in darkness and most planktonic Metazoa feed on these organisms. It is doubtful if planktonic organisms from the surface can reproduce underground; in any case, their multiplication is exceptional.

Although plankton does not persist underground, it is constantly being renewed from the surface and thus forms an important food-source for cavernicoles. The copepods are the principal food of cavernicolous fish and in particular of *Amblyopsis*.

Less information is available on the plankton of the subterranean rivers of Europe. It can be stated that, as in America, it contains a majority of cyclopids which belong to epigeous types (Claus, 1893; Schmeil, 1894).

A true underground plankton is present in the subterranean waters. It includes cyclopids, above all *Speocyclops;* and also, the thermosbenacean, *Monodella argentarii* (Stella).

3. *The hygropetric fauna:* The name "hygropetric fauna" (Thienemann) or "petrimadicolous fauna" (Vaillant) is given to that association of animals which lives in the thin layer of water over rocks, which issues from seepages. The underground hygropetric fauna has recently been studied by Orghidan, Dumitrescu and Georgescu (1961).

In the deep dark zones, the hygropetric fauna consists of turbellaria, nematodes, oligochaetes and amphipods. At the entrances of caves the hygropetric association includes numerous organisms containing chlorophyll: flagellates, algae and mosses. The hygropetric fauna is also very varied. It includes infusoria, rhabdocoeles, nematodes, oligochaetes (in particular representatives of the genus *Aelosoma*), rotifers, tardigrades, copepods and oribatids.

4. Torrenticolous fauna: The subterranean torrenticolous fauna contains only vertebrates. The most characteristic type is *Proteus* which lives in subterranean rivers of the Adriatic Karst. A race of the Pyrenean *Euproctus (Euproctus asper)*, which is being studied at the cave laboratory at Moulis, is found in the torrents which runs through the caves of the department of Ariège.

No hypogeous torrenticolous invertebrates have been reported (W. and A. Chodorowski, 1960). It is true that specimens of *Niphargus orcinus virei* and *Asellus cavaticus* have been collected in running water (Ginet, 1953). Also the amphipod, *Zenkevitchia admirabilis*, from the caves of Abkhasia (Caucasia) has also been reported as being a form characteristically found in running water (Birstein, 1940). However, experimental studies prove that fast-moving water is not the "preferendum" of the aquatic cavernicoles. *Niphargus orcinus virei* is a "rheophobe" or a "rheoxene" which is intolerant of running water, and when it can choose, settles in calm water (Ginet, 1960).

For this reason *Niphargus* can be found in forest pools and in springs, while they are always absent from running water. *Gammarus* are stream animals and are essentially rheophiles (Cousin, 1925). Therefore, temperature is not involved in the explanation of the differences of distribution in these two amphipods. It is based on a dynamic factor; running water for *Gammarus* and stagnant or slowly running water for *Niphargus*.

5. Subterranean waters: The true biotope of aquatic troglobia is either slow running or stagnant subterranean waters. This biotope has diverse aspects: slow running streams and rivers; lakes; pools containing silt; and small pools in the calcium deposits, the gours. It is impossible to establish rigid distinctions between the different types of aquatic media. Also at the time of the floods which periodically affect most caves these media are united. However, it must not be concluded that aquatic troglobia are uniformly distributed in these media. Certain harpacticids live in porous stalagmitic deposits, and appear only exceptionally in gours. Limivores such as oligochaetes populate only the silty pools. *Stenasellus* is found in water containing wood or dead leaves, and so on.

(f) Interstitial Medium and the Phreatobious Fauna†

Daubrée gave the name "nappe phréatique" (from ψρέαζ, a well) to the layers of soil completely and permanently impregnated with water (ground water). Ground water contains a specialised fauna which remained unknown for a long time because of the difficulty of access to observers.

Historical: The first information acquired on this fauna was obtained from the study of the fauna of wells. This work by F. Vejdovsky, *Tierische Organismen der Brunnenwässern von Prag*, which appeared in 1882, brought

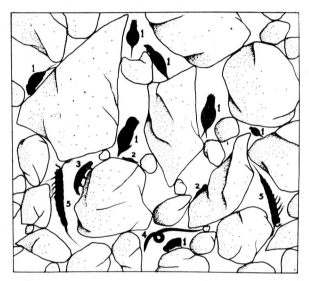

FIG. 68. The interstitial medium and its fauna. 1, Rotifera; 2, Gastrotricha; 3, Tardigrada; 4, Nematoda; 5, Harpacticidae (after Pennack).

the attention of zoologists to an until then entirely unknown biotope.

Several years later Moniez (1889) at Lille, Javorovsky (1893) at Krakow, and Chilton (1894) in New Zealand, carried out research on this subject. Later, P.A. Chappuis devoted his thesis appearing in 1920, to the fauna of the wells of Basle and its surrounding districts. More recently, Noll and Stammer (1953) have made an important contribution to our knowledge of this fauna by carrying out detailed study of the wells in the lower Main region.

It seems however, today, that the fauna of wells is a complex mixture of surface forms brought in by superficial streams, and subterranean species from the ground water.

† The hypotelminorheic biotope contains interstitial fauna but this is very different from the phreatobious fauna. The reader is referred to the first chapter of this book.

It is for this reason that zoologists have resorted to the use of "Norton" pumps which are connected to a tube placed directly in ground water. The fauna collected by this method is not mixed with surface material and represents the true population of ground water.† Hertzog (1933–1936) using this method obtained a rich and original fauna from the ground water of the Rhine valley around Strasbourg.

Previously, Sassuchin and his associates (1927, 1930) studied the flora and fauna of sands buried in the banks of the river Oka in Russia.

A few years later, P.A.Chappuis made use of a very simple method for easily and swiftly collecting the subterranean species of the ground fauna. This method was formerly applied by Stanko Karaman in the Vardar valley, in Macedonia (Motas, 1962).

A hole 1–1·5 metres deep is dug in the banks of a river. Ground water and the fauna it contains rapidly filter through the sand and gravel and fill the hole. The organisms present can be simply removed with a net.

This method has provided Chappuis with many original and unknown forms.

The interstitial fauna: The fauna which populates ground water (Fig. 68) consists of very small organisms which normally inhabit the interstices which separate grains of sand or fine gravel.†† This biotope has been given the name *interstitial medium* or *hyporheic biotope* (Orghidan, 1959). The fauna is termed the interstitial fauna (Nicholls, 1935) or the phreatobious fauna (Motas and Tanasachi, 1946).†††

The conditions existing in the interstitial medium are notably different from those in the subterranean medium. Thus, phreatobia differ in many characters from cavernicoles. The differences will be summarily considered below but for a detailed study the reader is referred to the book by Delamare-Deboutteville (1960).

The interstitial fauna is closely related to the size of the interstices which separate the gravels from each other, and the interstices are proportional to the size of the particles. Granulometry dominates the interstitial life (Prenant, 1932). The inhabitants of the ground water must be of a similar size to the cavities which they occupy. They are generally very small, measuring between 300 μ and 1–2 mm.* The interstitial forms are usually much smaller than the species living in other media. The exceptions to this rule are the infusoria which are larger than related epigeous genera in order to fill the interstices completely.

---

† Delamare-Deboutteville (1954a and b) has built a special apparatus to collect the fauna of ground waters or of marine shores.

†† This is why the ground waters can only support fauna when they are free of clay which would fill the interstices separating the sand particles.

††† The term "psammon" proposed by Sassuchin (1927) is less exact, because ground waters more often contain gravel and small pebbles than pure sand.

* A few species reach 5 mm: *Stenasellus skopljensis croaticus* and *Niphargus kochianus labacensis* (Meštrov, 1960).

The shape of the animals which inhabit the interstitial medium is also very characteristic. These organisms are extremely elongated and sometimes completely filiform. They are at the same time flattened. Such examples are found in numerous crustacea: *Angeliera* (Fig. 26), *Microcerberus* (Fig. 28), *Microcharon* and so on. They exhibit a pronounced positive thigmotaxis which stops them leaving a grain of sand when it is in isolation. They cannot therefore live a free life.

Ground waters, unless they are at considerable depth, do not possess the constant temperature of the water in caves. The temperature of ground waters which extend along rivers undergoes marked variations. Thus, representatives of the interstitial fauna are eurythermal in contrast to cavernicoles which are stenothermal. This fact has been recognised since 1915 by Schnitter and Chappuis.

Stages in the transition between marine, littoral, interstitial fauna and the freshwater underground fauna have been observed (p. 10). Thus most interstitial forms are euryhaline.

The interstitial medium, or at least the ground waters, are without light, as are all true subterranean media. Moreover, most representatives of the fauna of ground waters are depigmented and have reduced eyes, or sometimes no eyes at all.

BIBLIOGRAPHY

Subterranean Ecology

COIFFAIT, H. (1951) Quelques données actuelles sur l'écologie et l'éthologie des Arthropodes cavernicoles terrestres et endogés. *Bull. Soc. Hist. Nat. Toulouse*. **LXXXVI.**
COIFFAIT, H. (1956) La biocénose cavernicole du versant nord des Pyrénées. *Communic. 1er Congr. Intern. Spéléologie*. Paris. **III.**
DEROUET, L. and DRESCO, ED. (1955) Études sur la Grotte de Pèneblanque. I. Faune et Climats. *Notes biospéol.* **X.**
GINET, R. (1951) Étude écologique de la Grotte de la Balme (Isère). *Bull. biol. France. Belgique* **LXXXV.**
HEUTS, M. J. and LELEUP, N. (1954) La Géographie et l'Écologie des Grottes du Bas-Congo. *Ann. Mus. R. Congo Belge. Sc. Zool.* **XXXV.**
IVES, J. D. (1927) Cave Fauna with especial reference to ecological factors. *J. Elisha Mitchell Sci. Soc.* **XLIII.**
KENK, R. and SELIŠKAR, A. (1931) Études sur l'écologie de la faune cavernicole. I. Observations météorologiques et hydrologiques de la Podpeška jama, 1928–1931. *Prirod. Razpr.* **I.**
MEGUSAR, F. (1914) Ökologische Studien an Höhlentieren. *Carniola.*
TARMAN, K. (1961) Edafska Favna v Kraskih Jamah (Soil Fauna in Karst Caves). *Drugi Jugosl. Speleol. Kongres.* Zagreb.

Epigeous and Cavernicolous Life

COLAS, G. (1954) Note sur un *Oreonebria* cavernicole. *Notes biospéol.* **IX.**
CONDÉ, B. (1955) Matériaux pour une Monographie des Diploures Campodéides. *Mem. Mus. Hist. Nat. Paris N.S. Ser. A. Zool.* **XII.**

DEXTER, R.W. (1954) A record of the blind isopod, *Asellus tridentatus*, from an open stream habitat. *Amer. Midl. Nat.* **LII.**

DRESCO, ED. (1952) Répartition *d'Ischyropsalis luteipes* Simon (Opiliones, Ischyropsalidae). *Notes biospéol.* **VII.**

GODON, J. (1912) Les Crustacés Amphipodes des eaux douces de la région du Nord de la France. *Mém. Soc. Émulation. Cambrai.* **LXVII.**

JEANNEL, R. and LELEUP, N. (1952) L'Évolution souterraine dans la région Méditerranéenne et sur les montages du Kivu. *Notes biospéol.* **VII.**

KARAMAN, ST. (1950) Études sur les Amphipodes-Isopodes des Balkans. *Acad. Serbe. Sc. Monographies.* **CLXIII.** *Sect. Sc. Math. Nat. N.S.*

LEONARD, A.B. and PONDER, L.H. (1949) Crustacea of eastern Kansas. *Trans. Kansas Acad. Sc.* **LII.**

MINCKLEY, W.L. (1961) Occurrence of subterranean isopods in epigean waters. *Amer. Midl. Nat.* **LXVI.**

MÜLLER, G.W. (1914) Ist *Niphargus puteanus* ein typischer Höhlenbewohner? *Zool. Anz.* **XLIII.**

SCHÄFERNA, K. (1933) Über das Vorkommen von *Niphargus* im Teiche und in oberirdischen Lachen. *Verh. Intern. Ver. theor. angew. Limnologie* **VI.**

SCHELLENBERG, A. (1935) Der *Niphargus* des Thüringer Waldes und die Glazialrelikten-frage. *Archiv. f. Hydrobiol.* **XXIX.**

VANDEL, A. (1952) Isopodes terrestres (Troisième Série). *Biospeologica*, **LXXIII.** *Archiv. Zool. expér. géner.* **LXXXVIII.**

VANDEL, A. (1953) Isopodes terrestres récoltés dans les grottes de la Corse par le Professeur P.A. Remy, en 1942 et 1948. *Notes biospéol.* **VIII.**

VANDEL, A. (1955) La Faune isopodique cavernicole de l'Afrique du Nord (Berbérie). *Notes biospéol.* **X.**

VANDEL, A. (1960) Isopodes terrestres (Première Partie). *Faune de France.* No. 64. Paris.

Terrestrial Forms which are Amphibious

CHAPPUIS, P.A. (1927) Die Tierwelt der unterirdischen Gewässer. *Die Binnengewässer,* III. Stuttgart.

DELAMARE-DEBOUTTEVILLE CL. (1952) Données nouvelles sur la Biologie des Animaux cavernicoles. 4. Cheminement des Collemboles cavernicoles sous l'eau. *Notes biospéol.* **VII.**

DEROUET, L. (1953, 1956) Vie aérienne de quelques Crustacés aquatiques cavernicoles et épigés. *Premier Congr. Intern. Spéléol. Communic.* **III.** Paris.

DUDICH, E. (1932) Biologie der Aggteleker Tropfsteinhöhle "Baradla" in Ungarn. *Speläol. Monogr.* **XIII.** Wien.

FRANCISCOLO, M. (1951) La Fauna della "Arma Pollera", No. 24 Li, presso Finale Ligure. *Rassegna Speleol. ital.* **III.**

JEANNEL, R. and RACOVITZA, E.G. (1918) Énumeration des Grottes visitées, 1913–1917 (Sixième Serie). *Biospeologica,* **XXXIX.** *Archiv. Zool. expér. géner.* **LVII.**

PAVAN, M. (1940) La caverne della regione M. Palosso–M. Doppa (Brescia) e la loro fauna. *Suppl. Comment. Ateneo. Brescia.*

REMY, P. (1925) Contribution à l'étude de l'Appareil respiratoire et de la respiration chez quelques Invertébrés. Thèse, Nancy.

REMY, P. (1940) Sur le mode de vie des *Hadesia* dans la grotte Vjetrenica (Col. Bathys-ciinae). *Rev. franç. Entomol.* **VIII**

RIOJA, E. (1953) Estudios Carcinologicos, **XXIX.** Un nuevo Genero de Isopodo Triconiscido de la Cueva de Ojo de Agua Grande, Paraje Nuevo, Cordoba, Ver. *Anal. Instit. Biologia.* Mexico. **XXIII.**

UÉNO, S.I. (1955) Studies on the Japanese Trechinae (V) (Coleoptera, Harpalidae). *Mem. Coll. Sci. Univ. Kyoto Ser. B.* **XXII.**

VERHOEFF, K.W. (1908) Über Isopoden. *Androniscus.* 13. Aufsatz. *Zool. Anz.* **XXXIII.**

Aquatic Forms which are Amphibious

CARL, J. (1904) Materialien zur Höhlenfauna der Krim. I. *Zool. Anz.* **XXVIII.**
CHAPPUIS, P.A. (1920) Die Fauna der unterirdischen Gewässer der Umgebung von Basel. *Archiv. f. Hydrobiol.* **XIV.**
CHAPPUIS, P.A. (1933) Copépodes (Première Série). Avec l'énumeration de tous les Copépodes cavernicoles connus en 1931. *Biospeologica,* **LIX.** *Archiv. Zool. expér. géner.* **LXXVI.**
DEROUET-DRESCO, L. (1959) Contribution à l'étude de la Biologie de deux Crustacés aquatiques cavernicoles: *Caecosphaeroma burgundum* D. et *Niphargus orcinus virei* Ch. *Vie et Milieu* **X.**
VIRÉ, A. and ALZONA, C. (1902) Sur une nouvelle espèce de *Caecosphaeroma,* le *C. bericum. Bull. Mus. Hist. Nat. Paris* **VII.**

Conclusions

BOLE, J. (1961) Onekaterih Problemih Proucevanja subterane Malakofavne (Über einige Forschungsprobleme bezüglich der unterirdischen Molluskenfauna). *Drugi Jugosl. Speleol. Kongres.* Zagreb.
COMBAULT, A. (1909) Contribution à l'étude de la respiration et de la circulation des Lombriciens. *J. Anatomie Physiologie* **XLV.**
HNATENWYTSCH, B. (1929) Die Fauna der Erzgruben von Schneeberg im Erzgebirge. *Zool. Jahrb. Abt. System.* **LVI.**
SCHNEIDER, R. (1886) Amphibisches Leben in den Rhizomorphen bei Burgk. *Math. Naturw. Mitteil. Sitzb. k. Preuß. Akad. Wiss.* **VII.**

Niphargus foreli

SCHELLENBERG, A. (1934) Amphipoden aus Quellen, Seen und Höhlen. *Zool. Anz.* **CVI.**

Subterranean Biotopes

COIFFAIT, H. (1956) La biocénose cavernicole du versant nord des Pyrénées, *Comm. 1er Congr. Intern. Spéléol.* **III.** Paris.
JEANNEL, R. (1926) Faune cavernicole de la France. *Encyclopédie entomologique.* **VII.** Paris.
LERUTH, R. (1939) La Biologie du Domaine souterrain et la Faune cavernicole de la Belgique. *Mém. Mus. R. Hist. Nat. Belgique* No. 87.

The Parietal Association

CHOPARD, L. (1917) Note sur la biologie de *Dolichopoda palpata* Sulz. (Orth. Phasgonuridae). *Bull. Soc. Entomol. France.*
CHOPARD, L. (1959) Sur les mœurs d'un *Rhaphidophora* cavernicole. *Annal. Spéléol.* **XIV.**
CODINA, A. (1911) Sobre algunos Lepidopteros Heteroceros raros o curiosos de Cataluña. *Bol. Soc. Aragon. Cienc. Nat.* **X.**
DEAROLF, K. (1937) Notes on cave invertebrates. *Proc. Pennsylvania Acad. Sci.* **XI.**
DELEURANCE, S. (1959) Sur l'écologie et le cycle évolutif de *Choleva angustata* Fab. et *fagniezi* Jeann. (Col. Catopidae). *Annal. Spéléol.* **XIV.**
DEMAISON, L. (1907) Observations sur l'*Acrolepia granitella* Tr. *Bull. Soc. Entomol. France.*
DEMAISON, L. (1910) Observations sur quelques Lépidoptères des Pyrénées. *Bull. Soc. Entomol. France.*
DEMAISON, L. (1911) Lépidoptères des grottes de Catalogne. *Bull. Soc. Entomol. France.*
DEMAISON, L. (1915) Observations relatives aux *Acrolepia granitella, Cosmotriche potatoria* et *Callimorpha quadripunctaria* du littoral des Côtes du Nord. *Bull. Soc. Entomol. France.*
DOVER, C. (1929) Fauna of the Batu Caves, Selangor. XVI. Rhynchota. *J. Feder. Malay States Mus.* Singapore **XIV.**
GADEAU DE KERVILLE, H. (1905) Note sur la présence dans les cavernes du *Triphosa dubitata* L. et du *Scoliopteryx libatrix* L. *Bull. Soc. Entomol. France.*

300 BIOSPEOLOGY: THE BIOLOGY OF CAVERNICOLOUS ANIMALS

JEANNEL, R. (1912) Description d'un Réduviide (Hem. Heteroptera) troglophile nouveau de l'Afrique orientale. *Bull. Soc. Entomol. France.*
JEANNEL, R. (1919) Insectes Hemiptères. III. *Henicocephalidae* et *Reduviidae. In* Voyage de Ch. Alluaud et R. Jeannel en Afrique Orientale (1911–12). *Résultats scientifiques.* Paris.
KEMP, S. (1924) Rhynchota of the Siju Cave, Garo Hill, Assam. *Rec. Indian Mus.* Calcutta. **XXVI.**
LERUTH, R. (1939) La Biologie du Domaine souterrain et la Faune cavernicole de la Belgique. *Mém. Mus. R. Hist. Nat. Belgique* No. 87.
LHOSTE, J. (1939) Études biospéologiques. XI. Espèces africaines du genre *Bagauda* Bergroth (Hemiptera, Reduviidae). *Bull. Mus. R. Hist. Nat. Belgique.* **XV.**
MACATEE, W. L. and MALLOCH, J. R. (1926) Phillippine and Malayan Ploiariinae (Hemiptera). *Philippine J. Sci.* **XX.**
NICHOLAS, G. (1962) Nocturnal migration of *Hadenoecus subterraneus. Nat. Speleol. Soc. News.* **XX.**
PAIVA, C. A. (1919) Rhynchota from Garo Hills, Assam. *Rec. Indian Mus.* Calcutta. **XVI.**
ROUBAUD, E. (1933) Essai synthétique sur la vie du Moustique commun *(Culex pipiens).* L'évolution humaine et les adaptations biologiques du Moustique. *Annal. Sc. Nat. Zool.* (10) **XVI.**
TERCAFS, R. (1960) Notes préliminaires à propos de deux Trogloxènes réguliers des Cavernes de Belgique, *Scoliopteryx libatrix* L. et *Triphosa dubitata* L. *Annal. Féder. Spéleol. Belgique* **I.**
TOLLET, R. (1959) Contribution à l'étude des Diptères cavernicoles des grottes d'Italie et de Suisse, et description de deux *Mycetophilidae* nouveaux. *Bull. Annal. Soc. R. Entomol. Belgique* **XCV.**
VILLIERS, A. (1949a) Révision des Émesides africains. *Mem. Mus. Hist. Nat. Paris N.S.* **XXIII.**
VILLIERS, A. (1949b) Sur deux Emesines du Congo belge. *Rev. franç. Entomol.* **XVI.**
VILLIERS, A. (1953a) *Emesinae* cavernicoles du Congo belge (Hemiptera Reduviidae). *Rev. Zool. Bot. Afric.* **XLVII.**
VILLIERS, A. (1953b) Les Réduviides de Madagascar. IX–X. *Mém. Inst. Sci. Madagascar.* E. **III.**
VILLIERS, A. (1962) Les Réduviides de Madagascar. XX. *Physoderinae. Rev. franç. Entomol.* **XXIX.**

Guanobia

JEANNEL, R. (1943) *Les Fossiles vivants des cavernes.* Paris.
LELEUP, N. (1956) La Faune cavernicole du Congo belge et Considérations sur les Coléoptères reliques d'Afrique intertropicale. *Annal. Mus. R. Congo Belge. Sc. Zool.* **XLVI.**

Troglobia

GARMAN, H. (1892) The origin of the cave fauna of Kentucky, with a description of a new blind beetle. *Science* **XX.**
JEANNEL, R. (1926) Faune cavernicole de la France. *Encyclopédie Entomologique* **VII.** Paris.
LERUTH, R. (1939) La Biologie du domaine souterrain et la faune cavernicole de la Belgique. *Mém. Mus. R. Hist. Nat. Belgique* No. 87.
RACOVITZA, E. G. (1907) Essai sur les problèmes biospéologiques. *Biospeologica,* **I.** *Archiv. Zool. expér. géner.* (4) **VI.**

Aquatic Cavernicoles

BIRSTEIN, J. A. (1940) Über die Fauna der Höhlen-Amphipoden Abchasiens. *Biospeologica sovietica,* **III.** *Bull. Soc. Nat. Moscou Sect. Biol.* **XLIX.**

CHODOROWSKI, W. and A. (1960) Groupement de la Faune aquatique dans les cavernes des Montagnes des Tatras. *Biospeologica Polonica*, **V**. *Speleologia*. **II**.

CLAUS, C. (1893) Neue Beobachtungen über die Organisation und Entwicklung von *Cyclops*. Ein Beitrag zur Systematik der Cyclopiden. *Arbeit. Zool. Inst. Wien* **X**.

COUSIN, G. (1925) Un dispositif simple et démonstratif pour mettre en évidence le rhéotropisme des *Gammarus*. *Feuille des Naturalistes* **XLVI**.

GINET, R. (1953) Faune cavernicole du Jura méridional et des chaînes subalpines dauphinoises. I. Crustacés aquatiques. *Notes biospéol.* **VIII**.

GINET, R. (1960) Écologie, Éthologie et Biologie de *Niphargus* (Amphipodes Gammarides hypogés). *Annal. Spéleol.* **XV**.

KOFOID, C. A. (1900) The plankton of Echo River, Mammoth Cave. *Trans. Amer. Microsc. Soc.* **XXI**.

ORGHIDAN, TR., DIMITRESCU, M. and GEORGESCU, M. (1961) Sur le biotope hygropétrique de quelques grottes de Roumanie. *Die Höhle* **XII**.

PACKARD, A. S. (1886) The cave fauna of North America with remarks on the anatomy of the brain and origin of the blind species. *Mem. Acad. Washington* **IV**.

SCHMEIL, O. (1894) Zur Höhlenfauna des Karstes. *Zeit. f. Naturw.* **LXVI**.

SCOTT, W. (1909) An ecological study of the plankton of Shawnee Cave, with notes on the cave environment. *Biol. Bull.* **XVII**.

TELLKAMPF, T. A. (1845) Memoirs on the blind-fishes and some other animals living in Mammoth Cave in Kentucky. *New York J. Med.*

Interstitial Medium

CHAPPUIS, P. A. (1920) Die Fauna der unterirdischen Gewässer der Umgebung von Basel. *Archiv. f. Hydrobiol.* **XIV**.

CHAPPUIS, P. A. (1936) Subterrane Harpacticiden aus Jugoslawien. *Bull. Soc. Sci. Cluj.* **VIII**.

CHAPPUIS, P. A. (1942) Eine neue Methode zur Untersuchung der Grundwasserfauna. *Acta Sci. Math. Natur. Koloszvar.* **VI**.

CHILTON, CH. (1894) The subterranean Crustacea of New Zealand, with some remarks on the fauna of caves and wells. *Trans. Linn. Soc. London.* (2) **VI**. *Zool.*

DELAMARE-DEBOUTTEVILLE, CL. (1954a) Description d'un appareil pour la capture de la faune des eaux souterraines littorales sous la mer. *Compt. rend. Acad. Sci. Paris* **CCXXXVIII**.

DELAMARE-DEBOUTTEVILLE, CL. (1954b) Description d'un appareil pour la capture de la faune des eaux souterraines littorales sous la mer. Premiers résultats. *Vie et Milieu* **IV**.

DELAMARE-DEBOUTTEVILLE, CL. (1960) Biologie des eaux souterraines littorales et continentales. *Actual. sc. industr.* No. 1280. Paris.

HERTZOG, L. (1933) *Bogidiella albertimagni* sp. nov., ein neuer Grundwasseramphipode aus der Rheinebene bei Straßburg. *Zool. Anz.* **CII**.

HERTZOG, L. (1936) Crustaceen aus unterirdischen Biotopen des Rheintales bei Straßburg. I. Mitteilung. *Zool. Anz.* **CXIV**.

HERTZOG, L. (1936) Crustacés de biotopes hypogés de la vallée du Rhin d'Alsace. *Bull. Soc. Zool. France* **LXI**.

JAVOROVSKY, A. (1893) Fauna studzienna miasta Krakowa i Lwowa napisat. *Spravozd. Fizy.* Krakau.

MEŠTROV, M. (1960) Faunističko-ekološka i biocenološka istrazivanja podžemnih voda savske nizine. *Biološki Glasnik*. **XIII**.

MONIEZ, R. (1889) Faune des eaux souterraines du département du Nord, et, en particulier de la ville de Lille. *Rev. biol. Nord. France* **I**.

MOTAS, C. (1962) Procédés des sondages phréatiques. Division du domaine souterrain. Classification écologique des animaux souterrains. Le Psammon. *Acta. Mus. Macedon. Sci. Nat.* **VIII**.

MOTAS, C. and TANASACHI, J. (1946) Acariens phréaticoles de Transylvanie. *Notat. Biol* Bucuresti. **IV**.

NICHOLLS, A. G. (1946) Syncarida in relation to the interstitial habitat. *Nature, Lond.* **CLVIII.**

NOLL, W. and STAMMER, H. J. (1953) Die Grundwasserfauna des Untermaingebietes von Hanau bis Würzburg, mit Einschluß des Spessarts. *Mitt. Naturw. Mus. Aschaffenburg* *N.F.* **VI.**

ORGHIDAN, TR. (1959) Ein neuer Lebensraum des unterirdischen Wassers: der hyporheische Biotop. *Archiv. f. Hydrobiol.* **LV.**

PRENANT, M. (1932) L'analyse mécanique des sables littoraux et leurs qualités écologiques. *Archiv. Zool. expér. géner.* **LXXIV.**

SASSUCHIN, D. N., KABANOV, N. M. and NEIZWESTNOVA, K. S. (1927) Über mikroskopische Pflanzen und Tierwelt der Sandfläche der Okaufer bei Murow. *Russ. Hydrobiol. Zeit.* Saratov. **VI.**

SASSUCHIN, D. N. (1930) Materialien zur Frage über die Organismen des Flugsandes in der Kirgisensteppe. *Russ. Hydrobiol. Zeit.* Saratov. **IX.**

SCHNITTER, H. and CHAPPUIS, P. A. (1915) *Parastenocaris fontinalis* n. sp., ein neuer Süßwasserharpacticide. Zugleich ein Beitrag zur Kenntnis der Gattung *Parastenocaris*. *Zool. Anz.* **XLV.**

VEJDOVSKY, FR. (1882) *Thierische Organismen der Brunnenwässer von Prag.* Prag.

CHAPTER XVIII

PHYSICAL, CHEMICAL AND CLIMATIC FACTORS AND THEIR ACTION ON THE PHYSIOLOGY OF CAVERNICOLES

INTRODUCTION

The preceding chapter dealt with the distribution of cavernicoles in the subterranean world. The observations reported were made in the natural environment. To complete the investigations it is necessary to analyse the different external factors which affect the distribution, behaviour and physiology of the cavernicoles. This study cannot be carried out in the natural medium because the constant coexistence of different factors makes it impossible to isolate and clearly establish the influence of any one of them. Only experiments carried out in the laboratory can resolve these problems. However, there are still many gaps in our knowledge of this subject.

The experimenter must realise from the outset, that the same factor does not have an identical action on troglobia, troglophiles and trogloxenes. The mistake has too often been made of attributing to a factor whose action is very important with regard to a highly specialised troglobian, an analogous influence on the ancestors of this troglobian at the stage in evolution when they penetrated the subterranean environment.

In the last part of this book we shall show that subterranean evolution represents above all a specialisation which links the cavernicole more and more closely to the particular medium in which it lives. This evolution is the rule for ancient groups which have reached the last stage of their history. All true troglobia are relicts of a fauna which has very often disappeared from the surface of the Earth.

Four groups of factors which may affect cavernicoles can be distinguished.

Two of these influence all cavernicoles. They are:

A. Light, or more correctly the absence of light.
B. Temperature.

The other two factors are:

C. The atmosphere, which influences only terrestrial cavernicoles.
D. Water, in which the aquatic cavernicoles live.

A. LIGHT

(a) Light in the Subterranean World

Although the sun is the principal source of light on our planet, there are others which are not without significance. Certain rays of light are of biological origin. Many living creatures are capable of emitting light. These are the bioluminescent organisms.

The distribution of these organisms varies according to the medium in which they live. The immense majority are marine forms. These are the bacteria; Protista such as *Noctiluca;* medusae; alcyonaria; pennatulids; ctenophores; nemertines; polychaetes; molluscs and specially the cephalopods; copepods; ostracods; mysids; decapods; Enteropneusta; *Pyrosoma,* and fish (Harvey, 1952). Due to the presence of these animals the oceans are sometimes illuminated. However, the bioluminescent animals are particularly well represented in the abyssal fauna. Thus, even at a great depth these animals spread some light through marine waters.

There are far fewer freshwater and terrestrial bioluminescent forms. They are still more rare in cavernicoles. Thus, although there is a certain amount of light in the depths of the sea, the subterranean medium is, with rare exceptions, always dark. For this reason there are notable differences between abyssal organisms and cavernicolous forms with respect to pigmentation and the structure of the eye. We shall return to this point later.

Three examples of bioluminescence in the subterranean cavities can be given:

(1) W. G. Smith (1871) reported the presence of *Polyporus annosus* in the timbering of a coal mine at Cardiff, which produced a shining light.

(2) A glow worm, *Lychnocrepis antricola* has been observed in the Batu Caves (Malaya), at a distance of 270 metres from the entrance (Blair, 1929).

(3) The most remarkable case is that of the New Zealand "glow worm". This creature belongs to the Mycetophilidae, Diptera. The larvae of these flies feed on the spores of fungi. The scientific name of the "glow-worm" is *Arachnocampa luminosa* (Harrison 1961). Many publications have been devoted to this bioluminescent fly. These have recently been reviewed by A. M. Richards (1960).

The larvae are strongly luminous. The pupae and females emit less light, and the males are rarely luminous. Their light disappears and reappears (Gatenby and Ganguly, 1958). A loud noise will cause them to extinguish their light. The light emitted is greenish-blue in colour. The luminous apparatus belongs to a unique type which has not been found in any other animal. The light appears at the level of four spots situated on the eleventh abdominal segment. These four spots correspond to four terminal swellings of the malpighian tubules (Wheeler and Williams, 1915).

PLATE VI. The Waimoto Cave lit by "glow-worms" (after Goldschmidt).

The larvae of *Arachnocampa luminosa* are found by the millions in certain caves in New Zealand. They suspend themselves from the roofs of these caves and spread a very noticeable light. Because of the rich population of "glow-worms" in the Waimoto Cave in the Northern Island, some 200 miles north of Wellington, it has become a tourist attraction (Plate VI). "Glow-worms" are, however, known in other caves in New Zealand, and also in caves in Tasmania and Australia (New South Wales). This species is not a true cavernicole, and is at most a troglophile. They are also found in humid ravines under the fronds of tree-ferns.

The larvae build a cylindrical sac from mucus and silk, in which they enclose themselves. Numerous filaments bearing sticky globules hang from this sac (Plate VII).

An interesting condition is the carnivorous habit of these larvae. It is unique among the Mycetophilidae which usually eat fungi. In caves the prey of these larvae are chironomids belonging to the species *Anatopynia debilis*. These are attracted by the light from the posterior part of the body and are captured by the sticky filaments.

Thus the presence of glow-worms in subterranean cavities depends on a series of specialised characters which are interrelated to each other and of which it is difficult to explain the origin (Goldschmidt, 1948).

(b) Darkness

Although the facts given above are very interesting, they are very exceptional and one can state that the subterranean world is a totally dark medium. This is one of the most constant factors in the subterranean world.

This total darkness has results which are extremely important in relation to the physiology of cavernicoles.

1. The total darkness results in the absence of green plants in the subterranean habitat, with the exception of few rare Cyanophycae Rhodophyceae and Chlorophycae (pp. 57 and 58). Thus, one of the most important food sources available to surface animals is inaccessible to cavernicoles. Thus all strictly phytophagous animals are excluded from the subterranean world.

Green plants provide animals, not only with a source of energy, but also with vitamins, growth factors and other compounds which the Metazoa cannot synthesise. Cavernicoles must therefore exploit other sources of food in order to obtain them.

2. The absence of green plants brings about changes in the composition of the atmosphere: an increase in carbon dioxide and a decrease in oxygen.

Stimulation of the endocrine glands by light — and especially of the pituitary in the vertebrates — cannot occur in the subterranean world. The endocrine function of cavernicoles is therefore different in this respect from that of epigeous forms. We shall return to this problem in Chapter XXI.

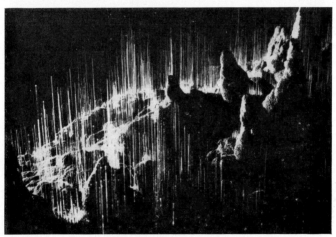

PLATE VII. The glutinous and luminous filaments secreted by the "glow-worms" (after Goldschmidt).

(c) Action of Light upon Cavernicoles

There is no doubt that light stimulates epigeous forms. However, as is the rule in biology, all types of stimulation become harmful above a critical level. The pigment which develops in the skin of most epigeous animals represents a screen which moderates the action of light. When light intensity exceeds a critical threshold the stimulatory effect disappears, and light becomes a lethal factor which induces necrosis. The ultra-violet part of the spectrum is the most harmful to organisms. These conclusions have been reached as a result of experiments on many different animals (Merker, 1925; Merker and Gilbert, 1932).

It is generally found that pigment is absent from the skin of cavernicoles (see Chapter XXV). It is thus to be expected that these animals would be particularly sensitive to the lethal action of light.

Epigeous planarians are rapidly killed by intense light (Merker, 1925) and more rapidly by ultra-violet light (Merker and Gilbert, 1932). Under the influence of light, planarians become necrotic in the same manner that they do under the effect of a toxic substance. Cavernicolous planarians are much more sensitive to light. We had the greatest difficulty in obtaining some pictures of cavernicolous planarians on film. The pictures shown in the film "Faune cavernicole" from the cave laboratory at Moulis, represent individuals which are already in a state of deterioration and are on the point of death.

The lethal action of light has also been demonstrated on cavernicolous Collembola and Acarina (Absolon, 1900) as well as on subterranean ostracods and copepods (Maguire, 1960).

However, one cannot generalise these conclusions and apply them to all cavernicoles. Higher Crustacea which possess a carapace, and vertebrates, are not killed by daylight and can be kept for a long time in normally illuminated aquaria. This has been shown for the troglophile crayfish, *Procambarus simulans* (Maguire, 1960); the thermosbenacean, *Monodella argenarii* (Stella, 1951); amphipods of the genus *Niphargus* (Gal, 1903; Ginet, 1960); the blind fishes: *Tiphlichthys subterraneus* (Verrier, 1929); *Anoptichthys jordani* (Breder and Gresser, 1941); *Caecobarbus geersti* (Petit and Besnard, 1937); and finally the amphibian, *Proteus*. In addition, the bathysciinae (Coleoptera) bear the action of light very well (S. Deleurance)

B. TEMPERATURE

(a) Temperature of Caves

The temperature of air and water. In speaking of the temperature of a cave it is necessary to describe the nature of the medium in which measurements are made, for example, in air, in water, or in the surrounding rocks.

When the water in a cave is of endogeous origin the temperature differences between air and water are very slight. The results of Polli (1958, 1962) and Renault (1961) both show that the temperature of water is on average 0·1 °C lower that that of air.

When a subterranean river arises from a surface river the temperature of the water undergoes extensive variation, corresponding to seasonal variation of the external temperature. For example:

Cave	Temperature variation	Author
The river Reka (Grotto of San Canzian)	0–27 °C	after Maritinitsch, reported by Martel, 1894
Timavo (outlet of the Reka)	4·6–17·8 °C	Boegan, 1921
Source of the River Orbe (a resurgent from the Lac de Joux, and Lac des Brenets)	3·5–14·7 °C	after Forel, reported by Martel, 1894

Absolute Temperature. It is generally stated that the temperature of a subterranean cavity corresponds to the mean temperature of the exterior. In fact, this is only a first approximation, as the temperature of a cave depends on numerous factors. The four principal ones are as follows:

1. The region. In France, the caves of the Jura and the Causses are on average 2 °C colder than those of the Alps and Pyrenees. In the Pyrenees, the caves on the Spanish side are 1–3 °C warmer than those on the French side (Jeannel, 1926).

In temperate Europe, the temperature of caves is normally between 8 and 12 °C (8–10 °C in England) and in southern Europe it is between 13 and 15 °C.

In the Ryugado Cave, hollowed from the island of Shikoku (southern Japan), the temperature is 15–16 °C (Torii, 1953).

Tropical caves are notably warmer. The temperature of caves in the Lower Congo, in the region of Thysville, is between 18 and 24 °C (mean temperature, 22 °C) (Heuts and Leleup, 1954). The caves of East Africa have still higher temperatures, 24–26 °C (Jeannel, 1926). The temperature of caves in Guinea is between 24 and 28 °C (Hiernaux and Villiers, 1955). Finally the Siju Caves, in Assam, have a temperature of 21–26·4 °C (Stadler, 1927).

2. Altitude. Caves become colder with increasing altitude. According to Jeannel (1926) the temperature decreases 2–3 °C for an increase of 100 metres. Numerous irregularities due to local causes (orientation, wooded or unwooded zones) impair the regularity of this relationship.

3. The Form of Caves. Descending caves or swallow-holes are cold caves because during winter they accumulate cold air which has a tendency to descend because of its greater density. The ice-holes or the snow-holes always belong to this type. Their temperature fluctuates around 0°C.

Ascending caves are warm and accumulate warm air during the summer.

4. Depth. It is rare to find a regular increase in geothermal temperature with depth in caves hollowed from calcareous rocks. This is because rocks, fissures and passages transport air from the exterior. However, the observations of Renault (1961) carried out in the cave at Moulis, show that the temperature at the entrance is lower by 1·0–1·5°C than that of the deepest part where the rock covering measures 400 metres in thickness.

Analogous conditions are found in other caves. The following figures from the Postojnska Jama reported by Martel (1894) are: 12·9°C at 400 metres from the entrance; 13·1°C at 600 metres; and 14°C at 2000 metres.
Temperature Variations. It is generally stated that the temperature of caves is constant, but this conclusion is not based on accurate information.

Caves which have several openings and are thus permanently transversed by currents of air from the exterior, must be considered apart. In this case, seasonal variations outside have a direct effect in the cave (Davies, 1960). Many caves in the Cevennes are of this type and annual variations of 4–5°C between summer and winter have been recorded (Jeannel, 1926). These variations even reach 12°C in the Grotte du Soldat (Balazuc, 1956).

Such deviations are exceptional; more generally seasonal temperature variations are much lower than at the exterior. However, it is only rarely that these variations are completely cancelled out.

The most accurate information available is that of Polli (1958), and of Ginet (1960). Polli studied the seasonal temperature range in the Grotta Gigente, near Trieste, during a period of five years. In summary the results are as follows:

Maximum difference

in air temperature	0·8°C
in water temperature	1·3°C
at 43 cm in the rock	0·7°C
at 117 cm in the rock	0·5°C

The observations of Ginet were made in three French caves:

Grotte de Corveissiat (Ain), amplitude of variations	2·2°C
Grotte de la Balme (Isère), amplitude of variations	1·2°C
Grotte de Moulis (Ariège), amplitude of variations	0·7°C

Temperature variations are particularly small in caves in tropical regions, because climatic variations are less than those of temperate or continental climates. According to Heuts and Leleup (1954), the seasonal temperature variations of the caves in the Thysville region (Lower Congo) never exceed 1°C.

(b) Temperature and the Cavernicoles

Absolute Temperature. From a review of the temperatures of regions populated by cavernicoles it can be stated there is considerable variation in the extreme values.

Lower Limits. A few cavernicoles can develop and exist at temperatures near to freezing point.

The Trou du Glaz, a huge subterranean system in the Dent de Crolles in the Alpes Dauphinoises is a true ice cave. It opens at an altitude of 1675 metres. The entrance still contains snow at the end of August and stalagmites and stalactites of ice form up to 100 metres from the entrance. The temperature of the caves in August is 1–2 °C, and 0 °C in winter. In these conditions, the life cycle of the Bathysciine beetle, *Isereus xambeui*, takes place (Fagniez, 1907).

The trechine beetle, *Arctaphaenops angulipennis*, found in the caves of the Dachstein Alps (Austrian Alps) has an analogous way of life (Vornatscher, 1961).

The Ghetarul de Scărisoara is an ice cave situated in Mont Bihar, in Transylvania. This cavity which has been studied by Racovitza (1927) is formed by an abyss at the bottom of which snow accumulates and is transformed to ice. In June the temperature is 0·1–0·7 °C; in August to October the temperature increases to 0·8 °C. It is probable that the temperature falls below 0 °C in winter but the cave is inaccessible at this time of the year. *Pholeuon glaciale* (Bathysciinae), lives in the deepest chamber of this cave. Its life cycle takes place between 0·1 and 0·8 °C (Jeannel, 1943).

Upper Limits: A few organisms which can be considered to be true cavernicoles have been captured in thermal springs at very high temperatures. "*Niphargus*" *thermalis* comes from a hot spring in Hungary where the temperature is 30 °C. *Thermosbaena mirabilis* lives in the warm springs of El Hamma (Tunisia), where the temperature lies between 45 and 48 °C (Monod, 1940). Finally, a syncarid, *Thermobathynella adami*, has been collected from the Kaziba resurgent in Katanga, where the waters have a temperature of 55 °C (Capart, 1951).

Are Cavernicoles Stenothermal or Eurythermal? Although the figures given in the preceding paragraph are remarkable, ecologists are more interested in the limits between which cavernicoles can live. That is, to know whether cavernicoles are stenothermal or eurythermal.

The question is easily answered in the case of the interstitial medium. This medium is characterised by often very extensive thermal variations. It is not surprising that the representatives of the phreatobious fauna are more or less eurythermal.

On the other hand, the temperature of caves although less stable than it was formerly believed to be, is still relatively constant. From this fact it has been concluded that cavernicoles are stenothermal. This idea was substantiated by the qualification, so often given to cavernicoles, that of "glacial relict". This is, however, based only on observations and is not confirmed by experimentation. The breeding of cavernicoles under experimental conditions has shown that they merit the designation "stenothermal", only approximately.

P. de Beauchamp (1932) wrote, concerning cavernicolous planarians, of which he raised many batches in the laboratory: "Strict stenothermia in the adults does not explain the causes of the distribution of these animals. In nearly all the species which I have kept, some specimens existed many months and others many years at temperatures of 18–20 °C, and resisted for a day or two 25–28 °C, provided that the rise in temperature was not too fast or the medium too restricted." Later he writes, "Cold-water planarians and obscuricolous planarians are not the same thing, as may be shown by the well-known species *Crenobia alpina* of which the biology has a few superficial similarities with cavernicolous species but which is basically considerably different."

Just after the beginning of this century, Gal (1903) showed that amphipods of the genus *Niphargus* could resist temperatures of 25 °C. More striking experiments were carried out by Ginet (1960). He wrote, "Although the optimum temperature for *Niphargus* varies between 8 and 14 °C experimental results show that the young can develop between 4–6 °C and 16–18 °C and that the adults survive temperatures between the lethal limits of -0.5 and 24·5 °C." These limits are close to those reported for epigeous species of *Gammarus*. However, *Niphargus* are less capable of adapting their metabolism to higher temperatures than *Gammarus*. These experiments establish the error of the interpretation, at first proposed by Thienemann (1908), that *Niphargus* species were glacial relicts. Schellenberg (1935) has put forward the arguments against this hypothesis.

With respect to terrestrial cavernicoles, the observations of Fagniez (1907) are particularly interesting. He found that the bathysciine, *Isereus xambeui*, lives in the Grotte du Guiers Vif, at 1150 metres altitude at a temperature of 5–6 °C; in the Grotte du Guiers Mort, at 1305 metres altitude at a temperature of 2·5–3 °C; in the Trou du Glaz at an altitude of 1675 metres and a temperature of 2–0 °C.

More detailed experiments were carried out by Sylvie Glaçon (1935) on another bathysciine, *Speonomus diecki*. This biospeologist experimentally established the limits of activity, and the upper and lower lethal temperature limits of this bathysciine. The first are between 0° and 20 °C and the second between -5 and $+25$ °C.

The optimum temperature for the troglophilic spider, *Meta menardi*, is between 6 and 8 °C. Escape reactions are induced by temperatures between

PLATE VIII. Measurements of the tensions of oxygen and carbon dioxide in the air at the bottom of the swallow-hole of Sainte Catherine at Balaguères (Ariège) (Photo by Bouillon).

28° and 29·5 °C (25° and 29·5 °C for a cavernicolous form from the Carpathians) (Szymczkowsky, 1953).

Conclusions. From these experiments it can be concluded that stenothermia in cavernicoles is relative. It is certainly much less marked than the stenothermia of organisms from cold springs.

It is probable that temperature played some part in the penetration of cavernicoles into the subterranean world. It is, however, certain that the temperature factor was not the only factor which has affected subterranean evolution.

C. THE ATMOSPHERE

This factor obviously influences only terrestrial cavernicoles.

(a) Chemical Composition

When the air of subterranean cavities can be regularly replenished its composition is the same as that of the external air.† However, this is not the general case. The preceding statement does not apply to galleries which end in a cul de sac and which open by a narrow hole, or to completely obstructed shafts (Plate VIII).

An increase in the CO_2 level of subterranean cavities is a general phenomenon. During the precipitation of limestone and the formation of concretions, carbon dioxide is liberated, and spreads through the subterranean atmosphere. Also, the carbon dioxide in the vegetable layer can penetrate to subterranean cavities by cracks in the soil.

In this manner, the carbon dioxide accumulates at the bottom of shafts and in the lower passages because of its higher density. This is why the atmosphere of caves, at least in some places, is richer in carbon dioxide than the external air. The CO_2 tension can reach and even exceed 10%. At this level it will put out a candle flame while the speleologist would begin to feel the first symptoms of uneasiness.

At the same time there is a decrease in oxygen content (Derouet and Dresco, 1955). Unfortunately no details are available on this point. A study of the composition of the air of caves and regional and seasonal variations would be of great interest to biospeologists.

Influence of the Composition of Air on Cavernicoles

The only detailed information on this subject is that provided by Derouet and Dresco (1955) on spiders. These biospeologists demonstrated the presence in the Grotte de Pèneblanque above Arbas (Haute-Garonne),

---

† Caves hollowed from volcanic areas where the air is polluted with gas are not included here.

of recesses containing carbon dioxide (as much as 5·4%) and a low oxygen level. However, these areas were populated by the troglophilic spiders: *Meta merianae*, *M. menardi* and *Tegenaria inermis*.

The experiments conducted by Louise Derouet (1956) on troglophilic spiders of the genus *Meta*, proved that these spiders can endure without any discomfort atmospheres containing 15% CO_2. An hour in an atmosphere containing 15 to 25% CO_2 caused anaesthetisation which ceased when the animal was placed in a normal atmosphere.

From these experiments it can be concluded, at least for the spiders, that cavernicoles can endure relatively high CO_2 tensions.

(b) Air Movements

A very large number of caves communicate with the exterior by one or more openings at the valley level, but are also connected with the surface of the calcareous masses in which they are hollowed, by ascending passages and galleries. If the temperature and pressure are the same in the cave and exterior the atmosphere is stagnant. However, this is an exceptional case. Generally, the temperatures of the interior and exterior are different. In summer the air in the cave is colder than that in the external atmosphere; it is thus more dense and tends to descend; cold air passes out through a lower opening. In winter this sequence is reversed. The air in the cave is warmer than that outside and, being lighter it moves upwards. During this stage air enters into the cave by lower openings and warm air passes out at higher levels.

When the openings which connect subterranean cavities with the exterior are very small the current of air is very strong. These have been called "blow holes" (remembering, of course, that they reverse during the course of the year). The presence of blow holes is one of the best indications to speologists of the existence of large subterranean cavities. The phenomena described above very briefly account for the fact that most caves are traversed, at least during certain times of the year, by air currents, sometimes weak, but occasionally very strong.

Air Movements and Cavernicoles: Movement of the air is of great importance to cavernicoles. A cave which is traversed by a permanent air current is usually azoic. However, the small isolated narrow chambers frequently have denser populations than the main passages. These observations, confirmed by biospeologists on many occasions, establish that movement of air is an unfavourable factor for cavernicolous life. This factor probably does not influence cavernicoles by its dynamic effect but by the drying which it induces. Terrestrial cavernicoles are extremely sensitive to this as will be shown in a following paragraph.

(c) Humidity

Humidity is one of the most characteristic features of the subterranean medium. Unfortunately, apparatus which is sensitive to small variations in humidity in the region of saturation and supersaturation are not at present available. Moreover, the most sensitive instruments are difficult to transport, and cannot therefore be used in caves.

A study of humidity has most often been carried out only by examining the behaviour of cavernicoles to humidity variations, as these animals provide a sensitivity greater than that of our instruments.

In general, the water-vapour tension of caves approaches saturation. The relative humidity is usually between 95 and 100% (Trombe, 1952).

In temperate countries the level of humidity of subterranean cavities is very different from that of the external medium. However, the difference is less in tropical regions. In the Congo the exterior atmosphere is always very humid; the relative humidity never decreases below 90% and can exceed 93% (Heuts and Leleup, 1954). Thus the atmosphere of caves does not differ very much from that of the exterior. This is probably the reason for the absence of true terrestrial cavernicoles in equatorial regions (Ridley, 1899).

The Action of Humidity on Cavernicoles: The level of humidity is the most important environmental factor in the life of cavernicoles, as has been stated for a long time. The importance of this factor was first recognised by Bedel and Simon (1875) and has been emphasised by Peyerimhoff (1906), Jeannel (1926, 1943) and Fage (1931).

A dry cave is an azoic cave. However, the bathysciinae cannot survive in an atmosphere of saturated or supersaturated humidity (S. Delerance). In all cases even the least amount of drying kills the troglobia. Drying is the factor to which they are most sensitive. This is the reason why Jeannel (1926) qualified them as "stenohygrobia".

It is certain that the humidity factor played an important role in the population of caves. Candidates for cavernicolous life were not only lucifugous but above all hygrophilic. For example, all Collembola, with the exception of tracheate forms, can live only in an atmosphere which possesses a saturated humidity (Davies, 1929).

All animals lose water by evaporation and transpiration. In a general way it may be stated that, among arthropods, the hygrophilic species lose more water than do xerophilic species under the same conditions (Palmen and Suomalainen, 1945). Very little is known of water balance in terrestrial cavernicoles. However, we know that a troglophilic spider *Meta menardi* is more sensitive to drying than the lucicolous spider *Araneus diadematus* (Derouet, 1955, 1960).

It would be interesting to know the reasons for the faster rate of water loss from hygrophiles, and in particular cavernicoles, than from xerophiles. It is known that:

(1) The cuticle of arthropods consists of three layers: an inner layer, the endocuticle; an external layer, the exocuticle, which is covered by a very thin layer, the epicuticle. The first layer is colourless and the second layer contains pigment in the superficial zone. The epicuticle comprises a waxy layer which prevents, or reduces, the evaporation of water from the body. Removal of this layer by abrasion, dissolution or melting makes the cuticle of the arthropod permeable to water. It is the monomolecular waxy layer which makes the integument of terrestrial arthropods impermeable to water. Thus transpiration is usually very slow in these invertebrates.

(2) The integuments of cavernicoles are generally thinner than those of related epigeous species. It would be useful to know whether this involves a decrease in the thickness, or even the total disappearance, of the epicuticular layer. If this should be so a cavernicole would not be protected from water loss in a non-saturated medium.

This hypothesis has recently been confirmed at least in part by the research of L. M. Semenova (1961) on the integument of myriapods. Although these animals are not cavernicoles the results are of great interest to biospeologists.

A normally structured cuticle is found in most chilopods (Lithobiomorpha, Scolopendromorpha, and Scutigeromorpha). On the other hand, endogeous myriapods such as the Symphyla and the Chilopods of the order Geophilomorpha, show a very definite simplification of these structures. The thickness of the exocuticle is reduced and the pigmented layer has disappeared. The yellow colour of endogeous myriapods is due to the normal colour of chitin. Endogeous myriapods are also without an epicuticle. As a result their integument becomes permeable to water. This is the reason why they can live only in an atmosphere of saturated humidity.

These observations concerning endogeous arthropods leads one to consider that analogous conditions are to be found in cavernicolous arthropods. If this is the case it accounts for the depigmentation of subterranean forms and their need to live in an atmosphere of saturated humidity.

(d) Ionisation and Conductivity of the Air

Measurements of ionisation and conductivity of the air have been carried out in the subterranean world by many physicists, in particular by Trombe and Henry de La Blanchetais (1947). The following are their conclusions:

(1) The air of caves can be a hundred times more conductive than air at the exterior.

(2) The radioactivity proceeds from the clay.

(3) The radioactive rays responsible for the high ionisation in caves, generally do not contain gamma-rays. There is one recorded exception, that of the Grotte de Saint Paul (Haute-Garonne) where significant gamma

radiation has been detected, probably in material brought in by deeply penetrating waters.

Atanasiu (1959) discovered the presence of radiocarbon C^{14} in stalactites and stalagmites, and also in the air of the caves.

No action of these factors on cavernicoles has been discovered up until now.

D. WATER

Water mainly concerns aquatic cavernicoles although it does have a certain influence on terrestrial cavernicoles.

A study of subterranean water involves an examination of its chemical composition and of its speed of flow (running or stagnant water). In fact, the later problem has already been considered in the preceding chapter (p. 294).

The chemical composition of subterranean water has now to be considered.

Oxygen Tension

The oxygen tension of subterranean waters is variable but on the whole it is comparable with that of surface waters (Chodorowski, 1959; Ginet, 1960). It is lower than that of mountain torrents but is higher than that in pools and in small areas of superficial water.

Inorganic Ions

It is known that pure water dissolves chlorides, sulphates and alkaline carbonates. When it contains carbon dioxide it dissolves calcium carbonate and decomposes silicates.

The calcium level of subterranean waters varies but as most caves are hollowed from calcareous rock formations, the level of calcium carbonate in subterranean waters is generally high. These waters are said to be "hard" compared with waters containing little calcium which are called "soft". The proportion of calcium in water is expressed by its hydrometric degree. Unfortunately the units of measurement vary according to the country. In France the degree corresponds to 0·0057 grams of CaO per litre.

The calcium level of water can also be found by pH measurements. With several rare exceptions subterranean waters are alkaline and their pH values are equal to 7 or higher. According to Ginet (1960) the pH of subterranean waters of France measures between 6·9 and 8·0.

There is very little information on the variations in composition of water and the calcium level according to the seasons. In Europe, there would appear to be only slight variations (Chodorowski, 1959). However, such variations are most marked in tropical and equatorial regions. The observations of Heuts and Leleup (1954) have shown that the calcium level of water in caves in the Congo is much higher during the dry season than that during the rainy season.

The Action of the Composition of Water on Cavernicoles

It would seem that the slight variations in composition of underground waters found in temperate countries can have little effect on aquatic cavernicoles. At least there is no information which shows that the composition of the water influences the distribution of aquatic cavernicoles.

BIBLIOGRAPHY

Light

BLAIR, K.G. (1929) Fauna of the Batu Caves, Selangor. XVII. Coleoptera. *J. Feder. Malay. States Mus.* Singapore **XIV.**

GOLDSCHMIDT, R.B. (1948) Glow-worms and evolution. *Rev. Sc.* Paris. **LXXXVI.**

GATENBY, J.B. and GANGULY, G. (195 8) On a possible explanation of the sudden dousing of the light by New Zealand glow-worm *(Arachnocampa luminosa)*. *J. R. Microsc. Soc. London* **LXXVI.**

HARRISON, R. A. (1961) Notes on the taxonomy of the New Zealand Glow-worm *Arachnocampa luminosa* (Skuse) (Dipt. Mycetophilidae). *Trans. R. Soc. New Zealand Zool.* **I.**

HARVEY, E.N. (1952) *Bioluminescence* New York.

RICHARDS, A.M. (1960) Observations on the New Zealand glow-worm, *Arachnocampa luminosa* (Skuse) 1890 *Trans. R. Soc. New Zealand* **LXXXVIII.**

SMITH, W.G. (1871) Luminous fungi. *J. Bot.* London **IX.**

WHEELER, W.M. and WILLIAMS, F.X. (1915) The luminous organ of the New Zealand glow-worm. *Psyche*, Cambridge, Mass. **XXII.**

Action of Light upon Cavernicoles

ABSOLON, K. (1900) Einige Bemerkungen über mährische Höhlenfauna (I. Aufsatz). *Zool. Anz.* **XXIII.**

BREDER, C.M. JR. and GRESSER, E.B. (1941) Further studies on the light sensitivity and behaviour of the Mexican blind characin. *Zoologica*, New York **XXVI.**

GAL, J. (1903) *Niphargus* et *Caecosphaeroma*. Observations physiologiques. *Bull. Soc. Études Sc. Nat. Nîmes* **XXXI.**

GINET, R. (1960) Écologie, Éthologie et Biologie de *Niphargus* (Amphipodes, Gammarides hypogés). *Annales de Spéléologie* **XV.**

MAGUIRE, B. JR. (1960) Lethal effect of visible light on cavernicolous ostracods. *Science* **CXXXII.**

MERKER, E. (1925) Die Empfindlichkeit feuchthäutiger Tiere im Lichte. *Zool. Jahrb. Abt. allg. Zool. Physiol.* **XLII.**

MERKER, E. and GILBERT, H. (1932) Die Widerstandsfähigkeit von Süßwasserplanarien in ultraviolettreichem Licht. *Zool. Jahrb. Abt. allg. Zool. Physiologie* **L.**

PETIT, G. and BESNARD, W. (1937) Sur le comportement en aquarium du *Caecobarbus geertsi* Blgr. *Bull. Mus. Hist. Nat. Paris* (2) **IX.**

STELLA, E. (1951) Notizie biologiche su *Monodella argentarii* Stella Thermosbenaceo delle acque di una grotta di Monte Argentario. *Boll. Zool. Ital.* **XVIII.**

VERRIER, M.L. (1929) Observations sur le comportement d'un poisson cavernicole: *Typhlichthys osborni* Eigenmann. *Bull. Mus. Hist. Nat.* Paris (2) **I.**

Temperature of Caves

BALAZUC, J. (1956) Spéléologie du département de l'Ardèche. *Rassegna Speleol. Ital. Soc. Speleol. Ital. Mem.* **II.**

BOEGAN, E. (1921) *La Grotta di Trebiciano.* Triest.

DAVIES, W.E. (1960) Meteorological observations in Martens Cave, West Virginia. *Bull. Nat. Speleol. Soc.* **XXII.**

GINET, R. (1960) Écologie, Éthologie et Biologie de *Niphargus* (Amphipodes, Gammarides hypogés). *Ann. Spéléol.* **XV.**

HEUTS, M.J. et LELEUP, N. (1954) La géographie et l'écologie des grottes du Bas-Congo. *Ann. Mus. R. Congo belge. Sc. Zool.* **XXXV.**

HIERNAUX, C.R. and VILLIERS, A. (1955) Speologia Africana. Étude préliminaire de six cavernes de Guinée. *Bull. Inst. franç. Afrique Noire* (A), **XVII.**

JEANNEL, R. (1926) Faune cavernicole de la France avec une étude des conditions d'existence dans le domaine souterrain. *Encyclopédie Entomologique* **VII.** Paris.

MARTEL, E.A. (1894) *Les Abîmes.* Paris.

POLLI, S. (1958) Cinque Anni di Meteorologia ipogea nella Grotte Gigante presso Trieste. *Atti d. VIII Congresso Naz. Speleologia.* Como.

POLLI, S. (1962) Tre anni di meteorologie ipogea nella grotta sperimentale "C. Doria" del Carso di Trieste. *Atti. Mem. Comm. grotte "Eugenio Boegan" Suppl. d. "Alpi giulie".*

RENAULT, PH. (1961) Première Étude Meteorologique de la Grotte de Moulis (Ariège). *Ann. Spéléol.* **XVI.**

STADLER, H. (1927) Fortschritte in der Erforschung der tierischen Bewohnerschaft der Höhlen Südasiens und Indonesiens. *Mitteil. Höhl. Karstf.*

TORII, H. (1953) Fauna der Ryugado Sinterhöhle in Kochi Präfektur (Die Berichte der speläobiologischen Expeditionen. VI). *Annot. Zool. Jap.* **XXVI.**

The Effect of Temperature on Cavernicoles

BEAUCHAMP, P. DE (1932) Turbellariés, Hirudinées, Branchiobdellidés (2ème Série). *Biospeologica,* **LVIII.** *Archiv. Zool. expér. géner.* **LXXIII.**

CAPART, A. (1951) *Thermobathynella adami* gen. et sp. nov. Anaspidacé nouveau du Congo belge. *Bull. Inst. R. Sc. Nat. Belgique* **XXVII.**

FAGNIEZ, CH. (1907) De l'influence de l'altitude et de la température sur la répartition des Coléoptères cavernicoles. *Bull. Soc. Entomol. France.*

GAL, J. (1903) *Niphargus* et *Caecosphaeroma.* Observations physiologiques. *Bull. Soc. Étud. Sc. Nat. Nimes* **XXXI.**

GINET, R. (1960) Écologie, Éthologie et Biologie de *Niphargus* (Amphipodes Gammarides hypogés). *Ann. Spéléol.* **XV.**

GLACON, S. (1953, 1956) Recherches sur la biologie des Coléoptères cavernicoles. *Comm. 1er Congr. Intern. Spéléologie.* Paris. **III.**

JEANNEL, R. (1943) *Les Fossiles vivants des cavernes.* Paris.

MONOD, TH. (1940) Thermosbaenacea. *Bronns Kl. Ordn. Tierreichs.* **V,** Bd. I.

RACOVITZA, E.G. (1927) Observations sur la glacière naturelle dite "Ghetarul de la Scarisoara". *Bull. Soc. Sc. Cluj* **III.**

SCHELLENBERG, A. (1935) Der *Niphargus* des Thüringer Waldes und die Glazialreliktenfrage. *Archiv. f. Hydrobiol.* **XXIX.**

SZYMCZKOWSKY, W. (1953) Preferendum termiczne jaskiniswego pajaka *Meta menardi* Latr. (Argiopidae). (The temperature preferendum of a cave spider, *Meta menardi* Latr.). *Folia Biologica Warszawa* **I.**

THIENEMANN, A. (1908) Das Vorkommen echter Höhlen- und Grundwassertiere in oberirdischen Gewässern. *Archiv. f. Hydrobiol.* **IV.**

VORNATSCHER, J. (1961) Die lebende Tierwelt der Dachsteinhöhlen. *Die Höhle* **XII.**

The Atmosphere

ATANASIU, G. (1959) Sur l'utilisation des méthodes radioactives en Spéologie. *Revue de Physique* Bucarest. **IV.**

BEDEL, L. and SIMON, E. (1875) Liste générale des Articulés cavernicoles de l'Europe. *Journal de Zoologie* **IV.**

DAVIES, W. M. (1929) The effect of variation in relative humidity on certain species of Collembola. *British J. exper. Biol.* **VI.**

DEROUET, L. and DRESCO, ED. (1955) Études sur la grotte de Pèneblanque. I. Faune et climats. *Notes biospéol.* **X.**

DEROUET, L. (1956) Action du CO_2 sur le métabolisme respiratoire de deux Araignées troglophiles. *Notes biospéol.* **XI.**

DEROUET-DRESCO, L. (1960) Étude biologique de quelques espèces d'Araignées lucicoles et troglophiles. *Archiv. Zool. expér. géner.* **XCVIII.**

FAGE, L. (1931) Araneae (5ème Série), précédée d'un essai sur l'évolution souterraine et son déterminisme. *Biospeologica,* **LV.** *Archiv. Zool. expér. géner.* **LXXI.**

HEUTS, M. J. and LELEUP, N. (1954) La Géographie et l'Écologie des Grottes du Bas-Congo. *Ann. Musée R. Congo belge Sc. Zool.* **XXXV.**

JEANNEL, R. (1926) Faune cavernicole de la France, avec une étude des conditions d'existence dans le domaine souterrain. *Encyclopédie Entomologique* **VII.** Paris.

JEANNEL, R. (1943) *Les Fossiles vivants des cavernes.* Paris.

PALMEN, E. and SUOMALAINEN, H. (1945) Experimentelle Untersuchungen über die Transpiration bei einigen Arthropoden, insbesondere Käfern. *Ann. Soc. Zool. Bot.Vanamo.* **XI.**

PEYERIMHOFF, P. de (1906) Recherches sur la faune cavernicole des Basses-Alpes. Considérations sur les origines de la faune souterraine. *Annal. Soc. Entomol. France* **LXXV.**

RIDLEY, H. N. (1899) Caves in the Malay Peninsula. The Batu Caves. *Rep. 68th Meet. Brit. Advanc. Sc.* Bristol.

SEMENOVA, L. M. (1961) Relation of cuticle structure in chilopods to the conditions of existence. *Zool. J.* **XL.**

TROMBE, F. (1952) *Traité de Spéléologie.* Paris.

TROMBE, F. and HENRY DE LA BLANCHETAIS, CH. (1947) Étude sur la conductibilité de l'air et la présence de radiations pénétrantes telluriques dans quelques souterrains des Pyrenées. *Ann. Spéléol.* **II.**

Water

CHODOROWSKI, A. (1959) Les Études biospéologiques en Pologne. *Biospeologica Polonica* **II.** *Speleologia* **I.**

GINET, R. (1960) Écologie, éthologie et biologie de *Niphargus* (Amphipodes Gammarides hypogés). *Ann. Spéléol.* **XV.**

HEUTS, M. J. and LELEUP, N. (1954) La Géographie et l'Écologie des Grottes du Bas-Congo. *Ann. Mus. R. Congo Belge. Sc. Zool.* **XXXV.**

PART 4

Physiology of Cavernicoles

Physiology can be divided into two sections:

(1) The study of the transformations of matter and energy which take place within the body of the organism and which are termed collectively, metabolism.

(2) Sensory and neural physiology.

These two aspects of physiology will be considered in the fourth and fifth parts of this book.

CHAPTER XIX

NUTRITION AND SOURCES OF FOOD OF CAVERNICOLES

A. ARE CAVERNICOLES ALWAYS STARVED ANIMALS?

If the entrances to caves, which are partly illuminated, are excluded, the subterranean environment possesses no green plants. Thus, the most important source of food for surface animals, is denied to the cavernicoles. Since such an important food source is lacking, early biospeologists concluded that there was a general scarcity of food for cavernicolous animals. Packard (1886) frequently insisted that the shortage of food available to cavernicolous animals is the reason for their small size. Verhoeff (1898) considered that cavernicolous animals were able to exist only because of their lower metabolic rates and their long periods of quiescence. Thienemann as recently as 1925, upheld the idea that sources of nutrition are scarce in the subterranean world.

Some very precise observations seem to strengthen the idea that there is a constant shortage of food in the underground world. These relate to the marked resistance to starvation shown by cavernicoles. In the first place several examples from the invertebrates may be quoted. Ginet (1960) kept a specimen of *Niphargus virei* for two years during which time the only food taken was one of its mates. A dytiscid, *Morimotoma phreatica* was maintained for eight months without food (Uéno, 1957). Bathysciinae can resist periods of fast of 6–8 months (Glaçon-Deleurance).

The examples of prolonged starvation observed in cavernicolous vertebrates are even more surprising. According to Packard (1886) a specimen of the blind fish *Amblyopsis spelaeus* from Mammoth Cave, remained alive for two years without receiving any food. Kammerer (1912) kept *Proteus* alive for two years even though they received no food. Gadeau de Kerville (1926) reported that a specimen of *Proteus* had been kept in captivity for fourteen and a half years and for the last eight of these had received no food.

Such examples should be accepted with some caution for the following reasons. Brode and Gunter (1959) have shown that specimens of *Amphiuma* and *Cryptobranchus* may survive and even grow, while apparently fasting. This anomalous situation may be explained by the following observations.

These animals eat the mucous layer impregnated with bacteria, diatoms and algae, which covers their bodies and detaches itself at intervals, rather like an exuvium.

Furthermore, it should not be forgotten that resistance to starvation is a general property of poikilotherms.

However, the views of early biospeologists concerning the food supply of the cavernicoles are certainly incorrect. Thus, although many cavernicoles have the ability to fast, they need some food, and even a great deal of nutriment. For example, individuals of the cavernicolous fish *Anoptichthys* consume more food than similar individuals of the surface-living genus *Astynax*, from which they have evolved (Rasquin, 1949). All workers who have cultured *Proteus* will know that these creatures consume large quantities of crustaceans and midges' larvae.

It is without significance to theorise on the general pattern of food sources in the subterranean world. Thus, as in other fields of biospeology, no generalisation concerning food can be made.

The most experienced workers (Hamann, 1896; Racovitza, 1907; Jeannel, 1926; 1943; Leruth, 1939) have recognised that sources of food are very diverse from cave to cave.

The nutritional factor plays a primary role in the distribution and abundance of cavernicoles. Those caves which have no food resources are azoic. Deep caves or shafts are generally without life. However, the factor of depth in itself is not important. Very deep abysses may be well populated provided that food supplies are able to reach them.

On the other hand certain caves contain Amphipoda, Isopoda, Myriapoda and Bathysciinae which can be captured in large numbers. This richness is always linked with an abundance of food. Brölemann (1923) reported that caves provide an ideal environment for the Blaniulidae because of the abundant food supply which exists there. According to Schwoerbel (1961) there is an abundant food supply in the interstitial medium.

There is a clear relationship between the abundance of food and the size of cavernicolous species. This is not because the amount of food has a direct effect on the size of the cavernicoles, but, it is only where food is abundant that they can successfully reproduce. *Proteus* is found only in certain large caves in the Adriatic areas of Karst which contain large food supplies brought in by rivers. The largest Isopoda, Amphipoda, Arachnida and Bathysciinae also live in the same caves.

B. THE FOOD OF CAVERNICOLES

In the preceding section only the quantitative aspects of food have been considered. In this section qualitative aspects will be discussed.

The quality of food has been a selective factor of prime importance by prohibiting the access of numerous organisms to the underground environment. In the second part of this book not only the animals which have colonised the subterranean world have been mentioned, but also those groups which have failed to do so. This absence is often due to the absence of the necessary types of food.

Animals which possess specialised feeding habits are obviously excluded from taking up cavernicolous life if their source of food is not available underground. This must be the case with strictly phytophagous species which cannot exist in the subterranean environment because of the absence of green plants. No strictly phytophagous group possesses cavernicolous representatives. Those species which capture green flagellates or algae, such as the anostracan branchiopods, the Cladocera and the diaptomids, cannot exist in caves.

Some species which appear perfectly preadapted to subterranean life fail to colonise this habitat simply because of the food-factor. This is the case with the Symphyla. Although they look the cavernicolous type, and are depigmented and anophthalmic they possess no cavernicolous representatives. The reason is that all Symphyla live on green plants. Before an animal can successfully colonise the underground habitat and reproduce there many barriers have to be overcome. If any of these is insurmountable then the transition to the troglobious existence cannot be made.

Those primitively phytophagous animals which have become successfully adapted to cavernicolous life have done so by a radical change in their diet. The case of the gasteropod mollusc *Oxychilus cellarius* is a good illustration of this point. Although this species is frequently found in caves it is also found in the epigeous medium, and is thus a troglophile. Tercafs (1960) has pointed out that epigeous specimens feed on dead leaves while the cavernicolous individuals feed on the debris of arthropods and even living Lepidoptera. Tercafs and Jeuniaux (1961) have measured the chitinase activity in the stomach contents and the hepatopancreas homogenates of *Oxychilus cellarius*. The level of activity is much higher than that which one finds in strictly phytophagous species such as *Helix pomatia*. It can thus be said that species of *Oxychilus* show a predisposition for becoming adapted to a cavernicolous diet. There can be little doubt that this predisposition is the reason why several species of the genus *Oxychilus* have been able to colonise the subterranean environment.

Wagner (1914) has reported that the molluscs which are best represented in the caves of Southern Dalmatia and Herzegovina, belong to the families Oleacinidae and Zonitidae, whose diets are mainly carnivorous. *Glandina algira*, an essentially carnivorous testacellid, has frequently been found in the caves of Southern Dalmatia.

Those animals with specialised dietary requirements have much less aptitude for colonising the subterranean environment than polyhagous species which can benefit from all types of food present in the caves. This is why most cavernicoles are more or less polyphagous and their dietary categories cannot be distinguished. It is generally impossible to classify a cavernicole as a humiphage, xylophage, mycophage, coprophage or necrophage, for example. Most underground animals can at some time or other be placed in one or more of these groups. The term detritivore is the one which best suits the majority of them. Only the guanobia can be rigidly defined according to their diet, but these are not true cavernicoles.

C. THE EXOGENOUS SOURCES OF FOOD

One of the most important problems in biospeology is to identify the various types of food available in the underground environment and to trace their origins.

Endre Dudich (1933) was the first worker to distinguish between the two sources of nourishment available to cavernicoles. Firstly, there is exogenous food or "allochtone", carried in from the exterior and secondly endogenous food or "autochtone" from the cave itself. The exogenous food can be divided into three categories:

(a) that carried in aerially,
(b) that carried in by water,
(c) material provided by living creatures.

(a) Material Brought in Aerially

The currents of air which pass through caves may introduce not only inorganic particles but also living creatures, bacteria, and fungal spores. Prat (1925) has shown that petri dishes containing culture media, placed inside newly opened galleries quickly become inoculated with various air-borne organisms. This phenomenon has led him to speak of an aeroplankton.

The penetration of pollen grains into caves has also been shown to occur. Van Campo and Leroi-Gourhan (1956) suspended glass slides covered with glycerine in the Grotte d'Arcy-sur-Cure, Yonne. They reported that pollen grains were deposited at a distance of 10 m from the cave entrance and even as far as 40 m when the air currents were stronger.

(b) Material Brought in by Water

1. *Plankton.* When subterranean rivers are fed by surface waters which have passed through ponds or marshes they contain true plankton (Kofoid, 1899; Scott, 1909). Most planktonic organisms disappear rapidly after entering the subterranean world but others persist and even reproduce. This is the case in the Cyclopidae.

2. *Organic Material*. An appreciable amount of organic material is brought into the subterranean environment by the leeching action of rain water on humus. Dudich (1932) has analysed the water dripping from one stalactite in the Baradla Cave. He found 0·009 g/l. organic material in this water. Ginet (1960) has reported for the cave of La Balme in the department Isère, levels of 0·7 mg of O_2/l. in winter and 1·35 mg in autumn. He gave the following figures for the caves of Corveissiat (Ain): 0·4 mg of O_2/l. in winter and 2 mg in summer.

3. *Branches of Trees, Wood and Dead Leaves*. Rivers carry, especially during floods, branches, twigs, masses of dead leaves and even whole trees. This is a very important source of food for both aquatic and terrestrial cavernicoles. It would obviously be very difficult to give a complete list of the materials which are brought into caves in this way. These materials are utilised for food by the scavengers and detritivores which populate the caves. Among these are *Stenasellus*, which feeds on dead leaves, and the terrestrial isopods which consume dead wood.

(c) Live Food

1. *The Fungi and Rhizomorphs* which utilise organic food materials are consumed by cavernicoles (Schneider, 1886). However, they are not commonly found underground. This is because the clay deposits are a medium with a reducing action which opposes the development of mycelia (*Aspergillus*, *Penicillium*, and the *Mucorales*), and brings about encystment (Caumartin).

2. *The Guano* of cave-dwelling bats is the source of nourishment for a large number of guanobia (p. 288). Guano provides food for both aquatic and terrestrial cavernicoles. According to Leleup (1956) *Stenasellus léleupi*, which lives in caves in the Congo, is a guanophage. The droppings of rodents and carnivores are also eaten by cavernicoles. *Stenasellus congolensis* feeds on the faeces of the porcupine (Leleup, 1956).

3. *The Corpses of Bats and Rodents* attract numerous cavernicoles.

4. *Man* occasionally provides food for cavernicolous animals. Chasms and swallow-holes are often used as rubbish-dumps and charnel-houses by local inhabitants. Caves which are opened to the public soon acquire various rubbish and excrement left by human beings. This provisionment is taken advantage of by numerous cavernicoles.

5. Finally, some cavernicoles are preyed upon by hypogeous carnivores. *Troglochaetus* consumes not only plant debris, but also testaceous amoebae (Delachaux, 1921; Stammer, 1937). Some Acarina feed on Collembola (Absolon, 1900). The gamasid mites are entomophagous (Carpenter, 1895). Cavernicolous opilionids of the genus *Ischyropsalis* eat molluscs (Verhoeff, 1900), and phryganids, (Dresco, 1947). Hypogeous spiders hunt flies, small Coleoptera, and also pseudoscorpions. According to Packard

the cavernicolous species of *Asellus* are the prey of the blind crayfish *Cambarus*. The Nematocera and Collembola are the main food-source of the trechine beetles. Both *Amblyopsis* and *Proteus* have varied diets. The former has been shown to consume small fish, *Cambarus*, *Crangonyx*, *Asellus* and various copepods, whilst the latter, according to Spandl, feeds on molluscs, oligochaetes and amphipods.

D. CLAY AND SILTS

The sources of food which have been considered above have been known for a long time. There are, however, other sources of which the importance has only recently been appreciated. These other sources are constituted by clays which are found always in caves populated by cavernicoles and silts, mixtures of sands and clays, where they are deposited in the beds of rivers or in gours.

(a) Early Observations

Several biospeologists have already indicated a possible nutritive value for clays and silts. Carpenter (1895) reported that Collembola of the genus *Lipura* ingest clay particles. Viré (1900) stated that one often discovers *Niphargus* feeding on the clay deposits at the bottoms of lakes like the geophageous savages until their digestive tracts are distended, a feature which explains the predilection of captured *Niphargus* for this material. Schreiber (1929, 1932) has observed the presence of nitrates in the slime deposited in the bed of the Piuka during its underground course, and has suggested that they act as a source of nourishment for limivores. Hertzog (1936) stated that the amphipod *Bogidiella albertimagni* feeds on silts rich in organic material, and regarded it as truly geophagous.

(b) The Work of R. Ginet and A. M. Gounot

The role played by clay and silt was not definitely established until after the work of R. Ginet (1955, 1960) and was confirmed by that of A. M. Gounot (1960). These experiments were carried out on amphipods of the genus *Niphargus*. Ginet has shown that species of *Niphargus* are polyphagous, falling at once into the categories of detritivore, coprophage, herbivore and carnivore, but most important in this context is the fact that during the early stages of their development they ingest particles of silt. These may be found by dissection in their masticatory stomachs or their intestines. Finally, to show the importance of clay in the diet of *Niphargus*, Ginet undertook four experiments.

(1) Young individuals of *Niphargus virei*, if deprived of clay and other food, survive for 3 to 6 months but they do not grow and finally die.

(2) When young *Niphargus* were deprived of clay but received regular exogenous nourishment (pieces of meat) they grew and moulted but their mortality was very high and no specimen lived for more than a year. Thus, exogenous nourishment alone is insufficient to ensure the successful development of *Niphargus*.

(3) Immature specimens of *Niphargus* were placed in tanks containing either silt or clay from caves. They received no exogenous nourishment. During the first three or four months of their existence they developed normally. However, A.M.Gounot has shown that if one continues the experiment for a longer period, growth begins to slow and finally ceases. If they are breeding under these conditions, no individuals survive for more than a year. Thus, clay by itself is sufficient for the early stages of growth but insufficient for its completion.

(4) The breeding of *Niphargus* in a medium including both clay and an exogenous source of food (meat), has allowed them to be maintained in good health, and to reproduce during three generations.

It is certain that the conclusions of these experiments are applicable only to *Niphargus* which are polyphagous. A parallel experiment carried out on exclusively limnivorous species would probably give different results.

(c) Importance of Clay to Proteus

Another rather less well-documented example is provided by *Proteus*. The young specimens of *Proteus* born in the cave laboratories at Moulis (Vandel and Bouillon, 1959), fed exclusively on the clay and silt which covered the bottom of the culture tank. This diet was continued for a year and then supplemented by regular meals of midges larvae. During the first year they increased their body length from 22 to 60 mm. It may therefore be concluded that silt normally insures the growth of *Proteus*, during the first year of life. As with *Niphargus* it is probable that such a diet would be insufficient to maintain the growth of older larvae, and an addition of exogenous food is indispensable for the attainment of the adult stage and sexual maturity.

(d) Importance of Clay to Terrestrial Cavernicoles

Our knowledge of the role of clay in this field is less advanced. However, it has been established that a large number of cavernicoles burrow into clay (Vandel, 1958). The significance of this habit is not clear but it is certainly different in the various groups studied. It appears most probable that a large quantity of clay would be ingested during these burrowings.

(e) Conclusions

The experiments and observations reported in the preceding lines clearly establish the importance of clay and silt in the diet and growth of cavernicoles. It is necessary to go more deeply into this question. The clay and silt deposits provide three kinds of nutritive elements:

(1) The silt deposited in gours or subterranean rivers contains organic matter, a fact which has been demonstrated chiefly by Giorgio Schreiber (1929, 1932). This biospeologist analysed the silt deposits in the subterranean section of the River Piuka and recorded a level of 0·237% organic nitrogen. This was made up by plant debris, guano, etc., brought in by the river. Schreiber stated, although no experimental data was given, that the exogenous organic material contained in silt was utilised by limivorous cavernicoles.

A. M. Gounot (1960) has reported that all the silt deposits which she had analysed contained quantities of organic matter. The levels of organic carbon varied between 0·026 and 0·66% dry weight of silt, that is 0·05–1·3% of organic matter, for the weight of carbon can be taken as making up about half the weight of the organic matter present. With regard to organic nitrogen, the proportion varies between 0·007 and 0·161%. The ratio C/N is between 3·6 and 5·0. These values are much lower than those obtained from lacustrine silts or from arable soil.

(2) These silts and probably also the clays contain a microfauna made up of Protista. Schreiber (1929, 1932), has denied the existence of protists in the section of the River Piuka which he has studied, but one should not generalise from this conclusion. Even if underground waters are lacking in green Protista they still contain numerous Protozoa as described previously (p. 61). Such slow-moving forms as *Amoeba*, Amoebina Testacea, *Bodo* and hypotrichous Ciliata frequently inhabit silt (Varga and Takats, 1960).

(3) Silts and clays contain a rich bacterial flora which will now be considered.

E. BACTERIA AND SPELEOBACTERIOLOGY

(a) Historical

For a long time subterranean waters were regarded as sterile. However, a large number of studies in particular those of the great speleologist E. A. Martel, have shown this to be incorrect. The rivers which disappear underground and the water passing through calcareous strata which feeds the subterranean reservoirs are not filtered and thus cannot be sterile. For this reason they are so frequently the cause of serious epidemic diseases. All the bacteria recorded by water-analysis in water from calcareous beds are the same as those found in surface waters.

The air of caves is generally sterile. This is because any small particles present in the subterranean atmosphere soon surround themselves by a droplet of water and are thus precipitated (Molnar, 1961). Man introduces bacteria into the atmosphere of caves during his visits. Also the air currents passing through subterranean galleries may become charged with bacteria from the outside world.

While our knowledge of the bacteria of subterranean cavities and waters was once limited to the few studies mentioned above, it should be realised that such studies were of greater interest to those concerned with public health than to biospeologists. But now the problem is viewed under an entirely different light, since the discovery of strictly subterranean bacteria. These are similar to those reported in the soil by S. Winogradsky (1949). It is thus legitimate today to speak of speleobacteriology.

Professor Endre Dudich was a pioneer in this field. In three most important publications (1930, 1932, 1933), he put forward the idea of the possible existence of an endogenous, or autochtonic, food-cycle based on the presence of autotrophic bacteria. He recognised the existence of three types of autotrophic bacteria in addition to the heterotrophic types. These were the Ferrobacteriales *(Leptothrix)*, (which were also reported by Magdeburg, 1933), the Thiobacteriales *(Beggiatoa)* and the nitrifying bacteria. These three groups of autotrophic bacteria obtain the energy they require for their metabolic processes by the oxidation of inorganic compounds, which enables them to synthesise organic compounds in the absence of light energy.

All papers on the bacteriology of subterranean cavities since the publications of Dudich's classic works recognise the two bacterial populations, one heterotrophic and the other autotrophic. Such papers are those of Birstein and Borutzky (1950), in Russia; Mason-Williams and Benson-Evans (1958), in Great Britain; Fischer (1959), and Chodorowsky in Poland; Varga and Takats (1960) in Hungary, and Caumartin (1957a and b, 1959, 1961) in France.

It may be added that the Actinomycetes, which are tentatively included in the bacteria, rather than in the fungi, are often abundant at the entrances of caves. They are responsible for the mouldy odour often characteristic of the entrances of caves.

(b) Methods of Investigation

The use of an optical microscope alone is insufficient to study the subterranean bacteria. When examining filtrates or foams, magnifications of 2,500 to 10,000 are necessary, such as can be obtained only with the electron microscope.

Finally, to complete the study of bacteria it is necessary to carry out culture techniques. Different bacterial types require particular nutritive media. The composition of these can be found in any work on bacteriological techniques.

Using these methods the bacteriologists have shown the presence of numerous bacterial types in the subterranean environment. However, the study of these is only in its infancy and to draw up a representative list of these forms will take many years.

(c) Heterotrophic and Autotrophic Forms

The heterotrophic forms are those bacteria which come from the exterior and are carried into subterranean cavities in air or water. These bacteria need organic material for their development but if this is present they grow as rapidly underground as they do in their normal habitat. They are thus an exogenous population.

Autotrophic bacteria are, on the other hand, indigenous to caves and are found in the clay deposits or in the crevices of rocks. Their complete development depends only on the presence of certain inorganic materials†. They form endogenous populations.

These two types of population, one exogenous and heterotrophic, and the other endogenous and autotrophic are not usually found together. Where bacterial populations are found in the presence of organic material it is the heterotrophic forms which predominate and the autotrophic forms, which are more sensitive to modifications in the chemical equilibrium, are eliminated. As a result autotrophs are never found in media contaminated with organic material. Gounot (1960) has reported that the number of heterotrophs in silt increases as a function of the level of water pollution, while the autotrophic populations do quite the reverse and decrease in numbers.

(d) Autotrophs

It is obvious that the endogeous bacteria are the most attractive group to biospeologists.

The autotrophs found in the underground habitat are always without the pigments which normally plays a role in photosynthesis. To use the terminology of Lwoff (1944) they are chemosynthetic autotrophs, which have no need of light for the synthesis of their own substance.

There are three categories of autotrophs, the Nitrobacteria, the Thiobacteria, and the Ferrobacteria.

(1) Nitrobacteria. It was Winogradsky who in 1890 first showed the presence of nitrifying bacteria in the soil. But they have also been shown to be present in the subterranean environment (Mason-Williams and Benson-Evans, 1958).

---

† This is not to say that the autotrophs are excluded from developing sometimes in a similar manner to the heterotrophs.

There are two categories of nitrifying bacteria: the *nitrous bacteria (Nitrosomas, Nitrococcus)* which carry out the oxidation of ammonia and its transformation to nitrous acid; and the *nitric bacteria (Nitrobacter)*, which carry out the oxidation of nitrous acid to nitric acid. The energy for these transformations comes from the oxidation of ammonia or nitrites. The nitrobacteria have the effect of mineralising proteins and thus tend to bring about an emptying of the medium of heterotrophic bacteria.

(2) Thiobacteria. The Thiobacteria or sulphur bacteria are abundant in water rich in hydrogen sulphide (H_2S). The best known of this type is *Beggiatoa.* They oxidise hydrogen sulphide and the sulphur liberated is deposited in colloidal form in the body of the bacterium. The Thiobacteria also oxidise the sulphur which they accumulate and produce sulphuric acid, which, in the presence of bases, produces sulphates. Finally, the Thiobacteria are capable of transforming sulphides into sulphates, in the presence of free oxygen.

(3) Ferrobacteria. The Ferrobacteria or iron bacteria, of which the best known type is *Leptothrix,* decompose iron carbonate and oxidise ferrous oxide into ferric oxide which is deposited as crystals of ferric hydroxide.

In 1957, V. Caumartin isolated a ferrobacterium which is widely distributed in the clay of caves, as well as in dolomite. He gave it the name *Perabacterium spelaei* (Caumartin, 1957b, 1959a and b). This form measures $1\cdot0$–$1\cdot5\,\mu$. It often takes the form of a double-wallet (from which the name *Perabacterium* is derived). Under normal conditions this bacterium is a microaerophilic and autotrophic form. It obtains carbon from the decomposition of iron carbonate and its energy from the oxidation ferrous oxide. It fixes the nitrogen of the air. This is one of the bacteria which is most sensitive to chemical equilibria and one of the types which suffers most from the presence of heterotrophs. Thus they find the optimum conditions for development in the clay of caves.

(e) Density of Bacteria

One must give some idea of the density of bacteria in the silts and clays of caves. Results obtained in France (Ginet, 1960; Gounot, 1960) showed that populations were of the order of several tens of millions per gram of dry silt (A. M. Gounot quotes 10 to 250 millions). These populations are much less dense than the bacterial populations found in arable soil which reach densities of hundreds or thousands of millions per gram.

It would also be extremely interesting to evaluate the proportions of the various types of bacteria which populate caves. Tentative studies have been carried out (E. Fischer, 1959; A. M. Gounot, 1960) but it would be difficult to draw any general conclusions at this early stage.

F. UTILISATION OF DIFFERENT FOOD SOURCES
AND THE FOOD CYCLES

From the preceding pages one can see how much in error the early biospeologists were to think of caves as deserted areas in the middle of which wandered a few constantly hungry cavernicoles. In fact, circumstances are very different. There is a wide variety of food available to cavernicoles, but it is very difficult to discover the ways in which hypogeous animals utilise the various food resources at their disposal in the underground world.

(a) Clays and Silts

The fact that clays and silts play a role in the nutrition of cavernicoles has been established by the experiments described earlier.

The clays and silts contain organic material, bacteria, Protista and small Metazoa (in particular Nematodes), which can serve as food for cavernicoles. But as these forms are so closely associated in this medium it is difficult to recognise the part that each of them plays in the feeding of hypogeous animals. It is known, however, that very many Protozoa feed on bacteria. We may also state that Amoebina, Amoebina Testacea, Flagellata and Infusoria in the subterranean environment also devour bacteria.

Certain Metazoa can feed on bacteria. Monakov and Sorokin (1960) gave autotrophic bacteria, labelled with carbon 14, to cyclopids (Mesocyclops leukarti and Acanthocyclops viridis). The adults do not ingest the bacteria. However, the nauplii absorbed large quantities of bacteria which insured normal growth. It would be extremely interesting to repeat these experiments using cavernicoles, in particular hypogeous harpacticids.

The marine abysses are not without resemblance to the subterranean medium. Abyssal animals are "silt eaters" which filter the marine deposits and feed on the bacteria which they contain. The bacterial populations of the abysses consist as do the subterranean media, of autotrophic and heterotrophic forms (Zo Bell). Certain abyssal fish eat bacteria which form their principal diet (Kriss and Asmane).

As for the terrestrial cavernicoles, Decu (1961) observed the presence of bacteria and Actinomycetes in the intestine of some Bathysciinae.

(b) Bacteria and Fungi as the Sources of Vitamins

In a study of the physiology of nutrition it is usual to distinguish the food material which provides energy on the one hand and the growth factors and vitamins on the other.

Animals have lost the ability to synthesise most vitamins. They rely on green plants which build up these compounds as a result of their photosynthesis. However, green plants cannot develop in the darkness of subterranean caves. Where can cavernicoles obtain these indispensable vitamins? Today the answer to this question is quite clear. It is known that bacteria and also fungi are capable of synthesising certain vitamins in the absence of light. The Thiobacteria can synthesise oligodynamic substances (nicotinic acid, pantothetic acid, riboflavin, pyridoxine, vitamin B_{12}). Actinomycetes can synthesise carotenes in complete darkness. It would seem that the importance of the autotrophic bacteria in the nutrition of cavernicoles lies mainly in their ability to synthesise vitamins.

It is obvious that the conclusions which can be drawn from the experiments carried out up until now, are only very superficial. The experimental methods which are used in research into nutritional physiology must be applied to cavernicolous animals.

For example, the rate and duration of growth under diets deficient in vitamins, or provided with them.

As has been said at the beginning of this book, the distinction between cavernicoles and endogeans is often delicate. One cannot establish definite limits between the clays and silts of caves and those of the surface soils. It is known that vitamins are regularly found in surface soils. "The principal vitamins seem to be normal constituants of most soils" (Schopfer, 1949). The vitamins present in the soil are derived from micro-organisms, in particular soil-bacteria. There is a striking analogy with subterranean deposits. It is pertinent to ask if the geophagia carried out by many primitive human societies is not an empirical method of combating avitaminosis.

(c) Food Cycles

In the studies of Dudich (1932), Ginet (1960) and A.M.Gounot (1960) the existence of food cycles in the subterranean environment has been recognised. Evidence has been given mainly for cycles involving aquatic cavernicoles, but it cannot be doubted that there are analogous cycles for terrestrial forms. It must not be forgotten however, that the categories which have been recognised must remain fluid because of the marked tendency of cavernicoles towards polyphagia.

(1) The bacterial populations can be placed at the lowest end of the scale. They include the autotrophs which are capable of synthesising organic material by utilising inorganic compounds, and the heterotrophic bacteria which use organic material of exogenous origin. The bacterial population furnish the clay and silt deposits with organic materials and vitamins.

(2) At a higher level are the microfauna constituted by the Protista, (Amoeba, Amoeba Testacea, Flagellates, Ciliates) which devour bacteria.

(3) The limivores feed on silt containing bacterial populations and Protista, and the compounds formed from these organisms. Certain cavernicolous forms belonging to the oligochetes and the nematodes, and perhaps also the molluscs, are pure limivores.

(4) Many cavernicoles are limivores in their developmental stages. This is the case in *Niphargus*, and in *Proteus* during the first year of its existence, as has already been stated.

However, the silt does not provide food material which is sufficient for the complete development of these animals and for sexual maturity. Food from exogenous sources is necessary during the second stage of growth. *Niphargus* and *Proteus* certainly need animal food to reach the adult state. Terrestrial isopods of the genus *Scotoniscus* can exist for a long time on clay deposits alone, but the diet must include wood or dead leaves before they can reproduce.

(5) At the summit of the food-pyramid are the carnivores, which prey on cavernicoles belonging to the other categories and also on the troglophiles. Among the aquatic forms the cavernicolous fish come into this category; a similar position is occupied by the carabid Trechinae among the terrestrial forms.

Thus, the nourishment of cavernicoles is assured by material from the surface environment. This was until recent years the only source of food known. It is however, now known that endogenous cycles exist which are based on the presence of autotrophic bacteria and largely independent of external phenomena. The endogenous cycles certainly play an important role in the subterranean world, as they do in the depths of the sea.

BIBLIOGRAPHY

The Food Supply of the Underground World

BRODE, E.W. and GUNTER, G. (1959) Peculiar feeding of *Amphiuma* under conditions of enforced starvation. *Science* **CXXX.**

BRÖLEMANN, H.W. (1923) Blaniulidae, Myriapodes (1ère Série). *Biospeologica*, **XLVII**; *Archiv. Zool. expér. géner.* **LXI.**

GADEAU DE KERVILLE, H. (1926) Note sur un Protée anguillard (*Proteus anguinus* Laur.) ayant vécu sans aucune nourriture. *Bull. Soc. Zool. France* **LI.**

GINET, R. (1960) Écologie, Éthologie et Biologie de *Niphargus* (Amphipodes, Gammarides hypogés). *Ann. Spéléol.* **XV.**

HAMANN, O. (1896) *Europäische Höhlenfauna*. Jena.

JEANNEL, R. (1926) Faune cavernicole de la France. *Encyclopédie Entomologique*, **VII.** Paris.

JEANNEL, R. (1943) *Les Fossiles vivants des Cavernes*. Paris.

KAMMERER, P. (1912) Experimente über Fortpflanzung, Farbe, Augen und Körperreduktion bei *Proteus anguinus*. Laur. Zugleich: Vererbung erzwungener Farbveränderungen. III. Mitteilung. *Archiv. f. Entwicklungsm.* **XXXIII.**

LERUTH, R. (1939) La Biologie du Domaine souterrain et la Faune cavernicole de la Belgique. *Mém. Mus. R. Hist. Nat. Belgique* No. 87.

PACKARD, A. S. (1886) The cave fauna of North America with remarks on the anatomy of the brain and origin of the blind species. *Mem. Nat. Acad. Sci. Washington* **IV.**

RACOVITZA, E. G. (1907) Essai sur les Problèmes biospéologiques. *Biospeologica*, **I.** *Archiv. Zool. expér. géner.* (4) **VI.**

RASQUIN, P. (1949) The influence of light and darkness on thyroid and pituitary activity of the characin *Astyanax mexicanus* and its cave derivatives. *Bull. Americ. Mus. Nat. Hist.* **XCIII.**

SCHWOERBEL, J. (1961) Subterrane Wassermilben (Acari: Hydrachnellae, Porohalacaridae und Stygothrombiidae), ihre Ökologie und Bedeutung für die Abgrenzung eines aquatischen Lebensraumes zwischen Oberfläche und Grundwasser. *Archiv. f. Hydrobiol. Suppl.* **XXV.**

THIENEMANN, A. (1925) Die Binnengewässer Mitteleuropas. Eine limnologische Einführung. Die Binnengewässer, **I.** Stuttgart.

UÉNO, S. I. (1957) Blind aquatic beetles of Japan, with some accounts of the fauna of Japanese subterranean waters. *Archiv. f. Hydrobiol.* **LIII.**

VERHOEFF, K. W. (1898) Einige Worte über europäische Höhlenfauna. *Zool. Anz.* **XXI.**

The Food of Cavernicoles

TERCAFS, R. R. (1960) *Oxychilus cellarius* Müll., un Mollusque cavernicole se nourrissant de Lépidoptères vivants. *Rassegna Speleol. Ital.* **XII.**

TERCAFS, R. R. (1961) Comparaison entre les individues épigés et cavernicoles d'un Mollusque Gastéropode troglophile, *Oxychilus cellarius* Müll. Affinités éthologiques pour le milieu souterrain. *Ann. Soc. R. Zool. Belgique* **XCI.**

TERCAFS, R. R. and JEUNIAUX, CH. (1961) Comparaison entre les individuals épigés et cavernicoles de l'espèce *Oxychilus cellarius* Müll. (Mollusque Gastéropode troglophile) au point de vue de la teneur en chitinase du tube digestif et de l'hépatopancréas. *Archiv. Intern. Physiol. Biochimie* **LXIX.**

WAGNER, A. J. (1914) Höhlenschnecken aus Süddalmatien und der Herzogewina. *Sitzb. K. K. Akad. Wiss. Math. Naturw. Kl. Wien. Abt. I.* **CXXXIII.**

Exogenous Food

ABSOLON, C. K. (1900) Einige Bemerkungen über mährische Höhlenfauna. III. Aufsatz. *Zool. Anz.* **XXIII.**

CARPENTER, G. M. (1895) Animals found in the Mitchelstown Cave. *Irish Naturalist* **IV.**

DELACHAUX, TH. (1921) Un Polychète d'eau douce cavernicole, *Troglochaetus beranecki* nov. gen. nov. sp. (Note préliminaire). *Bull. Soc. Neuchâtelois Sc. Nat.* **XLV.**

DRESCO, ED. (1947) Résultats biospéologiques. *In* TROMBE, F., DRESCO, ED., HALBRONN, G., HENRY DE LA BLANCHETAIS, CH. and NÈGRE, J., Recherches souterraines dans les Pyrénées centrales. Années 1945 à 1947. *Ann. Spéléol.* **II.**

DUDICH, E. (1932) Biologie der Aggteleker Tropfsteinhöhle "Baradla" in Ungarn. *Speläolog. Monogr.* **XIII.**

DUDICH, E. (1933) Die Klassifikation der Höhlen auf biologischer Grundlage. *Mitteil. Höhl. Karstf.*

GINET, R. (1960) Écologie, Éthologie et Biologie de *Niphargus* (Amphipodes, Gammarides hypogés). *Ann. Spéléol.* **XV.**

KOFOID, C. A. (1899, 1900) The plankton of Echo River, Mammoth Cave. *Trans. Americ. Micros. Soc.* **XXI.**

LELEUP, N. (1956) La Faune cavernicole du Congo belge et considérations sur les Coléoptères reliques d'Afrique intertropicale. *Ann. Mus. R. Congo belge. Sc. Zool.* **XLVI.**

PRAT, S. (1925) Das Aeroplankton neu geöffneter Höhlen. *Zentralbl. f. Bakteriol* 2. Abt. **LXIV.**

SCHNEIDER, R. (1886) Amphibisches Leben in den Rhizomorphen bei Burgk. *Math. Naturw. Mitteil. Sitzb. k. Preuß. Akad. Wiss.* **VII.**

SCOTT, W. (1909) An ecological study of the plankton of Shawnee Cave, with notes on the cave environment. *Biol. Bull.* **XVII.**

STAMMER, H.J. (1937) Der Höhlenarchannelide, *Troglochaetus beranecki*, in Schlesien. *Zool. Anz.* **CXVIII.**

VAN CAMPO, M. and LEROI-GOURHAN, A. (1956) Note préliminaire à l'étude des pollens fossiles de différents niveaux des grottes d'Arcy-sur-Cure. *Bull. Mus. Hist. Nat.* Paris. (2) **XXVIII.**

VERHOEFF. K.W. (1900) Zur Biologie von *Ischyropsalis*. *Zool. Anz.* **XXIII.**

Clay and Silts

CARPENTER, G.M. (1895) Animals found in the Mitchelstown Cave. *Irish Naturalist* **IV.**

GINET, R. (1955) Étude sur la Biologie d'Amphipodes troglobies du genre *Niphargus*. I.Le creusement des terriers; relations avec le limon argileux. *Bull. Soc. Zool. France* **LXXX.**

GINET, R. (1960) Écologie, Éthologie et Biologie de *Niphargus* (Amphipodes Gammarides hypogés). *Ann Spéléol.* **XV.**

GOUNOT, A.M. (1960) Recherches sur le limon argileux souterrain et sur son rôle nutritif pour les *Niphargus* (Amphipodes, Gammarides). *Ann. Spéléol.* **XV.**

HERTZOG, L. (1936) Crustacés de biotopes hypogés de la vallée du Rhin d'Alsace. *Bull. Soc. Zool. France.* **LXI.**

SCHREIBER, G. (1929) Il contenuto di sostenza organica nel fango delle Grotte di Postumia. *Atti Accad. Veneto-Trentino, Istria* (3) **XX.**

SCHREIBER, G. (1932) L'Azoto alimentare degli animali cavernicoli di Postumia (Considerazioni sul ciclo dell'Azoto). *Archiv. Zool. Ital.* **XVI.**

VANDEL, A. (1958) Sur l'édification de logettes de mue et de parturition chez les Isopodes terrestres troglobies et sur certains phénomènes de convergence observés dans le comportement des animaux cavernicoles. *Compt. rend. Acad. Sci.* Paris. **CCXLVII.**

VANDEL, A. and BOUILLON, M. (1959) Le Protée et son intérêt biologique. *Ann. Spéléol.* **XIV.**

VARGA, I. and TAKATS, T. (1960) Mikrobiologische Untersuchungen des Schlammes eines wasserlosen Teiches der Aggteleker Baradla-Höhle. *Biospeologica hungarica*, **VIII.** *Acta Zool. Acad. Sc. Hungar.* **VI.**

VIRÉ, A. (1900) *La Faune souterraine de France*. Paris.

Bacteriology and Speleobacteriology

BIRSTEIN, J.A. and BORUTZKY, E.W. (1950) Zizn v podziemnych vodach. *Zizn priesnych vod SSSR.* **III.**

CAUMARTIN, V. (1957a) La microflore des cavernes. *Notes biospéol.* **XII.**

CAUMARTIN, V. (1957b) Recherches sur une bactérie des argiles de cavernes et des sédiments ferrugineux. *Compt. rend. Acad. Sci.* Paris. **CCXLV.**

CAUMARTIN, V. (1959a) Quelques aspects nouveaux de la microflore des cavernes. *Ann. Spéléol.* **XIV.**

CAUMARTIN, V. (1959b) Aspect morphologique et position systématique du *Perebacterium spelaei*. *Bull. Soc. Botan. Nord. France* **XII.**

CAUMARTIN, V. (1961) La Microbiologie souterraine: ses techniques, ses problèmes. *Bull. Soc. Bot. Nord. France.* **XIV.**

CHODOROWSKI, A. (1959) Les Études biospéologiques en Pologne. *Biospeologica Polonica*, **II.** *Speleologica*, **I.**

DUDICH, E. (1930) Die Nahrungsquellen der Tierwelt in der Aggteleker Tropfsteinhöhle. *Allat. Közlem.* **XXVII.**

DUDICH, E. (1932) Biologie der Aggteleker Tropfsteinhöhle "Baradla" in Ungarn. *Speläol. Monogr.* **XIII.**

DUDICH, E. (1933) Die Klassifikation der Höhlen auf biologischer Grundlage. *Mitteil. Höhl. Karstf.*

FISCHER, E. (1959) Wyniki analizy bakteriologicznej drobnych zbiornikow wodnych groty Zimnej i Kasprowej. *Speleologia* III.

GINET, R. (1960) Ecologie, éthologie et biologie de *Niphargus* (Amphipodes Gammaridés hypogés). *Annal. Spéléol.* XV.

GOUNOT, A. M. (1960) Recherches sur le linon argileux souterrain et sur son rôle nutritif pour les *Niphargus* (Amphipodes, Gammarides). *Ann. Spéléol.* XV.

LIDDO, S. (1951) Ricerche batteriologiche nell'aria delle grotte di Castellana. Contributo allo studio della microflora cavernicola. *Boll. Soc. Ital. Biol. Speriment.* XXVII.

LWOFF, A. (1944) *L'Évolution physiologique. Étude des pertes de fonctions chez les Microorganismes.* Paris.

MAGDEBURG, P. (1933) Organogene Kalkkonkretionen in Höhlen. Beiträge zur Biologie der in Höhlen vorkommenden Algen. *Sitzb. naturf. Gesell. Leipzig.* LVI–LIX.

MASON-WILLIAMS, A. and BENSON-EVANS, K. (1958) A preliminary investigation into the bacterial and botanical flora of caves in South Wales. *Cave Res. Group. Great Brit. Public.* No. 8

MOLNAR, M. (1961) Beiträge zur Kenntnis der Mikrobiologie der Aggteleker Tropfstein-höhle "Baradla". *Biospeologica Hungarica*, XIV. *Annal Univ. Sc. Budapest. Rolando Eötvös nom. Sect. Biol.* IV.

VARGA, L. and TAKATS, T. (1960) Mikrobiologische Untersuchungen des Schlammes eines wasserlosen Teiches der Aggteleker Baradla-Höhle. *Acta Zool. Acad. Sc. Hungar.* VI.

WINOGRADSKY, S. (1949) *Microbiologie du Sol. Problèmes et Méthodes. Cinquante ans de recherches.* Œuvres complètes. Paris.

Food Utilisation and Food Cycles

DECU, V. (1961) Contributi la studiul morfologiei interne la coleoptere cavernicole din seria filetica *Sophrochaeta* Reitter (Catopidae-Bathysciinae). *Stud. Cerc. Biol. ser. Biol. animala. Romín.* XIII.

DUDICH, E. (1932) Biologie der Aggteleker Tropfsteinhöhle "Baradla" in Ungarn. *Speläol. Monogr.* XIII.

GINET, R. (1960) Écologie, Éthologie et Biologie de *Niphargus* (Amphipodes, Gammarides hypogés). *Ann. Spéléol.* XV.

GOUNOT, A. M. (1960) Recherches sur le Limon argileux souterrain et sur son rôle nutritif pour les *Niphargus* (Amphipodes, Gammarides). *Ann. Spéléol.* XV.

MONAKOV, A. V. and SOROKIN, J. I. (1960) The application of isotope methods to study the feeding of Cyclopes and their naupleated stages with bacterial forage. *Izvest. Akad. Nauk SSSR. Ser. Biol.* No. 6.

SCHOPFER, W. H. (1949) Les Vitamines du Sol. *In* Perspectives nouvelles dans la Chimie des Êtres vivants. *Actual. Sc. Industr.* No. 1073.

CHAPTER XX

THE METABOLISM OF CAVERNICOLOUS ANIMALS

THE TRANSFORMATIONS of material and energy which take place in living organisms and which are known under the name of metabolism, are extremely complex. An indication of the rate of metabolic activity as a whole can be obtained by investigating the respiratory exchanges, that is to say, those processes which provide the energy for life. The only knowledge we have of metabolism in cavernicoles concerns these phenomena. The first experiments carried out in this field were by Jules Gal in 1903.

A. RESPIRATORY METABOLISM

Respiratory metabolism can be expressed in terms of:
(1) The respiratory rate, that is the volume of oxygen absorbed (or of the carbon dioxide released) during one hour in relation to unit weight of animal.
(2) The respiratory quotient which is the ratio CO_2/O_2.
The measurement of respiration is carried out in respiratory chambers or micro-respirometers or eudiometers according to the size of the animal. The reader can find a description of these apparatus in the publications of L. Dresco-Derouet (1960 a and b).

B. RESPIRATORY METABOLISM IN DIFFERENT CAVERNICOLES

(a) Decapods

W. D. and M. P. Burbanck, and J. P. Edwards (1947, 1948), have compared the respiratory metabolism of two American crayfish, firstly the blind, cavernicolous *Cambarus setosus* from Smallin's cave, and *C. rusticus*, a river form.

These American biologists enclosed three specimens of each species in a jar containing one litre of boiled water covered with a layer of oil. Under such conditions the animals died of asphyxia. The duration of survival was measured. It was shown that the animals died when the oxygen tension

342

fell to a level which was the same for both species, $0·2$ cm³/litre. The duration of survival was however different for the two species. The average survival time for the epigeous species was 272 minutes and that of the cavernicole 892 minutes.

It must be concluded from these experiments that both animals were equally sensitive to oxygen shortage but respiratory metabolism was three times less intense in the cavernicoles than in the epigeans.

(b) Amphipods

The first experiments of Gal (1903) and Jeannel (1929) on *Niphargus*, have been repeated using more precise methods by L. Dresco-Derouet (1949, 1953, 1959). This physiologist compared the respiratory rate of an epigeous amphipod *Gammarus pulex* and a cavernicolous form *Niphargus virei*. At $10°C$ the respiratory rate of *G. pulex* was 10 to 15 times higher than that of *N. virei*.†

(c) Isopods

L. Dresco-Derouet (1953, 1959) has compared the respiratory rates of two spheromid isopods, a cavernicolous form, *Caecosphaeroma burgundum* and a marine littoral species, *Sphaeroma seratum*. The figures obtained were similar to those obtained with *Niphargus*.

From these experiments it can be seen that aquatic Crustacea resist asphyxia for a much longer time than do epigeous forms, not because they are less sensitive to low oxygen tensions but because their respiratory needs are less.

(d) Spiders

The excellent work of L. Dresco-Derouet (1960) on the physiology of spiders provides some precise information on their respiratory metabolism. The lucicoles, *Nephila*, and *Araneus* were compared with the troglophilic or obscuricole forms *Meta* and *Tegenaria*. Unfortunately the true troglobious genus *Leptoneta* could not be used owing to its small size.

The respiratory rates of the troglophiles were 5 to 7 times lower than those of the lucicoles. The respiratory quotients of the cavernicoles were low, of the order of $0·6$.

† Troïani (1954), and Wautier and Troïani, (1960), compared the respiratory rates of the two genera *Gammarus* and *Niphargus* but while the results are qualitatively similar to those of L. Dresco-Derouet it is difficult to compare them quantitatively. In most cases it seems that the respiratory rate of *N. virei* is clearly lower than that of *N. longicaudatus*.

(e) Fish

Eigenmann (1909) reported that the respiratory rhythm of *Amblyopsis spelaeus* was four times slower than that of epigeous genus *Chologaster*. Poulson (1961), using more precise methods, has established that the respiratory metabolism of the Amblyopsidae is very slow.

Professor Koch has measured the oxygen consumption of *Caecobarbus geertsi* (*in* Olivereau and Francotte-Henry, 1956). The oxygen consumption of this cavernicolous fish is reduced to the third of the consumption obtained in a tropical barb of the same size, and at the same temperature (23 °C).

The measurements of Schlagel and Breder (1947) appear to contradict the above results. These American biologists compared the respiratory rates of two characids, an epigean form *Astyanax mexicanus* and a cavernicolous form *Anoptichthys jordani*. However, the behaviour of these two fish is different (p. 386). In the first species periods of activity are separated by long periods of inactivity, while the second is continually active. Thus to obtain valid results it is necessary to compare fish whose behaviour is similar. If this condition is respected then it has been shown that measurements of respiratory rate yield similar results. Such a result is not surprising as *Anoptichthys jordani* is a recent cavernicole which is closely related to the original species represented by *Astyanax mexicanus*.

(f) Urodeles

In spite of the large size of its erythrocytes the oxygen consumption of *Proteus* is normal and of the same magnitude as the epigeous urodeles, newts, axolotls, and *Amblystoma*, (Korjoujev, 1950). These results can be explained by the fact that *Proteus* is a little-specialised cavernicole which remains very close to epigeous *Necturus*. It would be interesting to measure the respiratory rate of a very specialised type such as *Typhlomolge rathbuni*.

(g) Conclusions

The results obtained with crustaceans, spiders and fish, lead to the same conclusions. The respiratory metabolism and probably all other metabolic phenomena of ancient cavernicoles are much slower than those of related epigeous forms. It can be deduced that these cavernicoles lead a much less active life than lucicoles do.

These modifications are found only in ancient cavernicoles. Recent cavernicoles do not differ markedly from epigeous forms in their metabolic rates. It can be concluded that the physiological modifications of ancient cavernicoles have taken an extremely long time to develop.

C. THE ACTION OF EXTERNAL FACTORS
ON RESPIRATORY METABOLISM

The Neo-Lamarckians would have explained the very low respiratory rates and respiratory quotients of cavernicoles to be influenced by the external factors such as darkness and low temperature, and for terrestrial animals in the increased humidity. These interpretations were disproved when experimentation replaced the purely inductive method.

(a) Influence of Salinity

It has been known for a long time that subterranean amphipods such as *Niphargus* and related genera, are capable of living in brackish water. Fries (1879) reported the presence of *Niphargus* in wells on the island of Heligoland where the water was distinctly brackish. Chevreux (1901) stated that he had found *Pseudoniphargus africanus* in brackish water on some occasions.

The experimental studies of L. Dresco-Derouet (1952, 1959) have shown that *Niphargus virei* can live in 25% seawater and *Caecosphaeroma burgundum* in 33% seawater. In brackish water the respiratory rate increased during the first two days but by the end of 5–6 days some regulatory mechanism had intervened, and the level had returned to normal. In *Gammarus pulex*, which is an epigean form, the respiratory rate decreased rapidly when salt water was added to the tank.

It can be concluded from these experiments that the ancestors of the genera *Niphargus* and *Caecosphaeroma* must have been marine forms. This explains the tolerance of these animals to brackish water and that slightly salt water increases (even if temporarily) the respiratory rate.

(b) Influence of Light

Experiments carried out on the same Crustacea, *(Niphargus virei* and *Caecosphaeroma burgundum)* by L. Dresco-Derouet (1953, 1959) have shown that the effect of light on respiratory rate of these animals is weak. † It is much more noticeable when such illumination is accompanied by a rise in temperature.

(c) Influence of Temperature

L. Dresco-Derouet (1953, 1959), has shown that for the two Crustacea mentioned above, upon which she has carried out her experiments, a rise

† Wautier and Troïani (1960) would appear to have demonstrated a more definite effect of light on the respiratory metabolism of both epigeous and cavernicolous amphipods.

in temperature brings about an increase in respiratory rate. But the point of greatest interest noted by this physiologist is that the respiratory rate of these cavernicoles is less than that of similar epigeous species whatever the temperatures may be. At 10 °C the respiratory rates of the epigeous species *(Sphaeroma, Gammarus)* is 8 to 10 times greater than those of the caverni-coles *(Caecosphaeroma, Niphargus)*. At 17 °C the respiratory rate of these epigeans was 5 to 6 times higher. At 26 °C, which is the limiting temperature for these cavernicoles, the respiratory rate of the epigeans is still 3 to 5 times higher.

Thus respiratory metabolism is characteristic for each species. The differences in respiration are not the direct result of the action of external factors. L. Dresco-Derouet summed up in this way "the low rate of respira-tory metabolism of cavernicoles is not only the result of the effect of en-vironmental factors, but also of a long period of adaptation by the animal which would initially have been more or less preadapted for such conditions of life."

(d) Influence of Humidity

L. Dresco-Derouet studied the effect of humidity on the troglophilic spider *Meta menardi*. Lowering the humidity caused a decrease in respira-tory rate and irregularity of the respiratory quotient readings. Below 50% relative humidity the respiratory rate decreased by 20%. If the animal was maintained in a dry atmosphere for a long period the changes in respiration became irreversible and were followed by death of the animal. This was as a result of loss of water through the integument by evaporation.

Thus, *Meta menardi* appears to be a hydrophilic species which can sur-vive only in a humid atmosphere. It is to be expected that this effect would be emphasized in true troglobia, for example, species of the genus *Leptoneta*, but no data are available on this subject. Cavernicolous spiders can thus be considered as stenohygrobia, a category which also includes the troglobious Coleoptera.

(e) Conclusions

The experiments described above show that external factors influence metabolic processes in cavernicoles as well as in epigeans. However, there is no evidence that changes in the overall metabolic characteristics of an animal are caused by changes in the environment. The metabolism of caverni-coles remains of the cavernicolous type, and that of epigeans of the epi-geous type.

Thus, experimental studies do not confirm the statement which has so often been made, namely that the extremely reduced metabolic rate of cavernicoles represents an "adaptation" to conditions in the subterranean environment. The slower metabolism of ancient cavernicoles is a manifestation of phyletic senility. It corresponds to the end of a regressive evolutionary line which has been carried out in the subterranean world while such degenerative transformations would have rapidly been eliminated in the surface environment. The study of the morphology of cavernicoles also leads to this conclusion as will be seen when we return to the subject in the last part of this book.

BIBLIOGRAPHY

BURBANCK, W. D., EDWARDS, J. P. and BURBANCK, M. P. (1947) Toleration of lowered oxygen tension by cave and stream crayfish *Biol. Bull.* **XCIII.**

BURBANCK, W. D., EDWARDS, J. P. and BURBANCK, M. P. (1948) Toleration of lowered oxygen tension by cave and stream crayfish. *Ecology* **XXIX.**

CHEVREUX, ED. (1901) Amphipodes des eaux souterraines de France et d'Algérie. *Bull. Soc. Zool. France* **XXVI.**

DEROUET, L. (1949) Comparaison des échanges respiratoires chez *Gammarus pulex* L. et *Niphargus virei* Chevreux. *Compt. rend. Acad. Sci. Paris* **CCXXVIII.**

DEROUET, L. (1952) Influence des variations de salinité du milieu extérieur sur des Crustacés cavernicoles et épigés. I. Étude de l'intensité des échanges respiratoires. *Compt. rend. Acad. Sci. Paris* **CCXXXIV.**

DEROUET, L. (1953) Étude comparée du Métabolisme respiratoire chez certaines espèces de Crustacés cavernicoles et epigés. *Notes biospéol.* **VIII.**

DRESCO-DEROUET, L. (1959) Contribution a l'étude de la Biologie de deux Crustacés aquatiques cavernicoles: *Caecosphaeroma burgundum* D. et *Niphargus orcinus virei* Ch. *Vie et Milieu* **X.**

DRESCO-DEROUET, L. (1960a) Étude biologique de quelques espèces d'Araignées lucicoles et troglophiles. *Archiv. Zool. expér. géner.* **XCVIII.**

DRESCO-DEROUET, L. (1960b) Un micro-eudiomètre portatif pour l'analyse de petits volumes d'air. Technique simple de mesure de respiration dans les grottes. *Ann. Spéléol.* **XV.**

EIGENMANN, C. H. (1909) Cave vertebrates of America. A study in degenerative evolution. *Carnegie Inst. Washington Public.* No. 104.

FRIES, S. (1879) Mitteilungen aus dem Gebiete der Dunkelfauna. *Zool. Anz.* II.

GAL, J. (1903) *Niphargus* et *Caecosphaeroma.* Observations physiologiques. *Bull. Soc. Sc. Nat. Nîmes* **XXXI.**

JEANNEL, R. (1929) Le Vivarium du Jardin des Plantes pendant l'année 1928. *Rev. Hist. Nat.* **X.**

KORJOUJEV, P. A. (1950) Consumption of oxygen by erythrocytes of some Amphibia (*Molge cristata* and *Proteus anguinus*). *Compt. rend. Ac. Sc. U.R.S.S.* **LXXII.**

OLIVEREAU, M. and FRANCOTTE-HENRY, M. (1956) Étude histologique et biométrique de la glande thyroide de *Caecobarbus geertsi* Blgr. *Annal. Soc. R. Zool. Belgique* **LXXXVI.**

POULSON, T. L. (1961) Cave adaptation in amblyopsid fishes. *Thesis of the University of Michigan.*

SCHLAGEL, S. R. and BREDER, C. M. JR. (1947) A study of the oxygen consumption of blind and eyed cave characins in light and in darkness. *Zoologica* **XXXII.**

TROÏANI, D. (1954) La consommation d'oxygène de quelques *Gammaridae. Compt. rend. Acad. Sci. Paris* **CCXXXIX.**

WAUTIER, J. et TROÏANI, D. (1960) Contribution à l'étude du métabolisme respiratoire de quelques Gammaridae. *Ann. Stat. Centrale Hydrobiol. appliq.* **VIII.**

CHAPTER XXI

THE ENDOCRINE GLANDS OF CAVERNICOLOUS ANIMALS

A. INTRODUCTION

The measurements of the activity of the endocrine glands is another method which gives an indication of the metabolic rate of animals. In the vertebrates the pituitary and thyroid glands may be studied because it is known that these organs exert a multiple effect on metabolic and reproductive functions as well as on growth.

B. INVERTEBRATES

Very little is known of the morphology and physiology of the endocrine system in cavernicolous invertebrates. However, it cannot be doubted that the unusual modifications to reproduction and development found in cavernicolous Coleoptera, which will be described in the next chapter, are controlled by the endocrine organs (corpora allata and cardiaca and the prothoracic gland). A study of the endocrine system of these insects would certainly yield some interesting results, but it has not as yet been attempted.

Our present knowledge concerns cavernicolous Crustacea only, and so it is difficult to come to any general conclusions.

Husson and Graf (1961) reported that the androgenic gland of *Niphargus* is much smaller than that of *Gammarus* and shows reduced activity.

In some unpublished work Legrand and Juchault have shown the presence of two androgenic glands per gonad (one in the fifth segment and the other in the sixth or seventh), in the marine spheromids *Sphaeroma*, *Dynamene* and *Cymodocea*, while the cavernicolous species *Caecosphaeroma burgundum* possesses only one androgenic gland, situated in the fifth segment.

However, it must not be concluded from these observations that a reduced endocrine system is characteristic of cavernicolous Crustacea. In fact, the androgenic glands of the terrestrial, cavernicolous oniscoid *Scotoniscus macromelos* are normal and very similar to those of the epigous Trichoniscidae (Juchault, unpublished).

348

C. URODELA

The structure and function of the endocrine system has been studied in two urodela. These are *Proteus* and *Typhlomolge.*

(a) Proteus

The thyroid of *Proteus* has been studied by Leydig (1852), Oppel (1889), Bolau (1899), Versluys (1925) and Schreiber (1931). The most complete description of this organ has been given by Klose (1931). Contrary to the ideas of early anatomists, who described the thyroid of *Proteus* as reduced, Klose stated that this organ was not rudimentary at all. It was constructed in the same way as the thyroid of most other urodeles, consisting of four segments, two anterior, median and unpaired, and two posterior, paired and lateral ones. The follicles are well developed although certain of them appear to be abnormal following hypersecretion of chromophobic colloid. However, the mode of functioning of the thyroid of *Proteus* differs from that of other urodeles. In the latter, all the follicles develop at the same time and development ends with a massive discharge of colloid at the moment of metamorphosis. In *Proteus* the follicles develop out of phase with each other, and one may find in the same gland, vesicles in various stages of development. As a result there is no single, huge discharge of colloid. This phenomena is associated with the neotonous state of *Proteus* and not with the cavernicolous way of life. It can be concluded that the thyroid of *Proteus* does not represent a regressed organ, either on a histological or an anatomical level.

Physiological studies have shown that the thyroid is functional. Vialli (1931) implanted thyroids of *Proteus* in frog tadpoles and obtained metamorphosis. The activity of the thyroid has been shown to be under the control of the anterior pituitary (Schreiber, 1933) as is the case in other vertebrates.

The thymus of *Proteus* has been described by Leydig (1852), Siebold and Stannius (1856), and Schreiber (1931), but the best description of this organ was that made by Klose (1931). In *Proteus* the thymus consists of four pairs of glands situated one behind the other. The structure is similar to that of the thymus of the axolotol. The thymus persists in the adult *Proteus.*

We may conclude by saying that the endocrine glands of *Proteus* are about normal. The few changes in their structures result from the neotenous condition of this animal rather than from its cavernicolous mode of life.

(b) Typhlomolge

Typhlomolge rathbuni is a cavernicole showing many degenerate characters. In this species the thyroid is not absent as the early zoologists who studied this animal believed (Emerson, 1905; Uhlenhuth, 1923). The thyroid is almost normal but asymmetrical, and divided into two segments one of which is situated behind the branchial arches. However, it contains about a hundred follicles full of colloid. From all that can be estimated from the histological pattern the gland appears to be functional (Gorbman, 1957).

D. FISH

A distinction must be established in this group, between the recent and little modified cavernicoles, and the ancient, highly specialised ones. Examples of the first type are the genera *Chologaster* and *Anoptichthys*, and of the second the genera *Caecobarbus* and *Uegitglanis*. It is particularly interesting to study the first type as they demonstrate how the passage from the epigeous to the cavernicolous way of life was accomplished.

(a) Recent Cavernicoles

1. Chologaster. The Amblyopsidae of the genus *Chologaster* contain two species, one epigeous, *C. cornutus*, and the other troglophilic, *C. agassizi*. In spite of its epigeous way of life, *C. cornutus* is already preadapted to cavernicolous life. However, if it is maintained in darkness, some specimens show after three to four months, distortions of the vertebral column and erratic swimming movements (Poulson, 1961). These anomalies are probably the result of endocrine disorders.

2. Astyanax and Anoptichthys. Miss Rasquin (Rasquin, 1949; Rasquin and Rosenbloom, 1954) undertook the first experiments upon the Mexican Characidae. She experimented at first upon the river fish, *Astyanax mexicanus*, which represents the ancestor of the cavernicolous types. Miss Rasquin observed that when the river fish is maintained in permanent darkness, various anomalies appeared: accumulation of fat which deformed the body (Fig. 69); decrease of the rate of growth; sterilisation of the gonads, and in consequence the absence of reproduction. Miss Rasquin regarded these anomalies as the result of permanent darkness, and of the failure of the stimulation by light upon the endocrine glands, particularly upon the pituitary.

However, similar experiments undertaken by Mrs. Franck-Krahé (1962) give different results. No anomaly appeared in the fishes maintained in darkness. Otherwise, the rate of growth is normal and the gonads come to maturity in the river fish reared from the egg in permanent darkness.

Mrs. Franck-Krahé thinks that the anomalies observed by Miss Rasquin do not result from the darkness, but from old age of the fish. Fishes two or three years old show similar anomalies, although they were maintained by the light.

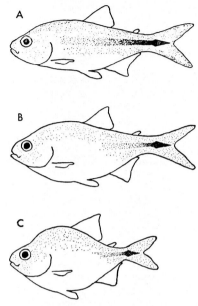

FIG. 69. The "River-fish", *Astyanax mexicanus*. A, a normal individual cultured in the light. B, an individual which has spent three months in darkness. C, an individual which has spent two years in darkness (after Rasquin).

Mrs. Franck-Krahé showed also that the river fish lay eggs at night. This behaviour represents a preadaptation to cavernicolous life.

As for the cavernicolous Characidae *(Anoptichthys jordani* and A. *hubbsi)*, their physiology and their reproduction are normal whether their eyes are regressed or not.

(b) Ancient Cavernicoles

1. Caecobarbus. The thyroid of *Caecobarbus geertsi* is reduced in volume and is probably of reduced activity (Olivereau and Francotte-Henry, 1956). The thyrotropic basophilic cells of the pituitary are very few and small (Olivereau and Herlant, 1954; Olivereau, 1960).

2. Uegitglanis. In another cavernicolous fish *Uegitglanis zammaronoi*, the majority of the follicles of the thyroid (except for those situated on the surface of the heart) are small and usually weakly differentiated and show signs of hypoactivity. They are also reduced in number and dispersed (Olivereau, 1960).

These phenomena must be interpreted as manifestations of regressive evolution rather than as an adaptation to cavernicolous life.

BIBLIOGRAPHY

Invertebrates

HUSSON, R. and GRAF, FR. (1961) Comparaison des glandes androgènes d'Amphipodes appartenant à des genres hypogés *(Niphargus)* et épigés *(Gammarus)*. *Compt. rend. Acad. Sci. Paris* **CCLII.**

HUSSON, R. and GRAF, F. (1961) Existence de la glande androgène chez le Crustacé troglobie *Niphargus*; comparaison avec celle du genre épigé *Gammarus*. *Ann. Spéléol.* **XVI.**

Urodela

BOLAU, H. (1899) Glandula thyroidea und Glandula Thymus der Amphibien. *Zool. Jahrb. Abt. Anat.* **XII.**

EMERSON, E. T. (1905) General anatomy of *Typhlomolge rathbuni*. *Proc. Soc. Nat. Hist. Boston* **XXXII.**

GORBMAN, A. (1957) The thyroid gland of *Typhlomolge rathbuni*. *Copeia.*

KLOSE, W. (1931) Beiträge zur Morphologie und Histologie der Schilddrüse, der Thymusdrüse und des postbranchialeu Körpers von *Proteus anguineus*. *Zeit. Zellf. mikr. Anat.* **XIV.**

LEYDIG, F. (1852) Über die Thyroidea und Thymus einiger Batrachier. *Froriep's Tagsber.*

OPPEL, A. (1889) Beiträge zur Anatomie des *Proteus anguineus*. *Archiv f. Mikrosk. Anat.* **XXXIV.**

SCHREIBER, G. (1931) La costituzione endocrina del *Proteus anguineus* Laur. ed il problema della Neotenia. *Atti. d. R. Ist. Veneto. d. Sc. Lett. Arti.* **XC.**

SCHREIBER, G. (1931) La struttura del timo nel Proteo. *Archiv. Zool. ital.* **XVI.**

SCHREIBER, G. (1933) Prove funzionale sulla tiroide del Proteo. *Boll. Soc. ital. Biol. Sperimentale* **VIII.**

SIEBOLD, C. TH. VON and STANNIUS, H. (1856) *Handbuch der Zootomie*. Berlin.

UHLENHUTH, E. (1923) The endocrine system of *Typhlomolge rathbuni*. *Biol. Bull.* **XLV.**

VERSLUYS, J. (1925) On the thyroid glands and the phylogeny of the perennibranchiate and derotremous salamanders. *Konink. Akad. Wet. Amsterdam Sect. Sc.* **XXVIII.**

VIALLI, M. (1931) Ricerche sulla metamorfosi degli anfibi. I. Innesti di tiroide di *Triton alpestris* neotenico e di *Proteus anguineus*. *Boll. Soc. ital. Biol. Sper.* **VI.**

Fish

FRANCK-KRAHÉ, A. C. (1962) Mexikanische Höhlencharaciniden im Vergleich zu ihren oberirdischen Vorfahren. *Staatsexamenarbeit der Universität Hamburg.*

OLIVEREAU, M. (1954) Hypophyse et glande thyroïde chez les Poissons. *Annal. Inst. Océanogr.* Paris. **XXIX.**

OLIVEREAU, M. (1960a) Quelques aspects anatomiques et physiologiques de la glande thyroïde des Poissons. *Annal. Soc. R. Zool. Belgique* **XC.**

OLIVEREAU, M. (1960b) Étude anatomique et histologique de la glande thyroïde d'*Uegitglanis zammaranoi* Gianferrari, Poisson aveugle et cavernicole. Comparaison avec un Clariidae voisin, *Clarias buthopogon* A. Dum. *Annal. Soc. R. Zool. Belgique.* **XC.**

OLIVEREAU, M. and HERLANT, M. (1954) Étude histologique de l'hypophyse de *Caecobarbus geertsi* Blgr. *Bull. Acad. R. Belgique* **XL.**

OLIVEREAU, M. and FRANCOTTE-HENRY, M. (1956) Étude histologique et biométrique de la glande thyroide de *Caecobarbus geertsi* Blgr. *Annal. Soc. R. Zool. Belgique.* **LXXXVI.**

POULSON, TH.-L. (1961) Cave adaptation in amblyopsid fishes. University of Michigan. *Phil. Dissert. Zoology.*

RASQUIN, P. (1949) The influence of light and darkness on thyroid and pituitary activity of the characin, *Astyanax mexicanus* and its cave derivates. *Bull. Amer. Mus. Nat. Hist.* **XCIII.**

RASQUIN, P. and ROSENBLOOM, L. (1954) Endocrine imbalance and tissue hyperplasia in teleosts maintained in darkness. *Bull. Amer. Mus. Nat. Hist.* **CIV.**

CHAPTER XXII

REPRODUCTION AND DEVELOPMENT
IN CAVERNICOLES

THE REPRODUCTION and development of cavernicoles shows some remarkable features. Their study is one of the most interesting aspects of biospeology, but one which has only recently been undertaken following the systematic breeding of these animals in the laboratory.

A. MODES OF REPRODUCTION

(a) Ovipary and Vivipary

The majority of cavernicoles are oviparous, only rarely are viviparous forms found. Vivipary has been reported in several parasites of cavernicoles but this condition is related more to the parasitic habit than to the subterranean one. Viviparous reproduction has been mentioned in *Proteus* but this is either an erroneous report or a case of abnormal and exceptional behaviour. It is now fully established that *Proteus* is normally oviparous (Vandel and Bouillon, 1959).

The fish belonging to the family Brotulidae are the only examples of viviparous cavernicoles. A few abyssal forms of this family, which is essentially marine, are viviparous. This is also the case with the three freshwater and cavernicolous members, as has been shown (primarily by Eigenmann, 1909), for *Stygicola* and *Lucifuga*. Fertilisation and gestation take place in the ovisac (Lane).

Eigenmann counted 4, 10 and 15 embryos, in three gravid females of *Lucifuga subterranea*. These figures are lower than those generally found in fish. This reduction is not, however, related to cavernicolous life, but to vivipary. A reduction in the number of offspring is a general characteristic of vivipary.

The work of H. Lane (in Eigenmann, 1909) which was devoted to oogenesis in the two cavernicolous Cuban fishes, *Lucifuga* and *Stygicola*, revealed some very interesting facts. The ovary of these fish is made up of "nests". Each of these contains many hundreds of cells. Of these cells only one develops and reaches a considerable size. This is the ovum which grows at the expense of the other cells of the "nest" which nourish it. This is an example of the phenomenon of adelphophagia.

(b) Forms which Incubate their Eggs

Although vivipary is rare among cavernicoles, incubation of the egg is found more frequently. Examples of this are known amongst cavernicolous fish and the Crustacea.

Fish. It was believed for many years that the Amplyopsidae resembled the related family Cyprinodontidae in being viviparous. This is not true. Eigenmann (1909) has shown that *Amblyopsis spelaeus* is an incubating form. This condition probably exists in the other species of the Amblyopsidae (Woods and Inger, 1957).

The branchial incubation which occurs in the Amblyopsidae is quite distinct from buccal incubation found in many fish belonging to the Cichlidae and Siluridae.

The urino-genital papilla of *Amblyopsis* is directed towards the anterior during post-embryonic development. In the adult it is situated behind the mouth and in front of the pectoral fins. Eggs pass from the papilla to the branchial cavity where they are incubated for two months. The Amblyopsidae have no copulatory organ and the mechanism of fertilisation is unknown.

Malacostraca. All the peracarids (Mysidacea, Amphipoda, Isopoda, and Spelaeogryphacea) are incubating forms. At the time of reproduction the females develop a marsupium which consists of osteostegites attached to the periopods. A reduction of the osteostegites is seen in highly specialised interstitial forms such as *Bogidiella brasiliensis* (Siewing, 1953).

Females of the Thermosbenacea possess a dorsal marsupium formed by the posterior part of the carapace (Stella, 1951, 1955, 1959; Barker, 1956, 1962). But the Bathynellidae do not incubate their eggs (Chappuis, 1948; Chappuis and Delamare-Deboutteville, 1954; Jakobi, 1954).

Copeopoda. In the copepods the eggs are not deposited in the external medium but in the ovisac. This structure is single in the Harpacticidae and double in the Cyclopidae. Normally, the eggs remain in the ovisac until the nauplii hatch.

This is not the case in some cavernicolous or obscuricolous forms where the female carries no ovisac. E. Graeter (1910), was one of the first zoologists to make some precise observations on the biology of the cavernicolous Copepoda. In the 76 females of *Moraria varica* which he inspected there was not one with an ovisac. Several years later Chappuis (1916, 1920), gave a reason for this state of affairs. The females of this harpacticid do form ovisacs but they are attached so loosely to the genital orifice that they very easily become detached. For this reason the sac can be seen only in animals reared in cultures and carefully observed.

Maupas (in Chappuis, 1916) established that the ovisac of *Viguierella caeca* has completely disappeared. The eggs are deposited in pairs on the substratum.

As *Moraria varica* and *Viguierella caeca* are only occasional caverni-coles there is no reason to think that these aberrant conditions of re-production are related to subterranean life. They should be regarded rather as a phenomenon of the regressive evolution peculiar to the very ancient lines, which involves the reduction of the cement glands of the oviduct.

B. NUMBER AND SIZE OF THE EGGS IN CAVERNICOLES

One of the most remarkable facts recorded by biospeologists is the re-duction in the number of eggs deposited by cavernicoles. This number is generally much less than that of epigeous species belonging to the same group. The decrease in the number of eggs is related to an increase in their size. This characteristic is more pronounced in those cavernicoles which are more specialised.

S. I. Smith reported in 1886 that abyssal decapods produce a small num-ber of large eggs. In the same year A. S. Packard showed that similar con-ditions are found in the cavernicolous spider, *Anthrobia mammouthia*. Raco-vitza (1910) for Sphaeromids and Fage (1913) for spiders, established a similar relationship between epigean and cavernicolous species with regard to the number and size of their eggs. Finally, it appears that this "embryogenic rule" is true for a great number of cavernicoles (Chappuis, 1920; Spandl, 1926; Hnatewytsch, 1929; Wächtler, 1929; Jeannel, 1943; Vandel, 1958). This rule applies similarly to members of the interstitial fauna (Remane, 1951; Delamare-Deboutteville, 1960).

A number of examples of this rule will shortly be given which have been taken from very different groups. However, the examples are taken only from the invertebrates. No examples are known in the vertebrates. This does not mean that this rule is not applicable to this group but only that for many reasons they cannot be included in the present section.

(1) Fish of the genus *Anoptichthys*, and *Proteus*, do not differ from epi-geous species as regards fecundity and the size of the eggs. This is not sur-prising as both genera represent little specialised cavernicoles.

(2) Nothing is known of reproduction in highly specialised cavernicolous vertebrates such as the hypogeous Cyprinidae and *Typhlomolge*.

(3) The Amblyopsidae seem to offer a good example of the conditions of ovulation in cavernicoles. In *Amblyopsis spelaeus* the number of eggs is reduced (about 70) and they are large (2·3 mm) (Eigenmann, 1909; Poul-son, 1961). The eggs of *Typhlichthys subterraneus* are larger than those of *Amblyopsis* and are less numerous (about 30 eggs according to Putnam, 1872)

A more remarkable example comes from the Brotulidae. As it has been said before, very few embryos are found in gestating females of *Lucifuga subterraneus* (4, 10 and 15) and the eggs are large due to the phenomenon of adelphophagia.

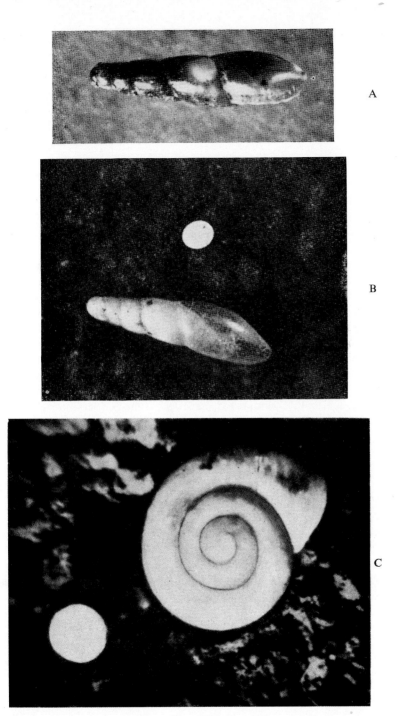

FIG. 70. *Caecilioides acicula*. A, whole animal; B, the animal and the egg. C, The animal and the egg of *Vallonia pulchella* (after Wächtler).

However, these examples cannot be taken as evidence for the embryogenic rule mentioned previously. The Amblyopsidae are incubating forms while the cavernicolous Brotulidae are viviparous. It is known that the number of eggs is always reduced in incubating or viviparous forms. The decrease in the number of eggs which one finds in cavernicolous fish belonging to these two families, must be considered as a consequence of incubation or vivipary rather than one of subterranean life.

Planarians. All that is known of the reproduction of cavernicolous planarians is due to the work of de Beauchamp (1932).The cocoons of obcuricolous planarians which were cultured by this zoologist, each produced 4–5 young while the coccoons of surface planarians *(Dendrocoelum lacteum, Planaria alpina)* contained about forty offspring. In the laboratory, obscuricolous planarians laid only once or twice while *Planaria polychroa* produced sixty cocoons during its lifetime.

Nematodes. According to Hnatewytsch (1929) the number of eggs produced by Rhabditidae living in the subterranean habitat is reduced compared with those produced by surface species.

Molluscs. Opeas cavernicola, a gasteropod from the caves of Assam, laid fewer and larger eggs than surface species belonging to the same genus (Annandale and Chopra, 1924).

Caecilioides acicula, a pulmonate gasteropod which is not a cavernicole but an endogean, gives us a good example of reproduction in hypogeous forms. Wächtler (1929) reported that this gasteropod laid a single egg at a time. It is of enormous size in relation to the size of the animal, measuring 0·75 mm in diameter and the shell of the mollusc 4–5 mm in length (Fig. 70). In captivity, the total number of eggs laid was 2 or 3 although it is probable that this number would be higher under natural conditions although not exceeding a dozen. These figures are very low when compared with those of epigeous molluscs: more than 200 eggs in *Helix pomatia;* 300–500 eggs in *Arion empiricorum*; 600–800 eggs in *Limax cinereo-niger*.

Opilionids. A comparison of the cavernicolous species *Ischyropsalis pyrenaea* and the muscicolous species *I. luteipes* shows that the number of eggs is much lower in the first species, while their size is greater. The figures reported by Juberthie (1961) are given below and clearly show these differences.

	Ischyropsalis luteipes	*Ischyropsalis pyrenaea*
Size of the eggs	0·8–0·9 mm	1·3–1·45 mm
Average number of eggs laid at one laying	16	10
Number of layings	8–15	3–6
Total number of eggs laid by one female	130–240	30–60

Palpigrades. Koenenia, which are endogenous or cavernicolous forms, only mature two eggs at a time. They are very large and almost completely fill the abdominal cavity (Millot, 1942).

Spiders. Emerton (1875) and Packard (1886) observed that the cavernicolous spider of the Mammoth Cave and neighbouring caves, *Anthrobia mammouthia*, built cocoons which contained only a small number of eggs (2–5), but these were very large.

The observations which have produced the most data in this field were carried out by L. Fage (1913, 1931). In his review of the Leptonetidae, he established that the number of eggs produced is related to the ecology of the animal. The coccoon of a sylvicolous form, *Leptoneta vittata* contains 6–8 eggs while that of the strictly cavernicolous form *L. leucophthalma* contains only two eggs. Finally, in the most specialised spider known, *Telema tenella* the cocoon contains a single egg of enormous size. It measures 0·4 mm in diameter while the whole animal is only a millimeter in length.

A fact can be reported here which will be considered again in the conclusions to this book. In certain spiders living in the great tropical forests, reproduction tends to be of the "cavernicolous type". A female of *Theotima fallax*, captured by Eugène Simon in the forest around Tovar Colony in Venezuela, carried only three eggs (Fage, 1912).

Copepods. A reduction in the number of eggs appears to be a general phenomenon in the cavernicolous copepods (Graeter, 1910; Chappuis, 1920; Spandl, 1926; Hnatewytsch, 1929; Rouch, 1961).

(1) A reduction in the number of eggs is seen in troglophilic cyclopids *(Acanthocyclops venustus)* and cavernicoles *(Eucyclops graeteri = macrurus subterranea)*, as well as in phreatobious types *(Acanthocyclops sensitivus)*. In these forms each ovisac contains only two eggs, that is to say four eggs are laid at one time. This figure is very low when compared with those of surface forms: 30–40 eggs per ovisac in *Macrocyclops fuscus* and 50–110 in *Cyclops viridis* (Röen, 1955).

(2) Similar conditions are to be found in the cavernicolous harpacticids. The ovisac carried by females of *Antrocampus catherinae* and *Nitocrella subterranea* contains 4–6 eggs while the average number in surface harpacticids is 15–25 (Rouch, 1961). Finally, the ovisac of *Moraria varica* contains only three eggs.

It has been reported that a similar reduction occurs in muscicolous forms. There are only two eggs in the ovisac of the anophthalmic species *Epactophanes muscicola* (Chappuis, 1916). It is the same with stenothermic forms of the genus *Arcticocamptus* from cold waters (Rouch, 1961). These facts are not surprising as it seems probable that cavernicolous harpacticids are related to muscicolous forms (p. 110).

Syncarids. Bathynella natans has two ovaries of approximately the same size. Each ovary contains a very large egg which is rich in yolk. Thus two eggs are laid at one time (Chappuis, 1915).

Asymmetry occurs in *Bathynella chappuisi*. This phenomenon is frequently seen in elongated forms which inhabit the interstitial medium (Remane, 1951; Delamare-Deboutteville, 1960). Only the left ovary is functional. In the right ovary the eggs abort. Thus one egg is laid at a time (Fig. 71) (Delachaux, 1919; Siewing, 1958).

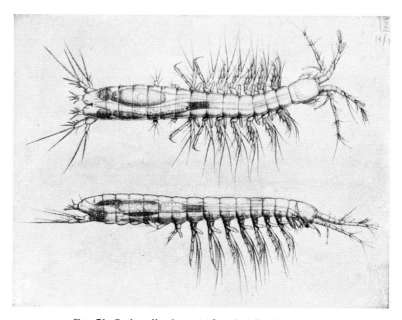

FIG. 71. *Bathynella chappuisi* female (after Delachaux).

Mysids. Females of the cavernicolous mysid from the Jameo de Agua, *Heteromysis cotti* possess a reduced incubating cavity which contains only two very large eggs (Fage and Monod, 1936).

Amphipods. The conclusions of a most interesting study carried out by R. Ginet (1960) devoted to members of the genus *Niphargus,* is summarised below:

(1) Small species of *Niphargus* and small epigeous gammarids measuring 4–6 mm contains approximately the same number of eggs (5–7).

(2) The ratio between the number of eggs and size of the females increases more rapidly in epigeans than in *Niphargus:* at equal size the large *Niphargus* have half as many eggs as the epigeans; with a similar number of eggs the females of *Niphargus* are much larger than epigeous females.

However, the volume of the eggs of large species of *Niphargus* such as *N. orcinus virei* is almost double that of epigeous species of the same size. It must be concluded that the volume of egg material produced is the same in cavernicoles and epigeans but its distribution is different. Analogous conditions are seen in the Bathysciinae (p. 364).

It is probable that a reduction in the number of eggs had begun in the epigeous ancestors of *Niphargus*. This large genus of cavernicolous amphipods is related, as has been mentioned earlier (p. 136) to the marine forms *Eriopisa* and *Eriopisella*. A marine species which lives in the interstitial

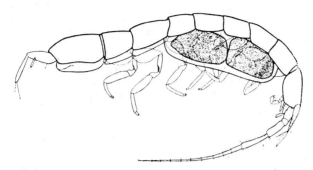

FIG. 72. An egg-bearing female of *Microcharon latus* (after Karaman).

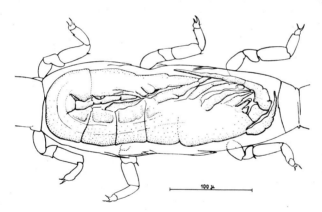

FIG. 73. An egg-bearing female of *Angeliera phreaticola* (after Delamare-Deboutteville and Chappuis).

medium on the coast of the island of Molene in Britainy, *Eriopisella pusilla*, incubates one egg at a time (Delamare-Deboutetville, 1960). It is the same in *Bogidiella brasiliensis*, another marine amphipod leading an interstitial mode of life (Siewing, 1953).

Anthuridae. The marsupium of females of *Microcerberus stygius* contains three eggs (Karaman, 1940).

Parasellidae. The marsupium of phreatobious parasellids contains a reduced number of eggs: three in *Pseudasellus* (Chappuis, 1951); two in *Microcharon* (Karaman, 1934, 1940; Chappuis and Delamare-Deboutteville, 1954) (Fig. 72); one egg in *Angeliera* (Gnanamuthu, 1954; Delamare-Deboutteville and Chappuis, 1956; Siewing, 1959). In the last example the egg is elongated and sausage-shaped; it is very large and reaches half the length of the mother's body (Fig. 73).

There is no doubt that the reduction in number of eggs seen in parasellids is associated with the phreatobious mode of life which these crustaceans lead.

Asellids. The marsupium of *Asellus cavaticus* contains 10–20 embryos (an average of 15) (Husson and Daum, 1955), while that of *Asellus aquaticus*, an epigeous form, contains 50–60 (Kaulbersz, 1913).

Sphaeromids. Marine sphaeromids lay numerous eggs but they are small. According to Hansen (1905) the number of eggs in the marsupium of *Sphaeroma serratum* is between 67 and 91. They were about 0·39 mm in diameter.

The detailed studies of the freshwater cavernicolous form *Caecosphaeroma burgundum*, (Husson and Daum, 1953; Daum, 1954; Husson, 1959) have established that the marsupium contains 4 to 15 eggs with an average of 7–8. The size is double that of the eggs of *Sphaeroma serratum* (0·75 mm).

The marsupium of *Monolistra caeca* contains about the same number of eggs as that of *Caecosphaeroma burgundum*, there being 6 to 8, but the size of the embryos is very large, measuring 2·75 mm while the females measure only 9 mm (Racovitza, 1910).

Decapods. The small number of eggs laid by females of *Munidopsis polymorpha* as well as their large size is opposite to the conditions seen in coastal galatheids, where the eggs are small and numerous (Calman, 1904). According to Fage and Monod (1936), the number of eggs carried by females of *Munidopsis polymorpha* varies from one to five. The eggs measure 1·5–1·8 mm.

Blattoids. In the small cavernicolous blattoid, *Alluaudella cavernicola* the ootheca contains 4–5 embryos (Chopard, 1932), a figure much lower than normal in epigeous species (6–50) (Chopard, 1938).

Coleoptera. The knowledge available on the conditions of ovulation in cavernicolous Coleoptera is due entirely to the remarkable work of Madame Sylvie Glaçon-Deleurance which she carried out for ten years with exemplary perseverence. This biologist systematically studied the conditions of oviposition as well as the structure of the egg in the two large groups of cavernicolous Coleoptera, the Bathysciinae and the Trechinae.

(a) Bathysciinae (Glaçon-Deleurance, 1953, 1954, 1957, 1958, 1959). Embryologically, the Bathysciinae can be divided into two groups.

(1) One group corresponds to muscicolous and little specialised cavernicolous forms. These species possess oligolecithal eggs containing a small quantity of yolk. These eggs are of small size but the number produced is large. Thus *Speonomus delarouzei* lays 15 to 18 eggs per month, *Bathysciola schiödtei* produces 8–10 per month and *Speonomus infernus* 7 per month.

(2) The second type corresponds to highly specialised cavernicoles, such as *Speonomus longicornis* and *S. diecki*, *Isereus serullazi* and *I. colasi* and *Cytodromus dapsoides*. A single egg is formed at a time; it is rich in yolk and very large. It entirely fills the abdomen of the female. The number of eggs produced is less than that in the first group, being 1 to 3 per month in *Speonomus longicornis*.

The species of the three Balkan genera *Sophrochaeta*, *Tismanella* and *Closania*, lay one egg at a time. The egg is very large and rich in yolk (Decu, 1961).

(b) Trechinae (Glaçon-Deleurance, 1954, 1958, 1960a and b). There is a remarkable parallelism in relation to ovogenesis between the Bathysciinae and the Trechinae, although these two groups are systematically far removed.

(1) It must be mentioned first that all the epigeous representatives of *Trechus* already show reproductive characteristics which are found in cavernicolous species. Only a small number of eggs are laid at a time (an average of 8) and these are of considerable size (0·5 mm) compared with the size of the insect (4 mm). The mature eggs completely fill the abdomen of the female (Leitner, 1943).

(2) In the second group which comprises the endogeans and cavernicoles belonging to the genera *Duvalius*, *Trichaphaenops* and *Geotrechus*, the eggs are oligolecithal and of relatively small size. In *Trichaphaenops* six eggs have been seen to reach maturity simultaneously in the one female.

(3) In the genus *Aphaenops* the females mature a single egg at a time as do the specialised Bathysciinae. This egg is enormous and completely fills the abdomen.

Females of the genus *Hydraphaenops* also produce one egg at a time. This egg is intermediate in size between that of *Trichaphaenops* and that of *Aphaenops*.

C. STRUCTURE OF THE OVARIES AND OOGENESIS IN CAVERNICOLES

It is evident that the reduction in number of eggs and the increase in their size results from a particular type of oogenesis which is different from the normal type. To find the origin of this embryonic manifestation which is so widely distributed among hypogeous types, it is necessary to consider the structure and function of the ovary of cavernicoles.

Information on this topic is unfortunately very fragmentary and concerns only a few species. However, from the few cases known it can be concluded that the reduction in the number of eggs arises by different ways or methods, according to the systematic group.

(a) Asymmetry of Ovaries

Among forms inhabiting the interstitial medium an increase in the length of the body is often accompanied by reduction, and then regression, of one of the gonads (Remane, 1951; Delamare-Deboutteville, 1960). The example of *Bathynella chappuisi* has already been given (p. 360).

This condition is related to a lengthening of the body and not with subterranean life. It has been known for many years that such manifestations are found in the serpentiform vertebrates.

(b) Variation in the Number of Ovarioles

In the Trechinae the number of ovarioles forming the ovary varies according to the genus (Glaçon-Deleurance, 1958, 1960). The ovary is formed by four ovarioles in *Duvalius*, three in *Trichaphaenops*, two in *Hydraphaenops* and *Geotrechus*, and one in *Aphaenops*.

Thus, the ovary of primitive forms, which produces small oligolecithal eggs is formed of many ovarioles, while the female of *Aphaenops* which lays only one large egg at a time has a single ovariole on each side of the body. However, the relationship between the number of ovarioles and the size of the eggs is not quite regular. The female of *Geotrechus* which possess only two ovarioles produces small oligolecithic eggs.

(c) The Ovary in the Bathysciinae

The ovary of Bathysciinae always contains 5–6 ovarioles (Glaçon-Deleurance, 1958, 1961). This number is constant; there is no reduction in the number of ovarioles as occurs in the Trechinae.

The differences in the ovaries of different types of Bathysciinae which give rise to large or small eggs, are insignificant. Yolk formation is very similar in all the Bathysciinae and the quantity of yolk produced remains the same. But, in the specialised types the yolk accumulates in one egg, while in primitive forms it is divided among many oocytes. The ovaries of specialised Bathysciinae differs from those of less evolved types in that the number of oogonial divisions is reduced and the rhythm of divisions is slower. Analogous conditions are probably found in large species in *Niphargus* (p. 361). The reduction of the number of oogonial divisions is probably theresult of the low metabolism which is the rule among the specialized cavernicoles.

D. PROCESSES OF DEVELOPMENT

It is certain that the reduction in the number of eggs characteristic of cavernicoles and associated with an in crease intheir size and yolk reserves, must influence their development.

Conditions of development have been followed systematically only in the cavernicolous Coleoptera. This is also the group in which one finds the most remarkable modifications.

(a) Historical

For a long time one of the most important gaps in biospeological knowledge, resided in the almost complete lack of information on the early stages of the cavernicolous insects. Professor Jeannel (1911, 1920, 1922, 1926, 1943) has on many occasions brought to the attention of entomologists the absence, or at least extreme rarity, of the larvae of cavernicolous Coleoptera. Although the larvae of little-evolved types belonging to the genera *Speonomus* and *Speotrechus* were known, one was ignorant of the early stages of the highly specialised genera such as *Aphaenops*, *Antrocharis* and *Leptodirus*.

Jeannel thought that the reason for this lack was because ovigerous females and larval stages hid in inaccessible crevices. This interpretation was shown to be incorrect (p. 290). The complete answer to the problem was found only after systematic breeding work had been carried out.

The problem can today be considered as completely solved with respect to the two large groups of cavernicolous Coleoptera, that is the Bathysciinae and the Trechinae. This is due to the work of Madame Glaçon-Deleurance (1953–1961) who has made one of the most significant discoveries in biospeology during the last few years.

(b) Development in the Bathysciinae

Entomologists (Brasavola de Massa, Cabidoche, Cerruti, Decou, Franciscolo, Jeannel, Menozzi, Peyerimhoff, Weber) had captured larvae of cavernicolous Bathysciinae belonging to the genera *Bathysciola*, *Neobathyscia*, *Parabathyscia*, *Speocharis*, *Breuilia*, *Speonomus*, *Speonesiotes*, *Hohenwartia*, and *Tismanella*, in caves. However, the larvae of the most specialised types such as *Antrocharis* and *Leptodirus* remained unknown. On the other hand, nothing was known of the duration of larval development in the Bathysciinae, and the number of larval stages was unknown. Our knowledge on the whole life cycle of cavernicolous Bathysciinae is due to the research of Madame Glaçon-Deleurance.

This worker (2954–1961) recognised many types of development among the Bathysciinae, ranging from a very specialised type to an almost normal one.

(a) Nothing was known of the most specialised Bathysciinae and their larvae have never been captured. The reason for this was given by Madame Glaçon-Deleurance in the work she published in 1953 concerning the life cycle of *Speonomus longicornis*. The cycle appeared to be so unusual that it

was at first thought to be anomaly resulting from the conditions of culture. However, it was not long before early results were confirmed and today there is no doubt that the cycle observed corresponds to the normal condition.

In this species, the female lays a single egg at a time. The egg is large and full of yolk reserves.

The egg gives rise to one larva whose free life is extremely short; a few hours or days. This is why larvae of this type have never been found in caves.

The larva very rapidly builds a clay capsule and becomes enclosed in it. The larva remains in the capsule in a state of diapause for five to six months, and then pupates. During the short phase of freedom the larva does not feed, grow, or molt. Thus the cycle of this insect comprises a single larval stage. Normally the Coleoptera pass through three successive larval stages. Larvae of this type possess a non-functional and completely empty digestive tract. But the embryonic digestive tract is functional and digest the stored yolk (Glaçon-Deleurance, 1961).

The same type of development is found in other species of the Bathysciinae: *Speonomus pyrenaeus*, *S. piochardi*, *S. stygius*, *S. diecki*, *S. hydrophilus*, *S. abeilli*, *Antrocharis querilhaci*, *Troglodromus bucheti*, *Cytodromus dapsoides*, *Isereus serullazi*, *I. colasi*, and *Leptodirus hohenwarti*.

(b) The second type is characteristic of less specialised species such as *Speonomus delarouzei*, *S. infernus*, *S. oberthuri*, *S. mazarredoi*, *S. bourgoini*, and *S. sioberi*. The eggs are smaller and more numerous than those of species of the first type although they are laid one by one. The larval phase comprises an early stage of free, active life. This period is long (fifty to sixty-five days in *S. delarouzei*). This is the reason why most larvae belonging to species of this second group were known as they had been captured in caves during the stage of free life.

The larvae of the second type feed and grow and they molt once inside a capsule of clay which is called the temporary or moult capsule. During the second stage the larva does not feed and the digestive system regresses (Glaçon-Deleurance, 1961). At the end of the second stage the larva encloses itself in a clay capsula which is called the final or pupation capsule. Inside this the larva undergoes a period of diapause which in *S. delarouzi* lasts for three to four months. Thus the growth cycle comprises two successive stages separated by a moult.

It must be mentioned that the construction of clay capsules, moult capsule and pupation capsule, is a constant phenomenon in cavernicolous Bathysciinae, but which is unknown in epigeous species.

(c) An example of intermediate type is given by *Diaprizus serrulazi* the larva of which passes through two successive stages, as do species of the second group. However, this larva does not feed at any stage of its existence and in this way resembles the first type.

(d) Finally, in the muscicolous Bathysciinae, e.g. *Bathysciola schiödtei*, the female lays small oligolecithal eggs at the rate of eight to ten per month. The larvae undergo two moults and pass through three successive stages. They feed throughout their life except for the last period of the third stage. In many respects this mode of development resembles the normal type but they behave like cavernicoles in that the larva construct clay capsules at the time of the molt.

Very briefly, cavernicolous evolution with respect to development involves a more and more pronounced contraction of the larval phase which, in the specialised types, comprises but a single stage.

(c) Development in the Trechinae

Larvae of the Trechinae have been described over many years by entomologists (Boldori, Coiffait, Decou, Dudich, Jeannel, Peyerimhoff, Strouhal, Uéno, Xambeu). They belong to weakly or partly differentiated genera *(Trechus, Neaphaenops, Geotrechus, Speotrechus, Paraphaenops, Allegrettia, Typhlotrechus, Duvalius, Trichaphaenops, Arctaphaenops, Anophthalmus, Trechiamia,* and *Rakantrechus).* The reader will find recent information on the larvae of the Trechinae in one of the last publications of Boldori (1958).

However, the larvae of *Aphaenops* remained unknown. Besides, the studies mentioned above did not vary from the strictly morphological plan. They concerned the systematics of Coleoptera more than biospeology. All that is known on the methods of development in Trechinae is due to Madame Glaçon-Deleurance.

The study of the development of the Trechinae is less advanced than that of the Bathysciinae, because the breeding of imagines and still more of the larvae, are difficult. Nevertheless, due to the culture-work carried out by Madame Glaçon-Deleurance (1959–1960) information is available concerning three species. The methods of development of these species correspond to different types which resemble those found in Bathysciinae.

In the two groups, cavernicolous evolution has followed the same direction: reduction in the number of eggs, increase in their size, morphological condensation of the larval stage, shortening of the free phase of larval life; partial and then complete suppression of feeding. This is an example of the remarkable phenomenon of convergence which is established between two groups which are systematically very distant.

Having said this, the information available on ontogenic evolution in the Trechinae will be given.

(a) Aphaenops cerberus. The larva does not feed but moults at least once. The cycle comprises at least two larval stages. A pupa has not been obtained. That which is known of the cycle indicates that it resembles that of the bathysciine species *Diaprysus serrulazi.*

(b) Hydraphaenops ehlersi. The larva feeds and must undergo at least one moult.

(c) Trichaphaenops gounelli. The larva undergoes two moults and passes through three successive stages. It feeds during the first part of the two early stages. The pupa has not been obtained.

E. FACTOR OF TIME

The different types of development seen in cavernicoles have been examined and they have been compared with epigeous ontogenesis. However, until now the factor of time has been omitted, that is to say, the duration of development. This will now be dealt with:

(a) the period of development, that is to say the embryonic phase and the postembryonic phase which extends from hatching to the adult stage.

(b) the period of the adult phase, that is to say a measurement of longevity.

(a) Growth during the Embryonic and Post-embryonic Phases

Planarians. The only information available in this respect is due to de Beauchamp (1932). Embryonic development in hypogeous types lasts 10–16 weeks while it lasts 2–6 weeks in *Dendrocoelum lacteum*, an epigeous species.

Oligochaetes. According to Ohfuchi (1941) the postembryonic development of cavernicolous species is much longer than in epigeous species.

Opilionids. Juberthie (1961) has compared the duration of embryonic and post-embryonic development in two related Opilionids, a muscicolous species, *Ischyropsalis luteipes*, and a cavernicolous species *I. pyrenaea.* Embryonic development lasts 44 days in the first species and 58 days in the second. Post-embryonic development extends from 4 to 4·5 months in the first species and from 6 to 7 months in the second.

Spiders. Madame Dresco-Derouet (1960) has compared the duration of growth of a troglophilic spider *Meta menardi* with that of epigeous species. The author concluded that:

"(1) The development of *Meta menardi*, a troglophilic species is very slow and takes at least 18 months.

(2) That of epigeous species is much more rapid and is completed in 8–10 months".

If one considers the troglobious species, the contrast between the epigeous and cavernicolous types is accentuated. Postembryonic growth in *Leptoneta microphthalma* takes a minimum of three years in spite of their small size (Glaçon-Deleurance, 1955).

Syncarids. In spite of their microscopic size, the adult stage of *Bathynella natans* is not reached until the end of nine months (Jakobi, 1954).

Amphipods. Ginet (1960) has reported that the moult-cycle of *Niphargus orcinus virei* is very slow compared with that of epigeous forms. The interval between two adult moults is eight months in cavernicolous species and from one to three weeks in the epigeous forms.

Myriapods. In the great work of Brölemann (1923), which he devoted to cavernicolous Blaniulidae, he established the following rule: "Today it is absolutely certain that the number of stages are slightly higher, or much higher, for cavernicolous forms, than for lucicolous forms". This is shown by the table given below taken from Brölemann's results.

Species	Mode of life	Number of larval stages	Maximum number of segments	Number of segments corresponding to sexual maturity in the male
Blaniulus guttulatus	Lucicole	9	59	38–40
Typhloblaniulus troglobius	Cavernicole	11	60	45–46
Archichoneiulus drahoni	Cavernicole	14	98	71–78

From the table above it can be concluded that development of cavernicolous diplopods is characterised by the three following conditions:

(1) The number of larval stages and as a result the number of moults, is higher in cavernicoles than in lucicoles. That is to say the post-embryonic phase of the first is longer than that of the second.

(2) It follows from this that the number of segments is always higher in cavernicoles than in epigeans.

(3) Sexual maturity manifests itself later in cavernicoles than in surface forms.

Coleoptera. The development of the cavernicolous Bathysciinae is very slow; it lasts 9 to 12 months (Glaçon-Deleurance).

Fish. Heuts (1952) has compared the growth of *Caecobarbus geertsi*, with that of an epigeous African barb, *Barbus holotaenia*. He reported that growth in the second species is twice as fast as that in the first. The duration of development of the Amblyopsidae is according to Poulson (1961) longer than that in epigeous species.

(b) Longevity

Information on the longevity of cavernicoles is rare and imprecise. Above all, when figures for cavernicoles are available, it is the comparison with related epigeans which is lacking. However, one can say generally that the longevity of cavernicoles is higher than that of epigeans.

Here are a few examples: Firstly that of *Niphargus orcinus virei* (Ginet, 1960). This amphipod does not reach the adult stage until after two and a half years and the average length of life is six years. However, specimens born in the Grotte-Laboratoire at Moulis have been kept alive for 8 years. Epigeous gammarids reach the adult state at the age of three months and the longevity is of the order of one, to one and a half years.

The adult Bathysciinae live on average 4 to 5 years (Glaçon-Deleurance) which is exceptionally long for insects.

We can find some information on the longevity of the Amblyopsidae in the thesis of Poulson (1961). *Chologaster* is sexually mature at one year old; *Typhlichthys*, at 2–3 years old; and *Amblyopsis* at 3–4 years old. The duration of life is about 3–4 years by *Typhlichthys* and 5–7 years by *Amblyopsis*.

Briegleb and Schwartzkopff (1961) estimated the length of life of *Proteus* to be thirty years. It is not to be ruled out that the figure could be greater than this.

F. CONCLUSIONS

Thus, the information available today constitutes substantial evidence in favour of a general slowing of ontogenesis in cavernicoles in relation to both embryonic development and post-embryonic growth. The increased length of life of cavernicoles is only the extension to the adult stage of the slowing which characterises the development of hypogeous animals.

This conclusion is not surprising. It simply translates to development the slowing of metabolism and endocrine activity which are manifestations common to all ancient cavernicoles.

BIBLIOGRAPHY

Modes of Reproduction

BARKER, D. (1956) The morphology, reproduction and behaviour of *Thermosbaena mirabilis* Monod. *Proc.* **XIV**. *Intern. Zool. Congr. Copenhagen.*

BARKER, D. (1962) A study of *Thermosbaena mirabilis* (Malacostraca, Peracarida) and its reproduction. *Quart. J. Microsc. Sci.* **CIII**.

CHAPPUIS, P. A. (1916) *Viguierella caeca* Maupas. Ein Beitrag zur Entwicklungsgeschichte der Crustaceen (unter Benützung eines Manuskriptes von E. Maupas). *Rev. suisse Zool.*. **XXIV**.

CHAPPUIS, P. A. (1920) Die Fauna der unterirdischen Gewässer der Umgebung von Basel. *Archiv f. Hydrobiol.* **XIV.**

CHAPPUIS, P. A. (1948) Le développement larvaire de *Bathynella*. *Bull. Soc. Sc. Cluj* **X.**

CHAPPUIS, P. A. and DELAMARE-DEBOUTTEVILLE, C. (1954) Recherches sur les Crustacés souterrains. *Biospeologica*, **LXXIV.** *Archiv. Zool. expér. géner.* **XCI.**

EIGENMANN, C. H. (1909) Cave vertebrates of America. A study in degenerative evolution. *Carnegie Inst. Washington Public. No.* 104.

GRAETER, E. (1910) Die Copepoden der unterirdischen Gewässer. *Archiv. f. Hydrobiol.* **VI.**

JAKOBI, H. (1954) Biologie, Entwicklungsgeschichte und Systematik von *Bathynella natans* Vejd. *Zool. Jahrb. Abt. System.* **LXXXIII.**

SIEWING, E. (1953) *Bogidiella brasiliensis*, ein neuer Amphipode aus dem Küstengrundwasser Brasiliens. *Kieler Meeresforsch.* **IX.**

STELLA, E. (1951) *Monodella argentarii* n. sp. di Thermosbenacea (Crustacea, Peracarida), limnotroglobio di Monte Argentario. *Archivio Zool. Ital.* **XXXVI.**

STELLA, E. (1955) Behaviour and development of *Monodella argentarii* Stella, a thermosbenacean from an Italian cave. *Verh. Intern. Verh. Limnologie* **XII.**

STELLA, E. (1959) Ulteriori osservazioni sulla riproduzione e lo svilluppo di *Monodella argentarii* (Peracarida, Thermosbenacea). *Riv. Biol. Ital.* **LI.**

VANDEL, A. and BOUILLON, M. (1959) Le Protée et son intérêt biologique. *Ann. Spéléologie* **XIV.**

WOODS, L. P. and INGER, R. F. (1957) The cave, spring and swamp fishes of the family *Amblyopsidae* of Central and Eastern United States. *Amer. Midl. Nat.* **LVIII.**

Number and Size of the Eggs in Cavernicoles

ANNANDALE, N. and CHOPRA, B. (1924) Molluscs of the Siju Cave, Garo Hills, Assam. *Rec. Ind. Mus.* **XXVI.**

BEAUCHAMP, P. DE (1932) Turbellariés, Hirudinées, Branchiobdellides (2ème Série). *Biospeologica*, **LVIII.** *Archiv. Zool. expér. géner.* **LXXIII.**

CALMAN, W. T. (1904) On *Munidopsis polymorpha* Koelbel, a cave dwelling marine crustacean from the Canary Islands. *Ann. Mag. Nat. Hist.* (7) **XIV.**

CHAPPUIS, P. A. (1915) "*Bathynella natans*" und ihre Stellung im System. *Zool. Jahrb. Abt. System.* **XLIV.**

CHAPPUIS, P. A. (1916) Die Metamorphosis einiger Harpacticiden-Genera. *Zool. Anz.* **XLVIII.**

CHAPPUIS, P. A. (1920) Die Fauna der unterirdischen Gewässer der Umgebung von Basel. *Archiv. f. Hydrobiol.* **XIV.**

CHAPPUIS, P. A. (1951) Un nouveau Parasellide de Tasmanie, *Pseudasellus nichollsi* n.g.n.sp. *Archiv. Zool. expér. géner.* **LXXXVIII.** *N & R.*

CHAPPUIS, P. A. and DELAMARE-DEBOUTTEVILLE, CL. (1954) Recherches sur les Crustacés souterrains. *Biospeologica*, **LXXIV.** *Archiv. Zool. expér. géner.* **XCI.**

CHOPARD, L. (1932) Un cas de microphthalmie liée à l'atrophie des ailes chez une Blatte cavernicole. *Soc. Entomol. France. Livre du Centenaire.*

CHOPARD, L. (1938) *La biologie des Orthoptères.* Paris.

DAUM, J. (1954) Zur Biologie einer Isopodenart unterirdischer Gewässer: *Caecosphaeroma (Vireia) burgundum* Dollfus. *Annal. Univ. Sarav.* **III.**

DECU, V. (1961) Contributi la studiul morfologiei interne la coleoptere cavernicole din seria filetica *Sophrochaeta* Reitter (Catopidae, Bathysciinae). *Stud. Cerc. Biol. Ser. Biol. animala. Romín.* **XIII.**

DELACHAUX, TH. (1919) *Bathynella chappuisi* n. sp., une nouvelle espèce de Crustacé cavernicole. *Bull. Soc. Neuchâtel. Sc. Nat.* **XLIV.**

DELAMARE-DEBOUTTEVILLE, CL. (1960) Biologie des eaux souterraines littorales et continentales. *Actual. Sc. Industr. No.* 1280.

EIGENMANN, C.H. (1909) Cave vertebrates of America. A study in degenerative evolution. *Carnegie Inst. Washington Public.* No. 104.

EMERTON, J.H. (1875) Notes on spiders from caves in Kentucky, Virginia and Indiana. *Amer. Nat.* **IX.**

FAGE, L. (1912) Études sur les Araignées cavernicoles. I. Revision des *Ochyroceratidae* (n. fam.). *Biospeologica,* **XXV.** *Archiv. Zool. expér. géner.* (5) **X.**

FAGE, L. (1913) Études sur les Araignées cavernicoles. II. Revision des *Leptonetidae.* *Biospeologica,* **XXIX.** *Archiv. Zool. expér. géner.* **L.**

FAGE, L. (1931) Araneae (5ème Série) precédée d'un essai sur l'évolution souterraine et son déterminisme. *Biospeologica,* **LV.** *Archiv. Zool. expér. géner.* **LXXI.**

FAGE, L. and MONOD, TH. (1936) La faune marine du Jameo de Agua, lac souterrain de l'île de Lanzarote. *Biospeologica,* **LXIII.** *Archiv. Zool. expér. géner.* **LXXVIII.**

GINET, R. (1960) Écologie, Éthologie et Biologie de *Niphargus* (Amphipodes, Gammarides hypogés). *Ann. Spéléol.* **XV.**

GLAÇON, S. (1953) Sur le cycle évolutif d'un Coléoptère troglobie, *Speonomus longicornis* Saulcy. *Compt. rend. Acad. Sci. Paris* **CCXXXVI.**

GLAÇON, S. (1954) Sur le cycle évolutif de quelques *Speonomus* (Coléoptères, Bathysciinae) cavernicoles. *Compt. rend. Acad. Sci. Paris* **CCXXXVIII.**

GLAÇON, S. (1957) Note sur la biologie et la morphologie larvaire de *Bathysciola schiödtei grandis* F. *Compt. rend. Acad. Sci. Paris* **CCXLV.**

GLAÇON-DELEURANCE, S. (1958) Biologie et morphologie larvaire d'*Isereus serullazi* F., *Isereus colasi* Bon. et *Cytodromus dapsoides* Ab. *Compt. rend. Acad. Sci. Paris* **CCXLVI.**

GLAÇON-DELEURANCE, S. (1958) La contraction du cycle évolutif des Coléoptères *Bathysciinae* et *Trechinae* en milieu souterrain. *Compt. rend. Acad. Sci. Paris* **CCXLVII.**

GLAÇON-DELEURANCE, S. (1959) Contribution à l'étude des Coléoptères troglobies. Sur la biologie des *Bathysciinae.* *Ann. Spéléol.* **XIV.**

GLAÇON-DELEURANCE, S. (1960) Sur la biologie de l'*Hydraphaenops ehlersi* Ab. (Coléoptère, Trechinae). Description de la larve du premier stade. *Compt. rend. Acad. Sci. Paris* **CCLI.**

GLAÇON-DELEURANCE, S. (1960) Biologie et morphologie larvaire du *Trichaphaenops gounelli* Bedel (Coléoptère, Trechinae). *Compt. rend. Acad. Sci. Paris* **CCLI.**

GNANAMUTHU, C.P. (1956) Two new sand dwelling isopods from the Madras sea-shore. *Ann. Mag. Nat. Hist.* (12) **VII.**

GRAETER, E. (1910) Die Copepoden der unterirdischen Gewässer. *Archiv. f. Hydrobiol.* **VI.**

HANSEN, H.J. (1905) On the propagation, structure and classification of the family *Sphaeromidae. Quart. J. microsc. Sci.* **XLIX.**

HNATENWYTSCH, B. (1929) Die Fauna der Erzgruben von Schneeberg in Erzgebirge. *Zool. Jahrb. Abt. System.* **LVI.**

HUSSON, R. (1959) Les Crustacés Péracarides des eaux souterraines. Considérations sur la biologie de ces cavernicoles. *Bull. Soc. Zool. France* **LXXXIV.**

HUSSON, R., and DAUM, J. (1953) Sur la biologie de *Caecosphaeroma burgundum. Compt. rend. Acad. Sci. Paris* **CCXXXVI.**

JEANNEL, R. (1943) *Les Fossiles vivants des Cavernes.* Paris.

JUBERTHIE, CH. (1961) Données sur la biologie des *Ischyropsalis* C.L.K. (Opilions; Palpatores; Ischyropsalidae). *Ann. Spéléol.* **XVI.**

KARAMAN, ST. (1934) Beiträge zur Kenntnis der Isopoden — Familie *Microparasellidae. Mitteil. Höhl. Karstf.*

KARAMAN, S. (1940) Die unterirdischen Isopoden Südserbiens. *Glasnik Skop. nauc. drustva.* **XXII.**

KAULBERSZ, G.J. VON (1913) Biologische Beobachtungen an *Asellus aquaticus,* nebst einigen Bemerkungen über *Gammarus* und *Niphargus. Zool. Jahrb. Abt. allg. Zool.* **XXXIII.**

LEITNER, E. (1943) Morphologische und entwicklungsbiologische Untersuchungen an Laufkäfern der Gattung *Trechus* (Ein Beitrag zur Frage der Artbildung). *Zool. Jahrb. Abt. Anat. Ont.* **LXVIII.**

MILLOT, J. (1942) Sur l'anatomie et l'histophysiologie de *Koenenia mirabilis* Grassi (Arachnida, Palpigradi). *Rev. franç. Entomol.* **IX.**

PACKARD, A.S. (1886) The cave fauna of North America, with remarks on the anatomy of the brain and origin of the blind species. *Mem. Nat. Acad. Sc. Washington* **IV.**

POULSON, TH. L. (1961) Cave adaptation in amblyopsid fishes. *University of Michigan. Phil. Dissert. Zool.*

PUTNAM, F.W. (1872) The blind fishes of the Mammoth Cave and their allies. *Amer. Nat.* **VI.**

RACOVITZA, E.G. (1930) Sphéromiens (Première Série) et Révision des *Monolistrini* (Isopodes, Sphéromiens). *Biospeologica,* **XIII.** *Archiv. Zool. expér. géner.* (5). **IV.**

REMANE, A. (1951) Die Besiedlung des Sandbodens im Meere und die Bedeutung der Lebensformtypen für die Ökologie. *Verh. d. deutsch. Zool. Gesell.*

RÖEN, U. (1955) On the number of eggs in some free-living freshwater copepods. *Verhandl. Intern. Ver. Theor. angew. Limnologie.* **XII.**

ROUCH, R. (1961) Le développement et la croissance des Copépodes Harpacticides cavernicoles (Crustacés). *Compt. rend. Acad. Sci. Paris* **CCLII.**

SIEWING, R. (1953) *Bogidiella brasiliensis,* ein neuer Amphipode aus dem Küstengrundwasser Brasiliens. *Kieler Meeresforsch.* **IX.**

SIEWING, R. (1958) Syncarida. *Bronn's Klass. Ordn. Tierreichs.* **V,** Abt. 1; 4. Buch; Teil II.

SIEWING, R. (1959) *Angeliera xarifae,* ein neuer Isopode aus dem Küstengrundwasser der Insel Abd-el-Kuri (Golf von Aden). *Zool. Anz.* **CLXIII.**

SPANDL, H. (1926) Die Tierwelt der unterirdischen Gewässer. *Speläol. Monogr.* **XI.**

VANDEL, A. (1958) Sur l'édification de logettes de mue et de parturition chez les Isopodes terrestres et sur certains phénomènes de convergence observés dans le comportement des animaux cavernicoles. *Compt. rend. Acad. Sci. Paris* **CCXLVII.**

WÄCHTLER, W. (1929) Anatomie und Biologie der augenlosen Landlungenschnecke *Caecilioides acicula* Müll. *Zeit. Morphol. Oekol. Tiere.* **XIII.**

Structure of the Ovaries

DELAMARE-DEBOUTTEVILLE, CL. (1960) Biologie des eaux souterraines et littorales et continentales. *Actual. Sc. Industr.* No. 1280.

GLAÇON-DELEURANCE, S. (1958) La contraction du cycle évolutif des Coléoptères *Bathysciinae* et *Trechinae* en milieu souterrain. *Compt. rend. Acad. Sci. Paris* **CCXLVII.**

GLAÇON-DELEURANCE, S. (1960) Sur la biologie de l'*Hydraphaenops ehlersi* Ab. (Coléoptère Trechinae). Descriptoin de la larve du premier stade. *Compt. rend. Acad. Sci. Paris* **CCLI.**

GLAÇON-DELEURANCE, S. (1961) Anatomie de l'appareil génital femelle des *Bathysciinae*. *Ann. Spéléol.* **XVI.**

REMANE, A. (1951) Die Besiedlung des Sandbodens im Meere und die Bedeutung der Lebensformtypen für die Ökologie. *Verh. deutsch. Zool. Gesell.*

Types of Development

BOLDORI, L. (1958) Larve di Coleotteri, I. Larve di *Trechini,* X. *Memorie d. Soc. Entomol. Ital.* **XXXVII.**

GLAÇON, S. (1953a) Sur le cycle évolutif d'un Coléoptère troglobie, *Speonomus longicornis* Saulcy. *Compt. rend. Acad. Sci. Paris* **CCXXXVI.**

GLAÇON, S. (1953b) Description des larves jeunes de *Speonomus longicornis* Saulcy et *Speonomus pyrenaeus* Lespès. *Notes biospéol.* **VIII.**

GLAÇON, S. (1954a) Sur le cycle évolutif de quelques *Speonomus* (Coléoptères Bathysciinae) cavernicoles. *Compt. rend. Acad. Sci. Paris* **CCXXXVIII.**

GLAÇON, S. (1954b) Morphologie de la nymphe de *Speonomus longicornis* S. (Bathysciinae) cavernicole. *Annal. Sc. Nat. Zool.* (11) **XVI.**

GLAÇON, S. (1955a) Remarques sur la morphologie et la biologie de quelques larves de *Bathysciinae* cavernicoles. *Compt. rend. Acad. Sci. Paris* **CCXL.**

GLAÇON, S. (1955b) Description de la larve jeune d'*Antrocharis querilhaci* Lespès. *Notes biospéol.* **X.**

GLAÇON, S. (1955c) Sur la morphologie et la biologie larvaire de *Speonomus infernus* D. *Compt. rend. Acad. Sci. Paris* **CCXLI.**

GLAÇON, S. (1956) Recherches sur la biologie des Coléoptères cavernicoles troglobies. *Premier Congr. Intern. Spéléol. Communic. III.* Paris.

GLAÇON-DELEURANCE, S. (1957a) Cycle évolutif des larves de *Troglodromus bucheti gaveti* (S.C.D.), *Bathysciella jeanneli* (Ab.) et *Diaprysus serullazi* (P.) *Compt. rend. Acad. Sci. Paris* **CCXLIV.**

GLAÇON-DELEURANCE, S. (1957b) Note sur la biologie et la morphologie larvaire de *Bathysciola schiödtei grandis* F. *Compt. rend. Acad. Sci. Paris* **CCXLV.**

GLAÇON-DELEURANCE, S. (1957c) Biologie larvaire et Morphologie externe nymphale d'*Antrocharis querilhaci* Lespès. *Notes biospéol.* **XII.**

GLAÇON-DELEURANCE, S. (1957d) Description de la Morphologie externe larvaire de *Diaprysus serullazi* Peyer. *Notes biospéol.* **XII.**

GLAÇON-DELEURANCE, S. (1958a) Morphologie de la larve de *Bathysciella jeanneli* Ab. (Coléoptère, Bathysciinae). *Compt. rend. Acad. Sci. Paris* **CCXLVI.**

GLAÇON-DELEURANCE, S. (1958b) Biologie et morphologie larvaire d'*Isereus serullazi* F., *Isereus colasi* Bon. et *Cytodromus dapsoides* Ab. *Compt. rend. Acad. Sci. Paris* **CCXLVI.**

GLAÇON-DELEURANCE, S. (1958c) La contraction du cycle évolutif des Coléoptères Bathysciinae et Trechinae en milieu souterrain. *Compt. rend. Acad. Sci. Paris* **CCXLVII.**

GLAÇON-DELEURANCE, S. (1958d) Larve de *Troglodromus bucheti gaveti* Deville. Biologie et Morphologie externe. *Notes biospéol.* **XIII.**

GLAÇON-DELEURANCE, S. (1959a) Contribution à l'étude des Coléoptères troglobies. Sur la biologie des *Bathysciinae*. *Ann. Spéléol.* **XIV.**

GLAÇON-DELEURANCE, S. (1959b) Sur *l'Aphaenops cerberus* Dieck (Insecte Coléoptère) et sa larve. *Compt. rend. Acad. Sci. Paris* **CCXLIX.**

GLAÇON-DELEURANCE, S. (1960a) Sur la biologie de l'*Hydraphaenops ehlersi* Ab. (Coléoptère Trechinae). Description de la larve du premier stade. *Compt. rend. Acad. Sci. Paris* **CCLI.**

GLAÇON-DELEURANCE, S. (1960b) Biologie et morphologie larvaire de *Trichaphaenops gounelli* Bedel (Coléoptère, Trechinae). *Compt. rend. Acad. Sci. Paris* **CCLI.**

GLAÇON-DELEURANCE, S. (1961a) Morphologie des larves de *Royerella tarissani* Bedel et *Leptodirus hohenwarti* Schm. *Ann. Spéléol.* **XVI.**

GLAÇON-DELEURANCE, S. (1961b) Sur le système digestif et le corps adipeux des *Bathysciinae* troglobies. *Compt. rend. Acad. Sci. Paris* **CCLIII.**

JEANNEL, R. (1911) Révision des *Bathysciinae*, (Coléoptères, Silphides). Morphologie, distribution géographique, systématique. *Biospeologica*, **XIX.** *Archiv. Zool. expér. gén.* (5) **VII.**

JEANNEL, R. (1920) Les larves des *Trechini* (Coléoptères, Carabidae). *Biospeologica*, **XLII.** *Archiv. Zool. expér. géner.* **LIX.**

JEANNEL, R. (1922) Les *Trechinae* de France (Première Partie). *Annal. Soc. Entomol. France.* **XC.**

JEANNEL, R. (1926) *Faune cavernicole de France.* Paris.

JEANNEL, R. (1943) *Les Fossiles vivants des Cavernes.* Paris.

The Time Factor

BEAUCHAMP, P. DE (1932) Turbellariés, Hirudinées, Branchiobdellidés (2ème Série). *Biospeologica*, **LVIII.** *Archiv. Zool. expér. géner.* **LXXIII.**

BRIEGLEB, W. and SCHWARTZKOPFF, J. (1961) Verhaltensweisen des Grottenolms (*Proteus anguineus* Laur.) und das Problem des Fortpflanzungsraumes. *Naturwiss.* **XLVIII.**

BRÖLEMANN, H.W. (1923) *Blaniulidae*, Myriapodes (1ère Série). *Biospeologica*, **XLVIII.** *Archiv. Zool. expér. géner.* **LXI.**

DRESCO-DEROUET, L. (1960) Étude biologique de quelques espèces d'Araignées lucicoles et troglophiles. *Archiv. Zool. expér. géner.* **XCVIII.**

GINET, R. (1960) Écologie, Éthologie et Biologie de *Niphargus* (Amphipodes Gammarides hypogés). *Ann. Spéléol.* **XV.**

GLAÇON-DELEURANCE, S. (1955) Sur la biologie de *Leptoneta microphthalma* Simon (Araignées cavernicoles). *Compt. rend. Acad. Sci. Paris* **CCXLI.**

HEUTS, M. J. (1952) Ecology, variation and adaptation of the blind cave fish, *Caecobarbus geertsi* Boulenger. *Annal. Soc. R. Zool. Belgique* **LXXXII.**

JAKOBI, H. (1954) Biologie, Entwicklungsgeschichte und Systematik von *Bathynella natans* Vejd. *Zool. Jahrb. Abt. System.* **LXXXIII.**

JUBERTHIE, CH. (1961) Données sur la biologie des *Ischyropsalis* C. L. K. *Ann. Spéléol.* **XVI.**

OHFUCHI, S. (1941) The cavernicolous Oligochaeta of Japan. I. *Sc. Rep. Tohoku Univ.* 4, *Biol.* **XVI.**

POULSON, T. L. (1961) Cave adaption in amplyopsid fishes. *Thesis of the University of Michigan.*

PART 5

The Behaviour of Cavernicoles; Sensitivity and Sense Organs

The fourth part of this book is devoted to that section of physiology which deals with the transformations of matter and energy which occur inside the organisms and which are known as metabolism. In this second physiological section the activities of the animals and the functioning of the sense organs will be considered. We shall reconsider the reactions of cavernicolous animals in relation to external factors, but this time with respect to sensory and nervous function. This is a wide field with many aspects but these are so interrelated that separation of them is difficult.

CHAPTER XXIII

THE BEHAVIOUR OF CAVERNICOLES

A. ACTIVITY RHYTHMS

The activities of animals are rarely continuous. More often they are cyclic. The cycles can be exclusively endogenous but more often they depend on the alternation of day and night (diurnal rhythm) or the succession of seasons (seasonal rhythm).

An actograph, of which there are many types, can be used to measure overall activity (Szymanski, Chauvin, Ginet, etc). With this apparatus it is possible to measure the time and duration of periods of activity of an animal.

(a) Diurnal Rhythms

Most animals living on the surface show a diurnal rhythm which extends over a period of twenty-four hours. Some show diurnal activity and others a nocturnal activity. It is thus the light or, alternatively the absence of light, which acts as a stimulus to activity in these animals.

The biospeologists must consider whether the diurnal rhythm is maintained in hypogeous forms. To answer this question correctly two groups of observations must be distinguished: those concerning troglophiles, and those concerning troglobia.

(1) Troglophiles. When the troglophilic gasteropod, *Oxychilus cellarius* colonises the subterranean environment it maintains its diurnal rhythm and nocturnal activity (Tercafs, 1961). Thus the acquired rhythm occurs in the absence of external stimuli.

(2) Troglobia. Three studies have been devoted to the examination of activity in troglobious invertebrates. Two of these concern aquatic forms and the third concerns a terrestrial form.

The first of these studies was carried out by Park, Roberts and Harris (1941). It relates to the behaviour of the blind crayfish of Mammoth Cave, *Orconectes pellucidus*. These biologists have shown that this crustacean shows no diurnal rhythm and is active during both day and night.

The second piece of work was carried out by Ginet (1960) on the troglobious amphipod *Niphargus virei*. From his observations he recognised the total absence of a diurnal rhythm in these cavernicoles. On the other hand, *Gammarus pulex* shows a definite diurnal rhythm with maximum activity during the night and minimal activity in the daytime. In *Niphargus*, the actograph revealed the existence of an alternation of periods of activity with those of rest. The total of the activity periods was about 5 to 6 hours and that of periods of rest was 18 to 19 hours. Thus the ratio of activity to rest, "A/R", was in the region of 0·25. In *Gammarus pulex* the ratio A/R was almost 1. The activity in *Niphargus* is therefore a quarter of that in *Gammarus*. This result agrees with the data relative to the metabolic rates of these two species (p. 343).

A third study concerns cavernicolous insects. S. Glaçon (1952, 1956) worked with the bathysciine beetle *Speonomus diecki*. In this species the diurnal rhythm is absent, and replaced by regularly alternating periods of activity and rest. The duration of the periods of activity in 24 hours varies according to the temperature and the sex. However, the ratio A/R remains at about one.

Conclusions. The results obtained from the three series of experiments are in agreement and lead to the conclusion that the diurnal rhythm is absent in troglobia. The alternation of periods of activity and rest persists but this depends on an endogenous cycle and is not controlled by light.

In addition, we notice the persistence among the cavernicoles, of a metabolic diurnal rhythm, linked with the earth's rotation, and which is present in all living organism, animal or plant (Brown, 1961).

(b) Seasonal Rhythms

All animals living on the surface exhibit seasonal rhythms which vary according to the climate and the latitude. These rhythms are shown in the following ways:

(1) A reproductive cycle which occurs once or more times per year, but always at definite times. Domestication tends to extend the period of reproduction without completely suppressing the cyclic nature of the process.

(2) In extreme climates, which are either very hot or very cold during the middle of summer or of the winter, the animal suspends activity or becomes lethargic. These periods of immobility or sleep have been called aestivation and hibernation, respectively.

Are similar conditions to be found in subterranean populations? This is a question to which biospeologists must find the answer.

Answers to this question have been distorted by the strong belief held by early biospeologists as to the constancy of the subterranean environment. It is true that climatic fluctuations are less marked in deep caves, but it is also certain that they are not completely abolished.

During his visit to the caves which open in the Popovo Polje in Herzegovina, R. S. Hawes (1939) was struck by the very different conditions in these caves when they were flooded and when they were dry. When the water which fills the Polje rushes into the subterranean cavities it carries with it all manner of debris which serves to feed the cavernicoles, and favours their reproduction. When the waters recede the aquatic cavernicoles enter a phase of reduced activity corresponding to aestivation.

These extreme conditions are obviously exceptional, but they are found to a lesser degree in all caves. All biospeologists who regularly visit certain caves know that these undergo noticeable changes during the course of the year. Professor Jeannel, whose knowledge of the subterranean world is unequalled, has described the various changes in the appearance of caves during the course of the year (Jeannel, 1943). He wrote that "seasonal modifications of temperature exist in many caves. It is true that they are often slight and scarcely exceeding an amplitude of 5–6 degrees. They are, however, quite real and regulated above all by air-currents, and accompanied by noticeable changes in the relative humidity. A seasonal influence is also manifested underground by the volume of water infiltrating through the crevices. When the snow melts in the spring the circulation of subterranean waters increase, and the caves remain very humid for the first part of the summer. Autumn and winter are the 'dry' period in the subterranean world."

Do these variations influence cavernicoles and do they impose on them a seasonal cycle? We shall attempt to answer this question by comparing the behaviour of cavernicoles and epigeans, in order to find out whether the cavernicoles show cyclic reproduction, hibernation, and aestivation.

(1) Cyclic Reproduction. Many biospeologists, convinced of the constancy of the subterranean environment have given a negative answer to this question and have concluded that seasonal reproductive cycles are absent in the cavernicoles. In such animals reproduction would thus occur without interruption throughout the year.

This belief which was first held by Bedel and Simon (1875) has been supported by many biospeologists. Some of them are listed below:

Nematodes. Hnatenwytsch, 1929.
Spiders. *Porrhoma thorelli* and *Lepthyphantes pallidus* (Hnatenwytsch, 1929); *Leptoneta microphthalma* (Glaçon-Deleurance, 1955).
Copepods. Chappuis, 1920; Hnatenwytsch, 1929.
Syncarids. *Bathynella natans* (Jakobi, 1954).
Isopods. *Caecosphaeroma burgundum* (Husson and Daum, 1953).
Polydesmids. Husson, 1937.
Blaniulids. *Typhloblaniulus* (Brölemann, 1923).
Staphylinids. *Quedius mesomelinus* and *Conosomus pubescens* (Paulian, 1937).
Bathysciines. *Speonomus longicornis* (Glaçon, 1956).
Fish. *Stygicola dentatus* (Eigenmann, 1909).

This list, which could be lengthened, seems to remove all doubt that a seasonal reproductive cycle is absent in cavernicoles. However, a more detailed examination of the data available does not warrant such a categorical statement. Most observations concerning reproduction in cavernicoles are very inaccurate. They are too often based on impressions rather than on detailed studies gained over a number of years.

It is certain that if the seasonal reproductive cycles are present in the underground they will be attenuated and more difficult to detect than the epigean cycles.

It has been established from more precise studies that in certain cavernicoles reproduction is carried out throughout the year but shows a seasonal maximum. Although this is less marked than in epigeous forms it is nevertheless unquestionably present.

Two troglophilic spiders, *Nesticus pallidus* and *N. carteri* deposit cocoons all the year round, but their production shows two definite maxima, one in April, the other in October (Ives, 1935, 1947).

Monodella argentarii (Thermosbenacea), reproduces at the beginning of the summer (Stella, 1955).

Finally, a seasonal reproductive cycle has definitely been shown for amphipods of the genus *Niphargus* by the detailed studies of Ginet (1960). He showed that reproduction in *Niphargus virei* continued throughout the year but was maximal at the beginning of summer and minimal at the end of autumn. The cycle is similar to that of epigeous species but is more extended than in surface forms, and the values of the maxima and minima are attenuated.

Hawes (1939) suggested that the flood factor, that is to say the huge overflows which submerge the karstic regions and the underground cavities when the snow melts, stimulates the reproduction of the cavernicolous animals, and in particular that of *Proteus*. Likewise, Poulson (1961) recognized the existence of an annual cycle of growth and reproduction among the Amblyopsidae. About 1909, Eigenmann reported that the reproduction in *Amblyopsis spelaeus* continued throughout the year but was slightly higher in March. Poulson believes that the reproductive cycle of the Amblyopsidae depends on the flood factor and on the great deal of exogenous food brought in with the flood waters. At this time the endocrine glands and reproduction are stimulated.

Recently, M. Cabidoche has proved that the reproductive cycle of the genus *Aphaenops* (Trechinae) is closely linked with the variations of the water level, and more exactly with floods and droughts.

In conclusion, the persistence of seasonal cycles of reproduction, although less apparent than in surface forms, definitely exists in certain cavernicoles. It appears extremely probable that new and carefully conducted studies will reveal other examples of annual cycles of reproduction while the continuous multiplication of subterranean organisms will be shown to occur only in exceptional cases.

(2) Hibernation and Aestivation. One may speak only of hibernation in

relation to the few forms which occupy the glacial caves of the Alpine system. However, these cavernicoles are always very rare and their biology remains unknown. However, a few aquatic cavernicoles are known whose behaviour could be referred to as aestivation. It allows them to resist the total or partial drying of caves during the summer, when gours and sub-terranean streams dry up. This desiccation is only relative in that it only concerns areas of water and not the atmospheric humidity.

This behaviour has been observed in Crustacea from two groups: the Amphipoda *(Niphargus)* and the crayfish *(Cambarus)*. It is possible that cavernicolous copepods also possess this means of defence against des-iccation, but evidence for this has been obtained up till now only for epigeous species.

Amphipods. The possibility of aestivation was suggested by Remy (1927) during a study of *Niphargus puteanus* and by Uéno (1957) for *Pseudo-crangonyx kyotonis*. However, the processes involved remained unknown for some time. The problem was finally solved following observations in caves and experimental duplication of the conditions of aestivation in the laboratory.

Michel Bouillon, Assistant-Biologist at the subterranean laboratory at Moulis, observed the behaviour in the cave of Arnac, near Audinac, (Ariège), of *Niphargus longicaudatus*, during the summer when the under-ground river was dry. During this season these amphipods are to be found in cells, hollowed in clay and covered by a large stone, which acts as a stopper.

Bouillon was able to duplicate the conditions for aestivation in the cave laboratory at Moulis. These experiments have been repeated and re-fined by Ginet (1960). When a specimen of *Niphargus longicaudatus* is enclosed in a glass tube half filled with clay and half with water it is not long before the animal digs the clay and hollows out a chamber which it enters. The water in the upper part of the tube is progressively lowered. When the water has been poured out a bubble of air appears in the cham-ber. The *Niphargus* completely closes the chamber which then has no connection with the exterior. The *Niphargus* aestivate in these cham-bers.

Ginet has observed the survival of *Niphargus* placed in humid sand. Half of the specimens died within 3 months. None existed for more than 5. However, all specimens of *Niphargus* aestivating in cells of clay ex-isted for 3 months, half were still alive after 5 months and the last ones did not die until the end of 10 months.

Thus, this behaviour allows aquatic amphipods to persist, with few los-ses, during the period of aestivation. This behaviour is further proof of the aptitude of aquatic amphipods to lead an amphibious life.

Crayfish. Hay (1902) recorded some observations made on *Cambarus pellucidus*. When the gours which they populate dry up, the crayfish dig holes in the silt or under stones and remain in these places in a semi-dormant condition. The duration of this period of inactivity is unknown.

Copepods. In epigeous copepods different methods of resistance to desiccation or other unfavourable conditions are known.

(1) Summer cysts containing the adults of stenothermic species are formed during the warm season. These cysts are secreted under water. These types of cyst have been observed in *Cyclops bicuspidatus* (Birge and Juday, 1908) and *Canthocamptus microstaphylinus* (Lauterborn and Wolf, 1909).

(2) Cysts which contain the adults can be found in the mud of marshes. They withstand drying and may be carried some distance by aquatic birds. This is the case with Cyclopidae, e.g. *Cyclops bisetosus* and *C. furcifer* and with the Harpacticoidea, e.g. *Canthocamptus staphylinus, Atheyella wulmeri* and *A. northumbrica* (Roy, 1932).

(3) Finally, some copepods lay resistant eggs (œufs de durée, Dauereier), which resist desiccation and can be carried by birds. These resistant eggs are known among the Centropagidae, such as *Diaptomus castor* and *D. vulgaris* (Wolf, 1905); the Cyclopidae, as *Cyclops furcifer* (Roy, 1932); and the Harpacticoidea, in particular those *Elaphoidella* with parthenogenetic reproduction, e.g. *E. grandidieri* and *E. bidens* (Roy, 1932; Chappuis, 1954a and b; 1956), and also in *Viguierella caeca* (Chappuis, 1920).

However, up till now no cysts or resistant eggs have been observed in hypogeous species, but such stages do probably exist among them.

B. SOME EXAMPLES OF BEHAVIOUR IN CAVERNICOLOUS ANIMALS

The behaviour of hypogeous animals is very varied and little is known about it. However, several modes of activity appear to be directly related to the subterranean life. A few examples will be given in the following lines.

(a) Behaviour of Cavernicoles in Relation to Crevices

The interstitial medium is formed essentially by the very small areas between the grains of sand. Delamare-Deboutteville (1960) has written that in the "interstitial media life is wholly made of contacts". The animals are retained in these small spaces by their special reactions which maintain them in direct contact with the surrounding solid-surfaces. These reactions have been called stereotropisms or thigmotropisms.

Analogous behaviour is shown by cavernicoles. Subterranean animals are sometimes to be seen running along the walls of caves or on the soil. These individuals are in search of food, and they are usually limited in number. If, however, one places some bait at any point in the cave, by the end of a few hours it is surrounded by hundreds of cavernicoles whose existence may not have been suspected. These animals were hidden under stones, in wall crevices and under the stalagmitic floor. This is as true for aquatic as for terrestrial forms. The behaviour of cavernicoles is thus similar to that of interstitial forms, although the need to be in direct contact with their surroundings is less imperative. This behaviour is the consequence of positive thigmotropism.

Very few studies have been devoted to thigmotropism and there is little information concerning it in cavernicoles. The best work in this field was carried out by R. Ginet (1960). With an actograph of his own design, he measured the time spent by a specimen of *Niphargus virei* in an artificial crevice. Ginet reported that the *Niphargus* spent $14\frac{1}{2}$ to $17\frac{1}{2}$ hours per day in the crevice. It moved freely around the aquarium during $6\frac{1}{2}$ to $9\frac{1}{2}$ hours per day. This crustacean spent twice as long in the crevice as it did in the open water. The thigmotactic reactions are controlled by rheoceptors which are represented in *Niphargus* by tactile hairs on the carapace.

Similar behaviour has been observed in the vertebrates; *Typhlotriton spelaeus* is strongly stereotropic (Eigenmann, 1909).

(b) Behaviour of Cavernicoles in Relation to Clay

The behaviour of cavernicoles with respect to the clay must be distinguished from other reactions to the surrounding media. They are much more complex and cannot often be considered as tropisms. It is rather a question of instinctive behaviour in this case. The life of cavernicoles is so closely connected with the clay that this material serves many purposes.

(1) Moult capsules and metamorphosis. Many arthropods do not moult in the air but enclose themselves in a moult capsule made from the clay which can be mixed with other materials. This occurs in terrestrial, cavernicolous isopods of the genus *Scotoniscus* (Vandel, 1958) but is exceptional for surface forms; it is found only in certain endogeans and humicoles *(Platyarthrus, Ligidium)*†.

Egg laying in isopods is always accompanied by a moult when the oostegites are released. During the laying the female of *Scotoniscus* is enclosed in a parturition chamber which is particularly spacious. The reason why ovigerous females of troglobious terrestrial isopods are so rarely collected in caves can certainly be explained by this behaviour whilst the ovigerous females of the epigeous species are very common.

Similar behaviour is shown by primitive opilionids belonging to the genus *Siro* which are endogeous (Juberthie, 1960) and in some diplopods

---

† This behaviour is not obligatory. If the materials necessary for building are not available or are destroyed during its construction the *Scotoniscus* can moult on the surface in a manner similar to epigeous species. Also, this behaviour is not universal in cavernicolous isopods. In the cave of Predjama near Postojna the author has seen dozens of *Titanethes albus* moulting on the clay which covered the walls and floor of the cave.

(Typhloblaniulus, Polydesmus). Finally, we have stated earlier (p. 366) that the construction of moult capsules is usual among cavernicolous Bathysciinae (Glaçon-Deleurance).

It is difficult to trace the origin of this behaviour. However, the following interpretation could be suggested. The construction of chambers is conditioned by the water requirements of the cavernicoles. These needs are always high but at the time of moulting are increased due to the increased permeability of the integument at this time (Wigglesworth). Enclosures in a covered chamber would certainly reduce evaporation.

(2) Burrows. Many aquatic cavernicoles hollow out burrows in which they spend a large part of their existence. These earthen chambers are regularly excavated by *Niphargus* (Racovitza, 1908; Jeannel, 1929; Ginet, 1955, 1960; Husson, 1959). Similar behaviour is shown by *Stenasellus virei* (Husson, 1959) and *Monolistra berica* (Viré and Alzona, 1902)†.

(3) Burrows of unknown significance. Apart from the above uses of clay there are others of which the significance remains unknown (Vandel, 1958). There are holes dug in the clay. They may be the work of Trechinae *(Aphaenops)*, Bathysciinae *(Speonomus)*, Diplopoda *(Typhloblaniulus)* or of terrestrial isopods *(Scotoniscus)*. The clay extracted from these burrows is thrown above the opening in the form of "sawdust" *(Aphaenops)* or cut into "ashlars" which are regularly deposited to for ma wall round the hole *(Scotoniscus)*. However, the terrestrial cavernicoles never remain in these holes for very long. They are therefore not resting chambers.

The burrows hollowed in the clay by *Aphaenops* are probably dug by way of a precaution against the floods. When the water rises, these insects sink into a hole in such a manner that only the extremity of the abdomen is left outside (Cabidoche).

Although the significance of the holes is obscure, they offer a good example of the convergence of behaviour in cavernicoles belonging to very different groups.

(c) The Feeding Behaviour of Subterranean Planarians

The surface planarians hunt their prey actively. The subterranean planarians lay a coat of mucus upon the surface of the clay. The prey are caught with this slime.

(d) Behaviour of Cavernicolous Fish

The behaviour of cavernicolous fish is very distinct from that of surface form s. While epigeous fish exhibit periods of intense activity separated by long periods of rest, cavernicolous fish swim continuously.

† The species of the genus *Caecosphaeroma* do not hollow burrows.

This behaviour was first reported for *Caecobarbus geertsi* by Petit and Besnard (1937), then in *Pimoledella kronei* by Pavan (1946) and finally confirmed in *Anoptichthys jordani* by Breder and Gresser, (1941), Breder (1943), Schlagel and Breder (1947), Hahn (1957) and Grobbel and Hahn (1958). This is the reason for the increased consumption of oxygen by the cavernicolous species compared with the river fish *(Astynax mexicanus)* when maintained in normal conditions. This result is at first surprising as it seems to contradict the general rule of lowered respiratory metabolism in cavernicoles. In fact it is an overactivity related to the absence of inhibitory reflexes determined by visual stimuli. When the river fish are placed in total darkness or blinded they behave in the same way as cavernicolous fish.

The epigeous characid, *Astyanax mexicanus* is a gregarious form. The cavernicolous species which are derived from these and placed in the genus *Anoptichthys* are solitary. That is to say that social behaviour is dependent in this case, on vision (Breder, 1943).

BIBLIOGRAPHY

Activity Rhythms

BEDEL, L. and SIMON, E. (1875) Liste générale des Articulés cavernicoles de l'Europe. *Journal de Zoologie.* **IV.**

BIRGE, E. A. and JUDAY, C. (1908) A summer resting stage in the development of *Cyclops bicuspidatus* Claus. *Trans. Wiscons. Acad. Sc.* **XVI.**

BRÖLEMANN, H. W. (1923) Blaniulidae, Myriapodes (1ère Série). *Biospeologica,* **XLVIII.** *Archiv. Zool. expér. géner.* **LXI.**

BROWN, F. A. (1961) Diurnal rythm in cave crayfish. *Nature. Lond.* **CXCI.**

CHAPPUIS, P. A. (1920) Die Fauna der unterirdischen Gewässer der Umgebung von Basel. *Archiv. f. Hydrobiol.* **XIV.**

CHAPPUIS, P. A. (1954a) Recherches sur la Faune interstitielle des sédiments marins et d'eau douce à Madagascar. IV. Copépodes Harpacticoides psammiques de Madagascar. *Mém. Instit. Sc. Madagascar Ser. A.* **IX**

CHAPPUIS, P. A. (1954b) Copépodes Harparticoides des Indes et de l'Iran. *Bull. Soc. Hist. Nat. Toulouse.* **LXXXIX.**

CHAPPUIS, P. A. (1956) Présence à Madagascar du genre *Echinocamptus: E. pauliani* n. sp. (Copépode, Harpacticoide). *Mém. Inst. Sc. Madagascar Ser. A.* **X.**

EIGENMANN, C. H. (1909) Cave vertebrates of America. A study in degenerative evolution. *Carnegie Inst. Washington Public.* No. 104.

GINET, R. (1960) Écologie, Éthologie et Biologie de *Niphargus* (Amphipodes, Gammarides hypogés). *Ann. Spéléol.* **XV.**

GLAÇON, S. (1952) Les Variations de l'activité de *Speonomus diecki* Saulcy en fonction de le température et du sexe. *Bull. Soc. Zool. France* **LXXVII.**

GLAÇON-DELEURANCE, S. (1955) Sur la biologie de *Leptoneta microphthalma* Simon (Araignées cavernicoles). *Compt. rend. Acad. Sci. Paris* **CCXLI.**

GLAÇON, S. (1956) Recherches sur la biologie des Coléoptères cavernicoles troglobies. *Premier Congrès Intern. Spéléol. Paris Communic.* **III.**

HAWES, R. S. (1939) The flood factor in the ecology of caves. *J. anim. Ecol.* **VIII.**

HAY, W. B. (1902) Observations on the crustacean fauna of Nickajack Cave, Tennessee and vicinity. *Proceed. U. S. Nat. Mus.* **XXV.**

388 BIOSPEOLOGY: THE BIOLOGY OF CAVERNICOLOUS ANIMALS

HNATENWYTSCH, B. (1929) Die Fauna der Erzgruben von Schneeberg im Erzgebirge. *Zool. Jahrb. Abt. System.* **LVI.**

HUSSON, R. (1937) Reproduction non saisonnière des Polydesmides des Galeries de Mines. *Bull. Soc. Sc. Nancy.*

HUSSON, R. and DAUM, J. (1953) Zur Biologie einer Isopodenart subterraner Gewässer: *Caecosphaeroma burgundum (Vireia burgunda)* Dollfus. *Annal. Univ. Sarav.* **II.**

IVES, J.D. (1935) A study of the cave spider, *Nesticus pallidus* Emerton, to determine whether it feeds seasonally or otherwise. *Jour. Elisha Mitchell Sc. Soc.* **LI.**

IVES, J.D. (1947) Breeding habits of cave spider, *Nesticus carteri* Emerton. *J. Elisha Mitchell Sc. Soc.* **LXIII.**

JAKOBI, H. (1954) Biologie, Entwicklungsgeschichte und Systematik von *Bathynella natans* Vejd. *Zool. Jahrb. Abt. System.* **LXXXIII.**

JEANNEL, R. (1943) *Les Fossiles vivants des Cavernes.* Paris.

LAUTERBORN, R. and WOLF, E. (1909) Cystenbildung bei *Canthocamptus microstaphylinus.* *Zool. Anz.* **XXXIV.**

PARK, O., ROBERTS, T.W. and HARRIS, S.J. (1941) Preliminary analysis of activity of the cave crayfish, *Cambarus pellucidus. Amer. Natural.* **LXXV.**

PAULIAN, R. (1937) Les larves des Staphylinidae cavernicoles. *Biospeologica,* **LXVII.** *Archiv. Zool. expér. géner.* **LXXIX.**

POULSON, T.L. (1961) Cave adaptation in amblyopsid fishes. *Thesis of the University of Michigan. Zool.*

REMY, P. (1927) Les *Niphargus* des sources sont-ils émigrés des nappes d'eau souterraine? *Compt. rend. Congr. Soc. Savantes.*

ROY, J. (1932) Copépodes et Cladocères de l'Ouest de la France. Thèse, Paris.

STELLA, E. (1955) Behaviour and development of *Monodella argentarii* Stella, a thermosbenacean from an Italian cave. *Verh. Intern. Ver. Limnologie* **XII.**

TERCAFS, R.R. (1961) Comparaison entre les individues épigés et cavernicoles d'un Mollusque Gastéropode troglophile, *Oxychilus cellarius* Müll. Affinités éthologiques pour le milieu souterrain. *Ann. Soc. R. Zool. Belgique* **XCI.**

UÉNO, S.I. (1957) Blind aquatic beetles of Japan with some accounts of the fauna of Japanese subterranean waters. *Archiv. f. Hydrobiol.* **LIII.**

WOLF, E. (1905) Die Fortpflanzungsverhältnisse unserer einheimischen Copepoden. *Zool. Jahrb. Abt. System.* **XXII.**

Examples of Behaviour in Cavernicoles

BREDER, C.M. JR. (1943) Problems in the behaviour and evolution of a species of blind fish. *Trans. New York Acad. Sc.* (2) **V.**

BREDER, C.M. JR. and GRESSER, E.B. (1941) Correlations between structural eye defects and behaviour in the Mexican blind characin. *Zoologica,* N.Y. **XXVI.**

DELAMARE-DEBOUTTEVILLE, CL. (1960) *Biologie des eaux souterraines littorales et continentales.* Paris.

EIGENMANN, C.H. (1909) Cave vertebrates of America. A study in degenerative evolution. *Carnegie Institution. Washington Public.* No. 104.

GINET, R. (1955) Étude sur la biologie d'Amphipodes troglobies du genre *Niphargus.* I. Le creusement des terriers; relations avec le limon argileux. *Bull. Soc. Zool. France* **LXXX.**

GINET, R. (1960) Écologie, Éthologie et Biologie de *Niphargus* (Amphipodes, Gammarides hypogés). *Ann. Spéléol.* **XV.**

GROBBEL, G. and HAHN, G. (1958) Morphologie und Histologie der Seitenorgane des augenlosen Höhlenfisches, *Anoptichthys jordani* im Vergleich zu anderen Teleosteern. *Zeit. Morphol. Oekol. Tiere.* **XLVII.**

HAHN, G. (1957) Ferntastsinn und Strömungssinn beim augenlosen Höhlenfisch, *Anoptichthys jordani* Hubbs and Innes, und das Problem der Rezeptionsorte von Lichtreizen. Dissertation, Köln.

Husson, R. (1959) Les Crustacés Peracarides des eaux souterraines. Considérations sur la biologie de ces cavernicoles. *Bull. Soc. Zool. France* **LXXXIV.**

Jeannel, R. (1929) Le Vivarium du Jardin des Plantes pendant l'année 1928. *Rev. Hist. Nat.* **X**

Juberthie, Ch. (1960) Sur la Biologie d'un Opilion endogé, *Siro rubens* Latr. (Cyphophthalmes). *Compt. rend. Acad. Sci. Paris* **CCLI.**

Pavan, C. (1946) Observations and experiments on the cave fish, *Pimelodella kronei* and its relatives. *Amer. Nat.* **LXXX.**

Petit, G. and Besnard, W. (1937) Sur le comportement en aquarium de *Caecobarbus geertsi* Blgr. *Bull. Mus. Hist. Nat. Paris* (2) **IX.**

Racovitza, E. G. *in* Jeannel, R. and Racovitza, E. G. (1908) Énumération des Grottes visitées (1906–1907) (Seconde Série). *Biospeologica,* **VI.** *Archiv. Zool. expér. géner.* (4) **VIII.**

Schlagel, S. R. and Breder, C. M. Jr. (1947) A study of the oxygen consumption of blind and eyed characin in light and darkness. *Zoologica* **XXXII.**

Vandel, A. (1958) Sur l'édification de logettes de mue et de parturition chez les Isopodes terrestres troglobies, et sur certains phénomènes de convergence observés dans le comportement des animaux cavernicoles. *Compt. rend. Acad. Sci. Paris* **CCXLVII.**

Viré, A. and Alzona, C. (1902) Sur une nouvelle espèce de *Caecosphaeroma,* le *C. bericum. Bull. Mus. Hist. Nat. Paris* **VII.**

CHAPTER XXIV

REACTIONS OF CAVERNICOLES TO EXTERNAL FACTORS

THE STUDY of the general activity of cavernicoles needs to be refined by some accurate research. It is necessary to examine the way in which hypogeous animals react to various environmental factors and the structure of the various organs which inform the animal of changes in the environment. There is no section of biospeology where the information available is so scant, so full of errors, or based upon hypothetical ideas without an adequate basis of fact, as this one.

The reactions of cavernicoles to the following factors will be examined:

(A) Mechanical agents
(B) Movement
(C) Running water
(D) Vibration
(E) Sound waves
(F) Chemical materials dissolved in water
(G) Chemical materials transported through the air
(H) Atmospheric humidity.

The reactions of cavernicoles to light and to darkness represent such a significant aspect of biospeology that they will be considered in a separate chapter.

A. BEHAVIOUR WITH RESPECT TO MECHANICAL FACTORS

All the animals are sensitive to blows or contact with foreign objects. This is called tactile sensitivity.

Do cavernicoles differ from epigeans with respect to tactile sensitivity? It is impossible to answer this question at present. The experiments in this field are very few and extremely rudimentary.

Banta (1910) compared the tactile sense of *Asellus stygius*, a cavernicole, and *Asellus communis*, an epigeous form. By touching individuals of each species with delicate or coarse needles he concluded that the tactile sense in the cavernicole was greater than in the epigean. However, accurate measurements were impossible with this method; and these conclusions, based on subjective impressions, are not conclusive.

There is little information on the tactile sensitivity of insects although it would be an interesting subject for research. The hairs which cover the integument are responsible for tactile sense in these animals. A comparative study of the trichobothria of lucicolous and cavernicolous arachnids would also yield some interesting results.

B. BEHAVIOUR WITH RESPECT TO MOVEMENTS

Many Crustacea possess statocysts which normally inform them of their position in space. According to Packard (1886), the statocysts of cavernicolous *Cambarus (C. hamulatus* and *C. pellucidus)* are small and degenerate compared with those of epigeous species.

C. BEHAVIOUR OF AQUATIC CAVERNICOLES WITH RESPECT TO RUNNING WATER

It has been stated earlier that there are no hypogeous, torrenticolous organisms. It would appear that the cavernicoles might react unfavourably to running water. Ginet (1960) was able to verify this by some accurate experiments. It is well-known that gammarids show a very strong positive reaction towards running water and swim energetically against it (Cousin, 1925). The name positive rheotropism has been given to this reaction. The behaviour is very different in *Niphargus*. Ginet observed that *Niphargus* began to orientate itself with the current of water (positive orientation) but exhibited a strong urge to move away from the current itself and to find calm water. *Niphargus* is thus a rheoxene or rheophobic type.

Asellus stygius reacted positively to the current but the reaction took longer than in the epigeous species *Asellus communis* (Banta, 1910).

Verrier (1929) stated that the fish, *Typhlichthys subterraneus* is probably rheophobic, whilst, according to Poulson (1961) *Amblyopsis spelaeus* is rheophilous.

D. BEHAVIOUR TOWARDS VIBRATION

Little work has been carried out in this field. Banta (1910) studied the sensitivity to vibrations of the epigeous species, *Asellus communis* and the cavernicole *A. stygius*. It appears that the second species is more sensitive to vibrations than the first. The most sensitive organs seemed to be the antennae.

The elytra of carabids, in addition to small erect hairs, possess large hairs which are articulated at the base and moveable. Jeannel has given them the name "fouets" (flagellae). They are few in number and have a definite position. They are longer in cavernicoles than epigeans. The "fouets" must fulfil a particular function, but this is unknown. For many years entomologists (La Brulerie, 1872; Chalande, 1881; Jeannel, 1943) attributed to them the role of receiving air-borne vibrations. However, this has not, as yet, been confirmed experimentally.

Not much more is known of aquatic cavernicolous vertebrates. However it is certain that the hypogeous vertebrates are sensitive to vibrations transmitted through the water and received by the neuromastes of the lateral system. This was proved for *Typhlomolge rathbuni* (Eigenmann, 1909) and for *Typhlichthys subterraneus* (Verrier, 1929). The *Amblyopsidae* are able to detect prey at a distance of from 1 to 5 cm (Poulson, 1961). *Anoptichthys jordani* does the same (Lüling). Detection results from the vibrations emitted by the prey.

Hahn (1960) has given us some information concerning *Anoptichthys jordani*. This fish is able to orientate itself in a medium containing obstacles more efficiently than a normal fish when blinded. Ablation experiments showed that this sense is a function of the lateral line. However, the lateral line system in fish with regressed eyes, shows the same morphological characteristics as that of epigeans.

E. BEHAVIOUR TOWARDS SOUND WAVES

Eigenmann (1909) studied the structure of the ear in *Amblyopsis spelaeus* and found it to be normal. Several experiments led him to conclude that these fish do not hear in the human sense. It is, however, difficult to distinguish in fish between hearing and the waves picked up by the lateral line system. The ear of *Anoptichthys jordani* clearly showed a degenerate structure (Breder, 1943).

F. BEHAVIOUR TOWARDS CHEMICAL MATERIALS DISSOLVED IN WATER

(a) Invertebrates

There are no experiments available to inform us of the chemical sensitivity of aquatic cavernicolous invertebrates. Only anatomical studies are available from which it is impossible to draw any physiological conclusions. The small structures on the antennules of Crustacea, which have been called "aesthetascs" may be chemosensory.

(b) Vertebrates

Two types of chemical receptor have been distinguished in the vertebrates, those of taste and those of smell. However, physiologically this distinction does not apply for aquatic forms and it is preferable to speak of a chemical sense. Nevertheless, anatomically the distinction between the two types of apparatus persists and will be examined successively:

1. the neurogemmae,   2. the olfactory apparatus.

(1) Neurogemmae. The chemical sense in fish is situated in small organs which have been called "Endknospen" or "taste buds". P. Gérard (1936) created the name neurogemmae ("neurogemmes") to describe them. Their structure seems to be closely related to the neuromasts which are the receptor organs of the lateral-line system.

Fig. 74. A cephalic neurogemma of *Caecobarbus geertsi* (after P. Gérard).

The neurogemmae are not limited to the buccal cavity but are also distributed over the surface of the head. They are particularly abundant on oral barbs, when these are present. Others are situated on the anterior surface of the body. In the gobiids and silurids they are spread across the whole body surface. Herrick estimated that in certain siluroids the skin contained up to 100,000 neurogemmae.

Neurogemmae are well developed in cavernicolous fish belonging to the Amblyopsidae *(Chologaster, Amblyopsis, Typhlichthys)* (Cox, 1905; Eigenmann, 1909).They are arranged in ridges which are particularly numerous on the head, but also extend along the body to the base of the caudal fin (Plate IX). A comparison of two species of *Chologaster*, one epigeous *(C. cornutus)* and the other cavernicolous, *(C. agassizi)* shows that the neurogemmae project in the second species and are buried under the skin in the first. Numerical comparisons are difficult to obtain as the number of neurogemmae varies with age and with size.

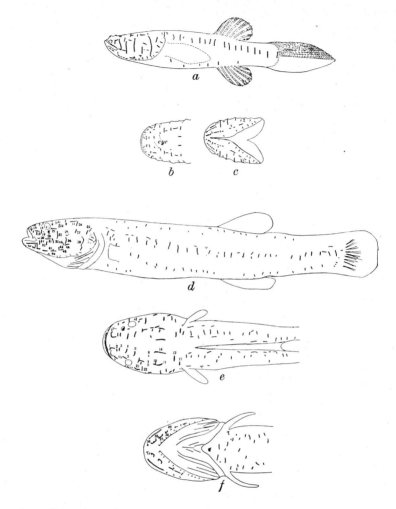

PLATE IX. *a, b, c,* Distribution of tactile ridges in *Troglichthys.* Side view of
entire fish, dorsal and ventral views of head.
d, e, f, Distribution of tactile ridges in *Chologaster papilliferus.* Side view of
entire fish, dorsal and ventral views of anterior part of body.

Similar arrangements have been observed in the blind fish *Typhleotris madagascariensis* from Madagascar (Angel, 1949).

The neurogemmae of *Caecobarbus geertsi* have been studied by Pol Gérard (1936). These organs are very abundant on the head and above all on the snout. They are arranged in transverse rows (Heuts). Each neurogemma is buried in a small cavity and covered with a muscous cup in which the sensory hairs are engaged (Fig. 74). Neurogemmae are also found on the trunk but here they are superficial and the mucous cup is reduced.

Gérard has pointed out that the development of neurogemmae is not specific to hypogeous species and cannot be considered as an adaptation to cavernicolous life. In *Barbus oligolepsis*, an epigeous form with eyes, there may be more neuromasts than in *Caecobarbus*.

The distribution of the neurogemmae in the cavernicolous Characidae is different (Breder and Rasquin, 1943). They are not set in lines, but scattered about the surface of the body. They are numerous on the head and the snout. The number of neurogemmae is larger, especially on the head, in the cavernicoles species, *Anoptichthys jordani* than the river fish, *Astyanax mexicanus*. The more specialized cavernicolous species from La Cueva de los Sabinos, *Anoptichtys hubbsi*, possess still more neurogemmae than *A. jordani*, from La Cueva Chica. This is a very rare example of progressive evolution among subterranean species (Breder and Rasquin).

In spite of the anatomical differences, the behaviour of the three species towards chemical materials, is very similar. However, the cavernicolous fish react more weakly than the river fish, *Astyanax mexicanus* (Breder and Rasquin).

Neurogemmae have been reported in the cavernicolous brotulid, *Stygicola dentatus*. They were first described by Putnam (1872) and later more accurately by Kosswig (1934). They are particularly abundant on the head. On the body they are arranged in a lateral line.

(2) Olfactory Apparatus. The olfactory system of fish is represented by two olfactory sacs, containing parallel lamellae and covered with olfactory cells.

Breder and Rasquin (1943) examined the olfactory organ of three Mexican characids, *Astaynax mexicanus*, the "river fish", *Anoptichthys jordani*, a cavernicole from the Cueva Chica, and *A. hubbsi* a highly specialised cavernicole from the Cueva de los Sabinos. The following numbers of lamellae were recorded:

Astyanax mexicanus:	8 pairs
Anoptichthys jordani:	7 pairs
A. hubbsi:	5 pairs.

There is thus a reduction in the number of lamellae in the cavernicolous types which is particularly marked in the highly specialised form. The figures are amongst the lowest recorded in fish.

G. BEHAVIOUR TOWARDS CHEMICAL SUBSTANCES CARRIED BY THE AIR

The chemical sensitivity of insects is localised in the antennae, in the small sensory apparatus, the sensilla basiconica (Slifer).

Nothing is known of the chemical sensitivity of cavernicoles except that many cavernicoles and in particular the Bathysciinae and Diplopoda, are attracted from long distances by bait. This behaviour indicates a highly developed olfactory sense.

Jeannel (1908, 1911) has described a structure situated at the interior of the seventh segment of the antenna in Bathysciinae. A vesicle is formed of hexagonal plates which are perforated by small holes through which the rods from the sensory cells protrude. This organ which closely resembles the "perforated plates" of Hymenoptera possibly have an olfactory function.

H. BEHAVIOUR TOWARDS HUMIDITY

Representatives of many orders of insect possess hygro-receptive organs on the antennae (the bibliography relative to these organs was given by Riegert, 1960). They are the sensilla coeloconica. Similar organs are found on the antennae of diplopods (Perttunen).

There has been no study of cavernicolous insects in this respect. This absence is regrettable as the humidity of the air plays an important part in the life of hypogeous terrestrial arthropods.

BIBLIOGRAPHY

ANGEL, F. (1949) Contribution à l'étude du *Tychleotris madagascariensis*, Poisson aveugle cavernicole du sud-ouest de Madagascar. *Bull. Mus. Hist. Nat. Paris* (2) **XXI.**

BANTA, A. M. (1910) A comparison of the reactions of a species of surface isopod with those of a subterranean species. Part II. *J. exper. Zool.* **VIII.**

BREDER, C. M. JR. (1943) Problems in the behaviour and evolution of a species of blind fish. *Trans. New York Acad. Sc.* (2) **V.**

BREDER, C. M JR. and RASQUIN, P. (1943) Chemical sensory reactions of the Mexican blind Characins. *Zoologica* **XXVIII.**

CHALANDE, J. (1881) De la sensibilité chez les Insectes aveugles cavernicoles. *Bull. Soc. Hist. Nat. Toulouse* **XV.**

COUSIN, G. (1925) Un dispositif simple et démonstratif pour mettre en évidence le rhéotropisme des *Gammarus*. *Feuille. Natural.* **XLVI.**

COX, U. O. (1905) A revision of the cave fishes of North America. *Rep. Bureau Fisheries for 1904.*

EIGENMANN, C. H. (1909) Cave vertebrates of America. A study in degenerative evolution. *Carnegie Inst. Washington Public.* No. 104.

GÉRARD, P. (1936) Sur l'homologie entre les appareils sensoriels du système latéral et ceux de système vestibulaire, chez les Teléostomes. *Bull. Acad. R. Belgique. Cl. Sc.* (5) **XXII.**

GINET, R. (1960) Écologie, Éthologie et Biologie de *Niphargus* (Amphipodes, Gammaridés hypogés). *Annal. Spéléol.* **XV.**

HAHN, G. (1960) Ferntastsinn und Strömungssinn beim augenlosen Höhlenfisch, *Anopitchtys jordani* Hubbs and Innes, im Vergleich zu einigen anderen Teleosteern. *Naturwissenschaften.* **XLVII.**

JEANNEL, R. (1908) Coléoptères (1ère Série). *Biospeologica,* **V.** *Archiv. Zool. expér. géner.* (4) **VIII.**

JEANNEL, R. (1911) Révision des *Bathysciinae.* Morphologie. Distribution géographique. Systématique. *Biospeologica,* **XIX.** *Archiv. Zool. expér. géner.* (5) **VII.**

JEANNEL, R. (1943) *Les Fossiles vivants des Cavernes.* Paris.

KOSSWIG, K. (1934) Über bislang unbekannte Sinnesorgane bei dem blinden Höhlenfisch, *Stygicola dentatus* (Poey). *Zool. Anz. 7. Supplementband.*

LA BRULERIE, CH. Piochard de (1872) Note pour servir à l'étude des Coléoptères cavernicoles. *Annal. Soc. Entomol. France* (5) **II.**

PACKARD, A. S. (1886) The cave fauna of North America, with remarks on the anatomy of the brain and origin of the blind species. *Mem. Nat. Acad. Sci. Washington* **IV.**

POULSON, TH. L. (1961) Cave adaptation in amblyopsid fishes. University of Michigan. *Phil. Dissert. Zool.*

PUTNAM, F. W. (1872) The blind fishes of the Mammoth Cave and their allies. *Americ. Nat.* **VI.**

RIEGERT, P. W. (1960) The humidity reactions of *Melanoplus bivittatus* (Say) (Orthoptera, Acrididae): antennal sensilla and hygro-reception. *Canad. Entomol.* **XCII.**

VERRIER, M. L (1929) Observations sur le comportement d'un Poisson cavernicole: *Typhlichthys osborni* Eigenmann. *Bull. Mus. Hist. Nat. Paris.* (2) **I.**

CHAPTER XXV

THE BEHAVIOUR OF CAVERNICOLES WITH RESPECT TO LIGHT

THE EFFECT of light on the behaviour of cavernicoles presents a particularly interesting topic, and has led to many experiments. For these reasons a whole chapter has been devoted to the study of this question.

A. GENERAL REACTIONS TO LIGHT

The effect of light on living organisms can be stimulating or harmful, depending on the intensity.

(a) Stimulating Action of Light

The stimulating action of light is clearly seen in the case of the seasonal reproductive cycles. The influence of light on the fertility of laying hens has been known for a long time. The action of light on the reproductive cycle has been studied experimentally in various birds (starling, sparrow, duck) and in mammals (ferret, *Peromyscus*).

However, the complexity of all biological phenomena must not be forgotten. The intervention of a chemical or physical factor on an organism is never simple or constant. The action of light on the reproductive cycle is marked in those animals whose sexual activity is strictly limited to certain times of the year. Light has little or no effect in animals which possess the faculty of continuous reproduction. Light has no action on reproduction in the guinea pig which can indeed take place in total darkness.

It cannot be stated *a priori* that light has an action on the physiological activity of cavernicoles, as it is known that seasonal variations have only a slight effect on these animals. In fact the results available do not settle this question.

P.H. Wells (1952, 1957) has stated that light has a stimulating action on the movements of the cavernicolous crayfishes: *Cambarus ayersi, C. setosus* and *C. hubrichti*. However, these results must be questioned. In Well's experiments, the crayfish were held down in a clamp. The behaviour of these animals when free is therefore unknown. It is probable that they would have shown escape reactions, that is to say negative phototropism which is more correctly termed photopathy (p. 400). But, the experimental procedure obscured the true nature of the reaction.

Some other experiments are also difficult to interpret. According to Wautier and Troïani (1960) the respiratory rate of *Niphargus longicaudatus* increased when the animals were placed in the light. On the other hand, L. Derouet-Dresco (1959) working with another species of the same genus, *N. virei*, came to opposite conclusions. She stated that light had little action on the respiratory rate.

It would be necessary to repeat these experiments while taking account of the interaction of many factors. The reaction of *Niphargus* to light is a photopathy which markedly increases motor activity (Ginet, 1960). As a result there is an increase in respiratory rate.

The only conclusion that we can draw from these experiments is as follows. The action of light on cavernicoles cannot be interpreted as a simple stimulus inducing general activity. Light induces complex reactions: escape reactions or photopathy whose adaptive value is evident. This will be considered later.

(b) Noxious Action of Light

It is easy to demonstrate the noxious action of light on cavernicoles. Cavernicolous planarians are particularly sensitive to light. They become necrotic so rapidly that it is extremely difficult to photograph or to film them.

Similarly, strong light is harmful to *Niphargus*. A light intensity of 20,000 lux causes the death of *N. virei* in two to three days while *Gammarus pulex* withstands 40,000 lux for 25 days (Ginet, 1960).

The cavernicolous sphodrids *(Sphodrus, Antisphodrus)* are killed by 15 minutes in sunlight (Bernard, 1932).

The noxious effects of strong light are not seen only among cavernicolous species. A large number of invertebrates and even lower vertebrates are effected, as has been shown by the experiments of Merker (1925). The action of ultra-violet rays is particularly harmful. The cavernicoles are more sensitive to the noxious action of light than epigeans because of the absence of a protective layer of pigment. While the normal, pigmented *Gammarus* can withstand a light intensity of 40,000 lux for 25 days, depigmented *Gammarus* are killed by the same light intensity in 4 to 5 days (Ginet, 1960).

Merker (1929) showed that the layer of chitin forms a protection against ultra-violet rays only when it is more than a millimetre in thickness. This condition is never found in arthropods nor in cavernicoles, where the integument is always very thin. The pigment forms a normal barrier which protects the animal from the noxious action of light and in particular ultra-violet light. Merker stated that the amount of ultra-violet light passing through the chitin is inversely proportional to the quantity of pigment which it contains. An examination of the relationships between pigmentation and cavernicolous life will be undertaken in a later paragraph.

B. ORIENTATION REACTIONS TO LIGHT

Light very frequently attracts or repels animals and these conditions are given the names of positive and of negative phototropism. However, Gaston Viaud has shown in a remarkable study of the reactions induced by light, that two types of behaviour are included under the terms of phototropism.

Positive phototropism corresponds to a general behaviour of animals to light. It originates from a photic reaction of the cell and of protoplasm. "It is probably a primitive property of organised material" (G. Viaud).

Negative phototropism is more complex. It is an adaptive reaction designed to remove the animal from an unpleasant stimulus. More than a directional movement of a true tropism, it is the search for a certain light intensity. This type or reaction was named photopathy by Graber (1884).

(a) Positive Phototropism

It is obvious that cavernicoles would not exhibit a positive phototropism which would have the effect of attracting them out from the subterranean environment. This rule applies only to true cavernicoles and not to representatives of the parietal association which cannot in any respect be considered as inhabitants of the subterranean world. Leruth (1934, 1939) has shown that Diptera of the family Helomizidae which are one of the important groups which make up of the parietal association, are positive phototropic. This must also be the case with many other inhabitants of cave entrances which are attracted there by the increased humidity.

There is, however, a very curious exception to this general rule. Several specimens of the cavernicolous characid *Anoptichthys jordani*, particuarly those captured in the Sotano de la Tijana, were weakly, but clearly, positively phototropic. Breder and Rasquin (1947) who discovered this remarkable behaviour asserted that it must be unique among cavernicoles. Later an interpretation of this anomaly will be given.

(b) Negative Phototropism or Photopathy

The immense majority of cavernicoles have been shown to negatively phototropic†. It seems probable that this photopathic characteristic of

---

† These results were obtained after studying the effect of white light. Experiments carried out on Crustacea and on fish have shown that cavernicoles react differently to different wavelengths of light. This is, however, more a problem of sensory physiology than of biospeology, and we shall not consider it further.

cavernicoles is related to the disappearence of the protective layers of pigment in the integument. It would be necessary for experiments to be carried out in order to verify this interpretation.

The following are a few examples although the list is far from complete.

Phreaticolous hydracarina (Schwoerbel, 1961)
Spiders. Meta menardi (MacIndoo, 1910)
Amphipods. Niphargus (Plateau, 1868; Jeannel, 1929; Ginet, 1960)
Isopods. Cavernicolous species of *Asellus* (Banta, 1910; Janzer, 1937; Janzer and Ludwig, 1952)
Caecosphaeroma (Jeannel, 1929, 1943)
Decapods. Cavernicolous *Cambarus* (Park, Roberts and Harris, 1941)
Diplopods. Typhloblaniulus (Jeannel, 1943)
Cavernicolous Collembola (Pactl, 1956)
Cavernicolous Coleoptera. Aphaenops, Duvalius, Speonomus (Jeannel, 1943)
Characids. Anoptichthys jordani (Breder and Gresser, 1941; Breder and Rasquin, 1947; Lüling, 1954; Thinès, 1954)
Cyprinids. Caecobarbus (Thinès, 1954)
Amblyopisids. Amblyopsis, Chologaster (Payne, 1907; Eigenmann, 1909; Woods and Inger, 1957)
Brotulids. Stygicola, Lucifuga (Eigenmann, 1909)
Urodela. Proteus (Dubois, 1890; Hawes, 1946), Larvae of *Typhlotriton spelaeus* (Noble and Pope, 1928).

One must remark that the "candidates" for cavernicolous life show similar reactions to those of hypogeous types. *Chologaster cornutus*, an inhabitant of the marshes in the south-east of the United States, which is still near to the ancestor of the Amblyopsidae, has nocturnal habits; it remains concealed during the day and is negatively phototropic. The behaviour of the troglophilous species, *Ch. agassizi* is similar when this fish lives in the springs; then it is active only during the night (Poulson, 1961). These are examples of preadaptation.

(c) Indifference to Light

Although it is general to observe negative phototropic reactions in cavernicoles a certain number of them are indifferent to light and react only when the intensity passes the noxious threshold.

This indifference to light can be considered as the final stage in the regressive evolution of phototropic reactions. Here are a few examples:

Turbellaria. *Sphalloplana percaeca* (Buchanan, 1936),
Molluscs. *Vitrella quenstedti* (Seibold, 1904).
Isopods. *Faucheria faucheri* (Gal, 1903).
Coleoptera. *Antisphodrus, Ceutosphodrus* (Bernard, 1932).

B 14a

Characids. *Anoptichthys hubbsi* and *A. antrobius* (Breder, 1943, 1944; Breder and Rasquin, 1947).

Amblyopsids. *Typhlichthys subterraneus* (Jeannel, 1929; Verrier, 1929).

Cyprinids. *Typhlogarra widdowsoni* (Trewavas, 1955).

Urodeles. *Typhlomolge rathbuni* (Eigenmann, 1909),
   Haideotriton (Pylka and Warren, 1958).

C. PHOTORECEPTORS OF CAVERNICOLES

With the exception of those species indifferent to light, cavernicoles react to light by photopathic manifestations. The photoreceptors which induce these reactions must be considered. The photoreceptors belong to several different types.

(a) The Eye

The eye is the normal light receptor. However most cavernicoles possess a regressed ocular apparatus, which is sometimes quite rudimentary. In spite of the degenerative state of the eye, some cavernicoles — for instance, the fishes of the genus *Anoptichthys* — are sensitive to light. We had formerly believed that the regressed eye was still sensitive to the light (Breder and Gresser, 1941; Breder and Rasquin, 1947; Kuhn and Kähling, 1954). This was erroneous. The fishes of the genus *Anoptichthys* are always blind, even the young which possess eyes less regressed than those of the adult (Lüling, 1953a and b, 1954; Kähling, 1961). However, these fishes do react to light; they may be trained to take their food when a lamp is lighted.

(b) The Diencephalon

Payne (1907) thought that the stimulation of *Amblyopsis spelaeus* by light results from a direct excitation of the light upon the brain. But it was a few years later, when the problem was experimentally considered.

Von Frisch (1911) stated that the chromatophores of the blinded minnow could expand or contract according to the conditions of illumination. Light stimulates the diencephalon through a depigmented area which covers this part of the brain. This stimulation is transmitted to the chromatophores via the hormonal secretions of the pituitary[†].

Von Frisch entrusted one of his students, E. Scharrer, with the task of carrying out the experiments. Scharrer (1928) trained the blinded minnows

---

[†] Vilter (1960) showed that certain abyssal fish possessing luminous organs on the head have a translucent window over the diencephalic region. Non-luminous abyssal fish do not have this structure. Thus, in the first group of fish, the pituitary may respond to light stimulation.

to take earthworms when a light is projected onto them. This biologist showed that only the diencephalon is sensitive to light and that the ventral part (the hypothalamus) is more sensitive than the dorsal part (epiphysal part).

The eye which is simply a vesicle formed by the diencephalon only shows to a higher degree the photosensitive properties of this region.

Breder and Rasquin (1950) extended this inquiry to include numerous species of fish. A certain number of them (principally non-acanthopterygians) possess a pineal surface covered by transparent tissue while other fish (acanthopterygians) have a pineal surface covered with strongly pigmented tissue. According to these American biologists, the two conditions determine the phototropic reactions of fish. The first group are positively phototropic and the second group negatively phototropic. One can, in fact, make fish with transparent pineal surface negatively phototropic by injecting chinese black ink into them. Thus the role of the diencephalon in photoreception has been established.

Breder and Rasquin (1947) paid particular attention to different races of *Anoptichthys jordani*. The pineal region of specimens of the race from Cueva Chica is protected by guanine and melanic pigments. On the other hand, the pineal region of the race from Sótano de la Tijana is transparent. The first race is negatively phototropic while the second is positively phototropic. That this behaviour is a consequence of the state of the pineal depression has been proved by two experiments. If one scrapes away the pigment layer which covers the pineal region of specimens from the Cueva Chica, these fish become weakly positively phototropic. If the pineal region of individuals from the Sótano de la Tijana is painted with Chinese black ink they become negatively phototropic.

These experiments establish beyond doubt the role played by the diencephalon on the phototropic behaviour in both epigeous and cavernicolous fish.

These conditions are found not only among the fish. Other organisms, particularly the Crustacea show these reactions. The photosensitive reactions of cavernicolous species of *Cambarus* is probably due to the direct influence of light on the brain (Wells, 1952, 1957). The brain of *Niphargus* would also seem to be sensitive to light (Merker, 1929).

(c) The Dermatoptic Sensitivity of the Skin

The sensitivity of the skin to light was first shown in 1872 by G. Pouchet during his experiments on dipterous larvae. V. Graber (1883, 1884) reported that animals without eyes (either naturally or with them experimentally removed), can react to light. This is to say that the sensory cells of the skin are influenced by light. This is, however, only one particular example of a general phenomenon: that is the property that all protoplasm has of reacting to light. Graber called this property, *dermatoptic sensitivity*.

Plateau (1886) and Willem (1891) produced a long list of animals showing dermatoptic sensitivity. This list may be considerably lengthened today. The dermatoptic sensitivity is particularly frequent among aquatic forms. It is much rarer in terrestrial forms, where it is only certainly established in chilopods and in the larvae of some insects (Viaud, 1961).

The information available on the dermatoptic sense of cavernicoles is fragmentary. This property is probably widespread among hypogeous invertebrates, but its existence has been established definitely only in Crustacea of the genus *Niphargus* (Ginet, 1960).

There are more numerous examples among the cavernicolous vertebrates. The dermatoptic sensitivity of *Proteus* has been known for many years (Configliachi and Rusconi, 1819; R. Dubois, 1890). It has been shown for many cavernicolous fish: *Chologaster* (Eigenmann, 1909), *Amblyopsis* (Payne, 1907; Eigenmann, 1909), *Caecobarbus* (Thinès, 1954). It is, however, absent in *Anoptichthys jordani* (Breder, 1944; Kähling, 1961).

D. PIGMENTS AND PIGMENTATION

(a) The Pigment Layer as a Protection against the Noxious Action of Light

The dermatoptic sensitivity manifests itself only in the presence of weak light. At increased intensity light becomes noxious as has been stated earlier. It cannot be doubted that the development of the pigment layer covering the body of most animals represents an adaptive manifestation. An opaque barrier opposes the action of the light rays and in particular that of ultra-violet ones†.

Most pigments represent by-products of excretion. These substances which would normally be passed to the exterior are stored in a layer around the body. They are used by the animal to form a protective layer against light.

The loss of the pigment is a degenerative alteration. As we have already said (p. 317), the loss of the pigments in the arthropods results from the thinning and simplification of the cuticle.

The loss of pigment, or depigmentation, is one of the most classical and most constant characteristics of cavernicoles. It is easy to understand the reason for this. If the epigeans lost their pigment layer they would be exposed to the harmful effects of ultra-violet light and it would not be long

---

† This interpretation applies only to integumentary pigments. The viscera are often surrounded by a layer of pigment which has a different function. These pigment layers arise in different ways as is shown by the fact that some cavernicoles with no integumentary pigments still retain pigment deposits in the visceral regions. For example, the mesentries of *Typhlogarra* (Trewavas, 1955) or the peritoneum of *Eilichthys* (Pellegrin, 1929) or the deep tissues of *Amblyopsis* (Eigenmann, 1909), of *Typhlichthys* (Poulson, 1961) and of *Caecobarbus* (Heuts, 1952).

before they disappeared from the surface of the earth. However, the humicoles, endogeans and even more so, the cavernicoles, which live away from the action of light rays can withstand this regressive characteristic without harm.

(b) Chromatophores

Pigments are usually contained in special cells with extensive ramifications. These are the chromatophores. According to the nature of the pigment they contain they are called erythrophores, xanthophores, melanophores, leucophores.

(c) Different Types of Pigment

It is outside the scope of this book to provide a detailed study of pigments. We shall limit this section to a brief description of pigments belonging to three types:

(1) Carotenoid Pigments. These are non-saturated hydrocarbons, red or yellow in colour.

Carotenoids are necessary to all living beings. However, while green plants, certain fungi, Actinomycetes and bacteria can synthesise these pigments, animals are incapable of doing so. These pigments must be provided by their food. However the animals are capable of transforming alimentary carotenoid into specific carotenoids (Lenel, 1961). The carotenoid pigments are insoluble in water but soluble in fats. It is for this reason that they are often found in adipose tissue.

Among the cavernicoles, carotenoids have been reported in Amoebina Testacea. An asellid, *Stenasellus virei,* is more often coloured red or pink than completely white. A carotene is present in the liver of *Proteus* (Beatty, 1941).

From the many functions of carotenoids and vitamin A, the synthesis of rhodopsin, is obviously suppressed in anophthalmic cavernicoles. The role of vitamin A as a growth factor is however maintained.

(2) Purine Pigments. Purine bases, and in particular guanine (2 amino-6 oxypurine) are derived from degradation of nucleoproteins. This substance, which is white in colour, is particularly abundant in the integument of fish. Bernasconi discovered the presence of pterins (leucopterins and xanthopterins) in some cavernicolous Coleoptera (Bathysciiae, Trechinae).

(3) Melanin and Ommatins. The melanins which are brown or black pigments, are derived from tyrosine and dioxyphenylalanine (DOPA) following oxidation by diastases, tyrosinase or DOPA-oxidase.

In the arthropods, melanins are replaced by pigments called ommochromes (ommines and ommatines) which are derived from tryptophan.

(d) Formation of Pigments

The formation of pigments poses many difficult problems which we feel unable to discuss. We shall limit ourselves to a brief description of the classical research of Butenandt (1940–1950), Becker (1942) and Kühn and Becker (1942), on ommochromes. The associated research of geneticists and of biochemists, has proved that the formation of the ommatins from tryptophan goes through the following stages:

Tryptophan → Kynurenine → 3 hydroxy-kynurenine → ommochrome.

The passage from one stage to another is controlled genetically by a specific gene, and biochemically by special enzymes.

(e) Conditions of Pigment Formation

The chemical reactions resulting in the formation of pigments can take place only if certain conditions are fulfilled:

(1) Presence of Oxygen. The formation of melanin pigments involves oxidation, as has been noted previously.

(2) Hormonal Intervention. Intermedin, produced by the intermediate lobe of the pituitary controls not only the dispersion of pigment in the chromatophores of lower vertebrates but also stimulates melanin formation (Pickford and Atz, 1957).

The experiments of Lillie and Juhn (1932) and those of Montalenti (1934), on chickens, showed that the thyroid hormone also influences melanin formation.

Although cavernicolous crayfish have lost their integumentary and retinal pigments, they still conserve the ability to produce neurosecretions which act as activators of these pigments (Fingerman and Mobberly, 1960).

(3) Action of Light: (a) Carotenoids – The carotenoid pigments are not produced by animals, whether they are cavernicolous or epigeans. However, the pigment persists in darkness as is proved by the constant presence of these pigments in certain cavernicoles.

The only information available on the behaviour of pigments in cavernicoles relates to *Gammarus* and to crayfishes. The colouration of these Crustacea is due mainly to carotenoids pigments. The decolouration of certain races of *Gammarus* living in subterranean waters can be explained by the low level of carotenoids in the food, for example *Gammarus pulex subterraneus* from Erzbergwerke in the Harz; if these colourless forms are fed on carrot they take on a dark colouration. The light or the absence of light are without effect on the body colour of these crustaceans (Anders, 1956a and b).

The colouration of *Procambarus simulans*, which lives in the caves of Texas, does not result from the presence – or absence – of light, but depends on the amount of carotinoids present in their food. Maguire (1961) succeeded in deepening the colour of the cavernicolous crayfishes by adding green algae to the food.

(b) Guanine – The development of purine pigments does not appear to be affected by light, or at least it has only a weak effect (Rasquin, 1947).

(c) Melanin Pigments — It is when one is considering the melanin pigments that the question of the influence of light arises. It is such a complex problem that we must consider its different aspects.

(f) Different Types of Colouration with Respect to Subterranean Life

The colouration of the cavernicoles belongs to different categories (Maguire, 1961). We can recognise three types.

(1) Stable Pigmented State

Many inhabitants of caves, even the permanent inhabitants, are not decoloured. There are many examples among the guanobia and the trolophiles. A particular example is that of the staphilinid, *Atheta subcavicola* which is very common in caves, but rare in the epigeous environment. It is, however, strongly pigmented as are other species of this genus. This is also the case with innumerable Collembola which live on guano.

This category even contains some true cavernicoles. For example, the dipteran *Phora aptina* or the opilionid *Ischyropsalis pyrenaea* which is strictly cavernicolous, and which is but little less-pigmented than the related epigeous form *I. luteipes.*

These cavernicoles considered from the aspect of pigmentation behave, in short, like epigeans, that is to say forms in which the pigment does not change in complete darkness and where it can develop in the absence of light. Some individuals of the epigeous species, *Asellus aquaticus*, were maintained in darkness for seven and a half months. They did not lose any colour and the young, born in darkness, developed normal pigmentation (Baldwin and Beatty, 1941).

(2) Stable Depigmented State

True cavernicoles are usually without melanic pigments. The red or pink colour seen in certain cavernicolous vertebrates is due to blood which circulates through the skin. Chromatophores are absent or very rare in the skin of troglobia.

In true cavernicoles, light does not modify the white colour. When adult or immature *Amblyopsis spelaeus* and *A. rosae* are placed in light they do not develop pigmentation (Eigenmann, 1900, 1909; Poulson, 1961). This is also the case in *Caecobarbus geertsi* (Gérard, 1936) and for *Anoptichthys hubbsi* (Rasquin, 1947).

Cavernicolous arthropods, and, in particular, *Niphargus* and *Asellus cavaticus* behave in the same way as the vertebrates reported above. When these animals are placed in light they do not become pigmented.

Thus the absence of pigmentation in these animals is a stable condition which is not modified by light. Thus it is certain that in these species the loss of integumentary pigments is complete and definitive.

This characteristic is not restricted to cavernicoles. It is to be found in endogeans, myremecophiles and even epigeans. Thus, although no caverni-coles are known among the symphylids, all the species are decoloured and blind. All palpigrades are decoloured and blind whether they are lapidi-colous or cavernicolous. Among the chilopods, the *Geophilomorphes* and *Cryptops* are without pigment and have no ocelli. The myrmecophilic iso-pods of the genus *Platyarthrus* and the muscicoles of the genus *Miktoniscus* are perfectly white.

(3) Unstable State of Depigmentation

The two preceding states represented simple conditions. There are, however, others which are more complex.

(a) Troglophiles. Troglophiles are known which live in caves as well as at the exterior and whose colour varies according to their mode of life.

Elaphe taeniura is a snake with nocturnal habits which is found in Burma and Malaya. Certain individuals enter caves to search for bats on which they feed. These cave-snakes lose their red and brown pigment but conserve the black pigment. Butler described such individuals under the variety *rid-leyi* (Annandale, Brown and Gravely, 1914). Another variety of this species which is also partially decoloured has received the name *grabowskyi*. It has been captured in the Batu caves in the State of Selangor in Malaya (Kloss, 1929).

The fish *Stygicola dentatus* shows all the intermediate stages between pigmentation and non-pigmentation (Eigenmann, 1909; Kosswig, 1934).

The troglophilic diplopods *(Archiboreiulus pallidus, Boreoiulus tenuis)* are coloured or completely depigmented, according to whether they live at the entrances to caves or in their depths (Leruth, 1939).

Populations of *Asellus aquaticus cavernicolus* which live in the sub-terranean part of the Piuka, in Slovenia, show extreme polychromatism. They include normal individuals, entirely depigmented specimens and all intermediate types (C. and L. Kosswig, 1940).

(b) Cavernicoles which are pigmented when young. Many cavernicoles are known which are pigmented at birth but gradually lose their pigmentation with age. Many cavernicolous Orthoptera of the family Rhaphidophoridae, *Diestrammema* (Chopard, 1919), *Pachyrhamma* (Richards, 1961), *Dolicho-poda* (Chopard, 1932), are examples.

The same situation occurs in cavernicolous fish, *Typhlichthys* and *Ambly-opsis* (Eigenmann, 1909) and also in *Proteus* (Zeller, 1888, 1889; Vandel and Bouillon, 1959) (Plate IV).

(c) The persistance of pigmentation in the adult under the influence of light. The regressive evolution of pigmentation which occurs during development and which has just been described, is sometimes irreversible.

But, this condition is not general in the cavernicolous fish. *Typhlichthys subterraneus*, according to Woods and Inger (1957) and Poulson (1961) is normally white in colour and without pigment. If it is maintained in light pigment appears after several months, in a lateral band and in lines corresponding to the myosepta. Thus disappearance of pigment is not definitive in this fish.

Other analogous examples are that of *Chologaster agassizi* (Poulson, 1961) and of *Anoptichthys jordani* (Rasquin, 1947) which becomes more pigmented when placed in light.

Similar examples are found in the Amphibia of the sub-class Urodela. The classical example is that of *Proteus* which becomes violet or blackish in colour when cultured in an illuminated medium. This fact has been known for many years. In their famous monograph devoted to *Proteus*, Configliachi and Rusconi (1819) mentioned that specimens kept in daylight acquire a violet colour. The observations of these Italian naturalists was confirmed by modern biologists (Poulton, 1889; Kammerer, 1912, Spandl, 1926; Kosswig, 1937). This phenomenon can be seen in zoological gardens and in particular in the Vivarium du Jardin des Plantes in Paris where a specimen of *Proteus* brought in by Dr. Jeannel has lived in the light for many years and has acquired a very dark blackish colouration.

Typhlotriton nereus develops a large amount of pigment when it is kept in light but depigments when it goes underground (Bishop, 1944). Thus the behaviour of *Proteus* is similar to this American urodele.

The behaviour of *Typhlotriton spelaeus* is different. The maintenance of the larvae in light induces pigmentation but light has no affect on the adults (Nobel and Pope, 1928).

This list can be concluded by mentioning several examples relative to invertebrates or rather, Crustacea. The spheromid *Caecosphaeroma burgundum* rapidly gains pigment when placed in light (Paris, 1925). Also the population of *Caecosphaeroma virei* found in the Grotte de Poncin, in the department of Ain, is made up of more or less pigmented individuals (Remy, 1948). It can thus be stated that *Caecosphaeroma* is a recent cavernicole.

(4) Conclusions

The observations reported above show a series of stages from the epigeous to the cavernicolous state.

The *epigeous state* is a stable state in which pigmentation is maintained regardless of the conditions of the medium (light or darkness). Certain cavernicoles have conserved, as for the pigmentation, the epigean state.

In the *cavernicolous state* the chain of reactions which leads to the formation of the pigment is totally suppressed. In this case the loss of pigment is definitive and nothing can cause it to reappear.

Between these two extreme conditions is the *unstable state* which is the most interesting to the biospeologist because it can be used for experimentation. *Proteus* is the best example. The processes of pigment formation are not lost, they are only weakened. It is probable that the loss of the pigment elaborated during the juvenile stages, results from a decrease in the endocrine secretions of the pituitary and thyroid. But the loss of pigmentation can be overcome if a stimulant, such as light, is provided during the young stages. However this hypothesis ought to be demonstrated experimentally.

BIBLIOGRAPHY

ANDERS, F. (1956a) Über Ausbildung und Vererbung der Körperfarbe bei *Gammarus pulex subterraneus* (Schneider), einer normalerweise pigmentlosen Höhlenform des gemeinen Bachflohkrebses. *Zeit. indukt. Abstamm. Vererbl.* **LXXXVII.**

ANDERS, F.(1956b) Modifikative und erbliche Variabilität von *Gammarus pulex. Verh. d. deutsch. Zool. Gesell.* Hamburg.

ANNANDALE, N., BROWN, J.C. and GRAVELY, F.H. (1914) The limestone caves of Burma and the Malay Peninsula. *J. Proc. Asiatic. Soc. Bengal.* **IX.**

BALDWIN, E. and BEATTY, R.A. (1941) The pigmentation of cavernicolous animals. I. The pigments of some isopod Crustacea. *J. exper. Biol.* **XVIII.**

BANTA, A.M. (1910) A comparaison of the reactions of a species of surface isopod with those of a subterranean species. Part II. *J. exper. Zool.* **VIII.**

BEATTY, R.A. (1941) The pigmentation of cavernicolous animals. II. Carotenoid pigments in the cave environment. *J. exper. Biol.* **XVIII.**

BECKER, E. (1942) Über Eigenschaften, Verbreitung und die genetisch-entwicklungsphysiologische Bedeutung der Pigmente der Ommatin- und Ommingruppe (Ommochrome) bei den Arthropoden. *Z. f. indukt. Abstamm. Vererbl.* **LXXX.**

BERNARD, FR. (1932) Comparaison de l'œil normal et de l'œil régressé chez quelques Carabiques. *Bull. biol. France. Belgique* **LXVI.**

BISHOP, S. (1944) A new neotenic Plethodon salamander, with notes on related species. *Copeia.*

BREDER, C.M. JR. (1943) Problems in the behaviour and evolution of a species of blind fish. *Trans. New York. Acad. Sc.* (2) **V.**

BREDER, C.M. JR. (1944) Ocular Anatomy and light sensitivity studies on the blind fish from Cueva de los Sabinos, Mexico. *Zoologica* **XXIX.**

BREDER, C.M. JR. and GRESSER, E.B. (1941) Further studies on the light sensitivity and behaviour of the Mexican blind characin. *Zoologica* **XXVI.**

BREDER, C.M. JR. and RASQUIN, P. (1947) Comparative studies in the light sensitivity of blind characins from a series of Mexican caves. *Bull. Amer. Mus. Nat. Hist.* **LXXXIX.**

BREDER, C.M. JR. and RASQUIN, P. (1950) A preliminary report on the role of the pineal organ in the control of pigment cells and light reactions in recent teleost fishes. *Science* **III.**

BUCHANAN, J.W. (1936) Notes on an American cave flatworm, *Sphalloplana percaeca. Ecology* **XVII.**

BUTENANDT, A., WEIDEL, W., and BECKER, E. (1940) Oxytryptophan als "Prokynurenin" in der zur Augenpigmentbildung führenden Reaktionskette bei Insekten. *Naturwiss.* **XXVIII.**

BUTENANDT, A., WEIDEL, W., WEICHERT, R. and DERJUGINE W. von (1943) Ueber Kynurenin. Physiologie, Konstitutionsermittlung und Synthese. *Zeit. f. physiol. Chemie.* **CCLXXIX.**

CHOPARD, L. (1919) Les Orthoptères cavernicoles de Birmanie et de la péninsule malaise. *Mem. Asiatic. Soc. Bengal* **VI.**

CHOPARD, L. (1932) Les Orthoptères cavernicoles de la Faune paléarctique. *Biospeologica*, **LVII.** *Archiv. Zool. expér. géner.* **LXXIV.**

CONFIGLIACHI, P. and RUSCONI, M. (1819) *Monografia del Proteo anguineo di Laurenti.* Pavia.

DEROUET-DRESCO, L. (1959) Contribution à l'étude de la Biologie de deux Crustacés aquatiques cavernicoles: *Caecosphaeroma burgundum* D. et *Niphargus orcinus virei.* Ch. *Vie et Milieu* **X.**

DUBOIS, R. (1890) Sur la perception des radiations lumineuses par la peau, chez les Protées aveugles des grottes de la Carniole. *Compt. rend. Acad. Sci. Paris* **CX.**

EIGENMANN, C. H. (1900) The blind fishes. *Biol. Lectures Marine Biol. Labor. Woods Hole.* **VIII.**

EIGENMANN, C. H. (1909) Cave vertebrates of America. A study in degenerative evolution. *Carnegie Inst. Washington Public.* No. 104.

FINGERMAN, M. and MOBBERLY, W. C. JR. (1960) Trophic substances in a blind crayfish. *Science* **CXXXII.**

FRISCH, K. VON (1911) Beiträge zur Physiologie der Pigmentzellen in der Fischhaut. *Archiv. f. gesam. Physiol.* **CXXXVIII.**

GAL, J. (1903) *Niphargus* et *Caecosphaeroma*. Observations physiologiques. *Bull. Soc. Sc. Nat. Nimes* **XXXI.**

GÉRARD, P. (1936) Sur l'existence de vestiges oculaires chez *Caecobarbus geertsi. Mém. Mus. Hist. Nat. Belgique* (2) **III.**

GINET, R. (1960) Écologie, Éthologie et Biologie de *Niphargus* (Amphipodes, Gammarides hypogés). *Ann. Spéléol* **XV.**

GRABER, V. (1883) Fundamentalversuche über die Helligkeits- und Farbenempfindlichkeit augenloser und geblendeter Thiere. *Sitzungsber. Math. Naturw. Kl. d. Akad.* **LXXXVII,** 1. Abt. Wien.

GRABER, V. (1884) *Grundlinien zur Erforschung des Helligkeits- und Farbensinnes der Tiere.* Prag und Leipzig.

HAWES, R.S., (1946) On the eyes and reactions to light of *Proteus anguinus. Quart. Jour. Micr. Sc.* **LXXXVI.**

HEUTS, M.J. (1952) Ecology, variation and adaptation of the blind fish, *Caecobarbus Geertsi* Boulenger. *Ann. Soc. R. Zool. Belgique* **LXXXII.**

JANZER, W. (1937) Versuche zur Entstehung der Höhlentiermerkmale. *Naturwiss.* **XXXVII.**

JANZER, W. and LUWDIG, W. (1952) Versuche zur evolutorischen Entstehung der Höhlentiermerkmale. *Z. indukt. Abstamm. Vererbl.* **LXXXIV.**

JEANNEL, R. (1929) Le Vivarium du Jardin des Plantes pendant l'année 1928. *Rev. Hist. Nat.* **X.**

JEANNEL, R. (1943) *Les Fossiles vivants des Cavernes.* Paris.

KÄHLING, J. (1961) Untersuchungen über den Lichtsinn und dessen Lokalisation bei dem Höhlenfisch *Anoptichthys jordani. Biol. Zentralbl.* **LXXX.**

KAMMERER, P. (1912) Experimente über Fortpflanzung bei *Proteus anguinus* Laur. Zugleich: Vererbung erzwungener Farbenveränderungen. III. Mitteilung. *Archiv. f. Entwicklungsm.* **XXXIII.**

KLOSS, C.B. (1929) Fauna of the Batu Caves, Selangor. *J. Fed. Malay St. Mus.* **XIV.**

KOSSWIG, C. (1934) Über bislang unbekannte Sinnesorgane bei dem blinden Höhlenfisch *Stygicola dentatus* (Poey). *Zool. Anz.* 7. Supplementbd.

KOSSWIG, C. (1937) Betrachtungen und Experimente über die Entstehung von Höhlentiermerkmalen. *Züchter.* **IX.**

KOSSWIG, C. and L. (1940) Die Variabilität bei *Asellus aquaticus*, unter besonderer Berücksichtigung der Variabilität in isolierten unter- und oberirdischen Populationen. *Istanbul Univer. f. Faküll. Mecm. Ser. B.* **V.**

Kühn, A. and Becker, E. (1942) Quantitative Beziehungen zwischen zugeführtem Kynurenin und Augenpigment bei *Ephestia kühniella* Z. *Biol. Zentralbl.* **LXII.**

Kuhn, O. and Kähling, J. (1954) Augenrückbildung und Lichtsinn bei *Anoptichthys jordani* Hubbs and Innes. *Experientia* **X.**

Lenel, R. (1961) Sur le métabolisme des pigments carotinotinoïdes de *Carcinus maenas* Linné. *Bull. Soc. lorraine Sc.* **I.**

Leruth, R. (1934) Diptères: *Dryomyzidae* et *Helomyzidae.* Exploration biologique des cavernes de la Belgique et du Limbourg Hollandais. XVI. *Bull. Annal. Soc. Entomol. Belgique* **LXXIV.**

Leruth, R. (1939) La biologie du domaine souterrain et la faune cavernicole de Belgique. *Mém. Mus. R. Hist. Nat. Belgique* No. 87.

Lillie, Fr. R. and Juhn, M. (1932) The physiology of development of feathers. I. Growth-rate and pattern in the individual feather. *Physiol. Zool.* **V.**

Lüling, K. H. (1953 a) Über das Sehen jugendlicher *Anoptichthys jordani* (Hubbs und Innes). *Aquar. Terrar. Zeit.* **VI.**

Lüling, K. H. (1953 b) Über die fortschreitende Augendegeneration des *Anoptichthys jordani* Hubbs und Innes (Characidae). Vorläufige Mitteil. *Zool. Anz.* **CLI.**

Lüling, K. (1954) Untersuchungen am Blindfisch *Anoptichthys jordani* Hubbs und Innes (Characidae). II. Beobachtungen und Experimente an *Anoptichthys jordani* zur Prüfung der Einstellung zum Futter, zum Licht und zur Wasserturbulenz. *Zool. Jahrb. Abt. allg. Zool. Physiol.* **LXV.**

MacIndoo, N. E. (1910) Biology of the Shawne Cave spiders. *Biol. Bull.* **XIX.**

Maguire, B. Jr. (1961) Regressive evolution in cave animals and its mechanism. *Texas J. Sc.* **XIII.**

Merker, E. (1925) Die Empfindlichkeit feuchthäutiger Tiere im Lichte. *Zool. Jahrb. Abt. allg. Zool. Physiol.* **XLII.**

Merker, E. (1929 a) Lichtsinn und allgemeine Lichtempfindlichkeit. *Zool. Anz. 4. Supplementbd.*

Merker, E. (1929 b) Die Durchlässigkeit des Chitins für ultraviolettes Licht. *Zool. Anz. 4. Supplementbd.*

Montalenti, G. (1934) A physiological analysis of the barred pattern in Plymouth Rock Feathers. *Jour. exper. Zool.* **LXIX.**

Nagel, W. A. (1896) Der Lichtsinn augenloser Tiere. Eine biologische Studie. Jena.

Noble, G. K. and Pope, S. H. (1928) The effects of light on the eyes, pigmentation and behaviour of the cave salamander *Typhlotriton. Anat. Rec.* **XLI.**

Pactl, J. (1956) Biologie der primär flügellosen Insekten. Jena.

Paris, P. (1925) Sur la bionomie de quelques Crustacés troglobies de la Côte d'Or. *Compt. rend. Assoc. franç. Avanc. Sc.* **XLIX;** Grenoble.

Park, O., Roberts, T. W. and Harris, S. J. (1941) Preliminary analysis of activity of the cave crayfish, *Cambarus pellucidus. Amer. Nat.* **LXXV.**

Payne, F. (1907) The reactions of the blind fish, *Amblyopsis spelaeus,* to light. *Biol. Bull.* **XIII.**

Pellegrin, J. (1929) L'*Eilichthys microphthalmus* Pellegrin, Poisson cavernicole de la Somalie italienne. *Bull. Mus. Hist. Nat. Paris* (2) **I.**

Pickford, G. E. and Atz, J. W. (1957) *The Physiology of the Pituitary Gland of Fishes.* New York.

Plateau, F. (1886) Recherches sur la perception de la lumière par les Myriapodes aveugles. *J. Anat. Physiol.* **XXII.**

Poulson, T. L. (1961) Cave adaptation in amblyopsid fishes. University of Michigan. *Phil. Dissert. Zool.*

Poulton, E. B. (1889) *The Colours of Animals.* London.

Pylka, J. and Warren, R. D. (1958) A population of *Haideotriton* in Florida. *Copeia.*

Rasquin, P. (1947) Progressive pigmentary regression in fishes associated with cave environments. *Zoologica* **XXXII.**

REMY, P. (1948) Sur quelques Crustacés cavernicoles d'Europe. *Notes biospéol.* **III.**

RICHARDS, A. M. (1961) The life history of some species of *Rhaphidophoridae* (Orthoptera). *Trans. R. Soc. New Zealand Zool.* **I.**

SCHARRER, E. (1928) Die Lichtempfindlichkeit blinder Elritzen (Untersuchungen über das Zwischenhirn der Fische. I). *Zeit. f. vergl. Physiol.* **VIII.**

SCHWOERBEL, J. (1961) Subterrane Wassermilben (Acari: Hydrachnellae, Porohalacaridae und Stygothrombiidae), ihre Ökologie und Bedeutung für die Abgrenzung eines aquatischen Lebensraumes zwischen Oberfläche und Grundwasser. *Archiv. f. Hydrobiol. Suppl.* **XXV.**

SEIBOLD, W. (1904) Anatomie von *Vitrella quenstedti* (Wiedersheim) Clessin. *Jahresh. Ver. vaterl. Naturk. Württemberg.* **LX.**

SPANDL, H. (1926) Die Tierwelt der unterirdischen Gewässer. *Speläol. Monogr.* **XI.**

THINÈS, G. (1953) Recherches expérimentales sur la photosensibilité du Poisson aveugle, *Caecobarbus geertsi* Blgr. *Ann. Soc. R. Zool. Belgique* **LXXXIV.**

THINÈS, G. (1954) Étude comparative de la photosensibilité des Poissons aveugles, *Caecobarbus geertsi* Blgr. et *Anoptichthys jordani* Hubbs et Innes. *Annal. Soc. R. Zool. Belgique* **LXXXV.**

TREWAVAS, E. (1955) A blind fish from Irak related to *Garra. Ann. Mag. Nat. Hist.* (12) **VIII.**

VANDEL, A. and BOUILLON, M. (1959) Le Protée et son intérêt biologique. *Ann. Spéléol.* **XIV.**

VERRIER, M. L. (1929) Observations sur le comportement d'un Poisson cavernicole: *Typhlichthys osborni* Eigenmann. *Bull. Mus. Hist. Nat. Paris* (2) **I.**

VIAUD, G. (1961) Le sens dermatoptique, mode primitif de photoréception, envisagé spécialement chez les organismes aquatiques. *Progress in Photobiol. Proceed. third Intern. Congress Photobiol.* Copenhagen.

VILTER, V. (1960) Différenciation d'une "fenêtre crânienne" chez les Poissons abyssaux porteurs d'organes lumineux cutanés. *Compt. rend. Soc. Biol. Paris* **CLIV.**

WAUTIER, J. and TROÏANI, D. (1960) Contribution à l'étude du métabolisme respiratoire de quelques Gammarides. *Ann. Stat. Centrale Hydrobiol. appl.* **VIII.**

WELLS, P. H. (1952) Response to light by the eyeless white cave crayfish, *Cambarus ayersii. Anat. Rec.* **CXIII.**

WELLS, P. H. (1957) Additional studies on responses to light by cave crayfishes. *Anat. Rec.* **CXXVIII.**

WELLS, P. H. (1959) Response to light by cave crayfishes. *Occas. Pap. Nat. Speleol. Soc.* **IV.**

WILLEM, V. (1891) Sur les perceptions dermatoptiques. Résumé historique et critique. *Bull. Sc. France Belgique* **XXIII.**

WOODS, L. P. and INGER, R. F. (1957) The cave, spring and swamp fishes of the family Amblyopsidae of central and eastern United States. *Amer. Midl. Nat.* **LVIII.**

ZELLER, E. (1888) Über die Larve des *Proteus anguinus. Zool. Anz.* **XI.**

ZELLER, E. (1889) Über die Fortpflanzung des *Proteus anguineus* und seine Larve. *Jahreshefte d. Ver. f. Vaterl. Naturkunde i. Württemberg.* **XLV.**

CHAPTER XXVI

THE VISUAL SYSTEM OF CAVERNICOLES

A. INTRODUCTION

The eye is the most important of all photoreceptors and the following chapter will be devoted to this topic.

It is known that the visual system undergoes some degree of regression in most animals living underground. Jeannel (1954) has remarked that the regression of the visual system or anophthalmia, is associated frequently, but however erroneously with loss of sight or blindness. Certain Coleoptera of the family Scaritidae, and also the hypogeous dytiscid, *Siettitia balsetensis* still possess eyes but they are non-functional. These insects are therefore blind, but not anophthalmic.

B. THE DISTRIBUTION OF ANOPHTHALMIA
IN THE ANIMAL KINGDOM

Many naturalists cannot reject the false idea that living organisms are passively modelled by the medium in which they live. An animal reacts actively to changing situations but different animal species never react in exactly the same way to similar situations. This is the reason for the absence of universal laws of biology. Certain paths are more frequently followed than others but unique paths are unknown in the living world.

It would completely distort reality to state that all cavernicoles are anophthalmic and that all surface forms possessed eyes. In reality the situation is much more complex, as can be shown by a few examples.

(a) Anophthalmia and Subterranean Cavities

Most cavernicoles possess regressed eyes and in some, the visual apparatus is completely absent. However, it would be wrong to believe that this condition is found in all cavernicoles. De Lattin (1939) has compiled a list of species inhabiting caves and possessing eyes. This list includes copepods, isopods, decapods, spiders, Coleoptera, Copeognatha, Diptera and fish. The list is certainly not complete. Guanobia and numerous troglophiles have eyes.

In certain groups such as the Acarina and the Collembola, numerous genera contain species some of which are anophthalmic, others possess eyes, but there is no discernable relationship between the type of visual system and the habitat, be it a cave or the surface medium (Absolon, 1900).

Hydracarina of the hyporheic medium are rarely totally anophthalmic; *Acherontacarus* which is completely deprived of the visual apparatus, is an exception (Schwoerbel, 1961).

(b) Anophthalmia and the Profundal Zones of Lakes

Other dark media present to animal species conditions similar to those of the subterranean environment. The biotope which in many ways shows the most ressemblance to the hypogeous medium is that of the bottom of lakes. Depigmented forms with regressed eyes, such as *Niphargus foreli* and *Asellus foreli*, are found there.

However, as it is the case in cavernicoles, lacustrine and bathyal forms show a great variety of visual structures. The planarians can be mentioned in this context as they allow comparison of lacustrine and cavernicolous species.

De Beauchamp (1932) remarked that the Dendrocoelidae which are usually depigmented, generally become anophthalmic in dark media, while the Planariidae which are normally pigmented, exhibit very varied states of the visual system when they penetrate the subterranean medium.

However, even in the Dendrocoelidae, the condition of the eyes is far from being constant. Specimens of *Dendrocoelum lacteum* found in sub-terranean waters have kept their eyes (de Beauchamp, 1920), while the forms which populate the deep regions of the great subalpine lakes are sometimes anophthalmic and sometimes possess eyes (Duplessis, 1876). *Dendrocoelum cavaticum* is a species which is normally without eyes, but Enslin has captured multi-ocular specimens in the springs of Lauter. Finally, *Sorocelis americana* from a cave in Oklahoma, possesses numerous eyes (Hyman, 1939).

Analogous examples are seen in the Planariidae. *Planaria anophthalma* is only a polypharyngeal race of *Planaria alpina*. Both live in springs, but while the first is anophthalmic, as the name indicates, the second has eyes. *Phagocata cavernicola* which is widely distributed in many caves of eastern America, also has eyes.

(c) Anophthalmia and Marine Abysses

Species which live in marine depths have been particularly well studied regarding the visual system. It has been stated many years ago (Cunning-ham, 1893) that the proportion of anophthalmia is much lower in the abyssal zones than in the subterranean world.

In an extensive study devoted to abyssal Brachyura, Döflein, (1904) noted that out of 450 species of crab dredged form the great depths of the sea by different oceanographic expeditions, 380 or 85% possessed eyes which appeared to be normal; 65 or 14% had small eyes. Only 5 or 1% were completely without eyes. Certain abyssal crabs such as *Geryon affinis* possess large eyes. The persistence of the ocular apparatus in bathetic forms is almost certainly linked to the presence of numerous bioluminescent forms found in the sea. Forms which produce light are exceptional in subterranean media (p. 304).

In other groups there is a higher proportion of anophthalmia. Stammer (1936) estimated that of 368 species of mysid, 39 or 10·6% are anophthalmic or have regressed eyes.

In one of the most recent accounts of anophthalmic fish, Thinès (1955) mentions 86 species with regressed eyes. They are distributed as follows: 41 are freshwater forms; 16 live in brackish water and 29 in the sea. Among the latter 17 are bathetic forms and 12 are coastal species. Moreover, of the 17 bathetic and anophthalmic forms listed by Thinès, 13 belong to the Brotulidae and 4 to the Ceratiidae.

These figures lead to the conclusion that:

(1) the relation between anophthalmia and a dark medium is far from being constant;

(2) anophthalmia is particularly frequent in certain families.

This phenomenon must be considered as a characteristic of certain phyletic lines.

(d) Anophthalmia and the Epigeous World

In the same way that one finds completely depigmented species in the epigeous habitat, many anophthalmic forms are also to be found there. These are the same species which as usual are both depigmented and anophthalmic. Examples are the Symphyla, Palpigrada, Geophilomorpha, *Cryptops*, and various Crustacea (Sayce, 1901).

Packard (1886) compiled a long list (comprising 19 pages of his book on the cavernicolous fauna of North America), of non-cavernicolous species without eyes or with regressed eyes. It is not necessary to reproduce this list. Its length alone establishes without doubt the absence of cause and effect between anophthalmia and the subterranean life.

However, it should be mentioned that there is not a uniform distribution of anophthalmic forms on the surface of the soil. Certain biotopes contain a particularly high proportion of such forms. We shall briefly mention here an example which will be reconsidered later in more detail as it is important in the understanding of the origin of cavernicoles. The research of Leleup and the studies of Jeannel (Jeannel and Leleup, 1952; Leleup, 1956), have shown the existence in the great forests which cover the Kivu

Mountains (Eastern Congo) between the altitude of 2000 and 2900 m of a humicolous fauna which contains numerous anophthalmic types showing very varied examples of ocular regression.

(e) Conclusions

From the examples given in the preceding pages it should be concluded that the relation between the presence of eyes, or on the other hand anophthalmia, and the way of life are very varied. Cavernicolous species are generally anophthalmic but there are many exceptions to this rule. Also, epigeous species without eyes are far from being rare.

It can also be established that certain phyletic lines are characterised by the frequence of anophthalmic forms (or even by the constant anophthalmia, as it is the case in the Geophilomorpha), whatever may be the mode of life, epigeous or hypogeous. Examples have been given from the planarians and the fish. Anophthalmia is thus characteristic of certain lines. It is a phyletic manifestation.

Finally, regression of the eyes is more marked when it is phyletically ancient. All the Bathysciinae are anophthalmic with the exception of a few rare species (Jeannel, 1911) and the reduction of the visual apparatus is a condition concerning epigeous as well as cavernicolous forms. It is certainly a very ancient condition in this line of Coleoptera. The situation is different in the Ptomophagini which are related to Bathysciinae, but which belong to the Catopidae. Certain species of the genus *Adelops* belong to this tribe. These are cavernicolous and populate the caves of North America. In this group ocular regression occurs sporadically and is always less accentuated than in Bathysciinae. In the Ptomophagini regressive evolution of the visual apparatus is in its early stages.

C. ANOPHTHALMIC CAVERNICOLES

Since the first studies of Newport (1855), Murray (1857), and Lespès (1868), published over a century ago, the number of anophthalmic or microphthalmic cavernicoles reported in the subterranean world has considerably increased. A list is given below of the more important publications devoted to a study of the ocular regression seen in hypogeous animals. The majority concern Crustacea. There are fewer publications concerning eyes of subterranean insects. The visual apparatus of fish and Amphibia have been the subject of some important studies.

A List of the Principal Publications Concerning the Ocular Structures
of Cavernicolous Animals

Planarians   Enslin, 1906; Berninger, 1911; de Beauchamp, 1920, 1932.

Molluscs   *Opeas:* Annandale and Chopra, 1924; Ghosh, 1929.
             Vitrella: Seibold, 1904.

Arachnids   *Pseudoscorpions:* Ellingsen, 1908; Beier, 1940.
             Opilionids; Ischyropsalis: Kratochvil, 1936; Caporiaco, 1947; Roewer, 1950.
             Spiders: MacIndoo, 1910; Simon, 1911; Fage, 1913, 1931; Kästner, 1926;
             Dresco, 1952.

Crustacea   *Copepods:* Chappuis, 1920; Borutzky, 1926; Hnatenwytsch, 1929.
             Ostracods: Vejdovsky, 1882; Hart and Hobbs, 1961.
             Mysidacea: Stammer, 1936.
             Amphipods: Leydig,1878; della Valle,1893; Hamann,1896; Vejdovsky, 1900,
             1905; Hay, 1902; Strauss, 1909; Jarocki and Krzysik, 1924; Schellenberg,
             1931; Wolsky, 1934, 1935; Szarski, 1935; Bartolazzi, 1937.
             Isopods; Asellids: Packard, 1885, 1886; Kosswig, 1935, 1936, 1937; de Lattin,
             1939.
             Oniscoids: de Lattin, 1939.
             Decapods; Troglocaris: Babič, 1922.
             Munidopsis: Harms, 1921; Bernard, 1937.
             Cambarus: Newport, 1855; Leydig, 1883; Packard, 1886; Parker, 1891.

Myriapods   Packard, 1886.

Insects   *Orthoptera:* Jörschke, 1914; Chopard, 1929, 1932; Bernard, 1937.
             Coleoptera: Lespès, 1867; Packard, 1886; Bernard, 1932, 1937; Jeannel, 1954.
             Diptera: Sadoglu, 1956.

Fish   *General list:* Thinès, 1955.
             Ambylopsis: Tellkampf, 1844; Wyman, 1854; Putnam, 1872; Eigenmann,
             1909.
             Chologaster: Eigenmann, 1909.
             Typhlichthys: Kohl, 1895; Eigenmann, 1909.
             Caecobarbus: Gérard, 1936.
             Typhlogarra: Marshall and Thinès, 1958.
             Phreatobius: Reichel, 1927.
             Anoptichthys: Gresser and Breder, 1940; Breder and Gresser, 1941; Breder,
             1944; Lüling, 1953, 1955; Stefanelli, 1954; Cahn, 1958; Frank, 1961; Käh-
             ling, 1961.
             Lucifuga and *Stygicola:* Eigenmann, 1909.

Urodeles   *American species:* Eigenmann, 1900, 1909; Noble and Pope, 1928; Hilton,
             1956.
             Proteus: Desfosses, 1882; Hess, 1889; Kohl, 1889, 1891, 1895; Schlampp,
             1891, 1892; Kammerer, 1912, 1913; Hawes, 1946; Gostojeva, 1949; Hilton,
             1956; Durand, 1962.

D. INSTABILITY OF THE OCULAR STRUCTURES
IN CAVERNICOLES

It has been known for many years that rudimentary organs, or those in the course of regression, usually show great variability. This instability shows itself in the existence in the same species, of varied visual structures according to the individual. Here are a few examples.

The visual structures in the mollusc *Opeas cavernicola* show this instability. The eye is sometimes pigmented and possesses crystalin and sometimes depigmented and without the dioptric apparatus (Annandale and Chopra, 1924).

Such variation is frequent among the spiders (Kästner, 1926) and in particular in the species belonging to the genera *Leptoneta* and *Troglohyphantes* (MacIndoo, 1910; Simon, 1911). Individual variation in the structure of the eye has been reported in amphipods of the genus *Bathyonyx* (Vejdovsky, 1905); in cavernicolous crayfish of the genus *Cambarus* (Parker, 1890); in myriapods of the genus *Antroherposoma* (Verhoeff, 1926–1928); and in various insects such as the cockroach *Alluaudellina cavernicola* (Chopard, 1932), the phasgonurid *Tachycines cuenoti* (Chopard, 1929); and in the pupiparous dipteran, *Nycteribosca africana* (Sadoglu, 1956).

There is however, no better example of this phenomenon than *Asellus aquaticus cavernicolus* from caves in Slovenia (De Lattin, 1939a and b; C. and L. Kosswig, 1940). A series of eye types have been described in this sub-species: normal, reduced, deformed, destroyed, blind, anophthalmic, which range from the normal eye to total disappearance of all eye structures (Fig. 75). The constitution of the eye in *Asellus stygius* from North America seems to be equally variable (Packard, 1885).

Arthropods are also known where the stages of ocular regression have progressed differently in the two sexes. Regression is always more rapid in the female than in the male. Thus in evolutionary terms the female is in advance of the male. These conditions are found in the opilionid, *Travunia vjetrenicae* (Hadži, 1933); in the cockroach, *Alluaudellina cavernicola* (Chopard, 1932); and in various pselaphids (Jeannel, 1952, 1954).

Variability in the size of the reduced eyes is also present in cavernicolous vertebrates, in particular in the hypogeous fish (Eigenmann, 1909). In *Lucifuga*, the size of the eye varies between 260 and 425μ. Frequently the size of the right eye is different from that of the left eye and the diameters can vary by as much as a proportion of 1 to 3. Finally, the variations in the regressed eyes of the mole may be noted (Tusques, 1954).

E. EYE PIGMENTS

Pigments normally form part of the visual apparatus. In the arthropods they consist, as do the integumentary pigments, of ommochromes. However, the eye pigments are generally more stable than those of the integument. In cavernicoles the disappearance of body colour normally precedes that of the visual pigments.

However, visual pigment is sometimes very sensitive to those factors which can cause its disappearance; for example, starvation (as shown by Tschugunoff (1913) on the cladoceran *Leptodora*); or darkness, (as

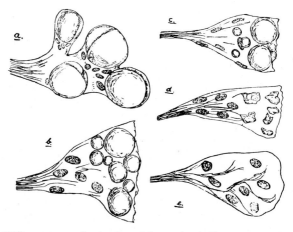

FIG. 75. Different types of reduction of the eye in *Asellus aquaticus cavernicolus*. *a*, normal eye. *b*, deformed with large nuclei. *c*, deformed with small nuclei. *d*, destroyed. *e*, blind (after de Lattin).

shown by Kapterew (1912) on various cladocerans); or starvation and darkness combined, (the experiments of Berninger (1911) on the planarians).

The visual pigments are in some respects independent of the other ocular structures. This is shown by the following facts:

(1) It has often been stated that troglophiles possess eyes when they live in illuminated zones, but are anophthalmic when they live in a completely dark region of the cave. Graeter (1910) maintained that three species of Cyclops (*C. teras*, *C. macrurus* and *C. zschokkei)* became "blind" when they lived in caves. This report of the abrupt disappearance and sudden reappearance of the visual apparatus according to the illumination of the habitat was dismissed by Chappuis (1910). He showed that the copepods which were studied by Graeter rapidly re-acquire their "eyes" when cultured in light. In fact, this visual apparatus persists under all conditions. It never disappears. However, the visual pigment rapidly degenerated in dark-

ness and reforms as rapidly in light. These variations represent a somation whose aspects depend upon the presence of light or darkness.

Hart and Hobbs (1961) observed similar changes in the ocular apparatus of the ostracod *Entocythere*, a commensal of the cavernicolous crayfishes.

The examples of populations of cavernicoles which live in the same cave and are represented by blind and eyed individuals, can be interpreted in the same way.

(2) Regression of the eye is sometimes accompanied, not by the disappearance of pigment, but by an excess of it. This is seen in species from the marine abysses and also in cavernicolous fish of the genus *Typhlichtys*. The excess pigment represents a waste deposit resulting from the degeneration of the visual structures.

(3) In a few species of hypogeous amphipods with regressed eyes one finds a brown pigment, examples are *Bathyonyx* (Vejdovsky, 1905) and *Niphargus tatrensis* of Lake Lunz, Austria (Brehm, 1955); or more generally yellow pigment such as in various species belonging to the genera *Crangonyx* (Spence Bate, 1859) and *Niphargus* (C. L. Koch, 1835; Leydig, 1878; della Valle, 1893; Garbini, 1895; Viré, 1900; Chevreux, 1901; Vejdovsky, 1905; Paris, 1925; Karaman, 1950; and observations on *Niphargus* captured around Paris by the author). This spot was formaly interpreted as the residue of the residue of the ocular pigment. However, this spot is made up of granules enclosed in the saccule of the antennal gland (Vejdovsky, 1901), or in the cephalic nephrocytes (Bruntz, 1904).

F. PATHS OF REGRESSION OF THE EYE IN CAVERNICOLES

(a) Introduction

The reduction of the visual structures in cavernicoles provides the zoologist with the most varied stages of regression of this organ, from eyes which are almost normal and certainly still functional, to total disappearance of all visual structures. This final stage is reached in the insects by certain Trechinae such as *Duvalius* and *Aphaenops*, where not only the eyes but also the optic ganglia disappear (Lespès, 1867; Bernard, 1937). Examples of total dispapearance of the visual apparatus are more rare among the vertebrates, but they have been reported in marine fish of the family Ipnopidae *(Ipnops murrayi, Bathymicrops sewelli)*. It seems, however, that certain cavernicolous fish are also completely deprived of eyes, although this has not been confirmed histologically. Such examples are *Trogloglanis pattersoni* and *Satan eurystomus*.

The paths of the regression of the eye depend:

(1) On the complexity of the eye. Racovitza (1907) stated, with good

reason, that the regression of the eye takes longer when its structure is more complex. The ocular system of a planarian, collembolan or copepod, would regress more rapidly than the compound eyes of an insect, or the eyes of a vertebrate.

(2) Although it is difficult to make general rules in biology, it can be seen that the state of regression of the eye generally depends on phylogenetic age (Verhoeff, 1930). The very ancient cavernicoles have usually very regressed eyes while the troglophiles, or the recent cavernicoles possess normal or slightly reduced eyes. It must be concluded that regression of the eye is a phenomenon which occurs very slowly and which is carried out for a long time before the final stages are reached.

There can be no question of describing the eyes of all cavernicoles in detail, as this would be a most tedious undertaking and would soon tire the reader. It is necessary, however, to establish the essential paths of regression in the different groups of cavernicoles.

In studies devoted to the regression of the eye it is the custom to use the terms "centrifugal regression" and "centripetal regression". We shall avoid using these terms in this text as they lead to confusion. In two otherwise excellent studies appearing within two months of each other and devoted to regression of the eye in isopods, de Lattin classed the eye of the same species, *Platyarthrus hoffmannseggi*, in the centripetal type (de Lattin, 1939a) and then as the centrifugal type (de Lattin, 1939b). There is, however, another reason which makes us reject these terms. It is that they lead to excessive simplification of an extremely variable and multiform phenomenon.

However, it can be stated that in most cavernicoles it is the peripheral structures, that is the dioptric apparatus, which first disappears while the central sensory and nervous structures are more stable and are those which persist the longest during the course of regressive evolution.

It is only in mutants obtained during laboratory breeding that one finds malformations of the sensory and nervous organisations appearing first. It is this which leads us to believe, and we shall return to this point later, that mutations can in no way give the answer to the problems raised by ocular regression in cavernicoles.

(b) Arachnids

Our study will be limited to Araneidae in which ocular regression has been extensively studied, in particular by Louis Fage (1913, 1931). The eyes of the spiders, as in all the arachnids, are simple eyes or ocelli. They are typically eight in number and arranged in an ocular platform in two transverse rows, one anterior, the other posterior. The two median and anterior eyes are called direct eyes because the retinal cells are directed towards the exterior. They are usually well pigmented. They were called diurnal eyes by the older arachnologists. The other eyes are termed in-

direct because the retinal cells are directed towards the interior. They usually contain a small amount of pigment and are whitish coloured. These
were earlier called nocturnal eyes.

Regressive evolution of the eyes of spiders always follows the same
path. This evolution appears in muscicolous forms or in the inhabitants
of great tropical forests, but is completed only in the subterranean
world.

(1) The median anterior eyes, the direct eyes, are the first to be affected
by regressive evolution. At first the amount of pigment decreases and then
disappears completely. Finally, the eyes themselves atrophy and disappear.

(2) This brings about a stage with six eyes of the indirect type. This is the
arrangement found in the genus *Leptoneta* which is composed of lucifugous
or cavernicolous species. In *Leptoneta* the six eyes are arranged in two
groups: an anterior group of four ocelli and a posterior group of two
ocelli. This condition is followed by dissociation of the ocular complex and
increased separation of the anterior and posterior groups. The posterior
ocelli may completely disappear in certain species of *Leptoneta*, e.g.
L. proserpina manca, and leave a type with only four eyes.

(3) Finally, the remaining eyes may degenerate. This is seen in totally
anophthalmic types such as *Telema*, *Stalita*, *Typhlonesticus*, *Iberina*.

(c) Crustacea

Cavernicolous Crustacea would certainly provide excellent examples of
regressive evolution. However, our information is very insufficient and it is
thus impossible to retrace the stages of regressive evolution in this group.

The excellent work of G.H.Parker provides much information on the
structure of the eye in American crayfish belonging to the genus *Cambarus*.
The animals studied were *C. bartoni*, a surface species (Parker, 1891) and
three cavernicolous species, *C. setosus*, *C. pellucidus* and *C. hamulatus*
(Parker, 1890) (Fig. 76). The eye-stalk of hypogeous forms terminates at the
cone because of the reduction of size of the eye. In *Cambarus setosus* there
are no cornea or facets. The retina is maintained in the condition of an
undifferentiated hypoderm although there are more nuclei in this than in
other parts of the body. The optic nerve and ganglia subsist. The retina
in *C. pellucidus* is represented by a very thick layer of hypodermal cells
which contain granular masses which Parker considered to be the remains
of the cones. Unfortunately, the processes of regression are not known in
these Crustacea because of the absence of structures intermediate between
the normal eye and the degenerated eyes described.

The eyes of terrestrial and aquatic isopods which are the ancestors of
cavernicoles, are already in a regressed state (de Lattin, 1939); thus the
very early stages of regressive evolution are absent.

The eyes of amphipods provide the best example of ocular regression.
Stages of regression in amphipods of the marine abysses are well known

following the work of Erich Strauss (1909). He clearly showed that the path of regression of the eye begins in the periphery of the organ and moves to the more central regions. Döflein (1904) earlier reported similar con-

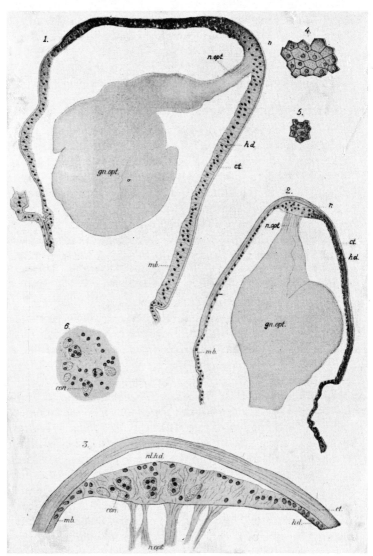

FIG. 76. *1, Cambarus setosus; 2, C. pellucidus. 1* and *2,* A longitudinal section through the right optic stalk. *3, 4, 5, 6, Cambarus pellucidus. 3,* Retinal enlargement of the hypodermis. *4, 5, 6,* Tangential section of the hypodermis (*4*), of the superficial (*5*) and of the deep portion (*6*) of the retinal thickening. *con.,* cone; *ct.* cuticula; *gn. opt.,* optic ganglion; *hd.,* hypodermis; *mb.,* basement membrane; *nl-con.,* nucleus of cone-cell; *nl-hd.,* nucleus of hypodermis-cell; *n. opt.,* optic nerve; *r.,* retina (after Parker).

ditions in the Brachyura. Strauss divided the different types studied into five categories which correspond to successive stages in regressive evolution.

(1) The first regressive changes seen in amphipods affect the dioptric apparatus. The crystalline cones dissolve and then disappear.

(2) *Liljeborgia Stage.* The dioptric apparatus has completely disappeared but the retinula cells and rhabdomes remain. This type of eye is still sensitive to light.

(3) *Tryphosa Stage.* This stage which is widely found, is characterised by the regression of the rhabdomes, followed by that of the retinula cells. Thus this eye cannot be sensitive to light, although the nervous pathways are unaffected.

(4) *Harpinia Stage.* The retinula cells have completely disappeared and are replaced by a layer of normal cells. However, a thickening of the hypodermis adjoining the optic nerve indicates the remains of the visual apparatus.

(5) *Andaniexis Stage.* The thickening of the hypodermis has disappeared. The nervous elements are altered and finally disappear.

There must be a similar evolution in cavernicoles but very little is known of it. The majority of descriptions of eyes of cavernicolous amphipods (Leydig, 1878; della Valle, 1893; Hamann, 1896; Vejdovsky, 1900, 1905; Strauss, 1909; Wolsky, 1935) are based on the examination of badly-fixed specimens or mediocre sections or are even incorrect interpretations.†

The same regressive series as that in the bathophilous amphipods can probably be found in hypogeous types. Brehm (1955) proposed the following agreement:

Niphargus puteanus	*Tryphosa stage*
Niphargus tatrensis	*Harpinia stage*
Niphargus aggtelekiensis	*Andaniexis stage*

However this parallelism can be established definitely only after new research using improved techniques.

The work of Wolsky has provided the best information on this topic. During their studies of the genetics of *Gammarus chevreuxi,* Allen and Sexton obtained mutants with *albino* eyes and others with *colourless* eyes. Both types of eyes show a degenerated structure. The eyes of these mutants were examined by Wolsky and Huxley (1934). During development, regression of the eye begins in the central region of the eye and proceeds to the periphery. The most modified structures are the distal parts of the optic tract (medulla externa and lamina ganglionaris) and retinula cells, that is to say the nervous and sensory elements. Wolsky (1935) has shown that

† The nerves of the pseudofrontal organ and the integumental nerve, described by Gräber (1933) were often thought by older workers to be optic nerves, or part of the optic tract. This mistake was pointed out by Wolsky (1935).

in the eye of *Niphargus aggtelekiensis*, evolution follows exactly the opposite pattern to that in the *Gammarus chevreuxi* mutants. In *Niphargus* regressive evolution proceeds from the periphery towards the central regions as was shown by Strauss for bathophilous forms. Thus, as has been concluded by Wolsky, regressive evolution of cavernicoles is entirely different from the appearance of mutants with abnormal eyes.

(d) Vertebrata

Much important work has been devoted to a study of the eyes of cavernicolous fish and urodeles. A list of publications has been given in an earlier table (p. 418).

Regressive evolution of the vertebrate eye shows in detail innumerable variations which cannot all be given here. However, degenerative evolution follows very similar paths in fish and urodeles and these will be now considered.

(1) The eye is more or less developed in the young stages (p. 429). It is situated on the surface, pigmented and often quite apparent.

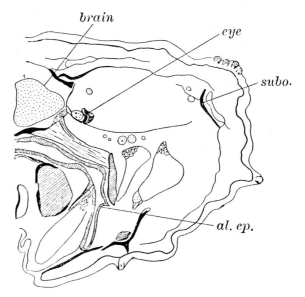

FIG. 77. Transverse section through the head of *Amblyopsis rosae* (after Eigenmann). *subo.*, suborbital; *al.ep.*, epithelium.

(2) In the adult the eye seems to have disappeared. This can be explained partly, by a thickening of the skin which covers it. Then, the eye migrates from the surface towards the brain and becomes buried in the orbit. This migration is very pronounced in *Amblyopsis rosae* (Fig. 77).

(3) At the same time as the eye migrates, its structure is altered. The rate of growth of the eye decreases. It is affected by a negative allometry instead of the positive allometric growth rate which is observed in surface

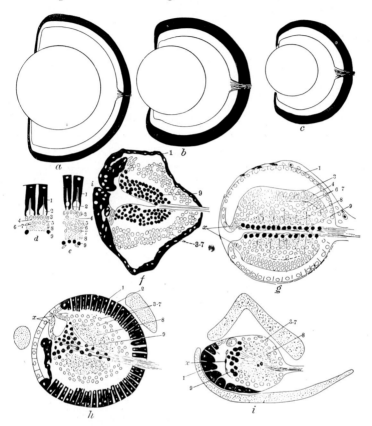

FIG. 78. The different types of regressed eyes in the vertebrates. *a, Chologaster cornutus; b, Chologaster papilliferus; c, Chologaster agassizi; d*, retina of *C. cornutus; e*, the retina of *C. papilliferus; f, Typhlomolge rathbuni; g, Typhlichthys subterraneus; h, Amblyopsis spelaeus; i, Amblyopsis rosae* (after Eigenmann).
1, Pigment epithelium; *2*, Rods and cones; *3*, Outer nuclear layer; *4*, Outer reticular layer; *5*, Horizontal cells; *6*, Inner nuclear layer; *7*, Spongioblastic layer; *8*, Inner reticular layer; *9*, Ganglion layer; *10*, Optic-fiber layer; *x*, Flattened cells beneath inner nuclear layer, of doubtful significance.

fish. Thus, the eye of the adult is smaller in relation to the body that it is in the young (Fig. 77).

(4) The dioptric region is the first to regress. The lens is most sensitive to the regressive tendencies and in most cases it disappears (Fig. 78). More rarely a rudimentary lens remains. However, the lens fibres are always missing in the cavernicolous vertebrates.

We may presume that the degeneration of the lens results from the deficiency of the induction, normally produced by the ocular vesicle (Cahn, 1958). But the experimental proof of this hypothesis is still lacking.

Filling tissue develops in place of the pupil *(Anoptichthys)* or a lump of pigment may do so *(Amblyopsis, Typhlomolge)* and this completely fills the anterior chamber of the eye. The eye is finally surrounded by a continuous layer of pigment.

(5) The vitrous body disappears in such a way that the eye cavity is filled by the folded-up retina (Fig. 78).

(6) The retinal structure regresses. Only rarely do the ten layers of the normal retina remain. The layer of rods and cones is the first to disappear and seems to be the most sensitive to regressive evolution. The median layers fuse and become indistinct from each other. The ganglion-cell layer persists for the longest period.

(7) The optic nerve persists in most cavernicoles (p. 434).

(8) The eye muscles remain, but their development and differentiation are stopped. In *Typhlomolge*, some small masses of connective tissue represent the eye muscles (Hilton, 1956).

G. ONTOGENETIC EVOLUTION OF REGRESSED EYES

(a) Introduction

The ocular apparatus of most cavernicoles is more reduced in the adult than it is in the young. This is the most frequent event. There are, however, very ancient cavernicoles in which not even rudiments of the eyes are to be seen at any stage. This is the case with copepods belonging to the genera *Viguierella*, *Epactophanes* and *Parastenocaris* in which the nauplii are already anophthalmic (Borutzky, 1926). The embryos of *Niphargus* are also anophthalmic (Vejdovsky, 1890).

(b) Insects

With respect to ontogenetic evolution the regressed eyes of insects correspond to a type of which there is no equivalent in any other group in the animal kingdom. The two most important studies devoted to regression of the eyes in insects are those of Bernard (1932, 1937). We shall consider here the regression of the compound eyes of cavernicolous insects.†

The regressed eye of hypogeous insects is the result of arrested development. The eye develops normally but growth is extremely slow compared with that of the normal eye. Thus, the animal does not produce a complete structure. Development finishes when the eye is still in a rudimentary state. This is why the eyes of larvae or nymphs of cavernicolous insects are less

† The ocelli are less stable structures than compound eyes and regress before them. This condition has been reported in numerous cavernicolous Orthoptera.

developed than those of the adults. This was seen by Chopard (1946) in cockroachs of the genus *Nocticola*.

In insects, reduction of the dioptric or peripheral region always precedes that of more central regions (Picard, 1923; Bernard 1937).

In the cavernicolous Coleoptera studied by Bernard regressive evolution shows the following pattern. The corneal facets and crystalline cones are usually the structures most markedly affected. These are followed by the retinal cells and the lamina ganglionaris; then the rhabdomes and finally the medulla externa and interna.

If one examines the ontogenesis of the eye it can be seen that development follows the exactly opposite pattern to that of regressive evolution. The retinal and pigment cells are the first to appear and later the corneagen and crystalline cells.

(c) General

The case of the insects is exceptional. In most cavernicoles early development of the eye is normal although growth is slower. Then, after a certain stage a turning point occurs. The structures developed in the larvae and in the young become disorganised and regressed and this reduction of the eye continues until the death of the individual. In these cases which are the reverse of those in insects, the eyes of the larvae or young are closer to the normal state than those of the adults. This is firstly a progressive ontogenesis, which is arrested and followed by degenerative organogeny.

These conditions recall those found during the development of integumentary pigmentation, which was mentioned earlier. Just as young cavernicoles show more or less fully developed pigmentation which disappears or decreases in the adult, the eyes of hypogeous animals are generally more developed and better differentiated in the young than in the adults.

These conditions have been reported for a number of cavernicoles, of which a few examples are given below:

Molluscs	*Vitrella* (Seibold, 1904)
Crustacea	Ostracods: *Cypris eremita* (Vejdovsky, 1882)
	Isopods: *Asellus foreli* (Blanc, 1879)
	Asellus cavaticus (de Lattin, 1939)
	Decapods: *Troglocaris schmidti* (Joseph, 1892)
	Orconectes pellucidus (Tellkampf, 1844; Cope, 1872; Packard, 1886)
Fish	*Typhloglobius californensis* (Eigenmann, 1909)
	Anoptichthys jordani (Lüling, 1955; Cahn, 1958; Frank, 1961; Kähling, 1961).
	Caecobarbus geertsi (Leleup, 1956)
	Amblyopsis spelaeus (Eigenmann, 1909)
	Stygicola dentatus (Eigenmann, 1909)
	Lucifuga subterraneus (Eigenmann, 1909)

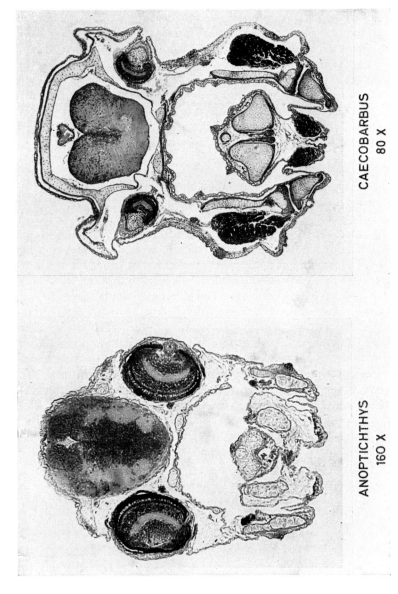

ANOPTICHTHYS
160 X

CAECOBARBUS
80 X

PLATE X. Transverse sections through the cephalic region of *Anoptichthys* and of *Caecobarbus* (Original photographs by Professor C. Koch, Louvain).

Urodeles.   *Typhlotriton spelaeus* (Eigenmann, 1909)
                    Proteus anguineus (see below).

(d) Vertebrates

Proteus. It has been known for many years that the eyes of young *Proteus* are more obvious and better developed than those of the adult, which are clearly regressed (Michahelles, 1831; Zeller, 1888, 1889; Schlampp, 1892; Schmeil, 1893; Hawes, 1946). These results have been confirmed by Vandel and Bouillon (1959) who followed the development of *Proteus* from the egg (Plate IV).

J. P. Durand (1962) used larvae bred in the Cave Laboratory of Moulis in order to examine the ontogenetic evolution of the eye of the *Proteus*. At birth, the eye possesses a structure which is like that of the embryos of epigeous urodeles. Thus, the reduction in speed of ocular development occurs from the embryonic stage.

The size of the eye increases slowly during the first two months. Afterwards, the size diminishes. At the same time the extreme retardation in the rate of development maintains the eye in an embryonic state. The lens degenerates and disappears when the larvae reach 10–12 cm in length.

After a year the eye is apparent as a small black spot, but after three years the eye is no longer visible externally.

H. EFFECTS OF REGRESSION OF THE EYE ON THE STRUCTURE OF THE BRAIN IN CAVERNICOLES

The sensory and nervous system are so closely linked functionally that regression of the first must affect the second. This is especially apparent in the effect of regression of the eye on the optic centres.

In the present state of our knowledge these changes with regard to the cavernicoles can be considered only in the arthropods or, more exactly, in the Crustacea and the insects, and also in the vertebrates. American biologists were the first to work in this field. Arthropods have been studied by A. S. Packard and vertebrates by C. H. Eigenmann.

(a) Arthropods

In the arthropods the optic centres are arranged in an elongated lobe fixed on each side of the protocerebrum. It is joined to the eye by the optic nerve and has been given the name optic lobe, (the optic ganglion of earlier authors). The optic lobe in Crustacea and insects is formed by three distinct regions which are united by two chiasmata, one external, and the other internal (Fig. 79 A). Hanström has proposed the following names for homologous structures in insects and crustacea.

Crustacea	Insects
Lamina ganglionaris	Periopticum
Medulla externa	Epiopticum
Medulla interna	Optic lobule

The regression of the optic centres is correlated with the reduction in the visual apparatus. Evolution of the lamina ganglionaris or periopticum directly follows that of the retina. On the other hand, there is a certain divergence between regressive evolution of the eye and that of median and internal centres, the first preceding the second. This is because formation

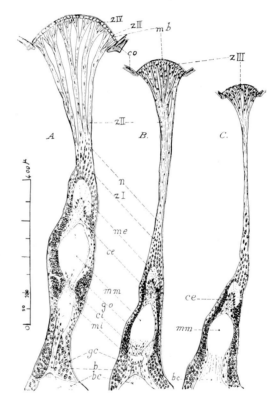

FIG. 79. The optic centres of three carabids; a normal form, *Steropus madidus* *(A)*, and two cavernicolous species, *Antisphodrus boldorii (B)* and *Ceutosphodrus navarricus(C).b*, optic tract which unites the optic centres with the protocerebrum *bc; ci*, internal chiasma; *ce*, external chiasma; *co*, periocular capsule; *gc*, grey matter of cerebrum; *go*, nucleus of visual neurones of the third order; *mb*, basal membrane of the eye; *me*, periopticum; *mi*, optic lobule; *mm*, epiopticum; *n*, fibres of the optic nerve; Z1, 2nd order giant neurones; Z2, first order sensory fibres; Z3, zone of anastomosis; Z4, fibres coming from the retina (fenestral zone) (after Bernard).

of the optic centres which embryologically precedes that of the eyes results in an earlier and more stable structure (Bernard, 1937).

Without establishing a definite parallelism between the different regressive series one can, however, recognise a general line of evolution comprising several successive stages, four of which are given below:

(1) Eyes Slightly Reduced in Size. Normal or slightly reduced optic lobes. This first stage is reached by troglophilic Orthoptera and has also been reported in a galatheid.

Orthoptera	*Ceutophilus* and *Hadenoecus* (Packard, 1886).
	Troglophilus (Jörschke, 1914)
	Dolichopoda (Bernard, 1937)
Decapod Crustacea	*Munidopsis polymorpha* (Harms, 1921; Bernard, 1937).

(2) Clearly-reduced Eyes. The size of the optic centres decreases, and moreover, they become compressed towards the brain, while the optic nerve increases in length. The optic lobule or the medulla interna have a tendency to fuse with the protocerebrum and become indistinct. These conditions are found in several cavernicoles:

Coleoptera	Carabids. Cavernicolous Sphodridae (Fig. 79 B and C) (Bernard, 1932, 1937)
Mysids	*Troglomysis* (Stammer, 1936)
Isopods	*Asellus cavaticus* (de Lattin, 1939)
	Titanethes (de Lattin, 1939)
Decapods	*Troglocaris* (Babić, 1922).

(3) The Optic Lobe Consists of a Single Centre. Decapods: Cavernicolous species of *Cambarus* (Parker, 1890).

(4) The Optic Lobe has Completely Disappeared or if it persists it is without nervous elements and consists of a supporting ridge joining the brain to the integument.

Isopod Crustacea	*Asellus stygius* (Packard, 1885, 1886)	
Diplopods	*Scoterpes* (Packard, 1886)	
Coleoptera	*Trechinae:*	*Pseudanophthalmus* (Packard, 1886)
		Duvalius, Geotrechus (Bernard, 1937)
		Aphaenops (Lespès, 1868; Bernard, 1932, 1937)
	Catopidae	*Adelops* (Packard, 1886)
	Bathysciinae	*Antrocharis, Speonomus* (Lespès, 1868)
		Sophrochaeta (Decu, 1961).

These are the cavernicolous Coleoptera, which among the arthropods represent the last stage of regressive evolution of the eye.

B 15a

(b) Vertebrates

In the lower vertebrates, the fibres of the optic nerve end in the optic centres, which are represented in the lower vertebrates by the roof of the mesencephalon or tectum opticum (which is divided into two optic lobes) and by the corpus geniculatum which belongs to the thalamic region (Kappers, 1947).

It can be expected that in a cavernicolous vertebrate showing regressed eyes and in which atrophy of the eye is certainly very ancient, the optic fibres would have disappeared. It is known that enucleation of the eye brings about marked degeneration of the tractus and optic centres.

(1) Fish. Representatives of the family Amblyopsidae have been particularly well-studied with regard to the optic centres in cavernicoles. It is necessary to mention firstly the older studies of Wyman (1853) the results of which were completed by Putnam (1872); then the research of Ramsay (1901) whose results are reported in the work of Eigenmann (1909); and finally the work of Charlton (1933) which can be set on a level with modern neurology.

(i) Optic Nerve. Although earlier authors all believed that the optic nerve in the Amblyopsidae was absent, at least in the adult, Charlton has shown that the optic nerve persists during the whole life of species with very regressed eyes, such as *Typhlichthys subterraneus* and *Amblyopsis rosae.* Charlton has provided a very simple explanation for this persistance. The optic nerve of cavernicolous fish contains fibres which correspond to the tractus preoptico-opticus of Holmgren. They are efferent fibres and lead to the eye. They are not inserted on the tectum opticum but onto the nucleus preopticus, a diencephalic nucleus which is probably a visceral centre.

(ii) The Oculo-motor Nerves (III, IV, VI) are absent, as are the oculo-motor muscles.

(iii) Tectum opticum. Ramsay has already stated that the width of the optic tectum is a third to a half of that in surface fish. Charlton has compared the optic tectum of the Amblyopsidae and surface fish of a similar size. The volume of the first is fourteen times less than that of the second. This reduction results, as has been shown by Charlton, from a disappearance of the optic fibres. The position of the large bundle of optic fibres which are normally contained in the fibrous layer of the peripheral zone of the tectum, is occupied in cavernicolous species by a vast lacuna. The optic fibres contained in the stratum medulare profundus of Kappers, are also absent in those fish with regressed eyes.

(iv) The geniculate body is itself atrophied with respect to surface forms.

Analogous regressions are to be found in the optic lobes of other cavernicolous fish, but less research has been carried out on their structure. Such fish include *Phreatobius* (Reichel, 1927); *Anoptichthys* (Stefanelli, 1954; Cahn, 1958); *Typhlogarra* (Marshal and Thinès, 1958).

(2) Urodela. Only *Proteus*, among the cavernicolous urodeles has been studied with respect to the optic paths.

Although earlier authors maintained that the optic nerve was incomplete and did not reach the brain, the recent descriptions of Benedetti (1922), Hawes (1946), and Durand (1962) have established that the optic nerve is continuous and joins the eye to the brain. The nerve is, however, reduced and probably non-functional.

The oculo-motor nerve (III) persists, and is inserted on the mesencephalic tegmentum. Development varies according to the individual (Kreht, 1931).

The knowledge available on the optic centres of *Proteus* is contained in the old and incomplete publication of Hirsch-Tabor (1908) and in a more precise and detailed study by Kreht (1931), to which the reader is referred. It is sufficient to note here, that the optic part of the thalamus has completely disappeared in *Proteus*.

I. CORRELATION BETWEEN DEPIGMENTATION, ANOPHTHALMIA AND APTERISM

For many years entomologists have remarked on the relationships between depigmentation, regression of the eye and apterism (Fairmaire and Laboulbène, Massonnat, Peyerimhoff, Jörschke, Picard, etc.). This association is obviously not constant, as there are no absolute rules in biology, but it is very frequent. Examples of this association are to be found in all the orders of insects.

Among the cavernicoles a case already mentioned will be considered. This is that of the cockroach, *Alluaudellina cavernicola*, captured in the Kulumuzi Cave in East Africa (Chopard, 1932). The males of this insect are macrophthalmic but the eye is sometimes more or less reduced. The size of the wings varies in parallel with that of the eyes. The females are microphthalmic, the size of their eyes varies but they are always smaller than those in the males, and they are apterous and without elytra.

Biospeologists are also interested in the "subterranean evolution" which occurs in the humus of the great mountain forests of Central Africa. The conditions found in the pselaphids are of particular interest (Jeannel and Leleup, 1952; Jeannel, 1952). Certain micropselaphids living in humus show a strong sexual dimorphism. The females are apterous and anophthalmic while the males have wings and eyes. Other species show phenomena of poecilandry: in the same species, males with wings and eyes, and apterous and anophthalmic males can be found. In other species the males with wings and eyes have disappeared. Then the two sexes are apterous and anophthalmic. It is most remarkable to find this "subterranean evolution" not underground, but in the humus of tropical forests.

We shall again refer to this type of evolution in the last part of the book.

The correlation between depigmentation, anophthalmia and apterism can be accounted for only by attributing the control of development of these three characters to the same genetic complex.

J. CONCLUSIONS. GENESIS OF ANOPHTHALMIA

Regression of the eye in cavernicoles is not a unique and exceptional phenomenon, and it would be a mistake to confine this problem to the world of biospeology. It is another example of a rudimentary organ. The problems which are raised by regression of the eye in subterranean animals are not different from those which the gill-slits and the mesonephros of the amniotes, or the Mullerian duct of the male and the Wolfian duct of the female amniotes, raise for embryologists. In all these cases, and many others in the living world, it is a question of phyletic regression. However, the problem to be analysed here is the regression of the eyes in cavernicoles.

(1) The normal eyes of epigeous forms are very stable structures. It is certain that light is not necessary to maintain it or for its development. Animals kept in darkness for many generations have shown no modification of the visual structures. Payne (1911) raised 69 generations of *Drosophila* in complete darkness. At the end of the experiment there was no change in the structure of the eyes nor in their sensitivity to light. It has been verified for Amphibia by Uhlenhuth (1915) and for fish by Loeb (1915) that the eye undergoes normal development in the total absence of light.

(2) The embryonic and adult stages of highly specialised cavernicoles are without eyes (e.g. *Aphaenops, Bathysciinae, Niphargus*). Among these animals loss of the eye is definitive and no intervention could modify this state. This condition is as stable as that of the presence of eyes.

(3) Between these two extreme conditions lies the long regressive evolution which has taken place during the history of cavernicoles. During this transitory period the eye is in a state of physiological and ontogenetic instability. This is why cavernicoles which have reached this stage of their history show a usually large variability in structure of the eye. Examples have been given in the preceding pages (p. 419). This phase is of greatest interest for the biospeologist because it offers the possibility of successful experimentation.

(4) Many biospeologists have considered that the specific conditions in the subterranean environment account for the evolution of cavernicoles and in particular the regression of pigmentation and of the eyes.

Fage and Jeannel attributed the responsibility for degenerative evolution of the eye to cold and humidity. More often darkness has to be considered as the cause. The facts reported above make this interpretation improbable. It must be added that regression of the eye has certainly begun in the epigeous habitat. Irrevocable evidence for this is given by the muscicolous Bathysciinae and the humicolous Pselaphidae. In these cases it is not darkness which is responsible for anophthalmia.

As regards the action of light on the development of the rudimentary eye of cavernicoles the only positive result available is from the well-known experiment of Kammerer (1912) on *Proteus*. Young specimens maintained in light developed almost normal eyes. It would seem that the existence of *Proteus* with large eyes cannot be contested since the well-known zoologist E.W. MacBride (1925) saw the specimens presented by Kammerer at a Linnean Society meeting in May, 1923, and wrote that these *Proteus* were "the most wonderful specimens, in my judgement, which have been exhibited to a zoological meeting". However, the interpretation of these animals must remain enigmatic, as only Kammerer has obtained this result although *Proteus* have been raised in light in zoological parks for many years. However, one must not exclude the fact that certain races of *Proteus* have conserved better developed eyes than those of the normal type.

(5) The topic to which naturalists have not paid sufficient attention is that of the physiological and ontogenetic conditions of regression of the eye.

Let us return to the interesting and frequent cases of the eye of cavernicoles which begins development in a normal way although its rate of growth is slowed. Having reached a certain turning point the eye begins to regress and already formed structures dissociate and degenerate. This regression can only be the consequence of the intervention of ontogenetic and physiological factors acting at the time of the turning point in evolution. These factors are accessible to experimentation and the experimental biospeologist must search for the answer to this problem.

One of the factors which can be studied most easily is that of the action of hormones. Already certain indications make it possible to predict that this method of study would yield positive results capable of throwing light on the problems of ocular regression.

It appears that the hypertrophy of the eye of the eel at the moment of metamorphosis, is dependent on the hyperactivity of the thyroid (Fontaine). In the tadpoles of green frogs, extracts of thyroxin accelerate the growth and differentiation of parts of the eye and also of the optic regions of the brain (Tusques, 1949). The injection of thyroxin into young moles causes the opening of the eyelids and some development of the eye (Tusques, 1954, 1955). It thus appears that the state of regression in the eye of moles is related to a decrease in sensitivity of the eye bud to thyroid hormone but this sensitivity has not completely disappeared and is shown in the presence of concentrated doses of thyroxine.

These experiments allow the correct interpretation of certain results which have been given a too simple explanation so far. If light favours the development of a regressed eye this is not affected through the direct intervention of light but by the intermediary of several endocrine pathways. Noble and Pope (1928) stated that the larvae of *Typhlotriton spelaeus* possess eyelids which are separate one from the other, while they are touching or fused in the adult. If the larvae are kept in light the eyelids remain free after metamorphosis. These results are so similar to those obtained by Tusques for the mole, that it is natural to provide a similar explanation for the results of these two experiments. The separation of the eyelids in adult *Typhlotriton* would result from the stimulation by light on thyroid activity (probably via the hypophysial pathway). However, it must be added that culturing the animals in light does not prevent the regression and degeneration of the retina after metamorphosis. This indicated that the eyelids and the retinal layer are not at the same stage of phyletic regression.

(6) The case of *Proteus* merits special attention. It has been established that the eyes of most "perennibranches" are reduced and even regressed although all, with the exception of *Proteus* and some American forms *(Typhlomolge* and *Haideotriton)*, lead an epigeous life (Vandel and Bouillon, 1959). The eyes of another representative of the Proteidae, *Necturus maculosus* are reduced, the optic nerve shows signs of degeneration, and the optic chiasma is absent (Kingsbury, 1895). The eye of *Aphiuma means* is extremely small. The diameter is 1·5 mm while the length of the body can reach one metre. The eye is covered by skin and is probably non-functional (Davison, 1895). As far as one can judge from the Davison's figure the eye of the young animal is already without the lens. The eyes of *Cryptobranchus* and *Siren* are also reduced (Herter, 1941).

It therefore appears that reduction of the eye of "perennibranchs" is not related to cavernicolous life but to their neotenous state. It seems possible that, as in the mole, the decrease in the size of the eye and its regression, results from a lessening in the sensitivity of the tissues of the "perennibranchs" to thyroid hormone.

BIBLIOGRAPHY

ABSOLON, K. (1900) Einige Bemerkungen über mährische Höhlenfauna. *Zool. Anz.* **XXIII.**

ANNANDALE, N. and CHOPRA, B. (1924) Molluscs of the Siju Cave, Garo Hills, Assam. *Rec. Indian Mus.* **XXVI.**

BABIČ, K. (1922) Über die drei Atyiden aus Jugoslawien. *Glasnik.* **XXXIV.**

BARTOLAZZI, C. (1937) L'occhio dei Gammaridi. *Atti. Soc. Ital. Mus. Milano* **LXXVI.**

BATE, C. Spence (1859) On the genus *Niphargus* Schiödte. *Dublin University Rev.* **I.**

BEAUCHAMP, P. DE (1920) Turbellariés et Hirudinées (1ère Série). *Biospeologica,* **XLIII.** *Archiv. Zool. expér. géner.* **LX.**

BEAUCHAMP, P. DE (1932) Turbellariés, Hirudinées, Branchiobdellidés (2ème Série). *Biospeologica,* **LVIII.** *Archiv. Zool. expér. géner.* **LXXIII.**

BEIER, M. (1940) Phylogenie der troglobionten Pseudoscorpione. *Sixth Intern. Congr. Entomol. Trans.*

BENEDETTI, E. (1922) Intorno all'esistenza del nervo ottico e del cervelletto nel *Proteus anguineus* Laurenti. *Rend. Unione Zool. Ital.* Trieste (1921) 1922. *In Monit. Zool. Ital.* **XXXII.**

BERNARD, FR. (1932) Comparaison de l'œil normal et de l'œil régressé chez quelques Carabiques. *Bull. biol. France Belgique* **LXVI.**

BERNARD, FR. (1937) Recherches sur la morphogenèse des yeux composés d'Arthropodes. Développement. Croissance. Réduction. *Bull. biol. France Belgique. Suppl* **XXIII.**

BERNINGER, J. (1911) Die Einwirkung des Hungers auf Planarien. *Zool. Jahrb. Abt. Zool. Physiol.* **XXXIII.**

BLANC, H. (1879) Über den *Asellus* aus der Tiefenzone des Genfer Sees. *Zool. Anz.* **II**

BORUTZKY, E. V. (1926) Les formes larvaires des Harpacticoida (Copepoda) d'eau douce. *Trav. Station biol. Kossino.* No. 3 (in Russian).

BREDER, C. M. JR. (1944) Ocular anatomy and light sensitivity studies on the blind fish from Cueva de los Sabinos, Mexico. *Zoologica* **XXIX.**

BREDER, C. M. JR. and GRESSER, E. B. (1941) Correlations between structural eye defects and behaviour in the Mexican blind Characin. *Zoologica* **XXVI.**

BREHM, V. (1955) *Niphargus*-Probleme. *Sitzb. d. Öster. Akad. Wiss. Math. naturw. Kl. Abt. I.* **CLXIV.**

BRUNTZ, L. (1904) Contribution à l'étude de l'excrétion chez les Arthropodes. *Archiv. Biol.* **XX.**

CAHN, P. H. (1958) Comparative optic development in *Astyanax mexicanus* and in two of its blind derivatives. *Bull. Americ. Mus. Nat. Hist.* **CXV.**

CAPORIACO, L. DI (1947) Seconda Nota su Aracnidi cavernicoli veronesi. *Mem Mus. Civ. Stor. Nat. Verona* **I.**

CHAPPUIS, P. A. (1920) Die Fauna der unterirdischen Gewässer der Umgebung von Basel. *Archiv. f. Hydrobiol.* **XIV.**

CHARLTON, H. H. (1933) The optic tectum and its related fibre tracts in blind fishes. *J. comp. Neurol.* **LVII.**

CHEVREUX, ED. (1901) Amphipodes des eaux souterraines de France et d'Algérie. *Bull. Soc. Zool. France* **XXVI.**

CHOPARD, L. (1929) Fauna of the Batu Caves, Selangor. XII, Orthoptera and Dermaptera. *J. Fed. Malay States Mus.* **XIV.**

CHOPARD, L. (1932) Un cas de microphthalmie liée à l'atrophie des ailes chez une Blatte cavernicole. *Soc. Entomol. France Livre du Centenaire.*

CHOPARD, L. (1946) Note sur quelques Orthoptères cavernicoles de Madagascar. *Rev. franç. Entomol.* **XII.**

COPE, E. D. (1872) On the Wyandotte Cave and its fauna. *Amer. Nat.* **VI.**

CUNNINGHAM, J. T. (1893) Blind animals in caves. *Nature. Lond.* **XLVII.**

DAVISON, A. (1895) A contribution to the anatomy and phylogeny of *Amphiuma means* (Gardner). *J. Morphol.* **XI.**

DECU, V. (1961) Contributi la studiul morfologiei interne la coleoptere cavernicole din seria filetica *Sophrochaeta* Reitter (Catopidae-Bathysciinae). *Stud. Cercl. Biol. Ser. Biol. animalà. Romîn.* **XIII.**

DESFOSSES (1882) D l'œil du Protée. *Compt. rend. Acad. Sci. Paris* **XCIV.**

DÖFLEIN, F. (1904) Brachyura. *Ergebnisse deutsch. Tiefseeexped.* **VI.**

DRESCO, ED. (1952) Remarques sur les *Centromerus* du groupe *paradoxus* et description de deux espèces nouvelles (Araneae). *Notes biospéol.* **VII.**

DUPLESSIS, G. *in* FOREL, F. A. (1876) Matériaux pour servir à l'étude de la Faune profonde du Lac Léman. *Bull. Soc. Vaud. Sc. Nat.* **XIV.**

DURAND, J. P. (1962) Structure de l'œil et de ses annexes chez la larve de *Proteus anguinus*. *Compt. rend. Acad. Sci.* Paris **CCLV.**

EIGENMANN, C. H. (1900) The blind fishes. *Biol. Lectures. Marine Biol. Lab. Woods Hole* No. **VIII.**

EIGENMANN, C. H. (1909) Cave vertebrates of America. A study in degenerative evolution. *Carnegie Inst. Washington Public.* No. 104.

ELLINGSEN, E. (1908) Pseudoscorpions (2ème Série). *Biospeologica,* **VII.** *Archiv. Zool. expér. géner.* (4) **VIII.**

ENSLIN, ED. (1906) *Dendrocoelum cavaticum* Fries. *Jahresh. Ver. f. Vaterl. Naturh. Württemberg* **LXII.**

FAGE, L. (1913) Études sur les Araignées cavernicoles. II. Révision des *Leptonetidae. Biospeologica,* **XXIX.** *Archiv. Zool. expér. géner.* **L.**

FAGE, L. (1931) *Araneae* (5ème Série), précédée d'un essai sur l'évolution souterraine et son déterminisme. *Biospeologica,* **LV.** *Archiv. Zool. expér. géner.* **LXXI.**

FRANK, S. (1961) A morphological study about blind cave fish, *Anoptichthys jordani. Vest. Českosl. Spol. Zool.* **XXV.**

GARBINI, A. (1895) *Gammarus* ciechi in aque superficiale basse. *Accad. d. Agricolt. Arti. Commerc. Verona* (3) **LXX.**

GÉRARD, P. (1936) Sur l'existence de vestiges oculaires chez *Caecobarbus geertsi. Mém. Mus. Hist. Nat. Belgique* (2) fas. 3.

GHOSH, E. (1929) Fauna of the Batu Caves, Selangor. VI. Mollusca. *J. Fed. Malay States Mus.* **XIV.**

GOSTOJEVA, M. N. (1949) New data on the problem of reduction of the eye in *Proteus. Compt. rend. Acad. Sc. U.R.S.S. N. S.* **LXVII.**

GRÄBER, H. (1933) Über die Gehirne der Amphipoden und Isopoden. *Zeit. Morphol. Oekol.* **XXVI.**

GRAETER, E. (1910) Die Copepoden der unterirdischen Gewässer. *Archiv. f. Hydrobiol.* **VI.**

GRESSER, E. B. and BREDER, C. M. JR. (1940) The histology of the eye of the cave characin *Anoptichthys. Zoologica* **XXV.**

HADŽI, J. (1933) Beitrag zur Kenntnis der Höhle Vjetrenica. *Bull. Acad. Sc. Math. Nat. Belgrade. B. Sc. Nat.* No. 1.

HAMANN, O. (1896) Europäische Höhlenfauna. Jena.

HARMS, W. (1921) Das rudimentäre Sehorgan eines Höhlendecapoden, *Munidopsis polymorpha* Koelbel aus der Cueva de los Verdes auf der Insel Lanzarote. *Zool. Anz.* **LII.**

HART, C. W. and HOBBS, H. H. JR. (1961) Eight new troglobitic ostracods of the genus *Entocythere* (Crustacea, Ostracoda) from the Eastern United States. *Proc. Acad. Nat. Sc. Philadelphia* **CXIII.**

HAWES, R. S. (1946) On the eyes and reactions to light of *Proteus anguinus. Quart. Jour. Microsc. Sc.* **LXXXVI.**

HAY, W. B. (1902) Observations on the crustacean fauna of Nickajack Cave, Tennessee and vicinity. *Proc. U.S. Nat. Mus.* **XXV.**

HERTER, K. (1941) Die Physiologie der Amphibien. In Kükenthal, W., *Handbuch der Zoologie.* Berlin. **IV.**

HESS, C. (1889) Beschreibung des Auges von *Talpa europaea* und *Proteus anguinus. Archiv. f. Ophthalmol.* **XXXV.**

HILTON, W. A. (1956) Eye muscles of salamanders. *Herpetologia* **XII.**

HIRSCH-TABOR (1908) Über das Gehirn von *Proteus anguineus. Archiv. f. mikros. Anat.* **LXXII.**

HNATENWYTSCH, B. (1929) Die Fauna der Erzgruben von Schneeberg im Erzgebirge. *Zool. Jahrb. Abt. System.* **LVI.**

HYMAN, L. H. (1939) North American triclad Turbellaria. X. Additional species of cave planarians. *Ecology* **LVIII.**

JAROCKI, J. and KRZYSIK, S. M. (1924) Materialien zur Morphologie und Ökologie von *Synurella ambulans* (Friedr. Müller). *Bull. Acad. Polon. Sc. Lett. Cracovie* B.

JEANNEL, R. (1911) Révision des *Bathysciinae.* Morphologie. Distribution géographique. Systématique. *Biospeologica,* **XIX.** *Archiv. Zool. expér. géner.* (5) **VII.**

JEANNEL, R. (1952) Psélaphides recueillis par N. Leleup au Congo belge. IV. Faune de l'Itombwee et du Rugege. *Ann. Mus. Congo. belge. Sér. in 8 Sc. Zool.* **XI.**

JEANNEL, R. (1954) Anophthalmie et cécité chez les Coléoptères souterrains. *Notes biospéol.* **IX.**

JEANNEL, R. and LELEUP, N. (1952) L'évolution souterraines dans la région Méditerranéenne et sur les montagnes du Kivu. *Notes biospéol.* **VII.**

JÖRSCHKE, H. (1914) Die Facettenaugen der Orthopteren und Termiten. *Zeit. wiss. Zool.* **CXI.**

JOSEPH, G. (1892) L'influence de l'éclairage sur la disjonction des organes visuels, leur réduction, leur atrophie complète et leur compensation chez les animaux cavernicoles. *Bull. Soc. Zool. France.* **XVII.**

KÄHLING, J. (1961) Untersuchungen über den Lichtsinn und dessen Lokalisation bei dem Höhlenfisch, *Anoptichthys jordani. Biol. Zentralbl.* **LXXX.**

KAMMERER, P. (1912) Experimente über Fortpflanzung, Farbe, Augen und Körperreduktion bei *Proteus anguinus* Laur. Zugleich: Vererbung erzwungener Farbveränderungen. III. Mitteilung. *Archiv. f. Entwicklungsm.* **XXXIII.**

KAMMERER, P. (1913) Nachweis normaler Funktion beim herangewachsenen Lichtauge des Proteus. *Archiv. Ges. Physiol.* **CLIII.**

KAPPERS, C. U. A. (1947) *Anatomie Comparée du Système nerveux.* Haarlem and Paris.

KAPTEREW, P. (1912) Über den Einfluß der Dunkelheit auf Daphnienauge (Eine experimentelle Untersuchung). *Biol. Centralbl.* **XXXII.**

KARAMAN, ST. L. (1950) Études sur les Amphipodes-Isopodes des Balkans. *Acad. Serbe Sc. Monogr.* **CLXIII.** *Sect. Sc. Math. Nat.* No. 2 (*Nouvelle Sér.*).

KÄSTNER, A. (1926) Überblick über die in den letzten 20 Jahren bekannt gewordenen Höhlenspinnen. *Mitteil. Höhl. Karstf.*

KINGSBURY, B.F. (1895) On the brain of *Necturus maculatus. Jour. Compar. Neurol.* **V.**

KOCH, C.L. (1835) Deutschlands Crustaceen, Myriapoden und Arachniden. Regensburg; Fas. 5; No. 2.

KOHL, C. (1889) Einige Notizen über das Auge von *Talpa europaea* und *Proteus anguineus. Zool. Anz.* **XII.**

KOHL, C. (1891) Vorläufige Mitteilung über das Auge von *Proteus anguineus. Zool. Anz.* **XIV.**

KOHL, C. (1895) Rudimentäre Wirbeltieraugen. Teil III. Zusammenfassung. *Bibliotheca Zoologica* **V.**

KOSSWIG, C. (1935) Die Evolution von "Anpassungsmerkmalen" bei Höhlentieren in genetischer Betrachtung. *Zool. Anz.* **CXII.**

KOSSWIG, C. (1937) Betrachtungen und Experimente über die Entstehung von Höhlentiermerkmalen. *Der Züchter,* **IX.**

KOSSWIG, C. and KOSSWIG, L. (1936) Über Augenrück- und -mißbildung bei *Asellus aquaticus cavernicolus. Verh. deutsch. Zool. Gesell.*

KOSSWIG, C. and KOSSWIG, L. (1940) Die Variabilität bei *Asellus aquaticus,* unter besonderer Berücksichtigung der Variabilität in isolierten unter- und oberirdischer Populationen. *Rev. Fac. Sc. Univ. Istanbul* (N.S.), **V.**

KRATOCHVIL, J. (1936) *Ischyropsalis strandi* n. sp., un Opilion cavernicole nouveau d'Italie *Festschrift 60. Geburtstag v. Prof. Embrik. Strand.* **I.**

KREHT, H. (1931) Faserzüge im Zentralnervensystem von *Proteus anguinus. Zeit. mikrosk. anat. Forsch.* **XXV.**

LATTIN, G. DE (1939a) Untersuchungen an Isopodenaugen (Unter besonderer Berücksichtigung der blinden Arten. *Zool. Jahrb. Abt. Anat. Ontog.* **LXV.**

LATTIN, G. DE (1939b) Über die Evolution der Höhlentiercharaktere. *Sitzungsber. Gesell. naturfr. Freunde.*

LELEUP, N. (1956) La Faune cavernicole du Congo Belge et Considérations sur les Coléoptères reliques d'Afrique intertropicale. *Annal Mus. R. Congo belge Ser. in. 8°. Sc. Zool.* **XLVI.**

LESPÈS, C. (1867) Recherches anatomiques sur quelques Coléoptères aveugles. *Annal. Sc. Nat. Zool.* **XVII.**

LEYDIG, F. (1878) Über Amphipoden und Isopoden. Anatomische und zoologische Bemerkungen. *Z. f. wiss. Zool.* **XXX.** *Suppl.*

LOEB, J. (1915) The blindness of the cave fauna and the artificial production of blind fish embryos by heterogeneous hybridization and by low temperatures. *Biol. Bull.* **XXIX.**

LÜLING, K.H. (1953) Über die fortschreitende Augendegeneration des *Anoptichthys jordani* Hubbs und Innes (Characidae). Vorläufige Mitteilung. *Zool. Anz.* **CLI.**

LÜLING, K.H. (1955) Untersuchungen am Blindfisch *Anoptichthys jordani* Hubbs und Innes (Characidae). III. Vergleichend anatomisch-histologische Studien an den Augen des *Anoptichthys jordani. Zool. Jahrb. Abt. Anat. Ontog.* **LXXIV.**

MACBRIDE, E.W. (1925) The blindness of cave-animals. *Nature, Lond.* **CXVI.**

MACINDOO, N.E. (1910) Biology of the Shawne cave spiders. *Biol. Bull.* **XIX.**

MARSHALL, N.B. and THINÈS, G.L. (1958) Studies of the brain, sense organs and light sensibility of a blind cave fish *(Typhlogarra widdowsoni)* from Iraq. *Proc. Zool. Soc. Lond.* **CXXXI.**

MICHAHELLES (1831) Beyträge zur Naturgeschichte des *Proteus anguinus. Isis* von Oken.

MURRAY, A. (1857) On the insect vision and blind insects. *Edinburgh New Philos. Journal, N.S.*

NEWPORT, G. (1855) On the ocelli in the genus *Anthophorabia. Trans. Linnean Soc. Lond.* **XXI.**

NOBLE, G.K. and POPE, S.H. (1928) The effects of light on the eyes, pigmentation and behaviour of the cave salamander, *Typhlotriton. Anat. Rec.* **XLI.**

PACKARD, A.S. (1885) On the structure of the brain of *Asellus* and the eyeless *Caecidotea. Amer. Nat.* **XIX.**

PACKARD, A.S. (1886) The cave fauna of North America, with remarks on the anatomy of the brain and origin of the blind species. *Mem. Nat. Acad. Sc. Washington* **IV.**

PARIS, P. (1925) Sur la bionomie de quelques Crustacés troglobies de la Côte d'Or. *Compt. Rend. 49ème Sess. Assoc. franç. Avanc. Sc.* Grenoble.

PARKER, G.H. (1890) The eyes in blind crayfishes. *Bull. Mus. Compt. Zool. Harvard College* **XX.**

PARKER, G.H. (1891) The compound eyes in crustaceans. *Bull. Mus. Compt. Zool. Harvard College* **XXI.**

PAYNE, F. (1911) *Drosophila ampelophila* bred in the dark for sixty-nine generations. *Biol. Bull.* **XXI.**

PICARD, F. (1923) Recherches biologiques et anatomiques sur *Melittobia acasta. Bull. biol. France Belgique* **LVII.**

PUTNAM, F.W. (1872) The blind fishes of the Mammoth Cave and their allies. *Amer. Nat.* **VI.**

RACOVITZA, E.G. (1907) Essai sur les Problèmes biospéologiques. *Biospeologica*, **I.** *Archiv. Zool. expér. géner.* (4) **VI.**

RAMSAY, E.E. (1901) The optic lobes and optic tracts of *Amblyopsis spelaeus. Jour. comp. Neurol.* **XI.**

REICHEL, M. (1927) Étude anatomique du *Phreatobius cisternarum* Goeldi, Silure aveugle du Brésil. *Rev. suisse Zool.* **XXXIV.**

ROEWER, C.FR. (1950) Über *Ischyropsalidae* und *Trogulidae.* Weitere Weberknechte. **XV.** *Senckenbergiana.* **XXXI.**

SADOGLU, P (1956) Ocular reduction of the bat-fly, *Nycteribosca africana* (Dipt. Streblidae) and its comparison with those seen in some nycteribiids. *Istanbul. Univ. Fen. Fak. Mec.* B, **XXI.**

SAYCE, O.A. (1901) On three blind Victorian fresh-water Crustacea found in surface-water. *Ann. Mag. Nat. Hist.* (7) **VIII.**

SCHELLENBERG, A. (1931) Amphipoden der Sunda-Expeditionen Thienemann und Rensch. *Archiv. f. Hydrobiol. Suppl.* Bd. **VIII.**

SCHLAMPP, K.W. (1891) Die Augenlinse des *Proteus anguineus. Biol. Centralbl.* **XI.**

SCHLAMPP, K.W. (1892) Das Auge des Grottenolmes *(Proteus anguineus) Zeit. wiss. Zool.* **LIII.**

SCHMEIL, O. (1893) Zur Höhlenfauna des Karstes. *Z. f. Naturwiss.* **LXVI.**

Schwoerbel, J. (1961) Subterrane Wassermilben (Acari: Hydrachnellae, Porohalacaridae und Stygothrombiidae), ihre Ökologie und Bedeutung für die Abgrenzung eines aquatischen Lebensraumes zwischen Oberfläche und Grundwasser. *Archiv. f. Hydrobiol. Suppl.* **XXV.**

Seibold, W. (1904) Anatomie von *Vitrella quenstedti* (Wiedersheim) Clessin. *Jahreshefte. Ver. vaterl. Naturk. Württemberg.* **LX.**

Simon, E. (1911) Araneae et Opiliones (Troisième Série). *Biospeologica*, **XXIII.** *Archiv. Zool. expér. géner.* (5) **IX.**

Stammer, H.J. (1936) Ein neuer Höhlenschizopode, *Troglomysis vjetrenicensis* n.g. n.sp. Zugleich eine Übersicht der bisher aus dem Brack- und Süßwasser bekannten Schizopoden, ihre geographischen Verbreitung und ihrer ökologischen Einteilung sowie eine Zusammenstellung der blinden Schizopoden. *Zool. Jahrb. Abt. System.* **LXVIII.**

Stefanelli, A. (1954) The differentiation of the optic lobes neurons in a blind cave teleost. *Experientia.* **X.**

Strauss, E. (1909) Das Gammaridenauge. Studien über ausgebildete und rückgebildete Gammaridenaugen. *Wiss. Ergebn. d. deutsch. Tiefsee-Exped.* **XX.**

Szarski, K. (1935) Über das Augenrudiment von *Niphargus puteans* Koch. *Kosmos.* **LX.**

Tellkampf, Th.G. (1844) Über den blinden Fisch der Mammuthöhle in Kentucky, mit Bemerkungen über einige andere in dieser Höhle lebende Thiere. *Müller's Archiv. f. Anat. Physiol.* **IV.**

Thinès, G. (1955) Les Poissons aveugles (I). Origine; Taxonomie. Répartition géographique. Comportement. *Annal. Soc. R. Zool. Belgique* **LXXXVI.**

Tschugunoff, N. (1913) Über die Veränderung des Auges bei *Leptodora kindtii* (Focke) unter dem Einfluß von Nahrungsentziehung (Eine experimentelle Untersuchung). *Biol. Zentralbl.* **XXXIII.**

Tusques, J. (1929) Action de la thyroxine sur l'appareil visuel des têtards de *Rana esculenta. Compt. rend. Soc. Biol.* Paris. **CXLIII.**

Tusques, J. (1954) Ouverture palpébrale et développement oculaire sous l'action de la thyroxine chez la Taupe *(Talpa europaea* L.*). Compt. rend. Acad. Sci. Paris.* **CCXXXVIII.**

Tusques, J. (1955) Essai d'analyse du phénomène d'agénésie de l'œil de la Taupe. *Compt. rend. Acad. Sci. Paris.* **CCXL.**

Uhlenhuth, Ed. (1915) Are function and functional stimulus factors in producing and preserving morphological structure? *Biol. Bull.* **XXIX.**

Valle, A. della (1893) Gammarini del Golfo di Napoli. *Fauna und Flora des Golfes von Neapel.* **XX.**

Vandel, A. and Bouillon, M. (1959) Le Protée et son intérêt biologique. *Ann. Spéléol.* **XIV.**

Vejdovsky, F. (1882) Thierische Organismen der Brunnenwässer von Prag. Prag.

Vejdovsky, F. (1890) Note sur une nouvelle Planaire terrestre *(Microplana humicola)* nov. nov. gen. sp. suivie d'une liste des Dendrocoeles observés jusqu'a présent en Bohême. *Rev. biol. Nord. France.* **II.**

Vejdovsky, F. (1900) Über einige Süßwasser-Amphipoden. II. Zur Frage der Augenrudimente von *Niphargus. Sitzungsber. K. böhm. Gesell.*

Vejdovsky, F. (1901) Zur Morphologie der Antennen und Schalendrüse der Crustaceen. *Zeit. wiss. Zool* **LXIX.**

Vejdovsky, F. (1905) Über einige Süßwasser-Amphipoden. III. Die Augenreduktion bei einem neuen Gammariden aus Irland und über *Niphargus casparyi* Pratz aus den Brunnen von München. *Sitzungsber. K. böhm. Gesell.*

Verhoeff, K.W. (1926–1928) Diplopoda. *Bronn's Tierreich.* V. Bd. II. Abt.

Verhoeff, K.W. (1930) Arthropoden aus südostalpinen Höhlen, gesammelt von Karl Strasser. *Mitteil. Höhl.Karstf.*

Viré, A. (1900) *La Faune souterraine de France.* Paris.

Wolsky, A. (1934) Phylogenetische und mutative Degeneration des Gammariden-Auges. *Math. Naturw. Anz.* **LI.**

WOLSKY, A. (1935) Über einen blinden Höhlengammariden, *Niphargus aggtelekiensis* Dudich mit Bemerkugnen über die Rückbildung des Gammaridenauges. *Verh. intern. Ver. Limnologie* **VII.**

WOLSKY, A. and HUXLEY, J.S. (1934) Structure and development of normal and mutant eyes in *Gammarus chevreuxi*. *Proc. R. Soc. London. B.* **CXIV.**

WYMAN, J. (1853) On the eye and the organ of hearing in the blind fish (*Amblyopsis spelaeus* de Kay) of Mammoth Cave. *Amer. J. Sc. Arts.* **XVII.**

WYMAN, J. (1854) Eyes of *Amblyopsis spelaeus*. *Proc. Boston Soc.* **IV.**

ZELLER, E. (1888) Über die Larve des *Proteus anguinus*. *Zool. Anz.* **XI.**

ZELLER, E. (1889) Über die Fortpflanzung des *Proteus anguineus* und seine Larve. *Jahreshefte d. Ver. f. Vaterl. Naturkunde i. Württemberg* **XLV.**

CHAPTER XXVII

ECHOLOCATION

A. PRINCIPLE OF ECHOLOCATION

(a) Nocturnal Animals and Cavernicoles

The problem of how animals can find their way in total darkness such as that found in caves is of interest to both sensory physiologists and biospeologists. It is not the same problem as that posed for nocturnal animals. It is never completely dark during the night and nocturnal animals show an extreme sensitivity to illumination of low intensity; that is to say scotopic vision, and their eyes are highly developed.

The problem of orientation in cavernicolous invertebrates has not yet been tackled. It is not even very far advanced with respect to troglobious vertebrates. However, in vertebrates which are temporarily cavernicolous a very specialised method of orientation has been discovered based on the echo principle. This method is used only when they are underground. D.R.Griffin (1944) called this method of orientation, *echolocation*. Those animals which use this method can be called *echolocators*.

(b) Use of Echo as a Means of Orientation and Recognition of Objects

The methods by which animals find their way in darkness and avoid objects were not understood until man had invented apparatus based on the same principles. It has been known for a long time that an echo produced by the reflection of sound waves from an object allowed a measurement to be made of the distance between the object and the observer. If a transmitter and a receiver are placed side by side it is easy to estimate the distance from this point to the obstacle, assuming one knows the velocity of the sound waves. The time between transmission and receiving is equal to $t = 2D/V$.

The properties of echos were used for the first time by the French physicist Langevin during the First World War, for the detection of submarines. He used ultra-sonic waves as they had a better performance than audible waves.

Radar, which is the abbreviation of "Radio Detection and Ranging", is based on the same principles and was developed during the Second World War, 1939–1945. Radar makes use of electro-magnetic waves and not of

445

sound waves. The electro-magnetic waves have the same velocity as light waves, namely 300,000 km/sec. Thus the interval between transmission and reception is very short, about one thousandth of a second.

As the reflected waves are very weak compared with direct waves they cannot be picked up if transmission is continuous. Thus, transmissions are very short (of the order of a millionth of a second) and are separated by comparatively long periods of silence. Also, the receiver is blocked during emission so that it picks up only reflected waves. The receiver is usually an oscilloscope which records vibrations and displays them on a screen. The shorter the wave used, the more accurate are the details of obstacles. Wave lengths of 3 cm are generally used.

(c) Origin of the Phenomenon of Echolocation

Echolation was first discovered among bats. This system was at first thought to be wonderful as the fact that it resulted from long evolution was unknown. It is the distant heritage of devices and functions going back to the origin of the vertebrates.

The Radar system which has become known to man only recently, has been used since the very earliest times by vertebrates. It results from the development of the lateral line, an organ already present in the most ancient vertebrates known, the ostracoderms of the Palaeozoic.

The lateral line system allows slow vibrations which are propagated through water to be recorded. The moving fish produces vibrations which can be reflected in the manner of an echo from obstacles in the medium, and thus inform the fish on the topography of the medium. This echolocation system relies on the physiological properties of the lateral line system. This was probably also the situation in early vertebrates. Thus, the echolocator system must go back as far as the origin of the vertebrates.

Fish use other methods of orientation and of avoiding objects. It has been known for many years that some fish produce electricity, the electric, or better, the electrogenic fish. Electric organs are used by fish to attack prey and defend themselves. However, H.W.Lissmann has put forward the hypothesis that the electricity produced by the electric organs is used in the recognition of objects. The electrogenic fish creates around itself, as the result of rhythmic discharges, an equipotential electric field. The introduction of a strange object in this electrostatic field would cause a deformation of the field. With specialised receptor organs derived from the lateral system (such as the organs found in the Mormyridae, and called "Mormyromastes" by Cordier) the fish is capable of picking up any changes in the electrostatic field and can thus localise the object.

However, the electric system does not prevail in the aerial vertebrates; the method used is that of echolocation. The mechanism is more complex. Firstly, the waves transmitted are not the result of body movements. They arise from a specialised organ, the phonator organ. Moreover, the waves are more rapid than those induced by the swimming of fish; they are generally ultra-sonic. The organ which receives the waves is no longer the lateral line system, which has disappeared in terrestrial vertebrates, but the cochlea and the organ of Corti, the essential elements of the inner ear. However, it must be remembered that the inner ear and the lateral line system have a common origin.

(d) Distribution of Echolocators in the Animal Kingdom

Echolocation was first discovered in bats, which are mainly nocturnal creatures and of which several species temporarily inhabit caves (p. 239). Later, the same method of orientation was found to exist in birds (Guacharo, Salanganes) which spend part of their life underground.

It would, however, be a grave error to believe in a constant relationship between echolocation and subterranean life. Indeed this method of orientation appears to be widespread in odontocete Cetacea. It is also possible but not certain, that echolocation is practised by certain rodents. However, echolocation has not yet been reported in any true cavernicoles.

B. ECHOLOCATION IN BATS

(a) Historical

The problem of how bats orientate themselves during the night or in caves, has been recognised for many years. The history of this topic was reviewed by R. Galambos (1942) and Dijkgraaf (1960).

The initiator: Abbé Spallanzani. In 1793 Abbé Lazzaro Spallanzani questioned the method by which bats took refuge in caves could direct their flight during nocturnal outings. He tried to solve the problem by experimentation. He did not keep the results to himself but wrote to several acquaintances: the Abbé Vassali at Turin; Pietro Rossi at Pisa; and Jean Senebier at Geneva. These letters and their answers were published in 1794 and also included in the *Works of Lazzaro Spallanzani* (1st Edition, 1825–1826; 2nd Edition, 1932).

Spallanzani wondered at first if vision played a part in flight orientation underground. He blinded bats and showed that their orientation was not affected by loss of sight. These results were confirmed by Vassali and Rossi. He suppressed taste and smell by removing the tongue and blocking the nares of bats, but orientation was again not affected. Finally by lacquering the body of the animal he showed that touch was not important in orientation.

From these negative results, Spallanzani concluded the existence of a sixth sense which had no equivalent among humans. In 1800 Cuvier mocked this explanation, saying that it was not based on any positive evidence He stated from purely anatomical data that orientation in bats was possible because of the exceptional sensitivity of tactile hairs which covered the alary surface.

Because of Cuvier's authority this hypothesis became an article of faith for naturalists and was reproduced for 120 years in all treatises on zoology. However, it appears today that this interpretation was entirely incorrect.

The Experiments of Louis Jurine. The naturalist Louis Jurine, from Geneva, was told of the experiments of Spallanzani by his compatriot Jean Senebier. He carried out some experiments of his own on bats. He stated that individuals in which the ears were completely blocked, crashed into obstacles. He concluded that in Chiroptera orientation is definitely a function of the ears and not of the eyes. The results obtained by Jurine were published in 1789 in the *Journal de Physique*. These results which, as we know today, would have led to the solution of the question were completely disregarded by naturalists and in particular by Cuvier, who held to his explanation based on a tactile sense. The reason for this mistake was that no one could hear the sounds emitted by bats nor suspect their existence.

During the nineteenth century no experiments were attempted on bats to find the secret of their orientation in darkness. It was not until 1900 that two French naturalists, Rollinat and Trouessart, and ten years later, an American zoologist W.L.Hahn, confirmed the results of Jurine and the role of hearing in orientation.

Hypothesis of Hartridge. In 1920 the English physiologist Hartridge, probably after learning of the research of Langevin on the detection of submarines, suggested an audacious hypothesis. He believed that bats directed themselves by emitting ultra-sonic waves which were reflected on neighbouring objects and informed the animals of their position.

(b) Verification of Hartridge's Hypothesis

The hypothesis of Hartridge had to be verified by detailed experimentation. This was carried out by three American biologists, G.W.Pierce, D.R.Griffin and R.Galambos. Their first results appeared in 1938 and were followed by numerous publications. Research was carried out on the same subject in Germany, England and Italy. For a detailed description of the subject the reader should refer to D.R.Griffin's book *Listening in the Dark*, which gives an account of the work carried out up till 1958.

Experiments. Rollinat carried out his experiments on orientation in bats in the ballroom at Argenton-sur-Creuse, in which he placed perches and large mesh nets.

The American biologists developed his methods further. They used a technique initiated in 1908 by Hahn. It consisted of suspending a row of iron wires of one millimetre diameter from the ceiling of a large room at a distance of 30 cm from each other. The bats are released in this room. If they collide with the wires during the flight an easily discernable metallic sound is produced. Normally bats can avoid these obstacles but if during experimentation certain faculties are inhibited they can no longer fly around the wires. This device provides a method for statistical recording of errors of orientation and the recognition of obstacles.

The Sense Used in Orientation. In their experiments Griffin and Galambos avoided injuring any organ. They limited themselves to suspending their functions.

(1) They repeated the experiments of Spallanzani but instead of using the barbarous method of blinding the bats they occluded the eyelids with a layer of colloidon. Such bats showed no anomalies of flight. Thus the statement of Spallanzani that sight was not involved in orientation, was confirmed.

(2) When the ears of the bats were perfectly occluded the bats were not only unable to avoid objects, but also hit the walls and the ceiling and finally fell to the ground. This was never observed in normal bats. These experiments confirmed those of Jurine, and established the role of hearing in the perception of obstacles.

(3) Finally, it was necessary to show that orientation involved the emission of sound by the bats. The American biologists closed the mouths of bats with a coat of colloidon. These "mute" bats were as unable to guide themselves as the deaf bats. Normally, bats (at least the vespertilionids, as it is not exactly true for other Chiroptera as will be shown later), fly with their mouths constantly open (Plate XI).

Sounds and Ultra-sounds. The nature of the sound emitted by bats must be studied. Bats emit a cry which is easily registered by the human ear and which is not very different from those of other mammals. These cries are composed of vibrations of the order of 7000 cycles per second. They are not used in orientation. Bats also emit ultra-sounds which are not perceptible to the human ear because they are above the maximum frequency registered by our auditory apparatus which is 20,000 vibrations per second. They can, however, be detected and analysed with the help of microphones and amplifiers associated with cathode ray oscilloscopes. These have shown that the sounds produced by bats consist of vibrations between 30,000 and 70,000 cycles per second and on average between 45,000 and 50,000 cycles per second.

The work of Dijkgraaf has shown that there are some transitions between audible sounds and ultra-sounds. The emission of ultra-sounds is preceded by an audible click. The clicks are particularly discernable in young bats or in bats coming out of winter lethargy.

The Sounds Emitted. The ultra-sonic cries emitted by bats are not continuous. They are separated by intervals. The duration of an ultra-sonic emission is very short, of the order of 1–5 msec.

PLATE XI. *Myotis lucifugus*, photographed in flight; note that the mouth is open (after Griffin).

The number of emissions in unit time is variable. The animal can increase or diminish the frequency "at will". The frequency increases as an object is approached which allows exact localisation of the obstacle. At rest a bat emits 5–10 ultra-sounds per second; when it is in flight 20–30 per second and when approaching an object 40–60 or even as many as 100 per second.

Apparatus Emitting Sounds. The larynx produces the sounds and the ultra-sounds of bats. The larynx of Chiroptera is constructed in the same way as in other mammals but it shows certain peculiarities (Robin, 1881; Elias, 1907; Motta Manno, 1951; Novick, 1955; Michel, 1961).

Indeed, the exact constitution of the larynx of the bats has been correctly understood only of late years (Fischer, 1961; Fischer, and Vömel, 1961). The larnyx of the bats works in the same manner as a sirene. The sound is produced by a strong draught flowing through a narrow slit which is alternately opened and closed.

Recording Apparatus. The method of echolocation must involve an extremely sensitive receiving apparatus. The tympanic bullae of bats are enormous and in the horseshoe-bats they almost join in the median line. The auditory centres of the brain are also well developed (Poliak, 1925).

Whilst the human ear can pick up only those sounds produced by vibrations of frequencies between 16 and 16,000–20,000 per second, Galambos (1942) showed that the ears of bats are sensitive to vibrations between 30 and 98,000 per second. Dijkgraaf (1957) stated that some bats can perceive ultra-sounds of a frequency of as much as 200,000 vibrations persecond (200 kc/sec).

Search and Capture of Prey. Bats which frequent caves must be able to orientate themselves in completely darkness. Besides all Chiroptera chase and capture often very small insects which are their source of food. The method employed by vespertilionids to locate their prey has been observed by Griffin (1960) who has studied the behaviour of *Myotis lucifugus*. These bats are capable of locating prey as small as *Drosophila* and mosquitoes and hunt them with unbelievable speed. Ten mosquitoes were captured within a minute, and fourteen *Drosphila* with the same time. The most extraordinary performance was that of a specimen of *Myotis* which captured two *Drosphila* separately in half a second.

What is the mechanism by which bats capture their prey? Certain bats locate insects by the humming sound produced by the latter's flight (Möhres, 1950; Kolb, 1959).

The experiments of Griffin have proved that location of prey is assured in *Myotis lucifugus* by a system of echolocation. Insects can be located at a distance of 21–135 cm and on average between 55 and 90 cm. The number of vibrations emitted increases as the bat approaches the prey. During the final stage vibrations are emitted without interruption and at an extremely rapid rate. This "buzz" is produced by repeated emissions which are separated from each other by intervals of only 4–7 milliseconds.

The capture of prey is possible only by employing sounds of very short wave length. The sounds can be accurately reflected only by surfaces whose dimensions are notably in excess of the wavelengths employed. As the length of the wave is inversely proportional to the frequency, only ultrasounds are capable of being reflected from objects as small as *Drosophila* or mosquitoes.

Thus bats appear to us as "a very efficient guided missile whose fuel is provided by the target" (Pye).

Biauricular Function. An accurate flight requires the intervention of the hearing. If the ears are blocked the animal can no longer avoid obstacles. What is the behaviour of bats with only one ear blocked?

We may first consider the vespertilionids. In these animals the ear lobes are small and immobile. A bat with one ear blocked can fly in the experimental room without crashing into the walls as a deaf animal, but it is incapable of avoiding the wires stretched across the room. Thus, there is no doubt that they would be unable to catch prey. The localisation of a small object situated a short distance away requires the utilisation of both ears. It is probably the very slight difference in time separating the arrival of the echo on one or other of the ears which allows localisation of the object.

The Mechanism of Echolocation. Hartridge (1945) stated that the echolocation system of bats would function in the manner of a Radar or Sonar, by estimation of the interval separating emission and reception of the echo. Hartridge imagined that in the ear of bats, as in the apparatus constructed by man, some mechanism would block reception (or hearing) during emission. This interruption would have to be induced by a reflex system which caused contraction on muscles of the tensor-tympani and the stapedius. This reflex would inhibit transmission of sound through the middle ear.

However, this hypothesis is not based on any experimental information. Also, it seems now very improbable. It fails to account for interference by external noises or the ultra-sonic cries produced by other bats.

Finally, it is certain that localisation of very small objects at a short distance away, that is those which require the most precision, implies a partial overlapping of the cry and the echo. This last condition led Pye (1961) and Kay (1961) to propose a new interpretation. They referred to the Doppler effect which is a difference in frequency between the emitted and the returning wave. In acoustics this difference is called the "beat-note". It can be stated that bats perceive neither the emitted cry or the e cho but only the beat-note, and this would exclude any problem of interference.

(c) Comparison of Echolocation in Chiroptera

Thus the problem originally posed by Spallanzani at the end of the eighteenth century can in general terms, today be considered to be resolved.

However, another problem has arisen meanwhile which has complicated the task of zoologists. It appears that the Chiroptera, which are structurally and behaviourally a very varied group, also show notable differences in their methods of orientation and their ability to recognise objects.

About thirty genera of the Chiroptera have been studied with respect to orientation and the list of these is given below:

Megachiroptera	Microchiroptera	
Pteropodidae	*Hipposideridae*	*Emballonuridae*
Rousettus	Asellia	Saccopteryx
Pteropus	*Rhinolophidae*	Rhynchonycteris
Hypsignathus	Rhinolophus	*Vespertilionidae*
Cynopterus	*Noctilionidae*	Antrozous
Nyctimene	Noctilis	Miniopterus
	Phyllostomatidae	Myotis
	Chilonycteris	Pizonyx
	Pteronotus	Plecotus
	Lonchorhina	Pipistrellus
	Phyllostomus	Nyctalus
	Glossophaga	Eptesicus
	Lonchophylla	Lasiurus
	Carollia	*Molossidae*
	Uroderma	Tatarida
	Artibeus	
	Desmodus	

From this list a few examples have been taken which will show the extent of the variations which have been discovered among the echolocation systems of bats. It seems probable that echolocation has developed independently in different phyletic lines. The system of location by sound is far from having reached the same degree of precision in all groups of the bats.

Vespertilionids. This truly cosmopolitan family contains the most common bats and thus those which have been the most widely studied. The general description of echolocation given in the preceding pages relates to the vespertilionids. It will be recalled that their system of echolocation is characterised by the following conditions:

(1) The sounds produced by the larynx and emitted by the mouth which is largely open during flight.
(2) The emissions are ultra-sounds with a frequency of 30–70 kc/s.

(3) The prey is located by echolocation.

(4) The ear lobes are small and immobile. Biauricular hearing is necessary to localise approaching objects.

Rhinolophidae. Echolocation in the horse-shoe bats has been studied in detail by Möhres (1953) and Möhres and Kulzer (1955). It shows the following characteristics:

(1) Contrary to that in the vespertilionids, the mouth of horseshoe bats is closed during flight. Sounds are emitted through the nares. This is made possible by the fact that the larynx is buried in the posterior region of the nasal fossae. This is why occlusion of the mouth does not prevent the horse-shoe bats from flying accurately. However, blockage of the nares interferes with flight. It is possible that the complicated appendages around the nasal region of these bats is related to nasal emission of ultra-sounds.

(2) Sounds are thus emitted from a double source, the two nares. The vibrations emitted by one or other of the nares interfere in the median region of the field of emission. Thus a narrow but powerful beam of sound waves is produced which is directed forward.

(3) The emissions are less numerous than those in the vespertilionids: 4 per second at rest and 5 - 6 per second during flight. They are of longer duration. The emissions coincide with respiratory movements and the beating of the wings. The frequency of the vibrations is higher than in the vespertilionids, being between 80 and 100 kc/s, but they are more constant.

(4) The effective range of emissions is much greater in horse-shoe bats: 8–9 metres for the greater horse-shoe bat *(Rhinolophus)* compared with 8–10 cm for vespertilionids.

(5) The ears of vespertilionids are immobile and thus the animals have to turn their heads to receive the echos. The horse-shoe bats can orientate their ear lobes, either forwards or backwards and alternatively for one or other ear. Localisation of an object probably results from the difference in intensity of sounds received by the left and right ears.

Rousettus. *Rousettus* belongs to the sub-order Megachiroptera, although the wing span of these bats does not exceed 60 cm. They are well-known and have received the name "flying-fox". The most well-known is *R. aegyptiacus*, the "tomb-bat" of Egyt and the Middle East. It was so named because it was found in tombs although it also frequents natural caves. These bats have a strong reaction to avoid light and take refuge in these dark places during the day. It is therefore a temporary cavernicole.

The flying-foxes are of particular interest here because they show an intermediate system of echolocation between the Microchiroptera and Megachiroptera.

These bats have large, well-developed eyes, as is the rule among Megachiroptera while the eyes of Microchiroptera are more or less reduced. The flying-foxes possess good scotopic vision. Thus one sees these animals emerging at twilight and accurately directing their flight in spite of the weak illumination.

However, the flying-foxes can fly in complete darkness, for instance in caves. This virtually implies some system of echolocation, and this has been studied during the last few years (Möhres and Kulzer, 1956; Kulzer, 1956, 1958, 1960; Griffin, Novick and Kornfield, 1958; Novick, 1958). *Rousettus aegyptiacus* has been studied principally but all species of this genus seem to behave in a similar manner.

Sound is emitted by movements of the tongue, the only case known among the Chiroptera. Tongue movements produce clicks which are partially audible to the human ear. Most of the sounds emitted are ultra-sounds however, with a frequency between 6·5 and 100 kc/s. The sounds are emitted via the nares as in horse-shoe bats.

The flying-foxes thus possess two methods of orientation. In dim light they can direct themselves because of their well developed eyes, with scotopic vision. In complete darkness they use a system of echolocation. The passage from one type to the other can be followed because of the audible clicks. The intensity of the clicks increases as the light intensity diminishes; inversely the animal produces no sound when it is in an illuminated zone. Similar characteristics are unknown among the Microchiroptera although they are found in the Guacharo (p. 456).

Rousettus provides the best example for understanding the development of echolocation. The echolocation system could develop in twilight because the optic control allows initial errors to be corrected.

Megachiroptera. Bats belonging to the Megachiroptera are generally very large; *Pteropus giganteus* measures 1·3 metres wing-span. *Pteropus* fly in twilight. They have large eyes and excellent scotopic vision. During the day these bats are found in trees; they appear not to be made uneasy by light.

Some research has been carried out on *Pteropus*, in particular on *Pteropus giganetus* of India, and *P. poliocephalus* of Australia (Möhres and Kulzer, 1955, 1956; Novick, 1958).

No echolocation system is found in *Pteropus* nor indeed in any Megachiroptera, with the exception of *Rousettus*. The eyes are the only organs used for orientation. *Pteropus* is incapable of flying in complete darkness. Blind individuals are disorientated. On the contrary, flight is not interrupted if the ears are blocked. *Pteropus* produces no ultra-sounds, only audible sounds which have no role in orientation.

Echolocation is therefore not a general property of Chiroptera, but is possessed by only some of them.

C. ECHOLOCATION IN BIRDS

(a) The Guacharo

The guacharo *(Steatornis caripensis)* nests in caves in the tropical regions of South America. In one of the best-known caves in which these birds nest, the Cueva del Guacharo, near Caripe in Venezuela, the nests are found some 650 metres from the entrance, in complete darkness. However, they direct themselves without hitting the walls of the cave.

D. R. Griffin (1953) has studied the method of orientation of the guacharo in this cave. He has shown that these birds find their way in the cave by echolocation. The sounds used are not ultra-sounds but are audible to the human ear. They are clicks with a frequency between 6,100 and 8,750 vibrations per second and with an average value of 7,300 vibrations per second. As in bats the emission of clicks is separated by intervals of a few milliseconds. If the ears of guacharos are blocked, the birds crash against obstacles.

In light, these birds direct themselves by vision and do not produce any clicks. Their behaviour is thus analogous to that of *Rousettus*.

(b) Salanganes

It seems probable that the salanganes and in particular those which produce the famous nests from which "bird's-nest soup" is made, that is *Collocalia fuciphaga*, find their way by echolocation. These birds nest in caves in Indonesia where they are found by the hundred thousand. They nest in completely dark sites which are very long distances from the entrance. This has led to the idea that they use echolocation in the way that the guacharo does.

This hypothesis, although plausible, had to be verified. The truth of this interpretation was shown almost simultaneously by A. Novick (1959) and Lord Medway (1959). Although the salanganes fly in silence in condition of light they produce audible clicks in condition of darkness. Specimens whose ears had been blocked could no longer direct themselves. The clicks are emitted at 5–10 per second with a frequency of 4–5 kc/s. This is the lowest frequency that has been recorded for animals possessing an echolocation system.

Salanganes penetrate caves to nest. They search for food, however, in light where they can use their eyes. Thus, the salanganes together with *Rousettus* and the guacharo possess two methods of orientation, one adapted to light and the other to darkness.

The echolocation system does not exist in all salanganes. It is found only in those which nest in caves. Those which nest above ground do not show this method of orientation, for example, *Collocalia esculenta* (Medway, 1962).

BIBLIOGRAPHY

Bats

A historical bibliography is given in great detail in the paper by Galambos (1942). All the bibliography concerning echolocation can be found in the work of Griffin (1958). Only the more important publications which have appeared since 1958 are given here.

DIJKGRAAF, S. (1960) Spallanzani's unpublished experiments on the sensory basis of object perception in bats. *Isis*, **LI.**

FISCHER, H. (1961) Der Kehlkopf der Fledermaus *Myotis myotis* ein primitiver Säuger-larynx mit der Sonderleistung der Ultraschallerzeugung. *Archiv. Ohren- Nasen u. Kehl-kopfheilk.* **CLXXVIII.**

FISCHER, H. and VÖMELS H.J. (1961) Der Ultraschallapparat des Larynx von *Myotis myotis.* Eine morphologische Studie über einen primitiven Säugerkehlkopf. *Gegen-baur's Morphol. Jahrb.* **CII.**

GALAMBOS, R. (1942) The avoidance of obstacles by flying bats. *Isis* **XXXIV.**

GRIFFIN, D.R. (1958) *Listening in the Dark.* New Haven.

GRIFFIN, D.R., NOVICK, A. and KORNFIELD, M. (1958) The sensitivity of echolocation in the fruit bat, *Rousettus. Biol. Bull.* **CXV.**

GRIFFIN, D.R., WEBSTER, F.A. and MICHAEL, C.R. (1960) The echolocation of flying insects by bats. *Animal Behaviour.* **VIII.**

KAY, L. (1961) Orientation of bats and men by ultrasonic echolocation. *Brit. Communic. Electron.* **VIII.**

KOLB, A. (1958) Die Nahrungsaufnahme einheimischer Fledermäuse vom Boden. *Verh. Deutsch. Zool. Gesell.* Frankfurt a.M.

KULZER, E. (1958) Untersuchungen über die Biologie von Flughunden der Gattung Rousettus Gray. *Z. Morphol. Ökol. Tiere.* **XLVII.**

KULZER, E. (1960) Physiologische und morphologische Untersuchungen über die Er-zeugung der Orientierungslaute von Flughunden der Gattung Rousettus. *Z. vergl. Physiol.* **XLIII.**

MICHEL, C. VAN (1961) Différenciation dans les cordes vocales de la Chauve-Souris en rapport avec l'émission d'ultra-sons? *Compt. rend. Soc. Biol.* Paris **CLV.**

NOVICK, A. (1958) Orientation in palaeotropical bats. II. Megachiroptera. *J. exper. Zool.* **CXXXVII.**

PYE, J.D. (1961) Localisation par écho chez les Chauves-Souris. *Endeavour* **XX.**

The Guacharo

GRIFFIN, D.R. (1953) Acoustic orientation in the oil bird, *Steatornis. Proc. Nat. Acad. Sc. Washington* **XXXIX.**

Salanganes

MEDWAY, Lord (1959) Echolocation among *Collocalia. Nature.* **CLXXXIV.**

MEDWAY, LORD (1962) The swiftletts *(Collocalia)* of Niah Cave, Sarawak-11. Ecology and the regulation of breeding. *Ibis* **CIV.**

NOVICK, A. (1959) Acoustic orientation in the cave swiftlet. *Biol. Bull.* **CXVII.**

PART 6

Evolution of Cavernicoles

CHAPTER XXVIII

THEORETICAL CONCEPTS

BIOSPEOLOGY today has become an autonomous science with its own requirements and following its own aims. However, in formulating its general conclusions, it of necessity reaches the stage to which all biological disciplines lead, namely that of the explanation of evolution of the living form. The origin of cavernicoles and their evolution will naturally be the subject of the last part of this book and will provide the conclusion.

At all stages, and in all fields, hypothesis has preceded experimentation. During the second half of the last century, when biospeology was in its infancy, the evolutionists on many occasions used cavernicoles as examples to support either one or other theory. Nothing remains of these speculations. They are only of historical interest today and therefore we shall consider them briefly.

A. NEO-LAMARCKISM

Although J.B. Lamarck attributed evolution of the organism to itself, be the usage or non-usage of the various organs, his successors, who ary called Neo-Lamarckists, altered the doctrine of their master. To account for biological evolution, they invoked the concept of the direct action of the medium on the organism. It is not without significance that the two first Neo-Lamarckists were E.D.Cope and A.S.Packard, the two pioneers of American biospeology.

Numerous biologists still hold Neo-Lamarckian views and of those who are concerned with biospeology we may mention E.W.MacBride (1925) in England; W.Peter (1922) in Germany; Louis Fage (1931) and R.Jeannel (1950) in France.

According to these biologists, cavernicoles have been modelled by the subterranean medium and this is responsible for their specialised characteristics.

B. MUTATIONISM

During the year 1893 a very curious controversy arose, some of which was published in *Nature*. This controversy opposed the views of many eminent naturalists of the day: E.R.Lankester, A.Anderson, J.T.Cunningham and G.A.Boulenger. It arose from a polemic between the celebrated zoologist E.R.Lankester and the biophilosopher Herbert Spencer. In his article which was published in *Nature* Lankester proposed an interpretation of the origin of cavernicoles which can be called the "Accident Theory". This postulates that if an animal with normal eyes accidentally strays into the subterranean medium, or into a marine abyss it would be attracted towards the surface by the light. If a similar accident happened to an animal with some malformation of the eye, it might remain in the subterranean world, and reproduce there. It would become a cavernicole. Given in this simple form such an interpretation is obviously inadequate.

However, this idea was recast in another form which was prompted by the development of genetics and the birth of the theory of mutation (Loeb, 1915; Banta, 1921; Nachtsheim, 1921; Kosswig, 1935, 1937, 1944, 1946, 1960; de Lattin, 1939; Ludwig, 1942; Komarek, 1955). These biologists argued that the existence of albino or eyeless mutants provided the usual method by which cavernicoles could appear. Later the arguments against this interpretation will be given. It is sufficient to recall here a fact given in a preceding chapter. The comparisons which have been established between albino or eyeless mutants and the cavernicolous state are superficial and frequently incorrect. Wolsky (1935) has provided some particularly clear proof of this. In hypogeous amphipods regression of the eye follows, as in most cavernicoles, a centripetal direction, that is to say the nervous tissues are the last to be modified. In mutants of *Gammarus chevreuxi* reduction of the eye follows a centrifugal direction, a type which is completely unknown in cavernicoles.

C. ORGANICISM

Contrary to a frequently expressed opinion, cavernicoles have not developed rapidly following some mutation. It can be stated today in all certainty that they have undergone a long period of development which can be measured only with the aid of the geological time-scale.

The origin of cavernicoles and their successive transformations present problems of the same order as those which held the attention of evolutionists. Firstly, it must be recalled that animal evolution follows a similar pattern although superficially showing great variety (Vandel, 1949, 1958). All phyletic lines pass through several successive stages: the stage of the creation of a new organic type (*typogenesis* of Schindewolf); the stage of expansion and diversification, and finally the stage of specialisation and senescence, resulting after a long period in the extinction of the group. The first stage of the cycle is a manifestation of progressive evolution and the last of regressive or gerontocratic evolution. The evolution of cavernicoles provides a particularly good example of regressive evolution which can occur only in the final stages of the evolution of animal lines.

Thus, all life, both individuals and phyletic lines, is fated to perish. As the phyletic lines age, the animal organisation degrades and thus all the zoological groups are led ineluctably towards their end and disappearance.

The living organism can exist only because it has devised processes of phyletic rejuvenation. One of these is that of paedomorphosis (W. Garstang), that is the development of sexual maturity in an otherwise juvenile organisation. This has often been thought to be exceptional but today it appears to be a very general character (Vandel, 1961).

Thus the origin of evolution must not be looked for in the organism itself and not at all in some external factor (medium or selection). This interpretation of life and evolution has been given the name of "organicism" (Vandel, 1958).

BIBLIOGRAPHY

BANTA, A. (1921) An eyeless daphnid with remarks on the possible origin of eyeless cave animals. *Science*. **LIII.**

FAGE, L. (1931) Araneae (5ème Série), précédée d'un essai sur l'évolution souterraine et son déterminisme. *Biospeologica*, **LV.** *Archiv. Zool. expér. géner*. **LXXI.**

JEANNEL, R. (1950) La Marche de l'Évolution. *Public. Mus. Hist. Nat. Paris* **XV.**

KOMAREK, J. (1955) Mutation und Adaptation bei dem Entstehungsprozeß von Höhlenwassertieren. *Zool. Anz.* **CLV.**

KOSSWIG, C. (1935) Die Evolution von "Anpassungs"-Merkmalen bei Höhlentieren in genetischer Betrachtung. *Zool. Anz.* **CXII.**

KOSSWIG, C. (1937) Betrachtungen und Experimente über die Entstehung von Höhlentiermerkmalen. *Der Züchter*. **IX.**

KOSSWIG, C. (1944) Zur Evolution der Höhlentiermerkmale. *Istanbul. Univ. Fen. Fak. Mec. Ser. B*. **IX.**

KOSSWIG, C. (1946) Bemerkungen zur degenerativen Evolution (vom genetischen Standpunkt). *Compt. rend. ann. et Archiv. Soc. Turque. Sc. phys. Nat.* **XII.**

KOSSWIG, C. (1960a) Darwin und die degenerative Evolution. *Abh. Verh. Naturw. Ver. Hamburg. N.F.* **IV.**

KOSSWIG, C. (1960b) Zur Phylogenese sogenannter Anpassungsmerkmale bei Höhlentieren. *Intern. Rev. ges. Hydrobiol*. **XLV.**

LANKESTER, E. R. (1893) Blind animals in caves. *Nature, Lond*. **XLVII.**

LATTIN, G. DE (1939) Über die Evolution der Höhlentiercharaktere. *Sitzb. Gesell. Naturf. Freunde Berlin*.

LOEB, J. (1915) The blindness of the cave-fauna and the artificial production of blind fish embryos by heterogeneous hybridization and by low temperatures. *Biol. Bull.* **XXIX.**

LUDWIG, W. (1942) Zur evolutorischen Erklärung der Höhlentiermerkmale durch Allelelimination. *Biol. Zentralbl.* **LXII.**

MacBride, E. W. (1925) The blindness of cave-animals. *Nature, Lond.* **CXVI.**

Nachtsheim, H. (1921) Augenlose Höhlentiere, Mutationstheorie und Lamarckismus. *Naturwiss. Wochenschr. N.F.* **XXI.**

Peter, W. (1922) Augenlose Höhlentiere, Mutationstheorie und Lamarckismus. *Naturw. Wochenschr. N.F.* **XXI.**

Vandel, A. (1949) *L'Homme et l'Évolution.* Paris, Gallimard.

Vandel, A. (1958) *L'Homme et l'Évolution.* Édition revue et augmentée. Paris, Gallimard.

Vandel, A. (1961) L'Origine des Vertébrés. *L'Année biologique.* **LXV.**

Wolsky, A. (1935) Über einen blinden Höhlengammariden, *Niphargus aggtelekinsis* Dudich, mit Bemerkungen über die Rückbildung des Gammaridenauges. *Verh. intern. Ver. Limnologie* **VII.**

CHAPTER XXIX

THE ANTIQUITY OF CAVERNICOLES

A. RELATIVE AGES OF CAVERNICOLES

It is certain that we cannot attribute the same phyletic age to all forms which populate the subterranean world. Some are relatively recent while others have originated in very ancient times.

However, there can be no doubt that population of the subterranean world is still taking place at the present time. The migration of surface forms to the depths of the soil represents a normal phenomenon which is accentuated in epochs in which profound climatic changes take place, but which has occurred throughout the length of the history of living organisms. Schwoerbel (1961) recognised that the population of the interstitial medium by surface-living Hydracarina is still occurring today. Certainly this statement also applies to other hypogeous forms.

One must take into account the following characteristics to fix the relative antiquities of cavernicoles.

(a) The forms which are not strictly subterranean and which are still to be found outside caves, are certainly recent cavernicoles. Thus Chopard (1936), considered *Gryllomorpha dalmatina* to be "a form showing before our eyes an example of the penetration of the subterranean world".

(b) Recent cavernicoles are usually united to surface types by intermediate forms, while ancient cavernicoles remain very isolated in the present world.

(c) Finally, the degree of physiological specialisation (see p. 346) allows a division of cavernicoles into recent and ancient types to be made.

Although these criteria and expressions are imprecise, it is indispensable to separate, at least approximately, the recent cavernicoles from ancient ones.

B. ANCIENT CAVERNICOLES. CONCEPT
OF RELICT FAUNAE

It is known today that the evolution of cavernicoles has taken place over a very long time period of time. As is to be expected, the fossils on which this assertion is based are rare but they are not totally absent as we shall show later. On the other hand, the manner of the distribution of the caverni-coles of which we have given examples in the preceding chapter, replaces in some measure the absence of palaeontological evidence. The geographical distribution of cavernicoles is sometimes capable of being superimposed onto palaeographical maps. This allows an estimate of the date of the period of their expansion to be made.

The true cavernicoles, the troglobia, are for the most part relics. We shall use the definition of Birstein (1947) that relics are those animal (or plant) types of which the evolution has stopped or is at least slowed †, and which have conserved the appearance of their distant ancestors. This recalls the expression used by Darwin in the fourth chapter of *The Origin of Species*, that of "living fossils".

Biospeologists are in almost complete agreement on this point. The relict character of the cavernicolous fauna has been clearly recognised by many zoologists and the following can be mentioned among the more recent authors: Spandl (1926), Stankovič (1932, 1960), Chappuis (1956) for aquatic forms; Fage (1931) and Jeannel (1943) for terrestrial animals.

It has been mentioned in a preceding chapter that the bathynellinids represent very ancient relics, the remains of a Carboniferous freshwater fauna.

The origin of certain North American and European troglobia must go back to the Tertiary. This is the case with cirolanids, whose Mesogean distribution is clear (p. 119). The distribution of cavernicolous spheromids and their commensal ostracods *(Sphaeromicola)* originated in the middle of the Tertiary because it coincides in a remarkable way with the extension of the Miocene seas. Differentiation of the annelid *Troglochaetus* must be dated in the same era. According to Wagner (1914) the molluscs which inhabit the caves of Southern Dalmatia and Herzegovina are in part the remains of an ancient fauna which dates back to the end of the Tertiary. These species no longer possess close relatives among the epigeous fauna.

Terrestrial cavernicoles are for the most part descendants of a tropical fauna which populated Europe and North America during the first half of the Tertiary.

Telema tenella, which is localised in a few caves in the Pyrénées Orientales is closely related to *Apneumonella* from Equatorial Africa (p. 96). *Prorhaphidophora* from the Baltic Amber, which is related to present day tropical Rhaphidophoridae, must be considered to be ancestral to *Troglophilus* and *Dolichopoda*, which today populate the caves of Europe (p. 179).

The tropical fauna which occupied Europe and North America during the Tertiary has disappeared from these regions because of the progressive

---

† Relicts confined to limited territories and isolated one from the other, undergo the phenomenon of micro-evolution which results in the appearance of numerous endemic species. The subterranean fauna provides numerous examples of this. However, this is only a particular case of a very general phenomenon: that of variation following the isolation of populations. The formation of endemic species is not a specifically biospeological problem and we shall not consider it further.

decrease in temperature which occurred with increasing severity just prior to the beginning of the Quaternary. What became of these tropical species? Some were no doubt destroyed following the climatic changes, while others migrated towards the present-day tropical regions. Only a few specimens of the primitive fauna remained in the same place, where they found refuge in the subterranean world. These have become relicts, abandoned by their related forms, redistributed in more suitable climates.

It may be asked what it was that led certain elements of the original fauna to abandon the migrating group and remain in their primitive habitat. This is a question which one cannot, and may never, answer correctly. The relicts may represent forms which had undergone regressive evolution for a long period, which had started before they entered caves. This may have been why they were incapable of undertaking remote migrations.

C. ORIGIN OF RELICTS IN RELATION TO CLIMATIC FACTORS

One must look for the origin of relict species and in particular that of cavernicoles in the climatic changes which took place on the surface of the Earth. True troglobia (at least the terrestrial forms) are found only in Europe, North America, and Japan, regions which underwent marked climatic modifications during the later geological periods. The climate, which was tropical during the first half of the Tertiary, became progressively colder until the Miocene and Pliocene. At the beginning of the Quaternary the Northern Hemisphere possessed a particularly cold and humid climate, which gave rise to immense glaciers. These climatic transformations were almost certainly one of the factors responsible for the evolution of subterranean life.

On the other hand, it would be incorrect to limit the intervention of climatic factors to the glacial periods alone. The origin of cavernicoles probably goes back to the middle of the Tertiary and perhaps even earlier (Jeannel). However, the glacial period hastened the penetration of subterranean environments by lines engaged in regressive evolution of the cavernicolous type.

Another factor influencing the retreat of the surface fauna concerns the Karstic evolution which often took place in calcareous regions, depriving them of water and plants and transforming them into deserts which thus drove the fauna underground.

The origin of the South African cavernicolous fauna results probably from the progressive drying of this region (Leleup).

D. CONCEPT OF REFUGE

From what has been said, it appears that the subterranean world represents a refuge which has allowed animal forms which have been driven from the surface of the earth by climatic alterations, to persist to the present day.

Other types of refuge which can be compared with the subterranean world are known, for example the abyssal world. This also contains ancient relicts (Zenkewitsch and Birstein, 1960). For example, the pedunculate crinoids; all representatives of the Pogonophora (with the exception of a few species of *Siboglinum*); molluscs of the genus *Neopilina*, the only survivors of a group which was expanding during the lower Paleozoic; the Porcellanasteridae the most primitive family of the Phanerozonia which are related to Cambrian types; the Lophogastridae, the most primitive mysids; the crabs of the family Eryonidae which are relicts of the Mesozoic, and many more examples.

E. DIFFERENT TYPES OF RELICTS

The refuges have acquired organisms of very diverse origins, which have been driven into the subterranean world for sometimes exactly opposite reasons. The underground world, with regard to climatic conditions, offers an intermediate stage between the extreme changes found on the surface. Thus both glacial and tropical relicts can be collected from caves at the same time. This has been reported for terrestrial isopods (Vandel, 1948), and Collembola (Yosii, 1956).

The relics which populate the subterranean world can be divided into three categories (Vandel, 1948, 1960): the thermophilic, glacial, and hydrophilic relicts. Marine relicts must be added for aquatic forms.

(a) Thermophilic Relicts

The majority of cavernicoles correspond to the remains of a tropical fauna which populate the Northern Hemisphere during the first part of the Tertiary. These are the most ancient troglobia whose evolution towards the cavernicolous type certainly began before the Ice Age.

(b) Glacial Relicts

The development of immense glaciers which covered a large part of Europe and North America at the beginning of the Quaternary† favoured the expansion of the nivicolous fauna, a fauna very similar to that which is found today on high mountains.

† The glacial extension appears to have been much more limited in other regions of the world.

However, the nivicolous fauna of the Quaternary lived at much lower altitudes than does the present-day fauna. The limit of continuous snow was in fact lowered during the ice age to 1200 to 1300 metres in the Alps, and to about 1000 metres in the Pyrenees (Penck).

The Quaternary nivicolous fauna still exists today almost in the same regions and at almost the same altitudes as in earlier times. However, when the great glacial period ended this fauna found refuge in order to escape from a climate which became progressively warmer and drier, in the

FIG. 80. Map of the distribution of the genus *Scotoniscus* (terrestrial Isopoda) (after Vandel).

depths of caves. These cavernicoles are thus glacial relicts. One can give the following examples of glacial relicts, oniscoids of the genus *Scotoniscus*, whose distribution indicates that they once lived on the edge of glaciers which covered the Central Pyrenees during the Quaternary (Fig. 80) (Vandel, 1948). The Collembola of the genus *Anurida* from the caves of Japan, belong in this category (Yosii, 1956). The Pyrenean genus *Aphaenops*, and thus the North American genus *Neaphaenops*, are probably also ancient nivicoles (Jeannel, 1928). The cavernicolous staphylinids of the genus *Apteraphaenops* are derived from the nivicolous genus *Paraleptus*; both genera are indigenous to the Djurdjura Mountains (Peyerimhoff, 1909).

(c) Hydrophilic Relicts

Other animal forms expanded during their distribution during Quaternary because of the extreme humidity which characterised the glacial periods. The development of the glaciers involved not only a lowering of the temperature, but also considerable snow-fall and therefore increased rainfall. Thus the glacial periods favoured the expansion of the hydrophilic fauna. The hydrophilic types differ from nivicoles in that they cannot endure low temperatures; they are thus forms confined to low altitudes. The decrease in humidity which induced the retreat of the glaciers resulted at the end of the Quaternary in southern regions in the disappearance of the hydrophilic types. They could exist only if they retreated underground. These cavernicoles are hydrophilic relicts. Numerous species belonging to the genus *Trichoniscoides* (terrestrial isopods) can be given as examples of such relicts (Vandel, 1948).

(d) Marine Relicts

The seas have undergone expansion and regression during geological times. These movements have been very slow. When the seas retreated their place was taken for a long period by brackish waters before freshwater conditions were established. These intermediate states allowed transformations of marine animals to the euryhaline and then the freshwater forms. The remains of the marine fauna became adapted to freshwater one called marine relicts (Chappuis, 1956). Numerous examples of marine relics have been given in the systematics section of this book: Nematodes (p. 69), Polychetes (p. 70 and 72), and above all, numerous Crustacea; copepods (p. 110), mysids (p. 118), Thermosbenacea (p. 116), spheromids (p. 123), microparasellids (p. 124), microcerberids (p. 128), and various amphipods (p. 132 and 135).

(e) Relicts from Abysses and from Underground

Some relict groups include together abyssal species and cavernicolous types, e. g. Bathynominae (p. 120), *Ingolfiella* (p. 131), Galatheidae (p. 142), Brotulidae (p. 229).

BIBLIOGRAPHY

BIRSTEIN, J. A. (1947) The conception of "relict" in Biology. *Zool. J.* Moscow. **XXVI.**

CHAPPUIS, P. A. (1956) Sur certaines reliques marines dans les eaux souterraines. *Communic. 1er Congr. Intern. Spéléol. Paris* **III.**

CHOPARD, L. (1936) Orthoptères et Dermaptères (1ère Série). *Biospeologica*, **LXIV.** *Archiv. Zool. expér. géner.* **LXVIII.**

FAGE, L. (1931) Araneae (5ème Série), précédée d'un essai sur l'évolution souterraine et son déterminisme. *Biospeologica*, **LV.** *Archiv. Zool. expér. gén.* **LXXI.**

JEANNEL, R. (1928) Monographie des *Trechinae*. 3ème Livr. *L'Abeille*, **XXXV.**

JEANNEL, R. (1943) *Les fossiles vivants des cavernes.* Paris.

PEYERIMHOFF, P. DE (1909) Position systématique et origine phylogénique du genre *Apteraphaenops. Bull. Soc. Entomol. France.*

SCHWOERBEL, J. (1961) Subterrane Wassermilben (Acari: Hydrachnellae, Porohalacaridae und Stygothrombiidae), ihre Ökologie und Bedeutung für die Abgrenzung eines aquatischen Lebensraumes zwischen Oberfläche und Grundwasser. *Archiv. f. Hydrobiol. Suppl.* **XXV.**

SPANDL, H. (1926) Die Tierwelt der unterirdischen Gewässer. *Speläol. Monogr.* **XI.**

STANKOVIČ, S. (1932) Die Fauna des Ohridsees und ihre Herkunft. *Archiv. f. Hydrobiol.* **XXIII.**

STANKOVIČ, S. (1960) *The Balkan Lake Ohrid and its living World.* Den Haag.

VANDEL, A. (1948) La Faune isopodique française (Oniscoïdes ou Isopodes terrestres). Sa répartition, ses origines et son histoire. *Rev. franç. Entomol.* **XV.**

VANDEL, A. (1960) Isopodes terrestres (Première Partie). *Faune de France.* No. 64.

WAGNER, A. J. (1914) Höhlenschnecken aus Süddalmatien und der Hercegovina. *Sitzb. k. k. Akad. Wiss. Math. Naturw. Kl. Wien. Abt.* I. **CXXIII.**

YOSII, R. (1956) Monographie zur Höhlencollembolen Japans. *Contrib. Biol. Lab. Kyoto Univer.* No. 3.

ZENKEVITSCH, L. A. and BIRSTEIN, J. A. (1960) On the problem of the antiquity of the deep-sea fauna. *Deep-Sea Res.* **VII.**

CHAPTER XXX

THE STAGES OF SUBTERRANEAN
EVOLUTION

THE HISTORY of cavernicoles extends over an immense number of years which are measured in geological time. Several stages can be distinguished in this long evolution:

(a) Period of preparation
(b) Period of instability
(c) Period of stabilisation.

A. PERIOD OF PREPARATION

(a) Historical

Theoreticians of biology have often imagined that cavernicoles appeared suddenly following several accidents. However, biospeologists have been convinced for many years, even more strongly so as more knowledge has become available, that the appearance of cavernicoles required a long period of preparation. Subterranean evolution comprises a first stage which is of considerable duration and occurs on the surface of the ground. This stage is indispensable in preparing the animal for life underground.

H. Garman in 1892, in an article appearing in *Science* was the first clearly to recognise this requirement. The idea was then developed by Eigenmann (1900, 1909), Banta (1907) and Cuénot (1909, 1911, 1914).

Davenport (1903), proposed the term "preadaptation" to describe the transformations which occurred during the preparatory stage. This term is badly chosen in the case of cavernicoles because here regressions and reductions prevail over adaptations. This is why we prefer to use the term "period of preparation".

(b) Alimentary Preadaptations

Although the food sources available to cavernicoles are varied, as has been described in Chapter XIX, there are certain categories of food which are completely absent underground. The principal nutritional deficiency of the subterranean medium is without doubt the complete absence of green plants.

Therefore strictly phytophagous forms are automatically eliminated from being candidates for subterranean life.

Only the polyphages, mycophages, detriticoles, guanobia and limivores have a chance of subsisting underground. Carnivores could also exist there when the populations on which they preyed had become sufficiently numerous.

A remarkable example of alimentary preadaptation has recently been discovered by Tercafs and Jeuniaux (1961). A gasteropod mollusc, *Oxychilus cellarius* can be considered to be a troglophile because it can be found both in caves and on the surface. Epigeous specimens feed principally on dead leaves while the cavernicolous ones feed on arthropod debris and even living Lepidoptera. From phytophagous forms they have become carnivores. If the rate of chitinase production of *Oxychilus* is compared with that of a strictly phytophagous gasteropod such as *Helix pomatia*, it is found that the rate is ten to twenty times higher in the first species. From this it can be concluded that *Oxychilus cellarius*, and more generally the representatives of the genus *Oxychilus*, are preadapted to a carnivorous diet, and thus preadapted to lead a cavernicolous life.

(c) Physiological and Ecological Preadaptations

On the surface of the Earth, forms can be found which are no longer perfectly adapted to live in daylight. They are more or less depigmented and lucifugous. They have a tendency to bury themselves in soil crevices or to hide in humus. They search for food not with their eyes, but by touch and smell. At the same time, a change in the impermeable layer of the integument makes them vulnerable to dessication. For this reason they search out humid media. All these modifications can in no way be regarded as "adaptations". They are regressive changes which alter the primitive organization. The animal can survive on condition that it confines itself to certain ecological niches. These forms are not yet cavernicoles but they already possess a behaviour and other requirements which direct them towards subterranean life. This fact was clearly recognised by Eugène Simon from 1872.

Many examples illustrating this idea can be given. In most groups of the animal kingdom forms are to be found which are already predisposed towards leading a cavernicolous life.

Many species of *Trechus* with reduced eyes and which are more or less depigmented, such as *T. microphthalmus*, *T. micros*, *T. fulvus*, are examples of the intermediate stages between surface and cavernicolous forms. Similar examples are those of *Tachys bisulcatus* and *Laemosthenus* of the sub-genus *Antisphodrus* (Jeannel, 1908, 1909).

The Bathysciinae provide excellent examples of preadaptation to a cavernicolous life, as has been shown earlier (p. 204). The non-cavernicolous

Bathysciinae are already depigmented and apterous and most of them are anophthalmic.

Many examples of transitional stages between surface and subterranean forms are provided by the spiders. They are particularly frequent among the genera *Leptoneta* (for example *L. infuscata*, *L. crypticola*, *L. vittata*) (Fage, 1913), and *Troglohyphantes*, as well as in the Ochyroceratidae (in particular, *Theotima fallax*) (Fage, 1912).

The silurids can be chosen as examples among the vertebrates. Reichel (1927) has stated that *Amiurus* uses its sense of taste rather than its sight (although their eyes are normal) to search for prey. Thus this cat-fish already shows cavernicolous behaviour. Thus, as has been stated by Hubbs (1936), the silurids are preadapted to cavernicolous life. This is particularly true for *Rhamdia guatemalensis* which represents "a very logical candidate for cave life" (Hubbs, 1938). It has been stated that fish rapidly learn to know their surroundings, for example the aquarium. This knowledge is acquired just as rapidly in forms without eyes as those with them. If obstacles are placed in an aquarium containing blind fish they very rapidly learn to avoid them. It is for this reason that the behaviour of cavernicolous fish does not differ fundamentally from that of epigeous forms. Many more examples could be given.

(d) Habitats in which Preparation for Cavernicolous Life may take Place

Certain media appear to be particularly favourable as habitats for the preparation of future cavernicoles. This is why the field of biospeology is not limited to caves. If the early stages of subterranean evolution are to be understood these surface biotopes must also be studied.

1. The Soil. The soil is populated by endogeans which differ from cavernicoles in many respects (p. 5). However, endogeans form part of the great group of hypogeans and there is hardly any doubt that certain of them, for example, *Duvalius*, represent the transitional stages of future cavernicoles.

2. Snow-pockets (Névés). The snow-pockets which persist on high mountains until the beginning of summer, constitute a cold and constantly humid medium. These climatic conditions are similar to those found in caves. Thus the nivicoles are prepared for cavernicolous life.

3. Mosses. The mosses which provide an excellent humid medium, contain muscicoles many of which will become converted into cavernicoles. An example is provided by the Bathysciinae (p. 204).

4. Humus. The masses of dead leaves which accumulate on the soil of great forests contain humicoles. These creatures are also candidates for cavernicolous life. This evolution is still occurring at the present time, at least in regions which have not been altered by man. An example is the Bihar Mountains in Transylvania (Jeannel, 1923, 1931). To impress this point on the reader we shall quote the vivid description given by Professor Jeannel (1943, p. 98). "The climatic conditions found today in large forests give little idea of those of the cold and humid great ancient forests of the Pliocene and glacial periods. There are however, still forested regions such as the Vercors (district to the south-east of Grenoble) in France and above all the Bihar Mountains in Transylvania where the very humid climate is a relic of the past. In the spruce forests of the High Bihar, which are under snow for six months and have six months with abundant rain, the soil ravines, facing north, remain sodden and cold during the major part of the year. *Duvalius* (Trechinae), *Pholeuon* and *Drimeotus* (Bathysciinae) populate the underground habitat in this area. *Pholeuon* species are strict troglobia; *Duvalius* and *Drimeotus* are also cavernicolous around 500 metres altitude, but the same species are found under stones at 1000 metres and on the surface at 1200 to 1400 metres.

Thus, certain species are still to be found on the surface but only at high altitudes. The climate allows them to live as they would have lived during the glacial periods. Lower down, at 1000 metres they have become endogeans and at 500 metres where the forest no longer exists the relatively dry conditions of the endogeous habitat have forced them to retire to caves and crevices."

5. *The Humus of the Mountains of Tropical Africa.* Leleup (1952, 1956), and Jeannel and Leleup (1952) have discovered a remarkable nursery of cavernicoles in humus which covers the forest soil of tropical Africa, to a depth of between 2·000 and 2·500 metres. The temperature of the humus is about 10–15°C; the annual variation does not exceed 2°C. The humus of these mountains possesses an isothermic climate. It is very different to that of Europe; but it is very much like the underground climate of the European caves.

These forests conceal a humicolous fauna, which is adapted to a cold and humid climate and comprises numerous anophthalmic and depigmented forms resembling cavernicoles: pterostichids (Straneo, 1951); scaritids (Basilewsky, 1951a); zuphiids (Basilewsky, 1951b); pselaphids (Jeannel, 1952); staphylinids; curculionids (Hoffmann, 1915). Leleup wrote that in Africa cavernicoles are not found in caves but in humus.

This rich fauna of cavernicoles "en puissance" offers a picture of the first stage of subterranean evolution while the hypogeous populations of Europe and North America provide us with the final stages. There seems to be no doubt that the ancestors of European cavernicoles lived in the Tertiary in humus of great humid mountain forests which would resemble the conditions in Africa today.

6. *Interstitial Medium.* The interstitial medium provides aquatic forms with a transitional habitat between the superficial biotopes and the subterranean environment (p. 297).

B. PERIOD OF INSTABILITY

Although the evolution of the cavernicoles began above ground it is in the subterranean world, where the animals are driven by climatic conditions, that cavernicolous evolution continues for the troglobia are restricted to the subterranean medium.

However, the troglobious state is not reached in one step. Between the epigeous stages and the truly cavernicolous stage there is an intermediate phase. This corresponds to the troglophile type. The troglophiles, or recent cavernicoles, are characterised by the great instability of their morphological characters and in particular of their pigment and ocular structures. Kosswig (1960a and b), attempted to provide a genetic interpretation.

Numerous examples of this have been given in the preceding chapters. Some forms, such as *Alluaudellina cavernicola* (p. 175); *Tachycines cuenoti* (p. 178); amphipods (p. 132); asellids (p. 126); and *Proteus* (p. 344), provide good examples of the state of instability of recent cavernicoles.

This stage of instability provides the most favourable material for the biospeological experimentalist because it is at this stage only that experimental intervention will have a chance of success.

C. PERIOD OF STABILITY

The troglobia or ancient cavernicoles are definitely fixed with regard to their morphology, physiology and behaviour. At the most the phenomena of micro-evolution may give rise to endemic species. The biospeologist can find among these some remarkable examples of "ultra-evolution", which is associated with an extraordinary physiological specialisation. Thus the idea of "phyletic senility" arises.

Because of their extreme specialisation and requirements the ancient cavernicoles provide material which can be used for experimentation only with great difficulty.

BIBLIOGRAPHY

BANTA, A. M. (1907) The fauna of the Mayfield's Cave. *Carnegie Inst. Washington Public.* No. 69.

BASILEWSKY, P. (1951a) Description d'un Scaritide aveugle du Kivu (Col. Carabidae, Scaritinae). *Rev. Zool. Bot. Afric.* **XLIV.**

BASILEWSKY, P. (1951b) *Leleupidia luvubuana* n.g.n.sp. (Col. Carabidae). *Rev. Zool. Bot. Afric.* **XLIV.**

CUÉNOT, L. (1909) Le peuplement des places vides dans la nature et l'origine des adaptations. *Rev. gén. Sc.*

CUÉNOT, L. (1911) *La Genèse des Espèces animales.* Paris.

CUÉNOT, L. (1914) Théorie de la préadaptation. *Scientia.* **VIII.**

DAVENPORT, C. B. (1903) The animal ecology of the Cold Spring sand spit, with remarks on the theory of adaptation. *The decenn. Public. Univ. Chicago* **X.**

EIGENMANN, C. H. (1900) The structure of blind fishes. *Popul. Sc. Mon.* **LVII.**

EIGENMANN, C. H. (1909) Cave vertebrates of America. A study in degenerative evolution. *Carnegie Inst. Washington Public. No.* 104.

FAGE, L. (1912) Études sur les Araignées cavernicoles. I. Révision des *Ochyroceratidae* (n. fam.). *Biospeologica,* **XXV.** *Archiv. Zool. expér. géner.* (5) **X.**

FAGE, L. (1913) Études sur les Araignées cavernicoles. II. Révision des *Leptonetidae*. *Biospeologica,* **XXIX.** *Archiv. Zool. expér. géner.* **L.**

GARMAN, H. (1892) The origin of the cave fauna of Kentucky, with a description of new blind beetle. *Science* **XX.**

HOFFMANN, A. (1951) Un nouveau Curculionide anophthalme du Congo belge et observations sur un genre voisin du Kenya. *Rev. Zool. Bot. Afric.* **XLIV.**

HUBBS, C. L. (1936) Fishes of the Yucatan Peninsula. *Carnegie Instit. Washington Public.* No. 457.

HUBBS, C. (1938) Fishes from the caves of Yucatan. *Carnegie Instit. Washington Public.* No. 491.

JEANNEL, R. (1908) Coléoptères (1ère Série). *Biospeologica,* **V.** *Archiv. Zool. expér. géner.* (4) **VIII.**

JEANNEL, R. (1909) Coléoptères (2ème Série). *Biospeologica,* **X.** *Archiv. Zool. expér. géner.* (5) **I.**

JEANNEL, R. (1923 a) Sur l'évolution des Coléoptères aveugles et le peuplement des grottes dans les monts du Bihor, Transylvanie. *Compt. rend. Acad. Sci. Paris* **CLXXVI.**

JEANNEL, R. (1923 b) Étude préliminaire des Coléoptères aveugles du Bihar. *Bull. Soc. Sc. Cluj.* **I.**

JEANNEL, R. (1931) Coléoptères nouveaux de la troisième campagne organisée par l'Institut de Spéologie dans les Carpathes méridionales. *Bull. Soc. Sc. Cluj* **V.**

JEANNEL, R. (1943) *Les Fossiles vivants des cavernes.* Paris.

JEANNEL, R. (1952) Psélaphides recueillis par N. Leleup au Congo belge. IV. Faune de l'Itombwee et du Rugege. *Ann. Mus. Congo belge. Sc. Zool.* **XI.**

JEANNEL, R. et LELEUP, N. (1952) L'évolution souterraine dans la région méditerranéenne et sur les montagnes du Kivu. *Notes. biospéol.* **VII.**

KOSSWIG, C. (1960 a) Darwin und die degenerative Evolution. *Abhandl. Verhandl. Naturw. Ver. Hamburg N.F.* **IV.**

KOSSWIG, C. (1960 b) Zur Phylogenese sogenannter Anpassungsmerkmale bei Höhlentieren. *Inter. Rev. ges. Hydrobiol.* **XLV.**

LELEUP, N. (1952) Réflections sur l'origine probable de certains Arthropodes troglobies. *Rev. Zool. Bot. Afric.* **XLV.**

LELEUP, N. (1956) La Faune cavernicole du Congo belge, et considérations sur les Coléoptères reliques d'Afrique intertropicale. *Ann. Mus. R. Congo belge. Sc. Zool.* **XLVI.**

REICHEL, M. (1927) Étude anatomique du *Phreatobius cisternarum* Goeldi, Silure aveugle du Brésil. *Rev. suisse Zool.* **XXXIV.**

SIMON, E. (1872) Notice sur les Arachnides cavernicoles et hypogés. *Annal. Soc. Entomol. France* (5) **II.**

STRANEO, S. L. (1951) Un Caeolostomide aveugle nouveau du Kivu (Col. Carabidae, Pterostichinae) *Rev. Zool. Bot. Afric.* **XLIV.**

TERCAFS, R. R. and JEUNIAUX, CH. (1961) Comparaison entre les individues épiges et cavernicoles de l'espéce *Oxychilus cellarius* Müll. (Mollusque Gastéropode troglophile) au point de vue de la teneur en chitinase du tube digestif et de l'hépatopancréas. *Archiv. Intern. Physiol. Biochimie* **LXIX.**

CHAPTER XXXI

THE PROCESSES OF SUBTERRANEAN EVOLUTION

A. CONCEPT OF ADAPTATION

The Darwinian concept resumed the ancient teleological theses, which proclaimed the universality of finalism, maintaining that anything which is not useful or adapted in an organism is automatically eliminated.

Workers became so obsessed with the idea of adaptation that depigmentation and anophthalmia have been regarded as adaptations to cavernicolous life. One could just as easily say that catarrh, rheumatism and presbyopia were adaptations to old age.

The theoretical ideas must be removed; and we must only take into account the factual evidence.

B. REGRESSIVE EVOLUTION

It is evident that the history of cavernicoles corresponds primarily to a regressive evolution, an idea which has been held for many years. For this reason, C. Eigenmann called his great work on the cavernicolous vertebrates of America *A study of degenerative evolution*. Many types of regression can occur in cavernicoles. They can be morphological (albinism, micro- or anophthalmia, apterism), physiological (decreased metabolism, hypofunction of endocrine glands, slowed development) or sensory or ethological (simplification of behaviour; weakening of tropisms).

C. SIGNIFICANCE OF REGRESSIVE EVOLUTION

Although the multiplicity of regressive characters has been recognised for many years their significance had often been interpreted in the wrong way. The Neo-Lamarckians considered these regressive attributes to be influenced by the medium. In fact, this conclusion was shown to be incorrect when biologists began to experiment rather than to merely speculate. The very detailed studies of Madame Dresco-Derouet (1959, 1960) have proved that external factors have a very similar effect on epigeans and cavernicoles. On

477

the other hand it is impossible to transform the epigeous type of metabolism into the cavernicolous type of metabolism and vice versa. Whatever the conditions of the experiment, epigeans conserve the metabolism characteristic of surface forms and cavernicoles that characteristic of cave animals.

There is a further argument against the Neo-Lamarckian interpretation, it is of great interest since it points the way to a solution of the problem. It is based on the existence of epigeous animals possessing the cavernicolous type of metabolism. The example of species of the genus *Grylloblatta* has been given earlier (p. 182). These very primitive Orthoptera live in the mountain forests of North America and seldom in caves (only in some western states, California, Oregon, Washington). However, the research of Miss Ford (1926), Walker (1937) and Mills and Pepper (1937) has established that *Grylloblatta* species show all the characters of cavernicoles: pale coloration, reduced and degenerate eyes, their humidity (atmosphere saturated with water vapour) and temperature requirements (optimum temperature between 3 and 7°C); marked stenothermia; slow rate of embryonic development (1 year), and post-embryonic development (five years); late appearance of reproductive phase (after seven years) and a slow metabolic rate.

Analogous conditions are found in another relic species which belongs to the vertebrates: *Sphenodon punctatus*. The metabolic rate of *Sphenodon* is very slow. Sexual maturity is reached at the age of twenty years which is much later than in any other recent reptiles. Growth continues for at least fifty years. It probably lives for a century or more (Dawbin, 1962).

A more remarkable fact is that metabolism of the cavernicolous type is found in certain mammals such as the sloth. Britton (1941) established that the metabolic rate of these animals is extremely slow; blood pressure is reduced; digestion is slow and the body temperature is three to six degrees lower than the average temperature of mammals.

From this it is clear that the regressive evolution of cavernicoles is not a result of subterranean life. It is related to the great age of the phyletic lines and their state of senescence. Regressive phenomena related to phyletic senescence have been observed in all living organisms. In the protists it has been shown by A. Lwoff (1944). The abyssal fish, which belong on the whole to primitive groups of teleosts, can be considered to be relics. They show very many degenerate characteristics (Bertin).

Regression of body structures and physiological functions is as directly related to senescence of phyletic lines as it is to the old age at the end of the life of the individual.

D. AUTOREGULATION

In the way that radar revealed the biological phenomenon of echolocation, cybernetics has shown the importance of autoregulation. The living organism appears today as an autoregulating system such that metabolism can be modified to suit the medium in which it lives. It is at the lower level of life that autoregulation is most clearly seen. A good example is that of the adaptive enzymes many of which have been reported in recent years by biochemists. The interpretations which have been proposed to account for autoregulation will not be considered here. It is sufficient to recall its existence.

E. AUTOREGULATION AND PHYLETIC SENESCENCE

Research carried out by the author (Vandel, 1959, 1960) has shown that young and expansive organisms have become widely distributed because of their ability to adapt to very varied conditions. These are the eurythermic, euryhygrobious and euryhaline organisms.

With the increasing age of a phyletic line the species tend to be confined to more specialised media. They became adapted to live in niches by auto-regulation and finally became imprisoned there. In the islands of the Atlantic, where the climate has hardly changed since the Tertiary, the expansive types which once colonised these islands have developed into small species or ecological races and localised in extremely specialised media which have become true ecological traps. They cannot move away and are destined to die there if the conditions change. These animals have become senescent species by reduction and then complete disappearance of the ability to autoregulate.

F. AUTOREGULATION IN CAVERNICOLES

(Period of Preparation)

The history of the cavernicoles comprises the same stages as does that of evolution of the epigeans. It is certain that an animal cannot live in the subterranean environment without having undergone a long period of preparation. This preparation involves a series of processes of autoregulation: adaptation to the absence of light stimuli; "adaptation" to life in a very humid medium, which may be only the result of the loss of the impermeable integumentary layers which protect epigeous arthropods from desiccation; construction of moulting, metamorphosis, and oviposition chambers which is perhaps related only to stenohygrobia, as has been said earlier (p. 385); a tendency towards polyphagia and the utilisation of "living clay" to provide vitamins and growth elements which epigeans normally obtain from green plants.

This long autoregulatory evolution occurs only on the surface and it is only when it is complete or at least very well advanced that the animal is capable of surviving in the subterranean world. An epigean which wanders into the subterranean medium is condemned to perish.

The strongest objection to the "mutation" theory is based on the need for this long period of preparation preceding the cavernicolous phase. The albino or eyeless individual is not preadapted through this anomaly to become a cavernicole, because a cavernicole is characterised more by its physiology than by its structure. The albino or eyeless mutant possesses the physiology of an epigean, and it is just as likely to die in a cave as a normal specimen. The mutation theory may satisfy the geneticist but it cannot be held by the biospeologist.

G. AUTOREGULATION IN CAVERNICOLES

(Final Period)

The period of preparation of a candidate for cavernicolous life takes place in any superficial medium where the climatic conditions are rather constant, e. g. under stones, in humus or mosses, at the edge of snow-pockets. When the metabolism of the animal is in accord with conditions in the hypogeous medium it can colonise caves. When the animal starts cavernicolous life, the ability to autoregulate is decreasing. This is, on the physiological level, a manifestation of regressive evolution.

Biospeologists are often asked about the factors which induce potential cavernicoles to enter caves. Some have given darkness as factor, but *Niphargus* and obcuricolous planarians are known which live in illuminated cold springs. *Typhlocaris galilae* has been found in a superficial pond. Other biospeologists have quoted humidity, which is certainly a major factor for terrestrial cavernicoles, but of course does not apply to aquatic forms. There is no point in considering low temperatures as we have already shown that the temperature conditions in which cavernicoles are found are extremely variable. Briefly, no factor so far suggested has satisfied the requirement of a general explanation.

The only condition which applies to all cavernicoles is the constancy of the medium which is the requirement of any biotope used as a refuge, whether the subterranean environment or marine abyss. The search for a constant medium is the direct result of a reduction in the ability to autoregulate. The degree of specialisation in a cavernicole can be measured by its capacity to autoregulate. *Niphargus* have conserved the ability to autoregulate with regards temperature (p. 312) and salinity (p. 345). However, *Aphaenops* are ultra-specialised forms as shown by their rigid requirement of high humidity. The cave is for them a prison before becoming their tomb.

In the preceding chapter we have considered the habitats in which preparation for cavernicolous life may occur. We may take the example of the forest soil. The classical ecological survey shows the presence of three superimposed layers.

(a) The superficial litter of dead leaves which are still intact. This biotope is populated by epigeous animals.

(b) Below the preceding layer is the humus, formed by the decomposition of dead leaves and containing the humicoles.

(c) The vegetative layer which constitutes the soil itself and contains endogeans.

Relative measurements of physical conditions, in particular temperature and humidity, show a stabilisation which is progressively more marked with increasing depth. In the litter, conditions differ very little from atmospheric conditions and there are considerable daily and seasonal variations. In the humus temperature variations decrease and there is little variation in humidity. In the soil temperature and humidity are stable. In forests of temperate countries the "subterranean climate" is practically stabilised at a depth of 0·5 metres.

It is thus normal for regressive lines with diminishing ability to autoregulate to abandon the superficial litter and become buried in the humus and finally the soil in order to find the stable conditions which become so necessary for them. These are the candidates for cavernicolous life which were considered in the previous chapter. The troglobia represent the last stage of regressive evolution which comprises a reduction and then the almost total loss of autoregulatory ability.

H. DOES PROGRESSIVE EVOLUTION OCCUR AMONG CAVERNICOLES?

Can one find anything contrary to the negative picture which has just been proposed? Is the evolution of cavernicoles entirely negative and do cavernicoles show some original structures and specialised behaviour?

Certain biospeologists (Racovitza, 1907, Banta, 1907) have concluded that new structures are completely absent in cavernicoles†. Even if this statement is correct the possibility of cavernicoles developing and extending structures present in their epigean ancestors must be considered. At this point in our argument we return to the old and teleological idea which was accepted by all ancient biospeologists, that of "the compensation for the loss of vision" in cavernicoles.

† The remarkable processes of echolocation have been developed in animals living in caves, but these are not true cavernicoles.

I. COMPENSATION FOR LOSS OF VISION AMONG CAVERNICOLES

The teleological concepts of biology which were universally accepted in the last century could not fail to have an influence on early biospeology. Naturalists had stated for many years that the reduction or loss of vision which was constantly found among cavernicoles of necessity brought about a compensatory hypertrophy of other sensory functions and organs. This interpretation was adopted by all biospeologists in the second half of the 19th century: Murray (1857), LaBrulerie (1872), de Rougemont (1876), Joseph (1882), Leydig (1883), Packard (1886), Hamann (1896), Viré (1900). However, it is very rare to find this interpretation held by contemporary zoologists. However, Verhoeff (1930) used it to explain the increase in the number of sensory organs in the antennae in cavernicolous diplopods and Matič (1957, 1958) to interpret the size of the organ of Tomösvary in an-ophthalmic species of *Lithobius*.

This theory led biospeologists to make numerous errors. An example was given by R.Jeannel (1926–1930) in his *Monographie des Trechinae*. Entomo-logists have stated that the large specialised hairs of carabids are more nu-merous in the cavernicolous Trechinae *(Aphaenops, Pheggomistes)* than in the lucicolous *Trechus*. But, Jeannel remarked very justly, that this is not an adaptation to the subterranean medium but the persistence of a primitive characteristic. The reduction in the number of hairs represent a specialisa-tion. Certain archaic lucicoles such as *Amblystogenium* from the Ile Crozet, and *Irechiama* from Japan, possess many hairs on the elytra, the disposition of which recalls that found in cavernicoles.

What are we to believe about this teleological interpretation? If the facts are examined impartially it can be stated that the answer is much less uni-vocal than is required by this theory. The negative cases are as frequent as the positive ones. Here are a few examples.

Positive cases: The most clear example which can be given in favour of "compensation to loss of vision" is that of cavernicolous fish. The sensory ridges containing neuromasts and neurogemmae are far more numerous and better developed in the *Amblyopsidae* (Fig. 62) than in representatives of related families, for instance the Cyprinodontodae. They extend along the whole length of the body and caudal fin, that is, the regions which are with-out sensory organs in surface fish (Woods and Inger, 1957). Also, the sen-sory organs are better developed in *Amblyopsis* and *Typhlichthys* which are strictly cavernicoles, than in *Chologaster* which comprises epigeous or troglo-philic species (Putnam, 1872; Cox, 1905; Poulson, 1961). *Stygicola dentatus* possesses a specialist type of neurogemma which is absent or exceptional in other brotulids (Kosswig, 1934).

However, the increase of the sensory elements in the lateral line system is not general in the hypogeous fishes. We may observe this increase in the ancient cavernicoles only, for instance in the Amblyopsidae. Among the recent cavernicoles, conditions are different. The lateral line system of *Anoptichthys* does not differ from that of ordinary fishes (Grobbel and Hahn, 1958). The underground fish *Typhlogarra* and the near epigeous form *Garra* possess a very similar lateral system (Marshall and Thinès, 1958).

Other positive examples are known among Crustacea. The size of sensory hairs or aesthetascs is sometimes greater in cavernicoles than in epigeans. According to Leydig (1878) the aesthetascs of *Niphargus puteanus* from the wells at Bonn are three times longer than those of *Gammarus pulex*. The sensory hairs of the cavernicolous crayfish, *Orconectes pellucidus* are better developed than those of epigeous forms (Eberley, 1960).

In other examples the number of aesthetascs of cavernicoles is higher than that found in epigeans. The antennae of *Asellus cavaticus* bear 6–7 each while those of *Asellus aquaticus* bear only 3 each in the female, and 4 each in the male (Weber, 1879).

Finally the antennae themselves are sometimes longer in cavernicoles than in epigeans. The *Gammarus* from Clausthal show an increase in the length of the antennae and their number of segments. (Mühlmann, 1938) The antennae of *Asellus aquaticus cavernicolus* from Carniola are much longer than those of epigeous species and there are double the number of segments (100 instead of 50) in the flagellum (Racovitza, 1925).

Negative Examples. There are many such examples among the Crustacea. The antennary sense organs of *Bathyonyx* are less numerous than those of *Gammarus* (Vejdovsky, 1905). The primitive and cavernicolous *Asellus* species (in particular *Asellus spelaeus*) have a lower number of sensory hairs than epigeous species belonging to the same line *(A. coxalis* and *A. meridianus)* (Racovitza, 1925). There is a decrease in the number of aesthetascs in cavernicolous terrestrial isopods of the genus *Androniscus* compared with surface species belonging to the same genus (Vandel, 1960).

The statocysts of cavernicolous crayfish belonging to the genus *Orconectes* are reduced compared with those of surface forms. Similar examples are known among the vertebrates. The ears of *Anoptichthys jordani* are regressed (Breder, 1943). There is a reduced number of olfactory lamellae in the cavernicolous Characidae compared with those of surface forms (Breder and Rasquin, 1943).

From these facts it can be seen that although unquestionable examples of "compensation for the loss of vision" can be given among cavernicoles, the inverse situation is just as common if not more frequent. The reduction of sensory organs is part of the regressive evolution which is reached by differing degrees by all cavernicoles. There is no better way to close this paragraph than to give the example of a cavernicolous planarian, *Sphalloplana percaeca* which exhibits almost no tropic reactions and does not react or reacts negatively or very slowly to those external agents which produce a lively response in epigeans (Buchanan, 1936).

J. FEATURES OF CAVERNICOLES

The multiplication of the sensory organs in some cavernicoles, considered in the previous section, seems linked, least in the arthropods, with the modifications to the shape of the body.

It has been recognised for many years that the inhabitants of the sub-terranean world have acquired a special cavernicolous appearance. The body is slender and the appendages elongated, especially the antennae, palps and limbs; sometimes there is a false, *(Leptodirus)*, or a true, *(Allopnyxia)*, physogastry.

These characteristics would be described by the neo-Darwinians as an adaptation to cavernicolous life and by the Neo-Lamarckists as the reaction of these species to the subterranean environment. It now appears that both these interpretations are inadequate. The cavernicolous appearance is to be found often among epigeans, humicoles and endogeans, long before they have ventured into the cave habitat. This prompted Jeannel (1923) to write "The subterranean world is more of a 'conservative' than 'modifying' medium for cavernicoles."

The significance of these characteristics will be made more clear by the following remarks.

(a) The slender body and elongated limbs are characteristic of some but not all cavernicoles. Also, they are not found among cavernicoles ex-clusively.

The elongation of the body and limbs is particularly well defined in caverni-colous Collembola belonging to the family Entomobryidae. The epigeous species are already remarkable in that they have a very slender appearance. However, subterranean Collembola belonging to the other families have conserved the characteristic short limbs and rounded bodies of epigeous species.

Among the Orthoptera the cavernicolous Rhaphidophoridae show an extraordinary elongation of their appendages but they are already very long among epigeous representatives of this family. On the other hand, cavernicolous species of *Troglophilus* and *Ceutophilus* possess limbs of nor-mal dimensions.

Cavernicolous Hemiptera of the family Emesinae possess limbs which are elongated out of all proportion to their bodies but this is only an ex-aggeration of a characteristic of all representatives of the family.

Among the Trechinae, *Aphaenops* species have exceptionally long limbs, but *Geotrechus* and *Duvalius* have the normal appearance of carabids.

Thus the slender body and elongated limbs are not cavernicole characteristic but the property of certain groups. Chopard (1929, p. 427) expressed this point very well when he wrote: "Contrary to that which has been believed for many years, the cavernicolous medium and the impossibility of vision appears to have no influence on the elongation of the limbs. It is true that certain cavernicolous Orthoptera have remarkably long limbs and palps but one can always find closely related lucicoles showing the same characteristics. If, on the contrary one examines species belonging to a group with heavy bodies and short limbs, it is found that those which have colonised caves have conserved exactly the same characters as those species from which they were derived".

(b) The form of the body of an animal and the relative dimensions of the parts are a result of the phenomenon of differential growth which is termed allometry. Heuts (1952) has already shown how the idea of allometry can be used in understanding the form of cavernicoles.

The exaggerated size of an appendage due to positive allometry is sometimes only the simple consequence of the overall increase in size of the animal. An example is that of the remarkably elongated antennae of *Heteromerus longicornis* (Fig. 36) which is one of the largest Collembola known (Absolon, 1900). However, the interpretation of the form of cavernicoles is not always as simple. One might envisage modifications in the rate of allometry of different parts with respect to the size of the animal. Biometric research on cavernicoles would be very useful in this respect.

(c) Thus, the cavernicolous type does not appear to be the consequence of subterranean life because a similar morphology can be found in many surface forms. It is a manifestation of orthogenesis, that is to say, directed evolution. Paleontology has familiarised the idea of orthogenesis and biometric methods have confirmed these views (Matsakis, 1962).

The phenomenon of orthogenesis corresponds to a property of the phyletic line. Orthogenesis which occurs among cavernicoles takes place in a remarkably constant medium. This fact establishes without doubt, that it originates within the animal itself and not from external influences such as the medium or natural selection. It thus forms sound proof in favour of organicism.

René Jeannel was the first to clearly recognise the role of orthogenetic evolution in the genesis of cavernicoles and its independence of environmental factors. He wrote (Jeannel, 1928, p. 33): "There is nothing to indicate why certain primitively lucicolous ancient lines have suddenly embarked in this orthogenetic direction which resulted in cavernicoles. Species have not undergone the morphological evolution called subterranean evolution because they entered caves. They entered caves because their degree of evolution made it impossible for them to exist among the epigeous fauna under the prevailing climatic conditions. It thus remains to establish the orthogenetic evolution which resulted in cavernicoles. It has been viewed as an accident during different geological areas among ancient lines and even among certain lines only. Perhaps it is very simply a result of senility of lines and is comparable to gigantism, established many times in paleontological evolution among groups on the point of disappearance".

Another French biologist, Bruno Condé, reached very similar conclusions. He wrote (Condé, 1955: p. 176), "It is necessary to interpret the remarkable modifications of certain troglobia as the result of orthogenesis which has occurred without interruption for several million years in the stable medium of the hypogeous habitat".

It must now be asked which factors allowed cavernicoles to undergo orthogenetic evolution which has been carried out beyond the stages reached by epigeans. We hope to answer this in the next paragraph.

K. "STRUGGLE FOR EXISTENCE" AND NATURAL SELECTION

Few subjects have caused more disagreement among biospeologists than the connected phenomena of the struggle for existence and natural selection. More than half a century ago Racovitza (1907) was already amused by their contradictory arguments.

In the paragraph devoted to cavernicoles in the *Origin of Species* Darwin emphasised the effect of non-usage. However the last sentence refers to another condition. "I am only surprised that more wrecks of ancient life have not been preserved owing to the less severe competition to which the scanty inhabitants of these abodes will have been exposed". †

The idea of the absence of natural selection in the subterranean environment was held by Packard (1886, 1894), Michaelsen (1933), Kosswig (1935), Hatch (1941), Chappuis (1956).

On the contrary, Lankester (1893), Chilton (1894), Verhoeff (1898), Absolon (1900a), Fage (1913), Jeannel (1943) have all recognised the ferocious character of competition in caves.

It appears however, that many of these contradictions are due to the confusion caused by imprecise usage of terms. Competition does not mean the same thing to all authors.

(1) There is no doubt that competition (in particular for food), between individuals of the same species or related species belonging to the same genus does occur. Eberley (1960) has mentioned that competition occurs between two American crayfish, *Orconectes pellucidus*, an ancient caverni-

---

† The interpretation proposed by August Weismann, although based on selection as well, is very different from that of Darwin. In one of his "Essays", "The Regression in Nature" which appeared in 1886 he accounted for all regressive characters which are found in organisms by the absence of selection and he applied this interpretation to the eyes of cavernicoles. He considered that the eyes of subterranean animals degenerate because in no longer having any function they are no longer influenced by selection.

cole and *Cambarus bartoni laevis,* a recent cavernicole. Both these animals attack the same prey, asellids. Christiansen (1960) speaks of the intense competition which is established between the species of Collembola belonging to the genus *Sinella.* Jeannel (1943) cited the competition which exists between different species of *Speonomus.*

The phenomenon of competition must be used to explain the extreme rarity of coexistence in the same cave of two species belonging to the same genus or same species group. Or, when two near species coexist in the same cave, one species is always more abundant than the other.

(2) On the other hand, by virtue of the mode of life, cavernicoles are sheltered from the many predators and parasites which swarm across the surface of the earth. They thus escape from one of the most active causes of destruction in the epigeous world. This condition is extremely important as it must not be forgotten that cavernicoles are for the most part ancient relics with regressed and unadapted structures for conditions in the world today. They would be eliminated at once if they emerged into the epigeous world. They can, however, exist in the subterranean world where access to surface forms is forbidden. Leruth (1939) and Birstein (1961) maintained very justly that *Niphargus* could live very well in springs which physically constitute the medium they require but they are rapidly eliminated from the springs through the gammarids which devour them.

Analogous conditions are found at the bottom of seas and on islands which have been isolated for many years from the continent and are sheltered from invasions by new-comers (Vandel, 1959, 1960).

This is the very reason why cavernicoles, sheltered from the intense competition which would be shown by epigeans are able to carry out orthogenetic evolution to great lengths.

L. CONCLUSION

Thus, subterranean evolution can easily be placed in the normal history of the animal world. Cavernicoles confirm the teachings of paleontology. Species have a history in the same way as phyletic lines: they undergo a period of youth, a stage of expansion, and finally a decline which terminates in the phase of senescence.

However, caves provide biospeologists with the chance to witness the terminal phase of phyletic evolutions which are never reached on the surface of the earth. For it appears that caves have always been for animals, as for man, a place of refuge and retreat.

BIBLIOGRAPHY

ABSOLON, K. (1900a) Einige Bemerkungen über die mährische Höhlenfauna. III. Aufsatz. *Zool. Anz.* **XXIII.**

ABSOLON, K. (1900b) Über zwei neue Collembolen aus den Höhlen des österreichischen Occupationsgebietes. *Zool. Anz.* **XXIII.**

BANTA, A. M. (1907) The fauna of the Mayfield's Cave. *Carnegie Inst. Washington Public.* No. 69.

BIRSTEIN, J. A. (1961) Biospeologica Sovietica. **XIV.** Les Amphipodes souterrains de Crimée. *Bull. Soc. Natural. Moscou Sect. Biol.* **LXVI.**

BREDER, C. M. JR. (1943) Problems in the behaviour and evolution of a species of blind fish. *Trans. New York Acad. Sc.* (2) **V.**

BREDER, C. M. JR., and RASQUIN, P. (1943) Chemical sensory reactions of Mexican blind characins. *Zoologica* **XXVIII.**

BRITTON, S. W. (1941) Form and function in the sloth. *Quart. Rev. Biol.* **XVI.**

BUCHANAN, J. W. (1936) Notes on an American cave flatworm, *Sphalloplana percaeca*. *Ecology* **XVII.**

CHAPPUIS, P. A. (1956) Sur certaines reliques marines dans les eaux souterraines. *Communic.* 1er *Congr. Intern. Spéléol.* Paris. **III.**

CHILTON, CH. (1894) The subterranean Crustacea of New Zealand, with some remarks on the fauna of caves and wells. *Trans. Linn. Soc. London* (2) **VI.** *Zool.*

CHOPARD, L. (1929) Note sur les Orthoptères cavernicoles du Tonkin. *Bull. Soc. Zool. France* **LIV.**

CHRISTIANSEN, K. (1960) The Genus *Sinella* Brook (Collembola; Entomobryidae) in nearctic caves. *Ann. Entomol. Soc. America* **LIII.**

CONDÉ, B. (1955) Matériaux pour une Monographie des Diploures Campodéidés. *Mém. Mus. Hist. Nat. Paris N.S. Zool.* **XII.**

COX, U. O. (1905) A revision of the cave fishes of North America. *Rep. Bureau Fisheries.*

DAWBIN, W. H. (1962) L'Hatteria dans son habitat naturel. *Endeavour* **XXI.**

DRESCO-DEROUET, L. (1959) Contribution à l'étude de la Biologie de deux Crustacés aquatiques cavernicoles: *Caecosphaeroma burgundum* D. et *Niphargus orcinus virei* Ch. *Vie et Milieu* **X.**

DRESCO-DEROUET, L. (1960) Étude biologique de quelques espèces d'Araignées lucicoles et troglophiles. *Archiv. Zool. expér. géner.* **XCVIII.**

EBERLEY, W. R. (1960) Competition and evolution in cave crayfishes of Southern Indiana. *System. Zool.* **IX.**

EIGENMANN, C. H. (1909) Cave vertebrates of America. A study in degenerative evolution. *Carnegie Inst. Washington Public.* No. 104.

FAGE, L. (1913) Études sur les Araignées cavernicoles. II. Révision des *Leptonetidae*. *Biospeologica*, **XXIX**, *Archiv. Zool. expér. géner.* **L.**

FORD, N. (1926) On the behaviour of *Grylloblatta*. *Canad. Entomol.* **XLVIII.**

GROBBEL, G. and HAHN, G. (1958) Morphologie und Histologie der Seitenorgane des augenlosen Höhlenfisches, *Anoptichthys jordani*, im Vergleich zu anderen Teleosteern. *Z. Morphol. Ökol. Tiere.* **XLVII.**

HAMANN, O. (1896) *Europäische Höhlenfauna.* Jena.

HATCH, M. H. (1941) An important factor in evolution. *Science* **XCIII.**

HEUTS, M. J. (1952) Ecology, variation and adaptation of the blind fish, *Caecobarbus geertsii* Boulenger. *Ann. Soc. Zool. Belgique* **LXXXII.**

JEANNEL, R. (1923) Sur l'évolution des Coléoptères aveugles et le peuplement des grottes dans les monts du Bihor, Transylvanie. *Compt. rend. Acad. Sci. Paris* **CLXXVI.**

JEANNEL, R. (1926–1930) Monographie des *Trechinae. L'Abeille* **XXXII–XXXVI.**

JEANNEL, R. (1943) *Les Fossiles Vivants des Cavernes.* Paris.

JOSEPH, G. (1882) Systematisches Verzeichnis der in den Tropfsteingrotten von Krain ein-heimischen Arthropoden nebst Diagnosen der vom Verfasser entdeckten und bisher noch nicht beschriebenen Arten. *Berliner Entomol. Zeit.* **XXV.**

KOSSWIG, C. (1934) Über bislang unbekannte Sinnesorgane bei dem blinden Höhlenfisch *Stygicola dentatus* (Poey). *Zool. Anz. 7. Supplementbd.*

KOSSWIG, C. (1935) Die Evolution von "Anpassungs"-Merkmalen bei Höhlentieren in genetischer Betrachtung. *Zool. Anz.* **CXII.**

LABRULERIE, CH. PIOCHARD DE (1872) Note pour servir à l'étude des Coléoptères caverni-coles. *Annal. Soc. Entomol. France.*

LANKESTER, E.R. (1893) Blind animals in caves. *Nature, Lond.* **XLVII.**

LERUTH, R. (1939) Notes d'Hydrobiologie souterraine. **VI.** Remarques écologiques sur le genre *Niphargus*. *Bull. Soc. R. Sc. Liège.*

LEYDIG, F. (1878) Über Amphipoden und Isopoden. Anatomische und zoologische Be-merkungen. *Zeit. wiss. Zool.* **XXX, Suppl. Bd.**

LEYDIG, F. (1883) Untersuchungen zur Anatomie und Histologie der Thiere. Bonn.

LWOFF, A. (1944) L'Évolution Physiologique. Étude des pertes de fonction chez les microorganismes. *Actual. Sc. Industr.* No. 970.

MARSHALL, N.B. and THINÈS, G.L. (1958) Studies on the brain, sense organs and light sensibility of a blind cave fish *(Typhlogarra widdowsoni)* from Irak. *Proc. Zool. Soc. Lond.* **CXXXI.**

MATIČ, Z. (1957) Description d'un nouveau *Lithobius* cavernicole des Pyrénées espa-gnoles (Myriapoda, Chilopoda). *Notes biospéol.* **XII.**

MATIČ, Z. (1958) Contribution à la connaissance des Lithobiides cavernicoles de France (Collection "*Biospeologica*" VII et VIII e séries). *Notes biospéol.* **XIII.**

MATSAKIS, J. (1962) Contribution à l'étude du développement postembryonnaire et de l'évolution de la forme chez quelques Crustacés Isopodes. *Bull. Biol. France Belgique* **XCV.**

MICHAELSEN, W. (1933) Über Höhlen-Oligochaeten. *Mitteil. Höhl. Karstf.*

MILLS, H.B. and PEPPER, J.H. (1937) Observations on *Grylloblatta campodeiformis* Walker. *Ann. Entomol. Soc. America* **XXX.**

MÜHLMANN, H. (1938) Variationsstatistische Untersuchungen und Beobachtungen an un-ter- und oberirdischen Populationen von *Gammarus pulex* (L.). *Zool. Anz.* **CXXII.**

MURRAY, A. (1857) On the insect vision and blind insects. *Edinburgh New Philos. Journal.* **N.S.**

PACKARD, A.S. (1886) The cave fauna of North America, with remarks on the anatomy of the brain and origin of the blind species. *Mem. Nat. Acad. Sc. Washington* **IV.**

PACKARD, A.S. (1894) On the origin of the subterranean fauna of North America. *Amer. Nat.* **XXVIII.**

POULSON, T.L. (1961) Cave adaptation in amblyopsid fishes. University of Michigan. *Phil. Diss. Zoology.*

PUTNAM, F.W. (1872) The blind fishes of Mammoth Cave and their allies. *Amer. Nat.* **VI.**

RACOVITZA, E.G. (1907) Essai sur les problèmes biospéologiques. *Biospeologica,* **I.** *Archiv. Zool. expér. géner.* (4) **VI.**

RACOVITZA, E.G. (1925) Notes sur les Isopodes. 13. Morphologie et phylogénie des an-tennes II. Le fouet. *Archiv. Zool. expér. géner.* **LXIII.**

RASQUIN, P. and ROSENBLOOM, L. (1954) Endocrine imbalance and tissues hyperplasia in teleost maintained in darkness. *Bull. Americ. Mus. Nat. Hist.* **CIV.**

ROUGEMONT, P. (1876) *Étude de la faune des eaux privées de lumière.* Paris.

VANDEL, A. (1959) Réflexions sur la notion d'espèce et sa signification. *Compt. rend. Acad. Sci. Paris* **CCXLIX.**

VANDEL, A. (1960) Les Isopodes terrestres de l'Archipel madérien. *Mém. Mus. Hist. Nat. Paris. N.S. Sér. A. Zool.* **XXII.**

VANDEL, A. (1960) Les espéces d'*Androniscus* Verhoeff 1908 appartenant au sous-genre *Dentigeroniscus* Aracangeli 1940 (Crustacés; Isopodes terrestres). *Ann. Spéléol.* **XV.**

VEJDOVSKY, F. (1905) Über einige Süßwasser-Amphipoden. **III.** Die Augenreduktion bei einem neuen Gammariden aus Irland und über *Niphargus casparyi* Pratz aus den Brunnen von München. *Sitzb. K. böhm. Gesell.*

VERHOEFF, K.W. (1898) Einige Worte über europäische Höhlenfauna. *Zool. Anz.* **XXI.**

VERHOEFF, K.W. (1930) Arthropoden aus südostalpinen Höhlen gesammelt von Karl Strasser. 5. Aufsatz. *Mitteil. Höhl. Karstf.*

VIRÉ, A. (1900) *La Faune souterraine de France.* Paris.

WALKER, E.M. (1937) *Grylloblatta,* a living fossil. *Trans. R. Soc. Canada. Sect. V. Biol. Sc.* (3) **XXXI.**

WEBER, M. (1879) Über *Asellus cavaticus* Schiödte in **I.** teste Leydig (*As. Sieboldii* de Rougement). *Zool. Anz.* **II.**

WEISMANN, A. (1886) Über den Rückschritt in der Natur. *Berichte d. Naturf. Gesell. Freiburg i. Br.* **II.** (Reprint *in: Aufsätze über Vererbung und verwandte biologische Fragen.* Jena, G. Fischer; 1892).

WOODS, L.P. and INGER, R.F. (1957) The cave, spring and swamp fishes of the family *Amblyopsidae* of Central and Eastern United States. *Amer. Midl. Nat.* **LVIII.**

AUTHOR INDEX

Simon, E. 24, 85, 87, 94, 95, 176, 271, 316, 359, 381, 418, 419, 472
Simroth, H. 76
Sjöstedt, J. 26
Skarbikovitch, T.S. 256
Sket, B. 74, 121, 123, 133
Skillen, S. 265
Slifer, E.H. 396
Smalley, A.E. 153
Smith, H.M. 233
Smith, S.I. 356
Smith, W.G. 304
Snodgrass, R.E. 154
Snow, D.W. 243
Sokolow, I. 100
Sonenshine, D.E. 257
Soos, L. 76
Sörensen, W. 91
Sorokine, J.I. 336
Sousa, M. 34
Spallanzani, L. 447, 448, 449, 453
Spandl, H. 221, 330, 356, 359, 409, 466
Speiser, P. 267
Spence Bate, C. 133, 421
Spencer, H. 462
Spooner, G.M. 114, 124, 132
Sprehn, C. 259
Stach, J. 165, 167, 169
Stadler, H. 26
Stammer, H.J. 69, 72, 73, 100, 117, 121, 126, 139, 248, 249, 250, 295, 329, 418, 433
Stankovič, S. 134, 466
Stannius, H. 349
Starostin, I.V. 148
Stefanelli, A. 418, 434
Steinmann, P. 107
Stejneger, L. 233, 234, 242
Stella, E. 116, 308, 355, 382
Stenn, Fr. 259
Stephensen, J. 73
Stephensen, K. 135, 152
Stiles, C.W. 265
Straneo, S.L. 474
Strassen, R. zur 208, 218
Strasser, K. 155
Strauss, E. 418, 424, 425
Streseman, E. 237
Strinati, P. 96
Strouhal, H. 121, 129, 249, 367
Stunkard, H.W. 251, 262
Suomalainen, H. 316
Synave, H. 183
Szalay, L. 12, 101

Szarski, K. 418
Szymanski, J.S. 379
Szymczkowsky, W. 314

Tabacaru, I. 155, 156
Tafall, B.F.O. 107, 143
Takashima, H. 155
Takats, T. 61, 333
Tanasachi, J. 10, 99, 100, 101, 102, 296
Tarman, K. 67
Tarsia in Curia, L. 165
Tellkampf, T.A. 24, 171, 227, 293, 418, 429
Tempère, C. 260
Tercafs, R.R. 76, 171, 286, 327, 279, 472
Ternetz, C. 61
Thaxter, R. 260
Theodor, O. 267
Theodoridès, J. 245, 252, 254
Thienemann, A. 17, 293, 312, 325
Thines, G.L. 220, 224, 225, 401, 404, 416, 418, 434, 483
Thomson, G.M. 113
Tichelman, G.L. 237, 243
Tiegs, O.W. 154
Tindale, N.B. 7
Tintant, H. 41
Tiwari, K.K. 150
Tollet, R. 286, 288
Tomaselli, R. 17, 57
Tomiyama, I. 230, 241
Torii, H. 26, 139, 142, 309
Torok, P. 113
Toschi, A. 244
Tosco, U. 57, 61, 84
Tragardh, I. 98
Trewavas, E. 224, 402, 404
Troïani, D. 343, 345, 399
Trombe, F. 316
Trouessart, E.L. 448
Tschugunoff, N. 420
Tusques, J. 419, 437
Tyne, J. van 244

Uchida, T. 101
Uéno, M. 27, 113, 135, 152, 261
Uéno, S.I. 25, 27, 69, 74, 139, 142, 154, 196, 202, 208, 237, 248, 254, 261, 282, 325, 367, 383
Uhlenhuth, E. 234, 350, 436
Uhlmann, D. 72
Ulrich, C.J. 171

SUBJECT INDEX

OTHER DIVISIONS IN THE SERIES IN
PURE AND APPLIED BIOLOGY

BIOCHEMISTRY

BOTANY

MODERN TRENDS
IN PHYSIOLOGICAL SCIENCES

PLANT PHYSIOLOGY